WITHDRAWN

AF680853

COMMUNICATIONS, COMPUTATION, CONTROL AND SIGNAL PROCESSING

a tribute to Thomas Kailath

COMMUNICATIONS, COMPUTATION, CONTROL AND SIGNAL PROCESSING

a tribute to Thomas Kailath

EDITED BY

Arogyaswami PAULRAJ
Stanford University
Stanford, California

■

Vwani ROYCHOWDHURY
University of California
Los Angeles, California

■

Charles D. SCHAPER
Stanford University
Stanford, California

KLUWER ACADEMIC PUBLISHERS
Boston/London/Dordrecht

Distributors for North America:
Kluwer Academic Publishers
101 Philip Drive
Assinippi Park
Norwell, Massachusetts 02061 USA

Distributors for all other countries:
Kluwer Academic Publishers Group
Distribution Centre
Post Office Box 322
3300 AH Dordrecht, THE NETHERLANDS

Library of Congress Cataloging-in-Publication Data

A C.I.P. Catalogue record for this book is available
from the Library of Congress.

Copyright © 1997 by Kluwer Academic Publishers

All rights reserved. No part of this publication may be reproduced, stored in a retrieval system or transmitted in any form or by any means, mechanical, photocopying, recording, or otherwise, without the prior written permission of the publisher, Kluwer Academic Publishers, 101 Philip Drive, Assinippi Park, Norwell, Massachusetts 02061

Printed on acid-free paper.

Printed in the United States of America

UNIVERSITY OF HERTFORDSHIRE
LIBRARY, HATFIELD CAMPUS,
HATFIELD, AL10 9AD
CONTROL
0792398157
CLASS
621 382 CON
COLLECTION/LOAN STATUS
OWL
ITEM
4404304300

Knowledge comes, but wisdom lingers.

– *Alfred, Lord Tennyson*

Thomas Kailath was born on June 7, 1935, in Pune, India, where he received a B.E. (Telecom) degree in 1956. In 1961, he became the first student from India to receive a doctorate in electrical engineering from the Massachusetts Institute of Technology. He was a researcher at the Jet Propulsion Labs, Pasadena, CA, and also taught part-time at Caltech, before joining Stanford University in 1963 as an Associate Professor of Electrical Engineering. He was promoted to Full Professor in 1968, served as Director of the Information Systems Laboratory from 1971 to 1980, as Associate Dept. Chair from 1981 to 1987, and since then as Hitachi America Professor of Engineering. He has also held shorter-term appointments at several institutions around the world, including Bell Labs, UC Berkeley, Cambridge University, K.U. Leuven, T.U. Delft, the Indian Institute of Science, Imperial College, the Weizmann Institute, and M.I.T.

Professor Kailath's research has spanned a large number of disciplines, emphasizing information theory and communications in the sixties, linear systems, estimation and control in the seventies, and VLSI design and sensor array signal processing in the eighties. Concurrently, he contributed to several fields of mathematics, especially stochastic processes, operator theory and linear algebra. While he maintains all these interests to varying degrees, his current research emphasizes their applications to problems of semiconductor manufacturing and wireless communications. In the course of his work, Professor Kailath has mentored an outstanding array of over a hundred doctoral and postdoctoral students. He is active in several engineering and mathematics professional societies and has served on various national and international boards and panels. He has been editor of the Prentice Hall Information and System Sciences Series since 1963. He is a cofounder and a director of Integrated Systems, Inc., a leading provider of software tools and services for the many products that now use embedded microcomputers.

Professor Kailath has received awards from the IEEE Information Theory Society, which he served as President in 1975, the American Control Council, the European Signal Processing Society, and the IEEE Societies in Signal Processing, in Circuits and Systems, and in Semiconductor Manufacturing. He is the recipient of the 1995 IEEE Education Medal and the 1996 IEEE Donald G. Fink Prize Award. He has held Guggenheim, Churchill and other fellowships and has been awarded honorary doctorates by Linköping University in Sweden and Strathclyde University in Scotland. He is a Fellow of the IEEE and of the Institute of Mathematical Statistics, and a member of the National Academy of Engineering, the American Academy of Arts and Sciences and the Third World Academy of Sciences.

Professor Kailath and his wife, Sarah, reside in Stanford, California. They have four children: Ann, Paul, Priya and Ryan.

CONTENTS

PART II
Mathematics and Computation

PART III

PART IV

Nonlinear Systems and Control 411

PART V

Signal Processing 493

1

INTRODUCTION

A. Paulraj*, V. Roychowdhury, and C. Schaper***

** Dept. of Electrical Engineering, Stanford University*
*** Dept. of Electrical Engineering, UCLA*

Innumerable conferences are held around the world on the subjects of communications, computation, control and signal processing, and on their numerous subdisciplines. Therefore one might not envision a coherent conference encompassing all these areas. However, such an event did take place June 22-26, 1995, at an international symposium held at Stanford University to celebrate Professor Thomas Kailath's sixtieth birthday and to honor the notable contributions made by him and his students and associates. The depth of these contributions was evident from the participation of so many leading figures in each of these fields. Over the five days of the meeting, there were about 200 attendees, from eighteen countries, more than twenty government and industrial organizations, and various engineering, mathematics and statistics faculties at nearly 50 different academic institutions.

They came not only to celebrate but also to learn and to ponder the threads and the connections that Professor Kailath has discovered and woven among so many apparently disparate areas. The organizers received many comments about the richness of the occasion. A distinguished academic wrote of the conference being "the single most rewarding professional event of my life." The program is summarized in Table 1.1; a letter of reflections by Dr. C. Rohrs appears a little later. This volume, dedicated to Professor Kailath, is an attempt to bring some of this wealth of material to a wider audience. It provides reviews, new results and perspectives on issues and problems of growing importance in the information era.

The editors did not venture to provide a commentary on the wide range of Professor Kailath's contributions. However, as an aid to students and researchers, we have presented the list of his publications under various categories rather than just sequentially. Table 1.2 is a graphical summary that also depicts the evolution of his research interests over time; for some light relief, see also Fig. 13.1 on page 253. There is a lot to be said about Professor Kailath's other contributions, e.g., his stellar array of about 70 Ph.D. graduates and 40 postdoctoral scholars, many of them already outstanding figures in their own right, his entrepreneurial ventures, and his awards and honors from a wide variety of institutions. We content ourselves with a list of his students and postdocs, and with reproducing an eloquent letter from Prof. J.F. Gibbons, Dean of Engineering at Stanford University.

Though Professor Kailath did not make any formal presentation at the meeting, the editors felt that a fine example of the way he has woven together ideas from a number of different fields could be provided by reprinting a paper he contributed to honor the birth centennial of one of his own personal heroes, and one of the greatest mathematicians of our time, Norbert Wiener.

Some final words of thanks. First, to Christine Lincke for the support she provided during the early planning stages of this book and, more importantly, her outstanding organizational efforts for the conference that led to this volume. Partial support for the conference was provided by the Army Research Office (Mathematics Division) and the National Science Foundation (Engineering Division). Many thanks also to all the conference participants and to everyone who helped with this happy and memorable event. We are especially grateful for the efforts of all the contributors to this volume.

Communication and Information Systems	Mathematics and Computation
A. Paulraj (Stanford)	*J. Helton (San Diego)*
R. Gallager (MIT)	P. Dewilde (Delft)
G.D. Forney (Motorola)	I. Gohberg (Tel Aviv)
J. Cioffi (Stanford)	C. Foias (Indiana)
W. Lindsey (USC)	*E. Parzen (Texas A&M)*
D. Messerschmitt (Berkeley)	G. Golub (Stanford)
A. Paulraj (Stanford)	G. Xu (UT, Austin)
D. Coll (Carleton)	*J. Chandra (ARO)*
N. Abramson (Hawaii)	B.D.O. Anderson (ANU)
T. Cover (Stanford)	J. Willems (Groningen)
B. Atal (Bell Labs)	S. Y. Kung (Princeton)
L. Kanal (Maryland)	*B. Widrow (Stanford)*
L. Zadeh (Berkeley)	V. Roychowdhury (Purdue)
A. Viterbi (Qualcomm)	D. Slock (Eurecom)
Informal presentations	*Informal presentations*

Systems and Control	Signal Processing
S. Sastry (Berkeley)	*M. Najim (Bordeaux)*
S. Mitter (MIT)	T. Durrani (Strathclyde)
P. Kokotovic (Santa Barbara)	A. Bruckstein (Technion)
D. Mayne (Davis)	E. Robinson (Columbia)
G. Zames (McGill)	*R. Baheti (NSF)*
K. Aström (Lund)	S. Meerkov (Michigan)
H. Kimura (Tokyo)	N. Viswanadham (Bangalore)
P. Khargonekar (Michigan)	*S. Rao (SDE, Inc., N.J.)*
L.Ljung (Linköping)	B. DeMoor (Leuven)
G. Verghese (MIT)	R.-W. Liu (Notre Dame)
M. Gevers (Louvain)	A. El Gamal (Stanford)
J. Baras (Maryland)	*S. Silverstein (GE)*
I. Landau (Grenoble)	R. Schafer (Georgia Tech)
S. Boyd (Stanford)	B. Ottersten (Stockholm)
Informal presentations	*Informal presentations*

Table 1.1 Program for the *International Conference on Communications, Computing, Control and Signal Processing*, Stanford, CA, June 22-26, 1995.

		TOTAL	60	61	62	63	64	65	66	67	68	69	70	71	72	73	74	75	76	77	78	79	80	81	82	83	84	85	86	87	88	89	90	91	92	93	94	95
1	*IEEE Info Theory*	42	2		3	1			3	2	1	2	2	1	2	3	2	4	2		1	1	1				2	1				2				1	2	1
2	*Info and Control*	2							1		1																											
3	*IEEE Comm Tech*	3								1																	1	1										
4	*Ann of Math Stat/Prb*	6								1				2			1	1			1																	
5	*Proc IEEE*	7								1			1						1	1									1	1	1							
6	*IEEE Auto Control*	40									2	1		2		3	5	1	2	1	4	1	1	2		2	1				1	1			1		6	3
9	*SIAM J App Math Sci*	3												1								1						1										
10	*J Math Anal & Appl*	2												1								1																
12	*Automatica*	9													1	1		1	1			1	1						1								1	1
18	*IEEE Circ & Sys*	13																		1	1	2		1			1	1						3	2		1	
19	*IEEE Sig Processing*	42																		1					1	3	3	6	2	3	3	2	1	4	5	1	6	1
21	*SIAM Review*	3																			1									1								1
22	*Indian J of Stat*	2																			1												1					
24	*Int Eqs & Op Theory*	5																			1				1			1									1	1
26	*Lin Alg & Appl*	15																				1							4	1	1	3		1	1	1	1	1
27	*Int J Control*	11																				1	1		1	1	1	1			1	1		2				1
31	*SIAM Sci Stat Comp*	2																											1	1								
32	*IEEE Computers*	5																											1			1	1	1		1		
33	*J Math Phys*	1																												1								
34	*IEEE Sig Pro Mag*	4																												1					1		2	
39	*J Acoustical Soc*	1																													1							
43	*IEEE Neural Net*	3																															1	1				1
47	*SIAM J Matrix Anal*	9																																2		1	4	2
49	*App Physics A*	1																																	1			
53	*IEEE Semi Manu*	3																																		1	2	
57	*J Optical Soc A*	1																																			1	
59	*IEEE PAMI*	1																																			1	
60	*Mach Vision & App*	2																																			2	
61	*J Par & Dist Comp*	1																																				1
64	*J Vac Sci Technol*	1																																				1
69	*Math of Comp*	1																																				1
	TOTAL Journals	282	2	0	3	1	0	0	4	5	4	4	3	9	4	8	8	7	7	5	11	12	4	3	3	7	9	12	10	11	11	12	8	17	11	10	35	22
	Books/Edited Books	7																		1			1	1				2					1					1
	Book Chaps (Unique)	32		3		1							1	1				1		1	1	1		1	3	1	1	3	2	1	2	1		2		2	1	2
	Conf Pubs (Unique)	95						2		1	2				1	2	4	2	3	3	1	5	1	1	1	5	7	4	5	5	2	2	1	3	1	11	15	5
	Overall TOTAL	416	2	3	3	2	0	2	4	6	6	4	4	10	5	10	12	10	10	10	13	18	6	6	7	13	17	21	17	17	15	15	10	22	12	23	51	30

Table 1.2 Publications profile (not all journals are shown, for space reasons).

SCHOOL OF ENGINEERING
STANFORD UNIVERSITY, STANFORD, CALIFORNIA 94305

JAMES F. GIBBONS
THE FREDERICK EMMONS TERMAN
PROFESSOR OF ENGINEERING
DEAN, SCHOOL OF ENGINEERING

TERMAN ENGINEERING CENTER
(415) 723-3938

June 16, 1995

Dear Tom,

I am very sorry that I cannot attend the symposium and dinner honoring you on your 60th birthday. I write this note as a poor substitute for telling you personally, in the presence of your many friends and admirers, how indebted we are to you for the marvelous contributions you have made to your profession, to Stanford, and to each of us personally. Your career has been an extraordinary success many times over, and for a different set of reasons each decade. I have never seen anything like it in 40 years of service at Stanford.

The quality and breadth of your many rewards illustrate the esteem that has properly come to you over the years. They do not, however, give any hint of your engaging personal characteristics, your infectious laugh, or your special, personal concern for the careers of students and young faculty members. I have the opportunity to see the results of this concern particularly well in my present position, and I thank you for it most sincerely.

Naturally, we do not expect you to stop work just because it is your 60th birthday. There are, after all, several decades left. The question is simply where your imagination will carry you next.

You always have our warmest, best wishes and our thanks for the life you have led among us.

Sincerely,

Jim

Reflections on the Kailath Conference

Charles E. Rohrs
Fellow
Tellabs Research Center
3740 Edison Lakes Parkway
Mishawaka, IN 46545

It was a great privilege to be able to attend Prof. Kailath's Birthday Conference and Dinner. There's a sign on an eating establishment near MIT that says, "Some of the smartest people in the world eat here." An extraordinary number of the smartest people in the world gathered at Stanford to honor Prof. Kailath. The list of speakers is in the record but what was truly astonishing was the list of those who were not called upon to speak but came great distances to listen, to share and to give testimony to Prof. Kailath through their presence. I personally felt like a kid who had somehow wandered into the Yankee clubhouse by mistake. While Mickey Mantle and Yogi Berra may have been absent, in their place I found the names from the cover pages of nearly every important textbook and seminal paper I had ever read. More importantly, the names came complete with the on-line wit and wisdom of the brains behind the papers. In this setting I was stimulated to listen, to conjecture, to muse, to ponder, to think deeply, to laugh, and to renew the joy of work we get to do.

The format of the conference allowed the full benefit of the assembled knowledge to be tasted and savored like a tasting of great wines. Though the speakers were instructed to present their ideas without equations, most speakers interpreted this in an engineering sense, producing zero equations for large values of zero. The balance that was struck produced thoughtful presentations accessible to all and revealing the true nature of the profoundly simple concepts needed to solve complex problems.

I participated through the four days and managed to hang in through almost every talk with significant understanding. Prof. Kailath, unbelievable but not unexpectedly, had something new and relevant to add to every talk. I understand that a great story teller had an act where the audience would shout out any subject and he would respond with an appropriate joke. I came to believe that Prof. Kailath could respond to any subject within the four fields covered by the conference with a conjecture, a theorem, or an application. It was truly an amazing display of virtuosity.

What had to be the most impressive part of the gathering was the depth of insight and philosophy that was offered. The presentations dealt not with technical extension and epsilon additions to existing principles but with serious, penetrating analyses of those fundamental principles and with philosophical discussions of how research does and should happen. This conference was not another of the all too often repeated sessions of, "Can you top this?". Egos were checked at the door as speakers searched to expose and present the fundamental concepts that they had learned with and through Prof. Kailath. Indeed, we were learning about learning.

I came to the conference also expecting to hear something of a roasting of Prof. Kailath with stories of projects that had flopped, ideas that had turned sour and general pratfalls that often come with overreaching. I was actually somewhat disappointed with this aspect. Outside of Prof. Kailath's apparent lack of understanding of the role of the man in the loop of an automotive system, one speaker after another seemed to strain to find an anecdote of a Kailath foible. The situation was resolved by the following explanation that speaks much of Prof. Kailath's personality. As best I can recall, the speaker related, "Tom has such an incredible sense of balance that he avoids the excesses that create funny stories even though he attacks life with spirit and a sense of adventure." For me, this is a lasting characterization of Tom Kailath, the man.

Perhaps the most satisfying impression of the conference is that Prof. Kailath seems to have passed many of his qualities on to his students, many of whom themselves are teachers with students. Indeed, Prof. Kailath represents a branching point of an exponentially growing tree of dedicated learners and teachers passing concepts, techniques and, most importantly, firmly rooted values to generations of researchers who can only hope to match Prof. Kailath's productivity but who will be out there passing on the fundamental values taught and practiced by Prof. Kailath.

I never have had the opportunity to work with Prof. Kailath but I am one of the many who have been affected by his writings and his sphere of influence and I am now one of the few who have had the opportunity to gain some insight into why he could succeed as he has. I thank the group that organized the conference for this opportunity.

1. T. Kailath
2. G. Sidhu
3. I. Gohberg
4. S.S. Rao
5. B. Anderson
6. M. Najim
7. E. Robinson
8. S. Mitter
9. C. Lincke
10. Y. Rosen
11. R-W. Liu
12. C. Foias
13. E. Parzen
14. S. Ratnavel
17. T-J. Shen
18. R. Gallager
19. G. Zames
20. G. Golub
21. A. Paulraj
22. R. Kosut
23. S. Parker
24. L. Tong
25. A. Venecek
26. H. Lev-Ari
27. P. Schalkwijk
28. K. Lashkari
29. O. Farotimi
30. J. Chandra
31. J. Helton
32. B. Prasada
33. K. Flamm
34. K. Aström
35. B. Friedlander
36. E. Veirriest
37. M. Gevers
38. W. Lindsey
39. B. De Moor
40. J. Chen
41. K. Senne
42. G. Gulak
43. T. Boros
44. V. Olshevsky
45. N. Blackman
46. B. Ottersten
47. P. Fine
48. L. Scharf
49. S. Bittanti
50. L. Lome
51. J. Willems
52. M. Hadidi
53. D. Slock
54. B. Lévy
55. D. Lee
57. Y. Pati
59. M. Viberg
60. V. Roychowdhury
61. K-Y. Siu
62. J. Baras
63. Y. Bresler
64. S. Silverstein
65. G. Xu
66. A. Swindlehurst
67. W. Gardner
68. I. Koltracht
69. R. Schmidt
71. L. Padulo
72. S.K. Rao
73. A. Bruckstein
74. G. Verghese
75. S. Meerkov
76. M. Deistler
77. N. Viswanadham
78. A. Ramakrishna
79. P. Algoet
80. A. Andrews
81. D. Forney
82. C. Rohrs
84. C. Papadias
85. P. Dewilde
86. G. Raleigh
87. A. van der Veen

Doctoral Students

Name	Years	Present Affiliation
J.P.M. Schalkwijk	1965	Tech. Univ. Eindhoven
Maung Gyi	1965	Ampex
R.F. Daly	1966	Off. Tele. Policy
H.M. Hall	1966	Bell Labs
M.J. Ferguson	1966	Quebec Univ.
J.K. Omura	1966	Cylink
A.J. Kramer	1967	GTE
T.E. Duncan	1967	Univ. of Kansas
R.J.F. Fang	1968	Comsat
P.A. Frost	1968	Xybion, Inc., N.J.
B. Gopinath	1968	Rutgers Univ.
J.H. Babcock	1968	Mitre Corp.
R.A. Geesey	1969	BDM Corp.
D.L. Duttweiler	1970	Bell Labs
H.B. Aasnaes	1971	Informations Kontrol, Oslo
M.R. Gevers	1972	Univ. Cath. Louvain, Belgium
A. Lender	1972	Lockheed Research
H.L. Weinert	1972	Johns Hopkins Univ.
A. Segall	1973	Technion
N.F. Krasner	1974	Probe Systems
M. Morf	1974	Canon Res. Labs
B.W. Dickinson	1974	Princeton Univ.
G.S. Sidhu	1975	Apple Computer
B. Friedlander	1976	UC Davis
A. Shah	1977	Amdahl
J.R. Nickolls	1977	Ampex
S.Y. Kung	1977	Princeton Univ.
A.G.C. Vieira	1977	LNC, Rio de Janeiro
G.C. Verghese	1978	MIT
J.R. Dobbins	1979	Bell Labs
E.I. Verriest	1980	Georgia Tech.
D.T.L. Lee	1980	Hewlett-Packard
B.C. Levy	1981	UC Davis

Name	Years	Present Affiliation
K. Lashkari	1982	Monterey Bay Res. Inst.
J-M. Delosme	1982	Yale Univ.
B. Porat	1982	Technion
M.T. Hadidi	1983	Mobil Research
H. Lev-Ari	1983	Northeastern Univ.
J.M. Cioffi	1984	Stanford Univ.
A.M. Bruckstein	1984	Technion
M. Wax	1985	Min. of Defence, Israel
S. Rao	1985	Silicon Design Experts
H.V. Jagadish	1985	Bell Labs
Y. Bresler	1985	Univ. of Illinois
J. Jover	1986	Silicon Design Experts
T.K. Citron	1986	Hughes Aircraft
T.J. Shan	1986	Bell Labs
R.H. Roy	1987	ArrayComm
V. Roychowdhury	1988	UCLA
J. Chun	1989	KAIST, S. Korea
D. Slock	1989	EURECOM, Sophia-Antipolis
B. Ottersten	1989	Royal Inst. Tech., Stockholm
D. Pal	1990	Amati Communications
O. Farotimi	1990	George Mason Univ.
K.Y. Siu	1991	MIT
A.L. Swindlehurst	1991	Brigham Young Univ.
G. Xu	1991	UT Austin
R. Ackner	1991	Canon Res. Labs
T. Varvarigou	1991	Bell Labs
M. Genossar	1992	ECI Telecom., Israel
A. Sayed	1992	UC Santa Barbara
J. Gisladottir	1993	Univ. of Illinois
M. Goldburg	1993	ArrayComm
Y.M. Cho	1993	United Technologies
H. Aghajan	1994	KLA Instruments
P. Park	1995	Pohang Inst. of Tech., S. Korea
T. Boros	1996	Watkins Johnson
B. Khalaj	1996	KLA Instruments
Y.T. Wang	1996	Numeritech
B. Hassibi	1996	Stanford Univ.

Postdoctoral Scholars and Research Associates

Name	Years	Present Affiliation
J.M.C. Clark	1965-'66	Imperial College, London
N.S. Jayant	1967-'68	Bell Labs
B. Gopinath	1968	Rutgers Univ.
P. Caines	1973-'74	McGill Univ., Canada
L. Ljung	1974-'75	Linköping Univ., Sweden
M. Morf	1974-'75	Canon Res. Labs
P. Van Dooren	1979-'80	Univ. Cath. Louvain, Belgium
B. Egardt	1980-'81	Chalmers Univ. of Tech., Sweden
V. U. Reddy	1981-'83	Indian Inst. of Science
C. Samson	1981-'82	IRISA, Rennes, France
A-Z. Di	1982-'84	Inst. of Systems Science, Beijing
I. Koltracht	1983-'84	Univ. of Connecticut
A. Paulraj	1983-'86	Stanford Univ.
H. Lev-Ari	1984-'86	Northeastern Univ.
Y. Bistritz	1984-'86	Tel-Aviv Univ.
P. Gulak	1986-'87	Univ. of Toronto
Y. Bresler	1986-'87	Univ. of Illinois
L. Thiele	1987-'88	ETH, Zurich
P. Comon	1987	Thomson Sintra, Nice
J-J. Chu	1987-'88	Tsinghua Univ., Beijing
A. Dembo	1987-'89	Stanford Univ.
R.H. Roy	1987-'90	ArrayComm, Sunnyvale
L. Chisci	1988	Univ. of Florence
B. DeMoor	1988-'89	Katholieke Univ., Leuven
V. Roychowdhury	1989-'91	UCLA
B. Ottersten	1989	Royal Inst. Tech., Stockholm
C. Schaper	1990-	Stanford Univ.
P. Algoet	1991	IBM
L. Tong	1991	Univ. of Connecticut
K.Y. Siu	1991	MIT
T. Constantinescu	1991-'93	UT Dallas
G. Xu	1991	UT Austin
M. Viberg	1992-'93	Gothenburg Univ., Sweden
A. Sayed	1992-'93	UCLA
Y.C. Pati	1992-'95	Harvard Univ.
V. Olshevsky	1994-	Stanford Univ.
P. Park	1995-'96	Pohang Univ., S. Korea
B. Hassibi	1996-	Stanford Univ.

Publications

Books

[1] T. Kailath, *Linear Systems*, Prentice-Hall, 1980.
(Reviews in *IEEE Trans. Inform. Thy.*, pp. 385-386, May 1981; *IEEE Trans. Autom. Contr.*, pp. 804-805, June 1981; *IEEE Circuits and Systems Magazine,* p. 21, 1981; *Proceedings IEEE,* pp. 1519-1520, 1981, p. 107 Jan. 1982; *Can. Elec. Eng. J.*, 1980; *Kybernetika,*1981; *Zentralblatt fur Mathematik,* 1981; *Control Systems Magazine,* 1982; *Automatica,* 1982; *Math. Reviews,* 1982; *SIAM J. Contr.*, 1984)

[2] T. Kailath, *Lectures on Wiener and Kalman Filtering*, Springer-Verlag, 1981.
(Reviews in *IEEE Trans. Inform. Thy.*, pp. 321-322, March 1983; *Zent. fur Math.*, 1983; *Control Systems Magazine,* 1984)

[3] K.Y. Siu, V. Roychowdhury and T. Kailath, *Discrete Neural Computation: A Theoretical Foundation*, Prentice-Hall, 1995.
(Review in *IEEE Trans. Neural Networks* p. 533, March 1996)

[4] T. Kailath, A. Sayed and B. Hassibi, *State-Space Estimation*, Prentice-Hall, 1997.

Edited Books

[1] T. Kailath, ed., *Linear Least-Squares Estimation*, Benchmark Papers in Elec. Eng. and Comp. Sci., 13, Dowden, Hutchinson & Ross, Stroudsburg, PA, 1977.

[2] T. Kailath, ed., *Modern Signal Processing*, Hemisphere and Springer-Verlag, 1985.

[3] S.Y. Kung, H. Whitehouse and T. Kailath, eds., *VLSI and Modern Signal Processing*, Prentice-Hall, 1985.

[4] L. Auslander, T. Kailath, S. Mitter, eds., *Signal Processing – Part I: Signal Processing Theory*, The IMA Volumes in Mathematics and Its Applications, Volume 22, Springer-Verlag, 1990.

IEEE Transactions on Information Theory

[1] T. Kailath, "Correlation Detection of Signals Perturbed by a Random Channel," **6**(3):361-366, June 1960.

[2] T. Kailath, "Solution of an Integral Equation Occurring in Multipath Communications Theory," **6**(3):412, June 1960.

[3] T. Kailath, "Review of 'Theory of random functions and its application to problems of automatic control,' by V.S. Pugachev, (Fizmatgiz, Moscow; 1957. 730 pages.)" **8**(1):69-70, January 1962.

[4] T. Kailath, "On Multi-link and Multi-Dimensional Channels," **8**(3):260-262, April 1962.

[5] T. Kailath, "Measurements on Time-Variant Communications Channels," **8**(5):229-236, September 1962.

[6] T. Kailath, "Report on Progress in Information Theory in the USA, 1960-1963, Pt. IV: Time-Variant Communication Channels," **9**(4):223-264, October 1963.

[7] J.P. Schalkwijk and T. Kailath, "A Coding Scheme for Additive Noise Channels with Feedback, Pt. I: No Bandwidth Constraint," **12**(2):172-182, April 1966.

[8] T. Kailath, "The Complex Envelope of White Noise," **12**(3):397-398, July 1966.

[9] T. Kailath, "Some Integral Equations with Nonrational Kernels," **12**(4):442-447, October 1966.

[10] T. Kailath, "A Simple Rule for Checking the Validity of Certain Detection Formulas," **13**(1):144-145, January 1967.

[11] T. Kailath, "A Projection Method for Detection in Colored Gaussian Noise," **13**(3):441-447, July 1967.

[12] T. Kailath, "A Simple Example of an Explicit Nonlinear Predictor," **4**(4):600, July 1968.

[13] T. Kailath, "A General Likelihood-Ratio Formula for Random Signals in Gaussian Noise," **15**(3):350-361, May 1969.

[14] T. Kailath, "Fredholm Resolvents, Wiener-Hopf Equations and Riccati Differential Equations," **15**(6):655-672, November 1969.

[15] T. Kailath, "Likelihood Ratios for Gaussian Processes," **16**(3):276-288, May 1970.

[16] T. Kailath, "A Further Note on a General Likelihood Formula For Random Signals in Gaussian Noise," **16**(4):393-396, July 1970.

[17] T. Kailath, "An RKHS Approach to Detection and Estimation Problems Pt. I: Deterministic Signals in Gaussian Noise," **17**(5):530-549, September 1971.

[18] T. Kailath, R. Geesey and H. Weinert, "Some Relations Between RKHS Norms, Innovation Representations and Fredholm Equations," **18**(3):341-348, May 1972.

[19] T. Kailath and D.L. Duttweiler, "An RKHS Approach to Detection and Estimation Theory, Pt. III: Generalized Innovations Representations and a Likelihood Ratio Formula," **18**(6):730-745, November 1972.

[20] D.L. Duttweiler and T. Kailath, "An RKHS Approach to Detection and Estimation Theory, Pt. IV: Non-Gaussian Detection Problems," **19**(1):19-28, January 1973.

[21] D.L. Duttweiler and T. Kailath, "An RKHS Approach to Detection and Estimation Theory Pt. V: Some Parameter Estimation Problems," **19**(1):29-37, January 1973.

[22] T. Kailath, "Some New Algorithms for Recursive Estimation in Constant Linear Systems," **19**(6):750-760, November 1973.

[23] T. Kailath, "A View of Three Decades of Linear Filtering Theory," **20**(2):145-181, March 1974.

[24] G.S. Sidhu and T. Kailath, "Development of New Estimation Algorithms by Innovations Analysis and Shift-Invariance Property," **20**(6):759-762, November 1974.

[25] T. Kailath and H. Weinert, "An RKHS Approach to Detection and Estimation Problems – Part II: Gaussian Signal Detection," **21**(1):15-23, January 1975.

[26] A. Segall and T. Kailath, "The Modeling of Random Modulated Jump Processes," **21**(2):135-143, March 1975.

[27] A. Segall,M. Davis and T. Kailath, "Nonlinear Filtering with Counting Observations," **21**(2):143-149, March 1975.

[28] T. Kailath, "Comment on 'A Note on the Use of Chandrasekhar Equations for the Calculation of the Kalman Gain Matrix,' by R.F. Brammer," **21**(3):336-337, May 1975.

[29] A. Segall and T. Kailath, "Orthogonal Functionals of Independent-Increment Processes," **22**(3):287-298, May 1976.

[30] L. Ljung and T. Kailath, "Backwards Markovian Models for Second-Order Stochastic Processes," **22**(4):488-491, July 1976.

[31] T. Kailath, B. Levy, L. Ljung and M. Morf, "Fast Time-Invariant Implementations of Gaussian Signal Detectors," **24**(4):469-477, July 1978.

[32] G.C. Verghese and T. Kailath, "A Further Note on Backwards Markovian Models," **25**(1):121-124, January 1979.

[33] T. Kailath, "Review of 'Factorization Methods for Discrete Sequential Estimation' by G. Bierman," **26**(1):130-131, January 1980.

[34] H. Lev-Ari and T. Kailath, "Lattice Filter Parametrization and Modeling of Nonstationary Processes," **30**(1):2-16, January 1984.

[35] H. Lev-Ari, J. Cioffi and T. Kailath, "Least-Squares Adaptive Lattice and Transversal Filters: A Unified Geometric Theory," **30**(3):222-236, March 1984.

[36] A.M. Bruckstein and T. Kailath, "Recursive Limited Memory Filtering and Scattering Theory," **31**(3):440-443, May 1985.

[37] H. Lev-Ari, S.R. Parker and T. Kailath, "Multi-Dimensional Maximum-Entropy Covariance Extension," **35**(3):497-508, May 1989.

[38] Y. Bistritz, H. Lev-Ari and T. Kailath, "Immittance-Domain Levinson Algorithms," **35**(3):675-681, May 1989.

[39] K.Y. Siu, J. Bruck, T. Kailath and T. Hofmeister, "Depth Efficient Neural Networks for Division and Related Problems," **39**(3):946-956, May 1993.

[40] K.Y. Siu, V. Roychowdhury and T. Kailath, "Rational Approximation Techniques for Analysis of Neural Networks," **40**(2):455-466, March 1994.

[41] L. Tong, G. Xu and T. Kailath, "Blind Identification and Equalization Based on Second-Order Statistics: A Time Domain Approach," **40**(2):340-349, March 1994.

[42] L. Tong, G. Xu, B. Hassibi and T. Kailath, "Blind Channel Identification Based on Second-Order Statistics: A Frequency-Domain Approach," **41**(1):329-333, Jan. 1995.

Other Papers on Communications

[1] T. Kailath, "Some Results on Singular Detection," *Information and Control*, **9**(1):130-152, April 1966.

[2] T. Kailath, "The Divergence and Bhattacharyya Distance Measures in Signal Selection," *IEEE Trans. Commun. Tech.*, **15**(1):52-60, February 1967.

[3] T. Kailath, "A Note on Least-Squares Estimates from Likelihood Ratios," *Information and Control*, **13**(6):534-540, December 1968.

[4] T. Kailath, "Some Extensions of the Innovations Theorem," *Bell Sys. Tech. J.*, **50**(4):1487-1494, April 1971.

[5] T. Kailath, "The Fundamental Theorem of Information Theory," *Science Today*, pp. 34-39, June 1971.

[6] J.M. Cioffi and T. Kailath, "An Efficient Exact-Least-Squares Fractionally Spaced Equalizer Using Intersymbol Interpolation," *IEEE Trans. Communications*, **2**(5):743-756, September 1984.

[7] J.M. Cioffi and T. Kailath, "An Efficient RLS Data-Driven Echo Canceller for Fast Initialization of Full-Duplex Transmission," *IEEE Trans. Communications*, **33**(7):601-611, July 1985.

[8] R. Roy and T. Kailath, "ESPRIT - Estimation of Signal Parameters via Rotational Invariance Techniques," *J. Optical Engrg.*, **29**(4):296-312, April 1990.

[9] A. Swindlehurst and T. Kailath, "Azimuth/Elevation Direction Finding Using Regular Array Geometries," *IEEE Trans. Aerospace Electron. Systems*, **29**(1):145-156, January 1993.

[10] C-Y. Chang, J-W. Liang, A. Paulraj and T. Kailath, "A Multiple Access Technique for Cellular Packet Networks with Admission Control," *Computer Science and Informatics*, **24**(1):1-10, March 1994.

[11] A. Sayed and T. Kailath, "A State-Space Approach to Adaptive RLS Filtering," *IEEE Signal Processing Magazine*, **11**(3):18-60, July 1994.

[12] T. Kailath "Some Challenges in Telecommunications," *IETE Technical Review*, **11**(4):249-253, July-August 1994.

[13] A. Naguib, A. Paulraj and T. Kailath, "Capacity Improvement with Base-Station Antenna Arrays in Cellular CDMA," *IEEE Trans. Vehicular Technology*, **43**(3):691-698, August 1994.

[14] T. Kailath and A. Sayed, "On Lossless Cascades in Structured Matrix Factorization," *Int'l. J. of Electronics and Communications*, **49**(5/6):307-312, Sept.-Nov. 1995.

IEEE Transactions on Automatic Control

[1] T. Kailath, "An Innovations Approach to Least-Squares Estimation, Pt. I: Linear Filtering in Additive Noise," **13**(6):646-655, December 1968.

[2] T. Kailath and P. Frost, "An Innovations Approach to Least-Squares Estimation, Part II: Linear Smoothing in Additive White Noise," **13**(6):655-660, December 1968.

[3] R. Geesey and T. Kailath, "Some Comments on 'The Relationship of Alternate State-Space Representations in Linear Filtering Problems'," **14**(1):113-114, February 1969.

[4] P. Frost and T. Kailath, "An Innovations Approach to Least-Squares Estimation, Pt. III: Nonlinear Estimation in White Gaussian Noise," **16**(3):217-226, June 1971.

[5] T. Kailath and R. Geesey, "An Innovations Approach to Least-Squares Estimation, Part IV: Recursive Estimation given the Covariance Functions," **16**(6):720-727, December 1971.

[6] R. Geesey and T. Kailath, "An Innovations Approach to Least-Squares Estimation Pt. IV: Innovations Representations and Recursive Estimation in Colored Noise," **18**(5):435-453, October 1973.

[7] M. Gevers and T. Kailath, "An Innovations Approach to Least-Squares Estimation, Pt. VI: Discrete-Time Innovations Representations and Recursive Estimation," **18**(6):588-600, December 1973.

[8] H. Aasnaes and T. Kailath, "An Innovations Approach to Least-Squares Estimation, Pt. VII: Some Applications of Vector Autoregressive Moving Average Models," **18**(6):601-607, December 1973.

[9] B. Dickinson, M. Morf and T. Kailath, "A Minimal Realization Algorithm for Matrix Sequences," **19**(1):31-38, February 1974.

[10] M. Morf, G.S. Sidhu and T. Kailath, "Some New Algorithms for Recursive Estimation in Constant, Linear, Discrete-Time Systems," **19**(4):315-323, August 1974.

[11] H. Aasnaes and T. Kailath, "Initial-Condition Robustness of Linear Least-Squares Filtering Algorithms," **19**(4):393-397, August 1974.

[12] T. Kailath, "Introduction to Special Issue on System Identification and Time-Series Analysis," **19**(6):638-640, December 1974.

[13] B. Dickinson, M. Morf and T. Kailath, "Canonical Matrix Fraction and State Space Descriptions for Deterministic and Stochastic Linear Systems," **19**(6):656-667, December 1974.

[14] M. Morf and T. Kailath, "Square-Root Algorithms for Linear Least-Squares Estimation," **20**(4):487-497, August 1975.

[15] T. Kailath and L. Ljung, "Asymptotic Behaviour of Constant-Coefficient Riccati Differential Equations," **21**(3):385-388, June 1976.

[16] H.L. Weinert and T. Kailath, "A Spline Theoretic Approach to Minimum-Energy Control," **21**(3):391-393, June 1976.

[17] L. Ljung and T. Kailath, "Efficient Change of Initial Conditions, Dual Chandrasekhar Equations and Some Applications," **22**(3):443-447, June 1977.

[18] N.K. Gupta and T. Kailath, "Corrections and Extensions to an 'An Innovations Approach to Least Squares Estimation, Pt. VII: Some Applications of Vector ARMA Models'," **23**(3):511-512, June 1978.

[19] B. Friedlander, T. Kailath, M. Morf and L. Ljung, "Extended Levinson and Chandrasekhar Equations for General Discrete-Time Linear Estimation Problems," **23**(4):653-659, August 1978.

[20] M. Morf, B. Levy and T. Kailath, "Square-Root Algorithms for the Continuous-Time Linear Least-Square Estimation Problem," **23**(5):907-911, October 1978.

[21] R.R. Bitmead, S.Y. Kung, B.D. Anderson and T. Kailath, "Greatest Common Divisors via Generalized Sylvester and Bezout Matrices," **23**(6):1043-1047, December 1978.

[22] B. Levy, T. Kailath, L. Ljung and M. Morf, "Fast Time-Invariant Implementations for Linear Least-Squares Smoothing Filters," **24**(5):770-774, October 1979.

[23] G.C. Verghese, B. Friedlander and T. Kailath, "Scattering Theory and Linear Least-Squares Estimation, Pt. III: The Estimates," **25**(4):794-779, August 1980.

[24] G.C. Verghese and T. Kailath, "Rational Matrix Structure," **26**(2):434-439, April 1981.

[25] G.C. Verghese, B. Levy and T. Kailath, "A Generalized State-Space for Singular Systems," **26**(4):811-831, August 1981.

[26] E.I. Verriest and T. Kailath, "On Generalized Balanced Realizations," **28**(8):833-844, August 1983.

[27] T. Kailath, L. Ljung and M. Morf, "Recursive Input-Output and State-Space Solutions for Continuous-Time Linear Estimation Problems," **28**(9):897-906, September 1983.

[28] T. Kailath and M. Wax, "A Note on the Complementary Model of Weinert and Desai," **29**(6):551-552, June 1984.

[29] Y. Baram and T. Kailath, "Estimability and Regulability of Linear Systems," **33**(12):1116-1121, December 1988.

[30] R. Ackner and T. Kailath, "Complementary Models and Smoothing," **34**(9):963-969, September 1989.

[31] R. Ackner and T. Kailath, "On The Ptak-Young Generalization of the Schur-Cohn Theorem," **37**(10):1601-1604, October 1992.

[32] D. Pal and T. Kailath, "Displacement Structure Approach to Singular Root Distribution Problems: The Unit Circle Case," **39**(1):238-245, January 1994.

[33] A. Sayed and T. Kailath, "Extended Chandrasekhar Recursions," **39**(3):619-623, March 1994.

[34] A. Sayed, T. Constantinescu and T. Kailath, "Time-Variant Displacement Structure and Interpolation Problems," **39**(5):960-976, May 1994.

[35] T. Constantinescu, A. Sayed and T. Kailath, "A Recursive Schur Based Approach to the Four-Block Problem," **39**(7):1476-1481, July 1994.

[36] Y.M. Cho, G. Xu and T. Kailath, "Fast Identification of State-Space Models via Exploitation of Displacement Structure," **39**(10):2004-2017, October 1994.

[37] A. Sayed, T. Kailath and H. Lev-Ari, "Generalized Chandrasekhar Recursions from the Generalized Schur Algorithm," **39**(11):2265-2269, November 1994.

[38] A. Swindlehurst, R. Roy, B. Ottersten and T. Kailath, "A Subspace Fitting Method for Identification of Linear State-Space Models," **40**(2):311-316, Feb. 1995.

[39] P. Park and T. Kailath, "New Square-Root Algorithms for Kalman Filtering," **40**(5):895-899, May 1995.

[40] P. Park and T. Kailath, "New Square-Root Bryson-Frazier Smoothing Formulas," **40**(4):761-766, April 1995.

[41] B. Hassibi, A. Sayed and T. Kailath, "Linear Estimation in Krein Spaces – Part I: Theory," **41**(1):18-33, Jan. 1996.

[42] B. Hassibi, A. Sayed and T. Kailath, "Linear Estimation in Krein Spaces – Part II: Applications," **41**(1):34-49, Jan. 1996.

[43] P. Park and T. Kailath, "New Square-Root Smoothing Algorithms," **41**(5):727-733, May 1996.

[44] T. Boros, A. H. Sayed and T. Kailath "A Recursive Method for Solving Unconstrained Tangential Interpolation Problems," to appear.

Automatica and **International Journal of Control**

[1] J. Rissanen and T. Kailath, "Partial Realization of Stochastic Processes," *Automatica*, **8**:389-396, July 1972.

[2] M. Gevers and T. Kailath, "Constant, Predictable and Degenerate Directions of the Discrete-Time Riccati Equation," *Automatica*, **9**:699-711, November 1973.

[3] T. Kailath, "Supplement to 'A Survey of Data Smoothing'," *Automatica*, **11**(1):109-111, January 1975.

[4] T. Kailath and L. Ljung, "A Unified Approach to Smoothing Formulas," *Automatica*, **12**(2):147-157, March 1976.

[5] M. Morf, J.R. Dobbins, B. Friedlander and T. Kailath, "Square-Root Algorithms for Parallel Processing in Optimal Estimation," *Automatica*, **15**:299-306, 1979.

[6] G.C. Verghese and T. Kailath, "Comments on 'Structural Invariants and the Root-Loci of Linear Multivariable Systems'," *Int'l. J. Contr.*, **29**(6):1077-1080, June 1979.

[7] S.Y. Kung and T. Kailath, "Fast Projection Methods for Minimal Design Problems in Linear System Theory," *Automatica*, **16**(4):399-403, July 1980.

[8] T. Kailath, "Some Alternatives in Recursive Estimation," *Int'l. J. of Control*, **32**(2):311-328, August 1980.

[9] T. Kailath and L. Ljung, "Two-Filter Smoothing Formulas by Diagonalization of the Hamiltonian Equations," *Int'l. J. of Control*, **36**(4):663-673, October 1982.

[10] W.H. Kwon, A.M. Bruckstein and T. Kailath, "Stabilizing State-Feedback Design via the Moving Horizon Method," *Int'l. J. of Control*, **37**(3):631-643, March 1983.

[11] M. Wax and T. Kailath, "Direct Approach to the Two-Filter Smoothing Formulas," *Int'l. J. of Control*, **39**(3):517-522, March 1984.

[12] T. Kailath and L. Ljung, "Explicit Strict Sense State-Space Realizations of Nonstationary Gaussian Processes," *Int'l. J. of Control*, **42**(5):971-988, November 1985.

[13] J.M. Jover and T. Kailath, "A Parallel Architecture for Kalman Filter Measurement Update and Parameter Estimation," *Automatica*, **22**(1):43-57, January 1986.

[14] A.M. Bruckstein, T.K. Citron, and T. Kailath, "On Inverse Scattering and Partial Realizations," *Int'l. J. Control*, **48**(4):1537-1550, October 1988.

[15] R. Ackner and T. Kailath, "Discrete-Time Complementary Models and Smoothing," *Int'l. J. Control*, **49**(5):1665-1682, May 1989.

[16] J. Chun and T. Kailath, "Systolic Array Implementation of the Square Root Chandrasekhar Filter," *Int'l. J. Control*, **54**(6):1385-1398, December 1991.

[17] L. Chisci, H. Lev-Ari, D. T. M. Slock, and T. Kailath, "Fast Parallel Self-Tuning Controllers," *Int'l. J. Control*, **54**(6):1353-1384, December 1991.

[18] Y. Cho, G. Xu and T. Kailath, "Fast Recursive Identification of State Space Models via Exploitation of Displacement Structure," *Automatica*, **30**(1):45-49, January 1994.

[19] Y.M. Cho and T. Kailath, "Fast Subspace-Based System Identification: An Instrumental Variable Approach," *Automatica*, **31**(6):903-905, June 1995.

[20] P. Park and T. Kailath, "Square-Root RTS Smoothing Algorithms," *Int'l. J. Control*, **62**(5):1049-1060, November 1995.

IEEE Transactions on Signal Processing

[1] M. Morf, B. Dickinson, T. Kailath and A. Vieira, "Efficient Solution of Covariance Equations for Linear Prediction," **25**(5):429-434, October 1977.

[2] V.U. Reddy, B. Egardt and T. Kailath, "Least-squares type algorithm for adaptive implementation of Pisarenko's harmonic retrieval method," **30**(3):399-405, June 1982.

[3] B. Porat and T. Kailath, "Normalized Lattice Algorithms for Least-Squares FIR System Identification," **31**(1):122-128, February 1983.

[4] M. Wax and T. Kailath, "Efficient Inversion of Toeplitz-Block Toeplitz Matrix," **31**(5):1218-1221, October 1983.

[5] M. Wax and T. Kailath, "Optimum Localization of Multiple Sources in Passive Arrays," **31**(5):1210-1218, October 1983.

[6] J.M. Cioffi and T. Kailath, "Fast, Recursive-Least-Squares Transversal Filters for Adaptive Filtering," **32**(2):304-337, April 1984.

[7] M. Wax, T-J. Shan and T. Kailath, "Spatio-Temporal Spectral Analysis by Eigenstructure Methods," **32**(40):817-827, August 1984.

[8] M. Wax and T. Kailath, "Detection of Signals by Information Theoretic Criteria," **32**(6), December 1984.

[9] M. Wax and T. Kailath, "Detection of Signals by Information Theoretic Criteria," **33**(2):387-392, April 1985.

[10] T-J. Shan and T. Kailath, "Adaptive Beamforming for Coherent Signals and Interference," **33**(3):527-536, June 1985.

[11] J.M. Cioffi and T. Kailath, "Windowed Fast Transversal Filters Adaptive Algorithms with Normalization," **33**(3):607-625, June 1985.

[12] T-J. Shan, M. Wax and T. Kailath, "On Spatial Smoothing for Direction-of-Arrival Estimation of Coherent Sources," **33**(4):806-811, August 1985.

[13] M. Wax and T. Kailath, "Decentralized Processing in Passive Arrays," **33**(5):1123-1129, October 1985.

[14] A.M. Bruckstein, T-J. Shan and T. Kailath, "The Resolution of Overlapping Echoes," **33**(6):1357-1367, December 1985.

[15] A. Paulraj and T. Kailath, "Eigenstructure Methods for Direction of Arrival Estimation in the Presence of Unknown Noise Fields," **34**(1):13-20, February 1986.

[16] R. Roy, A. Paulraj and T. Kailath, "ESPRIT – A Subspace Rotation Approach to Estimation of Parameters of Cisoids in Noise," **34**(5):1340-1344, July 1986.

[17] V.U. Reddy, A. Paulraj and T. Kailath, "Performance Analysis of Optimum Beamformer in the Presence of Correlated Sources and Its Behavior Under Spatial Smoothing," **35**(7):927-936, July 1987.

[18] I. Gohberg, T. Kailath and I. Koltracht, "A Note on Diagonal Innovation Matrices," **35**(7):1068-1069, July 1987.

[19] T-J. Shan, A. Paulraj and T. Kailath, "On Smoothed Rank Profile Tests in Eigenstructure Approach to Directions-of-Arrival Estimation," **35**(10):1377-1385, October 1987.

[20] Y. Bresler, V.U. Reddy and T. Kailath, "Optimum Beamforming for Coherent Signal and Interferences," **36**(6), June 1988.

[21] I. Koltracht and T. Kailath, "Linear Complexity Parallel Algorithm for Discrete-Time Wiener Filters with Optimum Lag," **36**(10):1680-1681, October 1988.

[22] Y. Yoganadam, V.U. Reddy and T. Kailath, "Performance Analysis of the Adaptive Line Enhancer for Sinusoidal Signals in Broadband Noise," **36**(11):1749-1757, November 1988.

[23] D.T. Slock and T. Kailath, "Fast Transversal Filters with Data Sequence Weighting," **37**(3):346-359, March 1989.

[24] R. Roy and T. Kailath, "ESPRIT – Estimation of Signal Parameters via Rotational Invariance Techniques," **37**(7):984-995, July 1989.

[25] B. Ottersten and T. Kailath, "Direction-of-Arrival Estimation for Wideband Signals Using the ESPRIT Algorithm," **38**(2):317-327, February 1990.

[26] D. Slock and T. Kailath, "Numerically Stable Fast Transversal Filters for Recursive Least-Squares Adaptive Filtering," **39**(1):92-114, January 1991.

[27] B. Ottersten, M. Viberg, and T. Kailath, "Performance Analysis of the Total Least Squares ESPRIT Algorithm," **39**(5):1122-1135, May 1991.

[28] H. V. Jagadish and T. Kailath, "Obtaining Schedules for Digital Systems," **39**(10):2296-2316, October 1991.

[29] M. Viberg, B. Ottersten, and T. Kailath, "Detection and Estimation in Sensor Arrays Using Weighted Subspace Fitting," **39**(11):2436-2449, November 1991.

[30] B. Ottersten, M. Viberg, and T. Kailath, "Analysis of Subspace Fitting and ML Techniques for Parameter Estimation from Sensor Array Processing," **40**(3):590-600, March 1992.

[31] R. Roy, B. Ottersten, L. Swindlehurst, and T. Kailath, "Multiple Invariance ESPRIT," **40**(4):867-881, April 1992.

[32] D. Slock, L. Chisci, H. Lev-Ari, and T. Kailath, "Modular and Numerically Stable Fast Transversal Filters for Multichannel and Multiexperiment RLS," **40**(4):784-802, April 1992.

[33] A. Swindlehurst and T. Kailath, "A Performance Analysis of Subspace-Based Methods in the Presence of Model Errors, Pt. I: The MUSIC Algorithm," **40**(7):1758-1774, July 1992.

[34] G. Xu and T. Kailath, "Direction-of-Arrival Estimation via Exploitation of Cyclostationarity – A Combination of Temporal and Spatial Processing," **40**(7):1775-1786, July 1992.

[35] A. Swindlehurst and T. Kailath, "A Performance Analysis of Subspace-Based Methods in the Presence of Model Errors, Pt. II: The MUSIC Algorithm," **41**(9):2882-2890, September 1993.

[36] G. Xu, R. Roy and T. Kailath, "Detection of Number of Sources Via Exploitation of Centro-Symmetry Property," **42**(1):102-112, January 1994.

[37] G. Xu, S.D. Silverstein, R. Roy and T. Kailath, "Beamspace ESPRIT," **42** (2):349-356, February 1994.

[38] M. Genossar, H. Lev-Ari and T. Kailath, "Consistent Estimation of Cyclic Autocorrelation," **42** (3):595-603, March 1994.

[39] G. Xu and T. Kailath, "Fast Subspace Decomposition," **42** (3):539-551, March 1994.

[40] A. Sayed, H. Lev-Ari and T. Kailath, "Time-Variant Displacement Structure and Triangular Arrays," **42**(5):1052-1062, May 1994.

[41] G. Xu, Y.M. Cho and T. Kailath, "Application of Fast Subspace Decomposition to Signal Processing and Communication Problems," **42**(6):1453-1461, June 1994.

[42] G. Xu, H. Liu, L. Tong and T. Kailath, "A Least Squares Approach to Blind Channel Identification," **43**(12):2982-2993, December 1995.

[43] B. Hassibi, A.H. Sayed, and T. Kailath, "H^∞ Optimality of the LMS Algorithm," **44**(2):267-277, February 1996.

Other Papers on Signal Processing

[1] M. Morf, A. Vieira, D. Lee and T. Kailath, "Recursive Multichannel Maximum Entropy Spectral Estimation," *IEEE Trans. Geoscience Elect.*, **16**(2):85-94, April 1978.

[2] B. Egardt, T. Kailath and V.U. Reddy, "High-Resolution Spectral Analysis using Multi-Step Adaptive Prediction," *Circuits, Systems and Signal Processing*, **2**(4):422-443, November 1983.

[3] A. Paulraj, V.U. Reddy and T. Kailath, "Analysis of Signal Cancellation Due to Multipath in Optimum Beamformers for Moving Arrays," *IEEE Trans. Oceanic Engineering*, special issue on Underwater Acoustic Signal Processing, **12**(1):163-172, January 1987.

[4] D. Spielman, A. Paulraj and T. Kailath, "Eigenstructure Approach to Directions-of-Arrival Estimation in IR Detector Arrays," *J. Appl. Optics*, **26**(2):199-202, January 1987.

[5] A. Bruckstein and T. Kailath, "An Inverse Scattering Framework for Several Problems in Signal Processing," *ASSP Magazine*, **4**(1):6-20, January 1987.

[6] D.T. Slock and T. Kailath, "A Fast RLS Transversal Filter for Adaptive Linear Phase Filtering," *Int'l. J. Adaptive Control and Signal Processing*, **2**:157-179, 1988.

[7] A. Paulraj and T. Kailath, "Direction of Arrival Estimation by Eigenstructure Methods with Imperfect Spatial Coherence of Wavefronts," *J. Acoust. Soc. of Amer.*, **83**(3):1034-1040, March 1988.

[8] Y. Bresler and T. Kailath, "Model Tracking Based on Signal Shift and Shape," *Traitement du Signal*, **22**:269-291, 1988.

[9] R. Roy and T. Kailath, "ESPRIT - Estimation of Signal Parameters via Rotational Invariance Techniques," *J. Optical Engrg.*, **29**(4):296-312, April 1990.

[10] D.T. Slock, L. Chisci, H. Lev-Ari and T. Kailath, "Modular and Numerically Stable Multichannel Fast Transversal Filters," *IETE Technical Review*, (special issue), **35**(2):41-51, March 1989.

[11] J. Gisladottir, H. Lev-Ari and T. Kailath, "Orthogonal Realization of First-Order Allpass Filters for Two-Dimensional Signals," *Multidimensional Systems and Signal Process*, **1**(1):39-50, March 1990.

[12] D.T. Slock, L. Chisci, H. Lev-Ari, and T. Kailath, "Modular and Numerically Stable Multichannel Statistical Signal Processing," *IETE Technical Rev.* **35**(2):41-51, March-April 1990.

[13] M. Viberg, B. Ottersten, and T. Kailath, "Subspace Based Detection for Linear Structural Relations," *J. Combin. Inform. System Sci.*, **16**(2&3):170-189, 1991.

[14] H. Aghajan and T. Kailath, "Sensor Array Processing Techniques for Super Resolution Multi-Line Fitting and Straight Edge Detection," *IEEE Trans. Image Processing*, **2**(4):454-465, October 1993.

[15] A. Sayed and T. Kailath, "A State-Space Approach to Adaptive RLS Filtering," *IEEE Signal Processing Magazine*, **11**(3):18-60, July 1994.

[16] H. Aghajan, B. Khalaj and T. Kailath, "Estimation of Skew Angle in Text Image Analysis by SLIDE: Subspace-based Line Detection," *Machine Vision and Applications*, **7**:267-276, 1994.

[17] P. Park and T. Kailath, "An Extended Inverse QR Adaptive Filtering Algorithm," *Signal Processing*, **40**:311-318, 1994.

[18] H. Aghajan and T. Kailath, "SLIDE: Subspace-based Line Detection," *IEEE Trans. Pattern Analysis and Machine Intelligence*, **16**(11):1057-1073, November 1994.

[19] Y-T. Wang, Y.C. Pati and T. Kailath, "Depth of Focus and the Moment Expansion," *Optics Letters*, 1841-1843, 1995.

[20] T. Kailath and A.H. Sayed, "On Lossless Cascades in Structured Matrix Factorization," *Archiv Für Electronik and Übertragungstechnik*, **49**:307-312, November 1995.

[21] A.H. Sayed, B. Hassibi and T. Kailath, "Inertia Properties of Indefinite Quadratic Forms," *IEEE Signal Processing Letters*, **3**(2):57-59, February 1996.

[22] H. Liu, G. Xu, L. Tong, and T. Kailath, "Recent Developments in Blind Channel Equalization: From Cyclostationarity to Subspaces," to appear in a Special Issue on Subspace Signal Processing in *IEE Signal Processing Journal*, 1996.

IEEE Transactions on Circuits and Systems

[1] A. Vieira and T. Kailath, "On Another Approach to the Schur-Cohn Criterion," **24**(4):218-220, April 1977.

[2] P. Dewilde, A. Vieira and T. Kailath, "On a Generalized Szegö-Levinson Realization Algorithm for Optimal Linear Prediction based on a Network Synthesis Approach," **25**(9):663-675, September 1978.

[3] B. D. Anderson and T. Kailath, "Passive Network Synthesis via Dual Spectral Factorization," **26**(10):866-873, October 1979.

[4] B.D. Anderson and T. Kailath, "Forwards, Backwards and Dynamically Reversible Markovian Models of Second-Order Processes," **26**(11):956-965, November 1979.

[5] V.U. Reddy, B. Egardt and T. Kailath, "Optimized Lattice-Form Adaptive Line Enhancer for a Sinusoidal Signal in Broadband Noise," **28**:542-550, June 1981.

[6] S.K. Rao and T. Kailath, "Orthogonal Digital Filters for VLSI Implementations," **31**(11):933-945, November 1984.

[7] S.K. Rao and T. Kailath, "VLSI Arrays for Digital Signal Processing: Pt. I – A Model Identification Approach to Digital Filter Realizations," **32**(11):1105-1117, November 1985.

[8] A. Dembo, O. Farotimi, and T. Kailath, "High Order Absolutely Stable Neural Networks," **38**(1):57-65, January 1991.

[9] H. Lev-Ari, Y. Bistritz, and T. Kailath, "Generalized (Bezoutians) and Families of Efficient Root-Location Procedures," **38**(2):170-186, February 1991.

[10] H. Lev-Ari, K-F. Chiang, and T. Kailath, "Constrained-Input/Constrained-Output Stability for Adaptive RLS Lattice Filters," **38**(12):1478-1483, December 1991.

[11] H. Lev-Ari, T. Kailath, and J. Cioffi, "Adaptive Recursive-Least-Squares Lattice and Transversal Filters for Continuous Signal Processing," **39**(2):81-89, February 1992.

[12] R. Ackner, H. Lev-Ari, and T. Kailath, "Transmission-Line Models for the Modified Schur Algorithm," **39**(4):290-296, April 1992.

[13] G. Xu, H. Zha, G. Golub and T. Kailath, "Fast Algorithms for Updating Signal Subspaces," **41**(8), August 1994.

Proceedings of the IEEE

[1] T. Kailath, "An Application of Shannon's Rate-Distortion Theory to Analog Communication over Additive Gaussian Channels," **55**(6):1102-1103, June 1967.

[2] T. Kailath, "The Innovations Approach to Detection and Estimation Theory," **58**(5):680-695, May 1970.

[3] L. Ljung, T. Kailath and B. Friedlander, "Scattering Theory and Linear Least-Squares Estimation, Pt. I: Continuous-Time Problems," **64**(1):131-138, January 1976.

[4] S.Y. Kung, B. Levy, M. Morf and T. Kailath, "New Results in 2-D Systems Theory, Pt. II: 2-D State- Space Models-Realization and the Notion of Controllability, Observability and Minimality," **65**(6):945-961, June 1977.

[5] A. Paulraj, R. Roy and T. Kailath, "A Subspace Rotation Approach to Signal Parameter Estimation," **74**(7):1044-1046, July 1986.

[6] H.V. Jagadish, S.K. Rao and T. Kailath, "Array Architectures for Iterative Algorithms," **75**(9):1304-1321, September 1987.

[7] S.K. Rao and T. Kailath, "Regular Iterative Algorithms and Their Implementations on Processor Arrays," **76**(3):259-282, March 1988.

Applied Mathematics-SIAM Journals

[1] T. Kailath, "Application of a Resolvent Identity to a Linear Smoothing Problem," *SIAM J. Control Optim.*, **7**(1):68-74, February 1969.

[2] B. Anderson and T. Kailath, "Some Integral Equations with Nonsymmetric Separate Kernels," *SIAM J. Appl. Math. Sci. Comp.*, **20**(4):659-669, June 1971.

[3] T. Kailath, "A Note on Least-Squares Estimation by the Innovations Method," *SIAM J. Comput.*, **10**(3):477-486, August 1972.

[4] T. Kailath, A. Vieira and M. Morf, "Inverses of Toeplitz Operators, Innovations, and Orthogonal Polynomials," *SIAM Review*, **20**(1):106-119, January 1978.

[5] T. Kailath, B. Levy, L. Ljung and M. Morf, "The Factorization and Representation of Operators in the Algebra Generated by Toeplitz Operators," *SIAM J. Appl. Math.*, **37**(3):467-484, December 1979.

[6] A.M. Bruckstein, B.C. Levy and T. Kailath, "Differential Methods in Inverse Scattering," *SIAM J. Appl. Math.*, **45**(2):312-335, April 1985.

[7] A.M. Bruckstein, I. Koltracht and T. Kailath, "Inverse Scattering with Noisy Data," *SIAM J. on Sci. Stat. Comput.*, **7**(4):1331-1349, October 1986.

[8] A.M. Bruckstein and T. Kailath, "Inverse Scattering for Discrete Transmission-Line Models," *SIAM Review*, **29**(3):September 1987.

[9] J. Chun, T. Kailath and H. Lev-Ari, "Fast Parallel Algorithms for QR and Triangular Factorization," *SIAM J. Scient. and Stat.*, **8**(6):899-913, November 1987.

[10] J.M. Jover, T. Kailath, H. Lev-Ari, and S. K. Rao, "On the Analysis of Synchronous Computing Systems," *SIAM J. Comput.*, **19**(4):627-643, August 1990.

[11] J. Chun and T. Kailath, "Divide-and-Conquer Solutions of Least-Squares Problems for Matrices with Displacement Structure," *SIAM J. Matrix Anal. and Appl.*, **12**(1):128-145, January 1991.

[12] Y. Bistritz, H. Lev-Ari, and T. Kailath, "Immittance-Domain Three-Term Schur and Levinson Recursions for Quasi-Toeplitz Complex Hermitian Matrices," *SIAM J. Matrix Anal. and Appl.*, **12**(3):497-520, July 1991.

[13] D. Pal and T. Kailath, "Fast Triangular Factorization and Inversion of Hermitian Toeplitz and Related Matrices With Arbitrary Rank Profile," *SIAM J. Matrix Anal. and Appl.*, **14**(4):1016-1042, October 1993.

[14] T. Kailath and J. Chun, "Generalized Displacement Structure for Block-Toeplitz, Toeplitz-Block, and Toeplitz-Derived Matrices," *SIAM J. Matrix Anal. and Appl.*, **15**(1):114-128, January 1994.

[15] R. Ackner, H. Lev-Ari and T. Kailath, "The Schur Algorithm for Matrix-Valued Meromorphic Functions," *SIAM J. Matrix Anal. and Appl.*, **15**(1):140-150, January 1994.

[16] D. Pal and T. Kailath, "Fast Triangular Factorization and Inversion of Hankel and Related Matrices With Arbitrary Rank Profile," *SIAM J. Matrix Anal. and Appl.*, **15**(2):451-478, April 1994.

[17] G. Xu and T. Kailath, "Fast Estimation of Principal Eigenspace Using Lanczos Algorithm," *SIAM J. Matrix Anal. and Appl.*, **15**(3):974-994, July 1994.

[18] T. Constantinescu, A. Sayed and T. Kailath, "Displacement Structure and Completion Problems," *SIAM J. Matrix Analysis and its Applications*, **16**(1):58-78, Jan. 1995.

[19] A. Sayed and T. Kailath, "A Look-Ahead Block Schur Algorithm for Toeplitz-like Matrices," *SIAM J. Matrix Anal. and Appl.*, **16**(2):388-414, April 1995.

[20] T. Kailath and A. Sayed, "Displacement Structure: Theory and Applications," *SIAM Review*, **37**(3):297-386, September 1995.

Journal of Linear Algebra and Its Applications

[1] B. Friedlander, M. Morf, T. Kailath and L. Ljung, "New Inversion Formulas for Matrices Classified in Terms of their Distance from Toeplitz Matrices," **27**:31-60, October 1979.

[2] A.M. Bruckstein and T. Kailath, "Some Matrix Factorization Identities for Discrete Inverse Scattering," **74**:157-172, February 1986.

[3] T. Kailath, A.M. Bruckstein and D. Morgan, "Fast Matrix Factorizations via Discrete Transmission Lines," **75**:1-25, March 1986.

[4] T. Kailath and I. Koltracht, "Matrices with Block Toeplitz Inverses," **75**:145-153, March 1986.

[5] I. Gohberg, T. Kailath and I. Koltracht, "Efficient Solution of Linear Systems of Equations with Recursive Structure," **80**:81-113, August 1986.

[6] I. Gohberg, T. Kailath, I. Koltracht and P. Lancaster, "Linear Complexity Parallel Algorithms for Linear Systems of Equations with Recursive Structure," **88/89**:271-315, 1987.

[7] Y. Bistritz and T. Kailath, "Inversion and Factorization of Non-Hermitian Quasi-Toeplitz Matrices," **98**:77-121, January 1988.

[8] J. Chun and T. Kailath, "A Constructive Proof of the Gohberg-Semencul Formula," **121**:475-489, August 1989.

[9] Y. Bistritz, H. Lev-Ari and T. Kailath, "Immittance Versus Scattering Domains Fast Algorithms for Non-Hermitian Toeplitz and Quasi-Toeplitz Matrices," **120**, August 1989.

[10] Y. Bistritz, H. Lev-Ari and T. Kailath, "Immittance Versus Scattering Domains Fast Algorithms for Non-Hermitian Toeplitz and Quasi-Toeplitz Matrices," **122/123**:847-888, September 1989.

[11] J. Chun and T. Kailath, "Displacement Structure for Hankel, Vandermonde and Related (Derived) Matrices," **151**:199-227, June 1991.

[12] H. Lev-Ari and T. Kailath, "State-Space Approach to Factorization of Lossless Transfer Functions and Structured Matrices," **162-164**:273–295, February 1992.

[13] A.H. Sayed, H. Lev-Ari and T. Kailath, "Fast Triangular Factorization of the Sum of Quasi-Toeplitz and Quasi Hankel Matrices," **191**:74-106, September 1993.

[14] T. Boros, A. Sayed and T. Kailath, "Structured Matrices and Unconstrained Rational Interpolation ," **203-204**:155-188, July 1994.

[15] A.H. Sayed and T. Kailath, "Fast Algorithms for Generalized Displacement Structures and Lossless Systems," **219**:49-78, April 1995.

[16] V. Olshevsky and T. Kailath, "Displacement structure approach to polynomial Vandermonde and related matrices," To appear, 1996.

[17] V. Olshevsky and T. Kailath, "Bunch-Kaufman pivoting for partially reconstructible Cauchy-like matrices with applications to Hermitian Toeplitz-like linear systems and rational matrix interpolation problems," To appear, 1996.

Other Math Journals

[1] T. Kailath, "On Measures Equivalent to Wiener Measure," *Ann. of Math. Statistics*, **38**(1):261-263, February 1967.

[2] T. Kailath and M. Zakai, "Absolute Continuity and Radon-Nikodym Derivatives for Certain Measures Relative to Wiener Measure," *Ann. of Math. Stat.*, **42**(1):130-146, February 1971.

[3] T. Kailath, "The Structure of Radon-Nikodym Derivatives with Respect to Wiener and Related Measures," *Ann. of Math. Stat.*, **42**(3):1054-1067, June 1971.

[4] B. Anderson and T. Kailath, "The Choice of Signal-Process Models in Kalman-Bucy Filtering," *J. of Math. Analysis & Appls.*, **35**(3):659-668, September 1971.

[5] H. Weinert and T. Kailath, "Stochastic Interpretation and Recursive Algorithms for Spline Functions," *Ann. of Stat.*, pp. 787-794, July 1974.

[6] A. Segall and T. Kailath, "Radon-Nikodym Derivatives with Respect to Measures Induced by Discontinuous Independent-Increment Processes," *Ann. of Probability*, **3**(3):449-464, June 1975.

[7] B. Friedlander, T. Kailath and L. Ljung, "Scattering Theory and Linear Least Squares Estimation, Pt. II: Discrete-Time Problems," *J. Franklin Inst.*, **301**(1&2):71-82, Jan.-Feb. 1976.

[8] M. Morf and T. Kailath, "Recent Results in Least-Squares Estimation Theory," *Ann. of Economic and Social Measurement*, **6**(3):261-274, 1977.

[9] T. Kailath, A. Segall and M. Zakai, "Fubini-Type Theorems for Stochastic Integrals," *Sankhya, The Indian J. of Stat.*, Ser. A, **40**(Pt. 2):138-143, April 1978.

[10] M. Morf, A. Vieira and T. Kailath, "Covariance Characterization by Partial Autocorrelation Matrices," *Annals of Stat.*, **6**(3):643-648, 1978.

[11] B.D. Anderson and T. Kailath, "Fast Algorithms for the Integral Equations for the Inverse Scattering Problems," *Integrals Eqs. & Operator Thy.*, **1**(1):132-136, 1978.

[12] B.D. Anderson and T. Kailath, "Forwards and Backwards Models for Finite-State Markov Processes," *Adv. Appl. Probl.*, **11**:118-133, 1979.

[13] T. Kailath, S.Y. Kung and M. Morf, "Displacement Ranks of Matrices and Linear Equations," *J. Math. Analysis & Appls.*, **68**(2):395-407, April 1979.

[14] T. Kailath, S.Y. Kung and M. Morf, "Displacement Ranks of a Matrix," *Bulletin Amer. Math. Soc.*, **1**(5):769-773, September 1979.

[15] T. Kailath, "Redheffer Scattering Theory and Linear State-Space Estimation Problems," *Ricerche di Automatica, Special Issue on Math. Physics and System Theory*, **10**(2):136-162, December 1979.

[16] H. Lev-Ari and T. Kailath, "Generalized Schur Parametrization of Nonstationary Second-Order Processes," *Integral Equations and Operator Theory*, pp. 321-329, 1982.

[17] A.M. Bruckstein and T. Kailath, "On Discrete Schrodinger Equations and Their Two Component Wave-Equation Equivalents," *J. Math. Physics*, **28**(12):2914-2924, December 1987.

[18] T. Kailath, "Remarks on the Origin of the Displacement Rank Concept," *J. Appl. Math. Computation*, **45**(2):193-206, September 1991.

[19] M. Viberg, B. Ottersten, and T. Kailath, "Subspace Based Detection for Linear Structural Relations," *J. Combin. Inform. System Sci.*, **16**(2&3):170-189, 1991.

[20] A.H. Sayed, T. Kailath, H. Lev-Ari, and T. Constatinescu, "Recursive Solutions of Rational Interpolation Problems Via Fast Matrix Factorization," *Integral Equations and Operator Theory*, **20**:84-118, September 1994.

[21] I.Gohberg, T. Kailath, and V. Olshevsky, "Fast Gaussian elimination with partial pivoting for matrices with displacement structure," *Math. of Computation*, 64:1557-1576, Oct. 1995.

[22] T. Kailath and V. Olshevsky, "Displacement Structure Approach to Chebyshev-Vandermonde and Related Matrices," *Integral Equations and Operator Theory*, **22**:65-92, May 1995.

Papers on Computing — VLSI and Neural Networks

[1] H.V. Jagadish, R.G. Mathews, T. Kailath and J.A. Newkirk, "A Study of Pipelining in Computing Arrays," *IEEE Trans. Computers*, **35**(5):431-440, May 1986.

[2] P.G. Gulak and T. Kailath, "Locally Connected VLSI Architectures for the Viterbi Algorithm," *IEEE J. in Selected Areas in Communication*," (special issue on VLSI in Communication), **6**(3):527-537, April 1988.

[3] H.V. Jagadish and T. Kailath, "A Family of New Efficient Arrays for Matrix Multiplication," *IEEE Trans. Computers*, **38**(1):149-155, January 1989.

[4] V.P. Roychowdhury and T. Kailath, "Subspace Scheduling and Parallel Implementation of Non-Systolic Regular Iterative Algorithms," *J. VLSI Signal Processing*, **1**:127-142, October 1989.

[5] A. Dembo and T. Kailath, "Model-Free Distributed Learning," *IEEE Trans. Neural Networks*, **1**(1):58-70, March 1990.

[6] V.P. Roychowdhury, J. Bruck, and T. Kailath, "Efficient Algorithms for Reconfiguration in VLSI/WSI Arrays," *IEEE Trans. Computers*, **39**(4):480-489, April 1990.

[7] O. Farotimi, A. Dembo, and T. Kailath, "A General Weight Matrix Formulation Using Optimal Control," *IEEE Trans. Neural Networks*, **2**(3):378-394, May 1991.

[8] T. Varvarigou, V. P. Roychowdhury, and T. Kailath, "New Algorithms for Reconfiguring VLSI/WSI Arrays," *J. VLSI Signal Process.*, **3**(4):329-344, October 1991.

[9] K-Y. Siu, V. Roychowdhury, and T. Kailath, "Depth-Size Tradeoffs for Neural Computation," *IEEE Trans. Computers*, **40**(12):1402-1412, December 1991.

[10] T. Varvarigou, V.P. Roychowdhury and T. Kailath, "A Polynomial Time Algorithm for Reconfiguring Multiple-Track Models," *IEEE Trans. Computers*, **42**(4):385-395, April 1993.

[11] K.Y. Siu, A. Dembo, and T. Kailath, "On the Perceptron Learning Algorithm on Data with High Precision," *J. Comput. System Sci.*, **48**(2):347-356, April 1994.

[12] V. Roychowdhury, K-Y. Siu and T. Kailath, "Classification of Linearly Non-Separable Patterns by Linear Threshold Elements," *IEEE Trans. Neural Networks*, **6**(2):318-331, March 1995.

[13] K. Siu, V. Roychowdhury and T. Kailath, "Toward Massively Parallel Design of Multipliers," *J. on Parallel and Distributed Computing*, **24**:86-93, 1995.

[14] V. P. Roychowdhury, K. Y. Siu, A. Orlitsky and T. Kailath, "Vector Analysis of Threshold Functions," *Information and Computation*, Vol. 120, No. 1, pp. 22-31, July 1995.

[15] V. P. Roychowdhury and T. Kailath, "Study of Parallelism in Regular Iterative Algorithms," to appear in *Algorithmica*, 1996.

[16] T. Varvarigou, V. P. Roychowdhury, T. Kailath and E. Lawler, "Scheduling In and Out Forests in the Presence of Communication Delays," to appear in *IEEE Trans. on Parallel and Distributed Systems*, 1997.

Applications in Semiconductor Manufacturing

[1] C.D. Schaper, Y.M. Cho, and T. Kailath, "Low-Order Modeling and Dynamic Characterization of Rapid Thermal Processing," *Applied Physics A: Solids and Surfaces*, **A:54**(4):317-326, April 1992.

[2] Y.M. Cho, G. Xu and T. Kailath, "On-line Subspace System Identification," *Control Engineering Practice*, **1**(4):681-690, 1993.

[3] H. Aghajan, C. Schaper and T. Kailath, "Machine Vision Techniques for Sub-Pixel Measurement of Critical Dimensions," *J. Optical Engineering*, **32**(4):828-839, April 1993.

[4] Y. Cho and T. Kailath, "Model Identification in Rapid Thermal Processing Systems," *IEEE Trans. Semiconductor Manufacturing*, **6**(3):233-245, August 1993.

[5] Y.M. Cho, A. Paulraj, T. Kailath and G. Xu, "A Contribution to Optimal Lamp Design in Rapid Thermal Processing," *IEEE Trans. Semiconductor Manufacturing*, **7**(1):34-41, February 1994.

[6] C. Schaper, M. Moslehi, K. Saraswat and T. Kailath, "Control of MMST RTP: Repeatability, Uniformity, and Integration for Flexible Manufacturing," *IEEE Trans. on Semiconductor Manufacturing*, **7**(2):202-219, May 1994.

[7] Y.C. Pati and T. Kailath, "Phase-Shifting Masks for Microlithography: Automated Design and Mask Requirements," *J. Opt. Soc. Am. A*, **11**(9):2438-2452, Sept. 1994.

[8] C. Schaper, M. Moslehi, K. Saraswat and T. Kailath, "Modeling, Identification, and Control of Rapid Thermal Processing Systems," *J. Electrochem. Socy.*, **141**(11):3200-3209, November 1994.

[9] B. Khalaj, H. Aghajan and T. Kailath, "Patterned Wafer Inspection by High Resolution Spectral Estimation Techniques," *Machine Vision and Applications*, **7**:178-185, 1994.

[10] Y. Wang, C. Schaper and T. Kailath, "Pattern Recognition of Trench Width Using a Confocal Microscope," *Journal of Vacuum Science and Technology B*, **13**(3):840-847, May/June 1995.

[11] C. Schaper and T. Kailath, "Thermal Model Validation for Rapid Thermal Chemical Vapor Deposition (RTCVD) of Polysilicon," *Jounal of The Electrochemical Society*, **143**(1):241-248, January 1996.

[12] K. El-Awady, C. Schaper and T. Kailath, "Improvements in C_{pk} Using Real-Time Feedback Control," *IEEE Trans. Semiconductor Manufacturing*, **9**(1):87-94, February 1996.

Contributions in Books

[1] T. Kailath, "Optimum Receivers for Randomly Varying Channels," pp. 189-122 in *Information Theory*, Ed. C. Cherry, Butterworths, London, 1961.

[2] T. Kailath "Channel Characterization: Dispersive Channels," pp. 95-123, in *Lectures on Communication System Theory*, Ed. E.J. Baghdady, McGraw-Hill Book Co., New York, 1961.

[3] T. Kailath "Estimation and Correlation Detection for Time-Variant Random Channels," pp. 306-320, in *Lectures on Communication System Theory*, Ed. E.J. Baghdady, McGraw-Hill Book Co., New York, 1961.

[4] T. Kailath, "Adaptive Matched Filters," Ch. VI, pp. 109-140 in *Mathematical Optimization Techniques*, Ed. R. Bellman, Univ. of California Press, 1963.

[5] T. Kailath and P. Frost, "Mathematical Modeling of Stochastic Processes," pp. 1-38 in *Stochastic Problems in Control*, ASME Publications, 1968.

[6] R. Geesey and T. Kailath, "Applications of the Canonical Representation to Estimation and Detection in Colored Noise," pp. 435-457, in *Proc. Symp. on Computer Processing in Communications*, Polytechnic Institute of Brooklyn Press, pp. 435-457, 1970.

[7] T. Kailath, "Coding for Feedback Channels – A Survey," chapter 3, pp. 53-68 in *Progress in Radio Science 1966-1969*, Eds. W. Tilston and M. Sauzade, URSI, Brussels, Belgium, 1971.

[8] M. Morf, T. Kailath and B. Dickinson, "General Speech Models and Linear Estimation Theory," pp. 157-182, in *Speech Recognition*, Ed. D. Raj Reddy, Academic Press, New York, 1975.

[9] T. Kailath and L. Ljung, "A Scattering Theory Framework for Fast Least-Squares Algorithms," pp. 387-406 in *Multivariate Analysis-IV*, North Holland Publishing Co., 1977.

[10] T. Kailath, L. Ljung and M. Morf, "Generalized Krein-Levinson Equations for Efficient Calculation of Fredholm Resolvents of Nondisplacement Kernels," pp. 169-184, in *Topics in Functional Analysis*, Eds. I.C. Gohberg and M. Kac, Academic Press, NY, 1978.

[11] T. Kailath, "Some Alternatives in Recursive Filtering," pp. 55-69, in *Appls. of Kalman Filter to Hydrology, Hydraulics & Water Resources*, Ed. by C-L. Chiu, Univ. of Pittsburgh, PA, 1978.

[12] T. Kailath, A. Vieira and M. Morf, "Orthogonal Transformation (Square-Root) Implementations of the Generalized Chandrasekhar and Generalized Levinson Algorithms," pp. 81-91, in *Int'l. Symp. on Syst. Optimization & Analysis*, Eds. A. Bensoussan and J.L. Lions, Springer-Verlag, New York, 1979.

[13] T. Kailath, "Some Topics in Linear Estimation," pp. 307-350 in *Stochastic Systems: The Mathematics of Filtering and Identification Applications, Proc. NATO Advanced Study Institute*, Eds. M. Hazewinkel and J.C. Willems, Les Arcs, France, 1981.

[14] T. Kailath and B. Porat, "State-Space Generators for Orthogonal Polynomials," pp. 131-163 in *Harmonic Analysis and Prediction Theory: Essays in Honor of P. Masani*, Eds. H. Salehi and V. Mandrekar, North Holland, 1982.

[15] T. Kailath, "Equations of Wiener-Hopf Type in Filtering Theory and Related Applications," pp. 63-64 in *Norbert Wiener: Collected Works, Vol. III*, Ed. P. Masani, MIT Press, 1982.

[16] T. Kailath, "Notes on the Szegö Unit Circle Orthogonal Polynomials in Least-Squares Prediction Theory," vol. 1, pp. 43-46 in *G. Szegö, Collected Works, Vol. 1, 1915-1927*, Ed. R. Askey, Birkhauser-Verlag, Boston, 1982.

[17] T. Kailath and H. Lev-Ari, "Constant-Parameter Lattice Filters for Nonstationary Processes," pp. 717-732 in *Outils et Modeles Mathematiques pour l'Automatique l'Analyse de Systemes et le Traitement du Signal*, Vol. 3, Ed. I.D. Landau, Editions du CNRS, Paris, France, 1983.

[18] T. Kailath and I. Koltracht, "Statistical Signal Processing," Ch. VI, pp. VI-1-VI-54, in *Foreign Applied Sciences Assessment Center* Tech. Report, TAR-2030 (Soviet Applied Math. Research: Math. Theory of Systems, Control, and Statistical Signal Processing), La Jolla, CA, July 1984.

[19] T. Kailath, "Linear Estimation," several articles for *IFAC Encyclopedia of Systems and Control*, Ed. by M. Singh, Pergamon Press, 1984.

[20] T. Kailath and A.M. Bruckstein, "Naimark Dilations, State-Space Generators and Transmission Lines," pp. 173-186, in *Operator Theory: Advances and Applications, Advances in Invariant Subspaces and Other Results of Operator Theory*, Birkhauser, 1986.

[21] T. Kailath, "Signal Processing in the VLSI Era," pp. 5-24 in *Modern Signal Processing and VLSI*, Eds. S.Y. Kung, H. Whitehouse, and T. Kailath, Prentice-Hall, 1985.

[22] T. Kailath "Linear Estimation for Stationary and Near-Stationary Processes," pp. 59-128, in *Modern Signal Processing*, Ed. T. Kailath, Hemisphere and Springer-Verlag, 1985.

[23] T. Kailath and H. Lev-Ari, "On Mappings Between Covariance Matrices and Physical Systems," pp. 241-252, in *Contemporary Mathematics*, Vol. 47, Ed. B. Datta, Amer. Math. Socy., Providence, R.I., 1985.

[24] H. Lev-Ari and T. Kailath, "Comments on [64a]: On the Oscillations of Nonlinear Systems ," in *Norbert Wiener: Collected Works*, Vol. IV, Ed. P. Masani, MIT Press, 1985.

[25] T. Kailath, "A Theorem of I. Schur and Its Impact on Modern Signal Processing," pp. 9-30 (special issue), *Operator Theory: Advances and Applications, I. Schur Methods in Operator Theory and Signal Processing*, Ed. I. Gohberg, Birkhauser, 1986.

[26] H. Lev-Ari and T. Kailath, "Triangular Factorization of Structured Hermitian Matrices," pp. 301-324 (special issue), *Operator Theory: Advances and Applications, I. Schur Methods in Operator Theory and Signal Processing*, Ed. I. Gohberg, Birkhauser, 1986.

[27] T. Kailath, "Signal Processing Applications of Some Moment Problems," Vol. 37, pp. 71-109, in *Moments in Mathematics*, Ed. by H. Landau, Amer. Math. Socy., Providence, RI, 1987.

[28] R. Roy, A. Paulraj and T. Kailath, "ESPRIT – Estimation of Signal Parameters via Rotational Invariance Techniques," pp. 235-265, in *Singular Value Decomposition and Signal Processing*, Ed. E.F. Deprettere, Elsevier Science Publishers, 1988.

[29] H. Lev-Ari, Y. Bistritz and T. Kailath, "Generalized Bezoutians and Efficient Root-Location," pp. 63-72, in *Linear Circuits, Systems and Signal Processing: Theory and Application*, Eds. C.I. Byrnes, C.F. Martin, R.E. Saeks, North Holland, New York, 1988.

[30] V.P. Roychowdhury, P.G. Gulak, A. Montalvo and T. Kailath, "Decoding of Rate K/N Convolutional Codes in VLSI," Chapter 33, in *Concurrent Computations: Algorithms, Architecture and Technology*, Eds. S.K. Tewksbury, B.W. Dickinson and S.C. Schwartz, Plenum Press, NY, July 1988.

[31] T. Kailath and J. Chun, "Generalized Gohberg-Semencul Formulas," pp. 231-246 in *Operator Theory: Advances and Applications, Gohberg Anniversary Volume*, Springer-Verlag, 1989.

[32] J. Chun T. Kailath and V. Roychowdhury, "Systolic Array for Solving Toeplitz Systems of Equations," pp. 547–560, in *Spectral Analysis in One or Two Dimensions*, Eds. S. Prasad and R. Kashyap, Oxford & IBH Publishing Co., India, 1990.

[33] J. Chun and T. Kailath, "Generalized Displacement Structure for Block-Toeplitz, Toeplitz-Block, and Toeplitz-Derived Matrices," pp. 215-236, in *Numerical Linear Algebra, Digital Signal Processing and Parallel Algorithms*, Eds. G. Golub and P. Van Dooren, Springer-Verlag, 1991.

[34] T. Kailath, "From Kalman Filtering to Innovations, Martingales, Scattering and Other Nice Things," pages 55–88, in *Mathematical System Theory: The Influence of R.E. Kalman*, Ed. A.C. Antoulas, Springer-Verlag, Communications and Control Engineering Edition, 1991.

[35] T. Kailath and A. H. Sayed, "Fast Algorithms for Generalized Displacement Structures," in *Recent Advances in Mathematical Theory of Systems, Control, Networks and Signal Processing II*, Ed. H. Kimura, S. Kodama, Mita Press, 1991.

[36] T. Constantinescu, A. Sayed and T. Kailath, "Structured Matrices and Moment Problems," pp. 25-43 in *Challenges of a Generalized Systems Theory*, Eds. P. Dewilde, M.A. Kaashoek and M. Verhaegen, North Holland, 1993.

[37] T. Kailath, "Root Distribution and Displacement Structure," pp. 93-101 in *Fundamentals of Discrete-Time Systems: A Tribute to Professor Eliahu I. Jury*, Eds. M. Jamshidi, M. Mansour, B.D.O. Anderson and N.K. Bose, TSI Press, Albuquerque, 1993.

[38] A. Paulraj, T. Kailath, B. Ottersten, A. Swindlehurst, G. Xu, and R. Roy, "Subspace Methods for Direction Finding and Parameter Estimation," pp. 693-740 in *Handbook of Statistical Signal Processing and Its Applications*, Vol. 10, Eds. N.K. Bose and C.R. Rao, Elsevier Science Publishers B.V., New York, June 1993.

[39] T. Kailath, B. Hassibi and A. Sayed, "H^∞ Filtering is Just Kalman Filtering in Krein Space," pp. 7-15 in *Computing and Intelligent Systems*, Eds. S. Keerthi, Y. Narhari, N. Viswanadham, Tata McGraw-Hill Publishing Co. Ltd., New Delhi, 1993.

[40] L. Tong, G. Xu and T. Kailath, "Blind Identification and Equalization using Spectral Correlation Measures, Pt. II: A Time-Domain Approach," in *Cyclostationarity in Communications and Signal Processing*, Ed. W.A. Gardner, IEEE Press, 1994.

[41] T. Kailath, "Encounters with the Berlekamp-Massey Algorithm," pp. 209-220 in *Communications and Cryptography: Two Sides of One Tapestry*, Ed. Ed. R. Blahut, D.J. Costello, Jr.,U. Maurer and T. Mittelholzer, Kluwer, 1994.

[42] B. Hassibi, A. Sayed and T. Kailath, "LMS is H^∞ Optimal," pp. 65-88 in *Adaptive Control, Filtering, and Signal Processing*, (IMA Volumes in Mathematics and Its Applications, Volume 74), Ed. K.J. Astrom, G.C. Goodwin and P.R. Kumar, Springer-Verlag, 1995.

[43] A. Sayed, T. Constantinescu and T. Kailath, "Square-Root Algorithms for Structured Matrices, Interpolation, and Completion Problems," pp. 153-184 in *Linear Algebra for Signal Processing* (IMA Volumes in Mathemathics and Its Applications, Volume 69), Ed. A. Bojanczyk and G. Cybenko, Springer-Verlag, 1995.

[44] T. Kailath, C. Schaper, Y. Cho, P. Gyugyi, S. Norman, P. Park, C. Davis, M. Moslehi, S. Boyd, G. Franklin, K. Saraswat, "Control for Advanced Semiconductor Device Manufacturing: A Case History", pp. 471-486 in *CRC Controls Handbook*, Ed. W. Levine, CRC Press, 1996.

[45] T. Kailath, "Nobert Wiener and the Development of Mathematical Engineering," in "The Legacy of Norbert Wiener: A Centennial Symposium (tentative)," Ed. D. Jerison, *et al, Proceedings of Symposium in Pure Mathematics*, 60 (E), American Mathematical Society, to appear in 1997,

[46] A. H. Sayed and B. Hassibi and T. Kailath, "Inertia Conditions for the Minimization of Quadratic Forms in Indefinite Metric Spaces," *Operator Theory: Advances and Applications*, Birkhäuser, to appear.

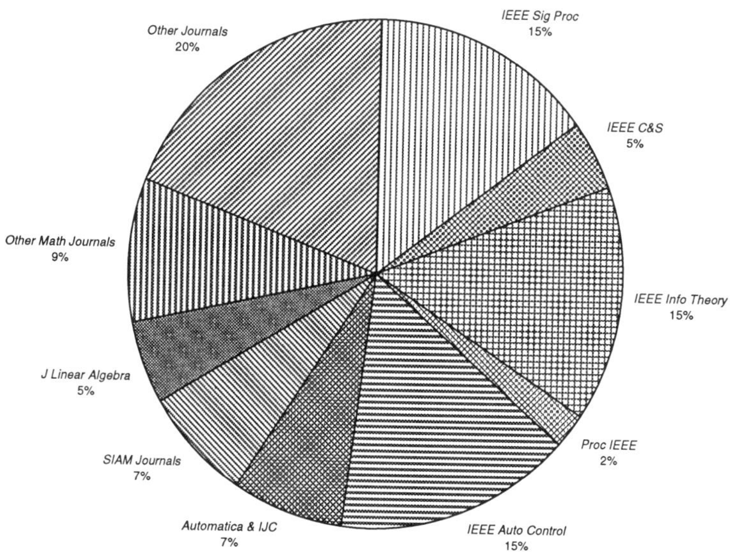

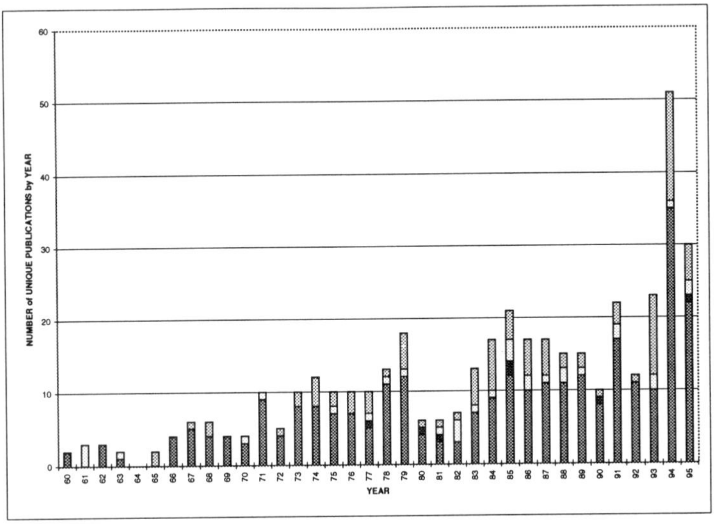

Number of unique publications by year grouped by journals (bottom), books, book chapters, and conferences (top).

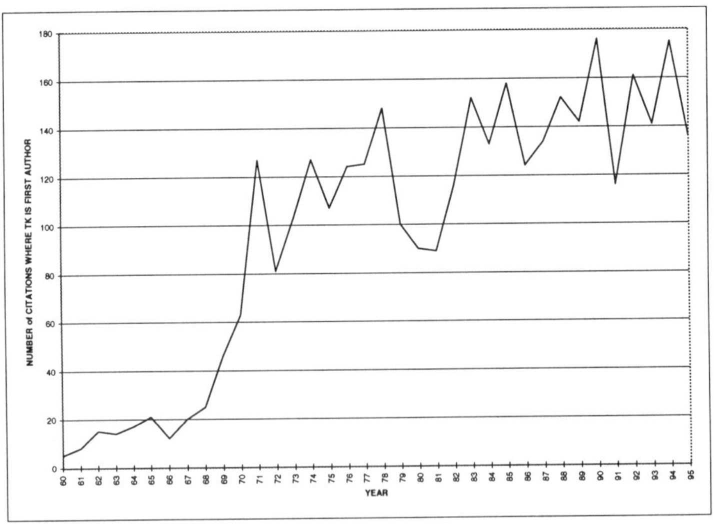

Citations to publications where TK is first author.
(There would be many more if all publications were included.)

2

NORBERT WIENER AND THE DEVELOPMENT OF MATHEMATICAL ENGINEERING

Thomas Kailath

Information Systems Laboratory
Department of Electrical Engineering
Stanford University
Stanford, CA 94305-4055

1 INTRODUCTION

"There is nothing better than concrete cases for the morale of a mathematician. Some of these cases are to be found in mathematical physics and the closely related mathematical engineering ..."

So wrote Norbert Wiener in 1949, in an obituary of G. H. Hardy. Here we shall describe how one particular concrete problem in Wiener's own work – solving the Wiener-Hopf equations encountered in astrophysics – led him, and then a vast host of followers, to chart out several new areas of investigation, and to develop a very significant body of knowledge, which can well go by the name Mathematical Engineering. In the era of the PC, the Internet and the World Wide Web, few of us can be unaware that mathematical engineering has come to play a major role in the world around us. And with this has come an increasing recognition of the seminal role of Norbert Wiener's ideas and influence in these developments.

It must be said that the term "Mathematical Engineering" does not enjoy the currency that the name "Mathematical Physics" does. Being a younger field, its proponents still focus on more specialized descriptions such as Information Theory, Communications, Computation, Control, Signal Processing, Image Processing, *etc.* The names 'System Theory' or even 'Mathematical System Theory' have been advanced but are not universally accepted. However this author believes that the increasing intermingling of the fields mentioned above, with many tools and techniques being successfully applied across them, as well as the tremendous opportunities ahead of them in the Information Era, could well lead to the adoption of Wiener's terminology. And be that as it may, Wiener's early vision and pioneering contributions will, as mentioned earlier, loom even larger with time. Already in 1962, in a special issue commemorating

This paper is reprinted from "The Legacy of Norbert Wiener: A Centennial Symposium (tentative)," edited by D. Jerison, *et al*, *Proceedings of Symposium in Pure Mathematics*, 60 (E), to appear in 1997, by permission of the American Mathematical Society.

the 50th anniversary of the effective existence of the IEEE (Institute of Electrical and Electronics Engineers), Lotfi Zadeh, winner of the 1995 IEEE Medal of Honor, wrote *"If one were asked to name a single individual who above anyone else is responsible for the conception of system theory, the answer would undoubtedly be "Norbert Wiener", even though Wiener has not concerned himself with system theory as such, nor has he been using the term "system theory" in the sense employed in this paper".*

There are many of Wiener's results that have come to be important in mathematical engineering. While Wiener was apparently never quite secure about his place in the pantheon of scientific innovators, he seemed to have no such doubts about the significance of the particular concrete problem that we shall focus on here. This is the Wiener-Hopf equation, which Wiener first encountered in 1931 in astrophysics.

We shall start, in the next section, with Wiener's beautiful technique of spectral factorization for solving it. In Sections 3 and 4, we shall note that Wiener encountered the equation again in solving a problem in anti-aircraft fire control and how his 1942 report on this project introduced two fundamental ideas that radically changed the way engineers approached important classes of problems. First is that the communication of information must be formulated as a statistical problem; the second, the introduction of optimization criteria to obtain limits of performance and replace the earlier 'trial and error' approach to design. From these two ideas has grown the huge flood of activity noted earlier. However to narrow the scope we shall return in the remainder of the paper to a specific problem studied by Wiener – filtering signals out of noisy observations. After describing his results, we shall turn to some of the mathematical developments following from it. First we shall show in Sec. 6, how, while Wiener was not quite successful in extending his results to the case of multiple time series, satisfactory computational solutions can be obtained by the introduction of state-space and Markov process descriptions. Somehow Wiener himself never really focused on the Markov property. As we shall see, the state-space description introduces the concept of recursive solution algorithms and enables straightforward extension to nonstationary/time-variant versions of the filtering problem, leading to the celebrated Kalman filter. In Sec. 7, we shall consider a finite-time prediction problem, which will not only bring in connections with the classical work of Carathéodory, Schur, Szegő and others on the moment problem, but also show the way to interesting extensions of these results through the concept of displacement structure. Finally in Sec. 8, we shall introduce the nonlinear filtering problem, which is still open, but for which a key tool is martingale theory, the prototype for which was the Wiener (Brownian motion) stochastic process. The presentation is sketchy and somewhat personalized; the aim is to indicate some of the richness of the work stimulated by just one of Wiener's many research interests.

2 THE WIENER-HOPF EQUATION

Given $\phi_{11}(\cdot)$ and $\phi_{12}(\cdot) \in \mathcal{L}_1(-\infty,\infty)$, $\phi_{11}(\cdot)$ even and positive definite, find $k(\cdot)$ such that

$$\phi_{12}(t) = \int_0^\infty \phi_{11}(t-\tau)k(\tau)d\tau, \qquad t \geq 0, \tag{2.1}$$

where $k(t) = 0$, $t < 0$. Equations of this type were apparently first introduced by Hvol'son (1894, St. Petersburg) while studying the scattering of light by milk glass, and later (ca 1920), by E. A. Milne, K. Schwarzschild and others in problems in astrophysics.

While such equations attracted considerable attention, explicit analytical solution was long thought to be impossible, a point of view that, in the words of M. G. Krein (1958), "was refuted in 1931 by the brilliant achievement of E. Hopf and N. Wiener". In fact, so striking in its ingenuity and simplicity was their solution that not only the technique, but the equation itself came to be known by the name "Wiener-Hopf". Moreover a decade later, Wiener encountered the equation again, with even more significant consequences!

The genesis of these events was apparently (see Pincus (1981)) a summer evening with E. Hopf at Wiener's country home in New Hampshire, where characteristically Wiener asked his visitor about challenging open problems in his field. The next morning Wiener came down for breakfast with a "solution". It in fact needed some reworking, which was done and the result published in 1931, with the unusual (for mathematicians) nonalphabetical ordering of names. We shall briefly outline the Wiener-Hopf idea (ignoring various technical assumptions, for which see, *e.g.*, Krein (1958)).

The tools used in the solution had in fact largely been developed by Wiener himself in collaboration with R. E. A. C. Paley. First were the properties in the complex plane of Fourier transforms of one-sided (or "causal") functions. One motivation for Wiener's interest in such topics may have been his study of electrical networks and his 1931 patent (with Y. W. Lee, Sc.D., Elec. Engg.) on a new cascade realization (based on Laguerre functions), of network impedance functions, known to mathematicians as Caratheodory (or positive-real) functions. In any case, for us the key property is that (again under certain conditions) a function $k(\cdot)$ such that $k(t) = 0$, $t < 0$, has a Fourier transform $\mathcal{K}(\omega)$ whose extension to the complex plane, $\mathcal{K}(s)$, $s = \sigma + i\omega$, is analytic in the RHP, $\sigma > 0$.

The second tool was the closely related fact that such extensions of nonnegative even functions

$$0 \leq \Phi(\omega) = \Phi(-\omega) \qquad -\infty < \omega < \infty$$

have unique canonical factorizations

$$\Phi(s) = \Psi(s)\Psi^*(-s^*), \qquad s = \sigma + i\omega, \tag{2.2}$$

where $\Psi(s)$ is analytic and bounded for $\sigma \geq 0$, while $\Psi^{-1}(s)$ is analytic for $\sigma > 0$. It may be noted that Wiener had recognized such functions, $\Phi(\cdot)$, as Fourier transforms of autocorrelation (real, even and positive definite) functions and described their significance as power spectral density functions. An equivalent result was discovered by Khinchin in 1934, leading to the famous Wiener-Khinchin theorem. [It turns out that they had been anticipated by no less than Albert Einstein, in a two-page 1914 sketch answering a question raised by a friend on measuring the "power" in meteorological time-series!] Now for the Wiener-Hopf technique.

We first extend the equation (2.1) to the whole line by introducing the function

$$g(t) = \phi_{12}(t) - \int_0^\infty k(\tau)\phi_{11}(t-\tau)d\tau \qquad -\infty < t < \infty.$$

By the fact that (2.1) only holds for $t \geq 0$, we know that $g(\cdot)$ is one-sided, but "anticausal",

$$g(t) = 0 \qquad t \geq 0$$

unlike the unknown "causal" function $k(\cdot)$,

$$k(t) = 0 \qquad t < 0.$$

Of course now we have two unknown functions, $g(\cdot)$ and $k(\cdot)$, but give us a moment. Take Fourier transforms in the complex plane to get

$$G(s) = \Phi_{12}(s) - \mathcal{K}(s)\Phi_{11}(s).$$

Now use the canonical factorization (2.2) to write

$$\frac{G(s)}{\Psi^*(-s^*)} = \frac{\Phi_{12}(s)}{\Psi^*(-s^*)} - \mathcal{K}(s)\Psi(s).$$

But by construction $G(s)$ and $1/\Psi^*(-s^*)$ are analytic in the LHP, $\sigma < 0$, while $\mathcal{K}(s)$ and $\Psi(s)$ are analytic in the RHP, $\sigma > 0$. Equivalently the time function obtained by inverse Fourier transformation (IFT) of $G(s)/\Psi^*(-s^*)$ will be zero for $t \geq 0$ (anticausal), while the time function corresponding to $\mathcal{K}(s)\Psi(s)$ will be zero for $t < 0$ (causal). This means that the latter time function must be equal to the $t \geq 0$ portion of the IFT of $\Phi_{12}(s)/\Psi^*(-s^*)$, leading to the famous formula for the solution of the Wiener-Hopf equation (2.1)

$$\mathcal{K}(s) = \frac{1}{\Psi(s)} \int_0^\infty dte^{-st} \oint \frac{\Phi_{12}(p)}{\Psi^*(-p^*)} e^{pt} \frac{dp}{2\pi i}. \tag{2.3}$$

We shall return to this equation in Sec. 5, after an apparent digression.

3 THE PROBLEM OF ANTI-AIRCRAFT FIRE CONTROL

Seeking eagerly to contribute to the War effort, Wiener submitted a proposal to the National Defense Research Council (NDRC) for a novel parallel computing machine for solving partial differential equations, as arising for example in aerofoil design problems. However, the proposal was deemed unlikely to be completed in a reasonable time frame. Wiener cast about for a more relevant problem and hit upon the problem of anti-aircraft fire control. In Dec. 1940, he sought and won a ($2350!) project on "*design of a lead or prediction apparatus in which, when one member follows the track of an airplane, another anticipates where the airplane is to be after a fixed lapse of time*". Under certain assumptions, the problem can be reduced to one of finding an effective method of approximating an exponential function by a rational function of suitable order, which Wiener proposed to do by using ideas from his earlier (ca 1930) work with Y. W. Lee on network synthesis. However working with the engineer hired for the project, Julian Bigelow, Wiener came to see in a few months, that because of noise and of model uncertainties, a satisfactory engineering solution demanded explicit consideration of the *statistical nature* of the problem, and also the use of an *optimization criterion*. Effectively assuming that the airplane trajectories were sample functions of a stationary and ergodic random process, and imposing the requirements of a *linear apparatus* and of the *mean-square-error criterion* led him again to the Wiener-Hopf integral equation. So the solution was at hand! But unlike the apocryphal mathematician, who promptly drops a problem when he reduces it to one that has already been solved, Wiener realized that much more was needed for a real engineering solution.

In fact, Wiener's project reports make a beautiful (and perhaps the first) case study in how to use the theory of *optimal* solutions in the real world. First, he confined himself to the practically most significant case: rational power spectral densities, necessary for "*the actual realization in the field by a finite electrical or mechanical structure.*" Here are a few other quotations: "*the detailed design of a filter involves ... choices ... which must be justified economically.*" "*Let us see ... the stages that are needed. The first stage determines the irreducible error, i.e., the error which cannot be reduced by any delay whatever. ... Next, ... , we can determine a reasonable delay, such that the delay error is not large in comparison with the irreducible error, the sense of "large" depending on the problem.*" Writing about the assumption of stationarity, he reassures the user that while the available "*statistical information will in fact never be complete, as our information does not run indefinitely far back into the past, it is a legitimate simplification of the facts to assume that the available information runs back much farther into the past than we are called upon to predict the future.*" And so on. Few mathematicians before Wiener had such an interest in the many issues arising in actual implementation.

Despite all this, however, the results actually obtained in field trials were not satisfactory, and the project was terminated; interesting accounts of this project have been written by Masani and Phillips (1985), Bennett (1994) and Mindell (1995). Wiener wrote up the theoretical results obtained in the course of the work, along with some mathematical preliminaries and his philosophical views on the nature of his new approach to the field of communication engineering, in a remarkable 1942 monograph "Extrapolation, Interpolation and Smoothing of Stationary Time Series". It was later declassified and published in 1949 by J. Wiley and the MIT Press. The classified report was widely circulated; while the engineers of the day were hard pressed to follow its contents, it was closely studied by mathematicians such as N. Levinson and R. Philips at MIT, and H. W. Bode, C. E. Shannon, R. Blackman, G. Stibitz and others at Bell Laboratories. They all wrote up expositions and variations of Wiener's results, some of which we shall mention later.

While preparing his report, Wiener was made aware (by W. Feller) that somewhat earlier, and apparently as a purely mathematical investigation, the famous Soviet mathematician A. N. Kolmogorov (1939,1941) had studied the prediction problem for general discrete-time stationary processes. The approaches were quite different: Wiener's concrete and focused on applications, Kolmogorov's more abstract and more general. Though later both the methodologies – concrete and abstract – became useful in applications, it was Wiener's work that had a greater impact on electrical engineering, far beyond the original problem and solution.

4 TWO NEW PARADIGMS

In fact, Wiener was quite aware (and perhaps too much so) of the revolutionary significance of his work and ideas, and he stated his views quite emphatically in his 1942 monograph and in various other talks and papers. Several passages from these works are quoted here, to show the clarity and prescience of Wiener's thinking.

We begin with his clear and forceful statement of the **statistical nature of the communication problem**. Thus on p.2 of his monograph:

> "*Communication engineering concerns itself with the transmission of messages. For the existence of a message, it is indeed essential that variable information be transmitted. The transmission of a single fixed item of information is of no communicative value. We must have a repertory of possible messages, and over this repertory a measure determining the probability of these messages.*"
>
> "*A message need not be the result of a conscious human effort for the transmission of ideas. For example the records of current and*

voltage kept on the instruments of an automatic substation are as truly messages as a telephone conversation."

And then on p.4:

"*A statistical method, as for example a method of extrapolating a time series into the future is judged by the probability with which it will yield an answer correct within certain bounds, or by the mean (taken with respect to probability) of some point,* i.e., *function or norm of the error contained in its answer.*"

"*No apparatus for conveying information is useful unless it is designed to operate, not on a particular message, but on a set of messages, and its effectiveness is to be judged on the way it performs on the average on messages of this set. ... The apparatus to be used for a particular purpose is that which gives the best result 'on the average' in an appropriate sense of the word 'average'.*"

To those now familiar with Information Theory, presented by Shannon in 1948, and Signal Detection Theory as presented by P. M. Woodward and others in the early 1950's, it is worth emphasizing that the above passages were written in 1942. As one major illustration of their influence, we may note Shannon's gracious acknowledgment [2] at the end of his magnificent 1948 paper founding Information Theory:

"*Credit must also be given to Professor Norbert Wiener, whose elegant solution of the problems of prediction and filtering has considerably influenced the writer's thinking in this field.*"

But there is more. Going beyond the statistical problems he had studied, Wiener stressed the possibility of formulating a variety of engineering problems as *optimization* problems, so that *performance limits* may be calculated and *reasonable* solutions sought; this stands in sharp contrast to many earlier *trial and error* methods:

"*These specifications give us an optimum filter to fit the situation exactly, whereas the earlier methods designed filters to certain specifications concerning passbands, sharpness of cut-off,* etc., *which stood in no obvious relation to the actual demands of a problem and had to be adjusted to these by the process of trial and error* [N.Y. Acad. Sci. lecture, Oct. 1946]."

[2]See also footnote 4 on p. 52 of that paper. A Feb. 1939 letter from Shannon (see Shannon (1994)) to Vannevar Bush gives more insight into Shannon's early (nonstatistical) thinking on the communications problem and to some of the reasons for the above remark.

and again from his 1942 monograph:

> "*Prediction and filtering do not exhaust the capacity of our methods. They may be applied whenever an ideally desirable linear operation ... is in fact not strictly realizable, although an approximation may be realized.*"

In particular, Wiener suggested that his ideas could be applied to the design of compensators for control systems. This suggestion was picked up in the Ph.D research of R. Newton at MIT, who, along with L. Gould and J. Kaiser, wrote a book on the topic entitled *Analytical Design of Linear Feedback Controls*, Wiley, 1957. From it, we may quote passages that clearly echo Wiener's early insights (in the fifties, authors were not as gender conscious as they are now):

> " ... *The analytical design procedure has several advantages over the trial and error method, the most important of which is the facility to detect immediately and surely an inconsistent set of specifications. The designer obtains a "yes" or "no" answer to the question of whether it is possible to fulfill any given set of specifications; he is not left with the haunting thought that if he had tried this or that form of compensation he might have been able to meet the specifications.*"
>
> "... *Even if the reader never employs the analytical procedure directly, the insight that it gives him into linear system design materially assists him in employing the trial and error design procedure.*"

However more than 50 years later, Wiener's message is apparently still worthy of repetition. In fact, the above quotation appears in Boyd and Barratt (1991). Or again, we may quote from a very recent text book by M. Green and D. J. N. Limebeer (1995):

> "*One does not want to waste time trying to solve a problem that has no solution, nor does one want to accept specification compromises without knowing that these are necessary. A further benefit of optimization is that it provides an absolute scale of merit against which any design can be measured – if a design is already all but perfect, there is little point in trying to improve it further.*"

We continue with some further remarks from this book on the Wiener ideas; From p.2, we quote (note the "free ride" that Hopf gets; he perhaps was completely unaware of these applications):

> "**Wiener-Hopf-Kalman Optimal Control**
> "*The first successes with control system optimization came in the*

1950's with the introduction of the Wiener-Hopf-Kalman (WHK) theory of optimal control. At roughly the same time, the United States and the Soviet Union were funding a massive research program into the guidance and maneuvering of space vehicles. As it turned out, the then new optimal control theory was well suited to many of the control problems that arose from the space program."

However, the major motivation for Green and Limebeer's 1995 book was the fact that they were introducing a new (so-called H_∞) formulation:

"*In contrast to experience with aerospace applications, it soon became apparent that there was a serious mismatch between the underlying assumptions of the WHK theory and industrial control problems. Accurate models are not routinely available and most industrial plant engineers have no idea as to the statistical nature of the external disturbances impinging on their plant.*"

"**Worst-Case Control: H_∞ Optimization**

"*H_∞ optimal control is a frequency-domain optimization and synthesis theory that was developed in response to the need for a synthesis procedure that explicitly addresses questions of modeling errors.*"

Ironically, the new H_∞ theory can be regarded as Wiener-Kalman theory, but in *Krein space* rather than in Hilbert space. More on this later.

Here, however, as a final quotation from Wiener's own monograph, let us present one that illustrates Wiener's keen insight into what mathematical refinements are relevant to engineers. Engineering textbooks are often still concerned about the level of rigor at which to present Fourier Theory – should one worry about pointwise convergence, L_2 convergence, *etc.*? Wiener cuts right to the heart of the matter:

"*Up to the present we have been treating the Fourier series as a purely formal expression without any regard to whether it converges or not. ... Now it is obvious that no physical quantity can be observed for a single precise value of t. small ranges of t rather than values at a precise point of t.*" [He goes on to very simply show that] "*all local averages of the formal series converge to the local average of* $f(t)$ [which] *is all that we need to make a practical employment of the Fourier series for* $f(t)$."

This is the approach that later in 1948 L. Schwartz elaborated in his theory of distributions which, among many other things, made rigorous the widespread use of impulsive functions by engineers. I cannot resist remarking that Wiener also uses the "Sampling Theorem for Bandlimited Functions", first derived by J. M. Whittaker in 1915, but to this day widely cited as Shannon's Sampling Theorem. There are several other little gems in Wiener's report, now fortunately also available in an MIT Press paperback entitled "Time Series". But let us now return to some more explicit mathematical problems.

5 THE WIENER FILTER

The appearance in 1949 of Wiener's monograph, and of Shannon's work on Information Theory, generated a huge wave of activity in, to use Wiener's phrase, "*mathematical engineering*". A striking range of mathematics has and is being used in these studies, as can be seen by glancing through, for example, the IEEE Transactions on Information Theory, on Automatic Control, on Signal Processing, on Circuits and Systems, on Image Processing, etc. Many of the algorithms were too complex for implementation till about five to ten years ago, but the pace of technology insertion is accelerating.

In the rest of this paper, we shall discuss one of the directions most closely related to Wiener's own work, focusing on some problems left open by him: spectral factorization of matrix-valued rational spectral densities; finite-time problems/nonstationary processes; nonlinear estimation/random signal detection. The idea is to give a glimpse of part of the wide range of consequences arising from just one of Wiener's mathematical contributions.

We begin with a review of the so-called (causal and noncausal) Wiener filters. A filter is a selective device, *e.g.*, one that allows the transmission of certain frequency regions (the passband) and rejects certain other regions (the stop band). In filtering signals out of a combination of the signals and additive noise, the conventional solution was a device that rejects as much of the noise as possible while passing the signal through undistorted. If as is usually the case, the signal is bandlimited (*i.e.*, occupies a limited frequency range) while the noise is wideband, the solution is the so-called Ideal Filter, with unit gain in the frequency regions of the signal and zero gain elsewhere. Wiener pointed out that if the signal is a random process, one might better be more concerned about regions where the power spectrum of the signal is much larger than that of the noise, with proper weighting to be assigned to different frequency regions depending upon the choice of an optimization criterion. The most fruitful has been Wiener's own choice of a least-squares criterion.

To be specific, consider the problem of estimating the value at a given time t of a stochastic signal process $s(\cdot)$ given the noisy observations

$$y(\tau) = s(\tau) + v(\tau) \qquad -\infty < \tau < \infty \tag{2.4}$$

by using a linear time-invariant filter $h(\cdot)$ such that

$$\hat{s}(t) = \int_{-\infty}^{\infty} h(t-\tau)y(\tau)d\tau$$

satisfies (E denotes expectation)

$$E\left|s(t) - \hat{s}(t)\right|^2 = \text{minimum}.$$

Under the assumption that the processes $\{s(\cdot), v(\cdot)\}$ are zero mean and jointly stationary, it is not hard to check that the optimum filter must satisfy the equation

$$\phi_{sy}(t) = \int_{-\infty}^{\infty} h(t-\tau)\phi_{yy}(\tau)d\tau,$$

where the $\phi(\cdot)$ are the autocorrelation functions

$$\phi_{sy}(t) = Es(\sigma+t)y^*(\sigma), \qquad \phi_{yy}(t) = Ey(\sigma+t)y^*(\sigma).$$

This equation is easily solved by Fourier transformation to yield

$$H(\omega) = \frac{\Phi_{sy}(\omega)}{\Phi_{yy}(\omega)}.$$

When the signal and noise processes are uncorrelated with each other, this reduces to

$$H(\omega) = \frac{\Phi_{ss}(\omega)}{\Phi_{ss}(\omega) + \Phi_{vv}(\omega)}. \tag{2.5}$$

To reflect the fact that the noise has a much higher frequency range than the signal, engineers often assume that the noise is *white*,

$$\Phi_{vv}(\omega) = R > 0, \qquad -\infty < \omega < \infty. \tag{2.6}$$

Such noise processes of course fall out of the scope of the usual theory of stationary processes, since for one thing, the IFT of $\Phi_{vv}(\omega)$ is a delta function. And, in fact, since to have finite power all physical processes must have spectra that decay to zero as $\omega \to \infty$, white noise is nonphysical as well. Yet it is widely used, as a model, for several important (mathematical <u>and</u> physical) reasons that we do not have time to explain here. However, it is interesting to see Wiener's free use of the white noise model, which is important for many reasons, both theoretical and practical (see Sec. 8):

> "*As for (the noise), we shall take a case which, although not formally contained in the theory we have given, constitutes the limiting case of it, and one of the greatest importance in practice. This is the case in which the noise input is due to a shot effect and has an equipartition of*

power in frequency. Theoretically, of course, this is not strictly realizable, as it would demand an infinite power; practically as in the case of Plank's law in optics, it may hold within the limits of observation up to frequencies of a magnitude so great that they are no longer of interest for our particular problem."

To return to our problem, with the white noise assumption, we can see that

$$H(\omega) = \frac{\Phi_{ss}(\omega)}{\Phi_{ss}(\omega) + R} R \to \infty.$$

On the other hand, as $R \to 0$, $H(\omega)$ tends to the ideal filter (with unit gain in the passband of the signal), but this is not the optimum filter when $R \neq 0$. As $R \to \infty$, we note that $H(\omega) \to \Phi_{ss}(\omega)/R$, so that the filter tends to reinforce frequencies where $\Phi_{ss}(\omega)/R > 1$, and to suppress the signal in regions where $\Phi_{ss}(\omega)/R < 1$. This fits, in hindsight, with our intuition, but the theory is necessary to tell us what to do for arbitrary values of R, or for nonwhite (often called colored) noise.

The above solution, though given by Wiener, does not use the Wiener-Hopf equation, because we are assuming that the process $y(\cdot)$ is observed over all time instants, past as well as future. When the observations of $y(\cdot)$ are restricted to the past, we have to solve a Wiener-Hopf equation, which when (2.4) and (2.6) hold, takes the form

$$Rk(t) + \int_0^\infty k(\tau)\phi_{ss}(t-\tau)d\tau = \phi_{ss}(t), \quad t \geq 0. \tag{2.7}$$

With the white noise assumption, the formula for the solution takes the striking form (apparently first noted by Yovits and Jackson (1960))

$$\mathcal{K}(\omega) = 1 - R^{\frac{1}{2}}\Psi^{-1}(\omega), \tag{2.8}$$

where $\Psi(\omega)$ is the canonical factor of $\Phi_{ss}(\omega) + R$ (cf. (2.2)). It is striking that the *canonical spectral factorization* completely defines the estimator in the additive white noise case!

One might justly wonder about the physical significance of the canonical factorization, and the stochastic problem allows a nice (and far reaching) interpretation, first given by Bode and Shannon (1950), and independently (and in somewhat more general form) in a lesser-known paper of Zadeh and Ragazzini, also appearing in 1950. These authors noted that passing the observations process $y(\cdot)$ through a linear filter with transfer function $\Psi^{-1}(\omega)$ gives us a process $e(\cdot)$ with power spectrum (using well-known formulas),

$$\Phi_{ee} = \Psi^{-1}(\omega)\Phi_{yy}(\omega)\Psi^{-*}(-\omega) = I.$$

Therefore, we can interpret the first term $\Psi^{-1}(\omega)$ in the general Wiener-Hopf formula (2.3)

$$\mathcal{K}(\omega) = \frac{1}{\Psi(\omega)}\left[\int_0^\infty dte^{-j\omega t}\oint \frac{\Phi_{12}(p)}{\Psi^*(-p^*)}e^{pt}\frac{dp}{2\pi i}\right],$$

as allowing us to replace the observations process $y(\cdot)$ by a much simpler stochastic process $e(\cdot)$, for which the problem of estimating a related stochastic process $s(\cdot)$ turns out to be much simpler: when $\phi_{11}(\cdot)$ in the Wiener-Hopf equation

$$\phi_{12}(t) = \int_0^\infty \phi_{11}(t-\tau)k(\tau)d\tau, \qquad t \geq 0$$

is a delta function, the solution is immediate: $k(t) = \phi_{12}(t), \quad t \geq 0$.
One might wonder if there is a "loss of information" in going from the original observed process $y(\cdot)$ to the white process $e(\cdot)$? The answer is no, because by the fact that the canonical factor $\Psi(s)$ and its inverse $\Psi^{-1}(s)$ have the property that they are analytic in the RHP, one can pass (recall the Paley-Wiener results quoted in Sec. 3) from $y(\cdot)$ to $e(\cdot)$ and from $e(\cdot)$ back to $y(\cdot)$ by causal and stable linear operations. Since

$$\mathcal{F}^{-1}\left\{\Phi_{ee}(\omega)\right\} = Ee(\tau+t)e^*(t) = \delta(\tau) = 0, \quad \tau \neq 0,$$

the value of $e(\cdot)$ at any instant is uncorrelated with its values at any other instant, and therefore every observation, $e(t)$, brings *new* information, which cannot be said about a (correlated) nonwhite process. The process $e(\cdot)$ is called the *innovations process* of $y(\cdot)$; it has been useful in extensions of Wiener filtering theory to nonstationary processes and to nonlinear problems, see Sec. 8. We may note that Kolmogorov's more general approach to the discrete-time prediction problem was based on the use of the innovations process, which avoids (or rather trivializes, as we noted earlier) the use of the Wiener-Hopf equation. Thus somewhat ironically, Kolmogorov's more abstract approach ultimately became more powerful than Wiener's more concrete approach, a phenomenon, mathematicians may be pleased to know, that is not uncommon in applications.

6 MATRIX SPECTRAL FACTORIZATION AND THE KALMAN FILTER

Wiener's monograph inspired various attempts at extensions – to finite-time nonstationary problems, and to vector-valued processes in particular. When observations are only available over a finite time, say $(0, t)$ rather than $(-\infty, t)$, the W-H equation is replaced by one of "W-H type",

$$h(t,s) + \int_0^t h(t,\tau)\phi(\tau - s)d\tau = \phi(t-s), \quad 0 \le s \le t. \tag{2.9}$$

No general methods were or are known for its solution, and a vast literature developed on various special cases, tricks, etc.; so much so that a 1958 editorial by P. Elias urged no more work on "*Two Famous Papers*". One generic title was "The Optimum Linear Mean Square Filter for Separating Sinusoidally Modulated Triangular Signals from Randomly Sampled Stationary Gaussian Noise, with Applications to a Problem in Radar". (The other: "Information theory, Photosynthesis and Religion").

The apparent mess was cleaned up by R.E. Kalman in 1960 by using the *state-space description* of processes with rational spectral densities. Such descriptions are actually of much older vintage. J. L. Doob wrote two long papers on them in 1944 and 1949 but, alas, did not mention them in his very influential 1953 book! Had he done so many developments might have occurred much earlier.

The so-called Kalman (or sometimes Kalman-Bucy) filter has been widely discussed in a host of papers and textbooks. *e.g.,* Davis (1977). It gains its power, as just noted, from the introduction of state-space models, which turns out to be equivalent to modeling stochastic processes as linear combinations of the components of a vector-valued Markov process. Briefly, we model a scalar process $s(\cdot)$ with an n-th order rational spectral density as

$$\begin{cases} s(t) = Hx(t) \\ \dot{x}(t) = Fx(t) + u(t) \end{cases} \quad t \ge t_0, \tag{2.10}$$

where $H \in \mathcal{C}^{1\times n}$, $F \in \mathcal{C}^{n\times n}$ are known matrices, $u(\cdot)$ is an $n \times 1$ vector-valued zero-mean white noise process, with

$$\langle u(t), u(s)\rangle \triangleq Eu(t)u(s)^* = Q\delta(t-s), \tag{2.11}$$

and the initial *state*, $x(t_0)$, is such that

$$\langle x(t_0), 1\rangle = Ex(t_0) = 0 \ , \ \langle x(t_0), x(t_0)\rangle = \Pi_0 \ , \ \langle x(t_0), u(t)\rangle = 0. \tag{2.12}$$

The matrices $Q \in \mathcal{C}^{n\times n}$ and $\Pi_0 \in \mathcal{C}^{n\times n}$ are also assumed to be known.

We use the inner product notation to follow Kolmogorov in assuming that (zero-mean) random variables defining a (second-order) stochastic process live in a Hilbert space (or Hilbert module, when the random variables are vector-valued). Of course we are stretching this formulation when we deal with white noise processes, but rigor can be regained by regarding a white noise process as the formal derivative of a process with orthogonal increments; see Doob (1953, p.435, 535) for the use of white noise processes precisely in this way. Actually Doob (1953, p.638) credits this practice to Wiener, which recalls our comments at the end of Sec. 5 on the sureness of Wiener's judgement as to the appropriate level of "rigor" in mathematical engineering.

Though the linear system relating the stochastic input process $u(\cdot)$ to the output stochastic process is time-invariant, the process $s(\cdot)$ will in general be nonstationary, because the "transients" arising from the fact that the input is switched on at time t_0 and does not begin in the remote past. In fact, it is not hard to see that

$$\Pi(t) \triangleq \langle x(t), x(t) \rangle$$

will obey

$$\dot{\Pi}(t) = F\Pi(t) + \Pi(t)F^* + Q, \quad \Pi(t_0) = \Pi_0. \tag{2.13}$$

However, when F is "stable", *i.e.*, all its eigenvalues have strictly negative real parts, then it turns out that the process $s(\cdot)$ will be stationary if the initial state variance is chosen as

$$\Pi(t_0) = \Pi,$$

where Π is the unique nonnegative definite solution of the (Lyapunov) equation

$$0 = F\Pi + \Pi F^* + Q. \tag{2.14}$$

Now since we have introduced matrix notation, we can as well take the step of regarding $s(\cdot)$ as a $p \times 1$ vector-valued process, so that $H \in \mathcal{C}^{p\times n}$. The filtering problem is now a "multichannel" problem of attempting to find

$$\hat{s}(t) = \text{the linear least-squares estimate of } s(t) \text{ given } \{y(\tau), t_0 \leq \tau < t\},$$

where

$$y(t) = s(t) + v(t), \quad t \geq 0 \tag{2.15}$$

and $v(\cdot)$ is a white noise process such that

$$\left\langle \begin{bmatrix} u(t) \\ v(t) \end{bmatrix}, \begin{bmatrix} u(\tau) \\ v(\tau) \end{bmatrix} \right\rangle = \begin{bmatrix} Q & S \\ S^* & R \end{bmatrix} \delta(t-\tau), \tag{2.16}$$

where $S \in \mathcal{C}^{n\times p}$ and $R \in \mathcal{C}^{p\times p}$ are also assumed to be known a priori. The presence of the additive white noise is essential to get useful results, and so it is assumed that R is strictly positive definite, $R > 0$.

It is widely believed that the reason for the greater scope of the Kalman theory (applying to vector-valued, finite-dimensional, stationary and nonstationary processes) is that it starts, as above, with a model for the process $s(\cdot)$ rather than with power-spectral or covariance data. However this is not true – the Wiener and Kalman approaches become equivalent if one carries over the state-space characterization of the process to its power-spectra and/or covariance functions.

We shall demonstrate this by using the state-space model to first solve a problem not satisfactorily resolved by Wiener and several later researchers – finding an effective way of computing the canonical factorization of a rational power spectral density function matrix. We start by noting that since the transfer function from the input white noise processes $\{u(\cdot), v(\cdot)\}$ to the output process $y(\cdot)$ is

$$\left[H(i\omega I - F)^{-1} \; I\right], \tag{2.17}$$

the power-spectral density function of $y(\cdot)$ can be computed as

$$\Phi_{yy}(\omega) = \left[H(i\omega I - F)^{-1} \; I\right] \begin{bmatrix} Q & S \\ S^* & R \end{bmatrix} \begin{bmatrix} (-i\omega I - F^*)^{-1}H^* \\ I \end{bmatrix}. \tag{2.18}$$

An alternative expression can be found by taking the Fourier transform of the covariance function:

$$\phi_y(\tau) = < y(t), y(t-\tau) >= R\delta(\tau) + He^{F\tau}N1(\tau) + N^*e^{F^*\tau}H^*1(\tau), \tag{2.19}$$

where

$$N = \Pi H^* + S, \tag{2.20}$$

Π is defined by (2.13), and $1(\cdot)$ is the (Heaviside) unit step function. The Fourier transform of $\phi_y(\cdot)$ is

$$\Phi_{yy}(\omega) = \left[H(i\omega I - F)^{-1} \; I\right] \begin{bmatrix} 0 & N \\ N^* & R \end{bmatrix} \begin{bmatrix} (-i\omega I - F^*)^{-1}H^* \\ I \end{bmatrix}. \tag{2.21}$$

Comparing (2.18) and (2.21) shows that different "central" matrices could be used to specify $\Phi_y(\omega)$, some nonnegative definite as in (2.18), while the one in (2.21) is indefinite. It is natural to ask how we can characterize the nonuniqueness of the central matrix? The answer is that we can use *any* central matrix of the form

$$M = \begin{bmatrix} Q + FZ + ZF^* & S + ZH^* \\ S^* + HZ & R \end{bmatrix}, \qquad Z = Z^*. \tag{2.22}$$

The choices $Z = 0$ and $Z = \Pi$ give the previous expressions (2.18) and (2.21). The fact that any such M yields $\Phi_{yy}(\omega)$ can be verified by a direct (but tedious) calculation. However, a nicer and more useful derivation can be obtained by allowing the random variables to live in an indefinite (Krein) space, rather than in Hilbert space. The inner product in such spaces has the usual linearity and reflexivity properties, except that there can be nonzero elements whose norm is indefinite or even zero; a nice (finite-dimensional) example is the Minkowski space used in special relativity. For example, we could have

$$\left\langle \begin{bmatrix} u^0(t) \\ v^0(t) \end{bmatrix}, \begin{bmatrix} u^0(\tau) \\ v^0(\tau) \end{bmatrix} \right\rangle \triangleq \begin{bmatrix} Q^0 & S^0 \\ S^{0*} & R^0 \end{bmatrix} \delta(t-\tau) = \begin{bmatrix} 0 & N \\ N^* & R \end{bmatrix} \delta(t-\tau). \tag{2.23}$$

Accepting this, we can add to the original $\{u(\cdot), v(\cdot)\}$, elements $\{u^0(\cdot), v^0(\cdot)\}$ such that (in an obvious notation)

$$\begin{cases} \dot{x}(t) + \dot{x}^0(t) = F(x(t) + x^0(t)) + G(u(t) + u^0(t)) \\ y(t) + y^0(t) = H(x(t) + x^0(t)) + v(t) + v^0(t), \end{cases} \tag{2.24}$$

where

$$\langle \begin{bmatrix} u(t) \\ v(t) \end{bmatrix}, \begin{bmatrix} u^0(\tau) \\ v^0(\tau) \end{bmatrix} \rangle \equiv 0 \text{ and } \langle y^0(t), y^0(\tau) \rangle \equiv 0 \text{ for all } t, \tau. \tag{2.25}$$

Now by using the formulas (2.20)-(2.21), we can write

$$\Phi_{y^0 y^0}(\omega) = \begin{bmatrix} H(i\omega I - F)^{-1} & I \end{bmatrix} \begin{bmatrix} 0 & \Pi^0 H^* + S^0 \\ H\Pi^0 + S^{0*} & R^0 \end{bmatrix} \begin{bmatrix} (-i\omega I - F^*)^{-1} H^* \\ I \end{bmatrix}, \tag{2.26}$$

where

$$F\Pi^0 + \Pi^0 F^* + Q^0 = 0. \tag{2.27}$$

From these we conclude that $\Phi_{y^0 y^0}(\omega)$ will be identically zero if we choose

$$S^0 = -\Pi^0 H^* \quad , \quad Q^0 = -F\Pi^0 - \Pi^0 F^* \text{ and } R^0 = 0. \tag{2.28}$$

Finally setting $Z = -\Pi_0$ gives

$$\langle \begin{bmatrix} u(t) + u^0(t) \\ v(t) + v^0(t) \end{bmatrix}, \begin{bmatrix} u(\tau) + u^0(\tau) \\ v(\tau) + v^0(\tau) \end{bmatrix} \rangle = \begin{bmatrix} Q + Q^0 & S + S^0 \\ S^* + S^{0*} & R + R^0 \end{bmatrix} \delta(t - \tau) =$$
$$\begin{bmatrix} Q + FZ + ZF^* & S + ZH^* \\ S^* + HZ & R \end{bmatrix} \delta(t - \tau) \triangleq M\delta(t - \tau) \tag{2.29}$$

exactly as claimed above (see (2.22)). Moreover, the arbitrary matrix Z can be interpreted as the (negative of) the state-variance matrix of a process with zero s-spectrum. However so far we only have a formal calculation. The significant theorem is the so-called KYP Lemma (see Willems and Trentelman (1991) for a recent discussion):

Theorem (KYP Lemma) *When $\Phi_{yy}(\omega) > 0$, then there exists a $Z = Z^*$ such that the central matrix is non-negative definite (i.e., it is the covariance matrix of a collection of genuine random variables).*

We do not need F to be stable; a weaker condition from linear system theory, a subject developed in the engineering literature of the last 30 years, will suffice: the pair $\{F, H\}$ should be detectable, *i.e.*, it should be such that $[H^* \; sI - F^*]$, $s = \sigma + j\omega$, should be full rank for all $\sigma \geq 0$. Here however we shall, for simplicity, stay with the assumption that F is stable so that we are dealing with a stationary process $y(\cdot)$. There are important corollaries of the above theorem characterizing matrix *positive-real* (Caratheodory) and *bounded-real*

(Schur) functions, which are widely encountered in applications. We note also that the Krein space interpretation introduced above can be used to give a simple geometric proof of the lemma as well as several other results, in particular a common framework for the H_2 and H_∞ theories (see, *e.g.*, Hassibi *et al* (1996), Sayed *et al* (1996)).

However while the KYP Lemma is an important result, which is why we mentioned it here, it is not necessary to use it to obtain a spectral factorization of $\Phi_{yy}(\omega)$. To this end, note that although we cannot make any assertions on the positivity of the central matrix M, defined in (2.28), the positivity of

$$\Phi_{yy}(\omega) = \begin{bmatrix} H(i\omega I - F)^{-1} & I \end{bmatrix} M \begin{bmatrix} (-i\omega I - F^*)^{-1}H^* \\ I \end{bmatrix} > 0 \tag{2.30}$$

shows that the $(n+p) \times (n+p)$ matrix M has at least p positive eigenvalues for all choices of Z. It is then interesting to ask whether Z can be chosen so that M has only p positive eigenvalues and no negative eigenvalues, *i.e.*, if Z can be chosen so that M has minimal rank p. To see that this is indeed possible, we use the easily verified decomposition (recall that we have assumed the invertibility of R),

$$M = \begin{bmatrix} Q + FZ + ZF^* & S + ZH^* \\ S^* + HZ & R \end{bmatrix} = \begin{bmatrix} I & K \\ 0 & I \end{bmatrix} \begin{bmatrix} \Delta & 0 \\ 0 & R \end{bmatrix} \begin{bmatrix} I & 0 \\ K^* & I \end{bmatrix}, \tag{2.31}$$

where

$$\Delta(Z) \triangleq Q + FZ + ZF^* - (S + ZH^*)R^{-1}(S + ZH^*)^* \tag{2.32}$$

$$K \triangleq (S + ZH^*)R^{-1}. \tag{2.33}$$

Therefore $\Phi_{yy}(\omega)$ in (2.30) can be written as

$$\begin{aligned}\Phi_{yy}(\omega) = {} & H(i\omega I - F)^{-1}\Delta(Z)(-i\omega I - F^*)^{-1}H^* + \\ & [I + H(i\omega I - F)^{-1}K]R[I + H(i\omega I - F)^{-1}K]^*\end{aligned} \tag{2.34}$$

The second term on the RHS is $p \times p$ and non-negative definite, so we can immediately obtain a factorization by choosing Z so that it satisfies

$$0 = \Delta(Z) = Q + FZ + ZF^* - (S + ZH^*)R^{-1}(S + ZH^*)^* \tag{2.35}$$

The only issue is whether the resulting spectral factor has a well-defined inverse, *viz.*, one that when extended into the complex plane is analytic in the right half plane (*cf.* (2.2)). There is an interesting result here. The nonlinear algebraic equation (2.35), which for reasons explained below is called an Algebraic Riccati Equation (ARE), has many solutions. However it can be shown (see, *e.g.*, Lancaster and Rodman (1991)) that when F is stable(or even just when $\{F, H\}$

is detectable) and $Q \geq 0$, $R > 0$, there is one and only one non-negative solution, say P; moreover, this solution is such that the spectral factor

$$\Psi(s) \triangleq [H(sI - F)^{-1}K + I]R^{\frac{1}{2}}, \quad K = S + PH^*R^{-1} \tag{2.36}$$

and its inverse

$$\Psi^{-1}(s) \triangleq [H(sI-F)^{-1}KR^{\frac{1}{2}}+R^{\frac{1}{2}}]^{-1} = R^{-\frac{1}{2}}[I-H(sI-F+KH)^{-1}K], \tag{2.37}$$

are both analytic in the right half plane. There are several computationally effective methods of finding the desired nonnegative definite solution of the ARE – a good source is the reprint volume edited by Patel *et al* (1994).

The point of the above discussion is that the introduction of the ARE, first done in the Kalman theory, overcame what had been regarded as one of the stumbling blocks to the Wiener theory. A minor quibble may be that the factorization is expressed in terms of the parameters $\{F, G, H, Q, R, S\}$ of a particular model for the process rather than in terms of the spectral data. Now for the state-space model, the covariance and the spectral density are fixed by (*cf.* (2.19)-(2.21)) by $\{H, F, N\}$. To use this data, all we need to do is to choose the central matrix M not as in (2.31), but as (*cf.* (2.19)-(2.22))

$$\begin{bmatrix} 0 + FZ + ZF^* & N + ZH^* \\ N^* + HZ & R \end{bmatrix} \tag{2.38}$$

Then proceeding as before, the rank of this matrix can be dropped by now choosing Z so that is satisfies a slightly different ARE,

$$0 = FZ + ZF^* - (N + ZH^*)R^{-1}(N + ZH^*)^* \tag{2.39}$$

which will lead to a factorization of the form

$$\Phi_{yy}(s) = [I + H(sI - F)^{-1}(N + ZH^*)]R^{-1}[\ \cdots\ ]^* \tag{2.40}$$

The particular choice that will give a factor with an inverse that is analytic in the right-hand plane can be shown to be the unique negative semidefinite, say $-\Sigma$, solution of the ARE (2.39). The corresponding factor is therefore

$$\Psi(s) = [I + H(sI - F)^{-1}K]R^{\frac{1}{2}}, \tag{2.41}$$

where we define

$$K \triangleq N - \Sigma H^*, \tag{2.42}$$

and

$$\Sigma \geq 0, \quad 0 = F\Sigma + \Sigma F^* + (N - \Sigma H^*)R^{-1}(N - \Sigma H^*)^* \tag{2.43}$$

The reader may have wondered that we used the *same* symbols $\Psi(s)$ and K as in the earlier formula (2.36) – the reason is that the canonical factorization is unique! [This implies the interesting identity $\Pi = P + \Sigma$, which we shall not explore here.]

To close the story of Wiener filtering, let us note that with the canonical factor in hand, we can really write down the optimal filter by using (2.8) and (2.37)

$$\mathcal{K}(s) = I - R^{\frac{1}{2}}\Psi^{-1}(s) = H(sI - F + KH)^{-1}K \tag{2.44}$$

where K can be found either from the model parameters, as in (2.36), or from the covariance/spectral parameters, as in (2.42). This is a reasonably explicit formula for the optimal filter, but another advantage of the state-space formulation is that we can readily write down a state-space model for the filter:

$$\dot{\hat{x}}(t) = (F - KH)\hat{x}(t) + Ky(t), \quad \hat{x}(t_0) = 0 \tag{2.45}$$

$$\hat{s}(t) = H\hat{x}(t) \tag{2.46}$$

as can be verified by checking that the transfer function from $y(\cdot)$ to $\hat{s}(\cdot)$ is exactly as in (2.44). We have used the notation $\hat{x}(\cdot)$ for the state-variable in (2.45), because in fact we have a bonus: $\hat{x}(\cdot)$ is the linear least-mean-squares estimate of the state $x(\cdot)$ itself.

We close with some remarks that, *inter alia*, will fulfill our promise to explain the name ARE. The first remark is obtained by going back to our state-space model, (2.10) *seq.* Observe that stationarity arose from a particular choice of initial condition, $\Pi(t_0) = \Pi$ defined as the unique matrix such that

$$\Pi \geq 0, \quad 0 = F\Pi + \Pi F^* + Q$$

For any other choice of $\Pi(t_0)$, or if F is unstable (so that (2.14) will not have a solution $\Pi \geq 0$), the process $s(\cdot)$ will be nonstationary, with covariance function

$$Es(t+\tau)s^*(t) = He^{F\tau}N(\tau), \quad \tau \geq 0$$

where

$$N(t) = \Pi(t)H^*, \quad \dot{\Pi}(t) = F\Pi(t) + \Pi(t)F^* + Q,$$

It turns out that the previous discussions can all be extended by now working in terms of covariance functions rather than power-spectral-density functions. The key change is that instead of the algebraic (Riccati) equation

$$P \geq 0, \quad 0 = Q + FP + PF^* - KRK^*, \quad K \equiv (S + PH^*)R^{-1}$$

we shall have the matrix Riccati differential equation,

$$\dot{P}(t) = Q + FP(t) + P(t)F^* - K(t)RK^*(t), \quad P(t_0) = \Pi(t_0)$$

$$K(t) \triangleq (S + P(t)H^*)R^{-1}$$

When the state is one dimensional, the resulting quadratically nonlinear equation is the one first studied by Jacopo Francesco (Count Riccati) and later introduced by Legendre and others into the calculus of variations. It was introduced into control theory by R.E. Bellman (1957) and the matrix version by R.E. Kalman (1960a).

Explicit analytic solution of the Riccati equation is impossible in the matrix case. But fortunately, this is a (nonlinear) initial value problem, so it can be solved via a discretization scheme, *e.g.,* in the naive way,

$$P(t+\delta) = P(t)+\delta[Q+FP(t)+P(t)F^* - K(t)RK^*(t)]+0(\delta^2), \quad t = 0, \delta, 2\delta, \dots$$

Now, an important observation is that once the need for some computer-based iterative algorithm is realized, one might further guess that there is no particular need to restrict oneself to time-invariant systems: one can just as easily consider time-variant models,

$$\begin{aligned} \dot{x}(t) &= F(t)x(t) + v(t), \quad t \geq t_0 \\ y(t) &= H(t)x(t) + v(t) \end{aligned}$$

with

$$\left\langle \begin{bmatrix} u(t) \\ v(t) \end{bmatrix}, \begin{bmatrix} u(\tau) \\ v(\tau) \end{bmatrix} \right\rangle = \begin{bmatrix} Q(t) & S(t) \\ S^*(t) & R(t) \end{bmatrix} \delta(t-\tau).$$

We just have to work with covariance matrices rather than spectral matrices - all the above analysis, e.g., (2.22), has appropriate time-domain analogs. One end result is that the (Riccati) iteration is essentially as before,

$$P(t+\delta) = P(t) + \delta[Q(t) + F(t)P(t) + \dots] + 0(\delta^2), \quad t = 0, \delta, 2\delta, \dots$$

except that we now need to store the values of the functions $\{F(\cdot), Q(\cdot), \dots\}$.

We have thus arrived at the Kalman-(Bucy) filtering algorithm, though via a very different route than in the vast literature on it. In retrospect we can see that the key idea was the introduction of state-space descriptions, a step whose significance was emphasized by Kalman, who demonstrated its value in several other fields as well, esp. control, circuits, computation, linear system theory.

By now, even the estimation studies have led to a very wide range of results and issues. For example, the Kalman filtering results can be connected to problems in radiative transfer and scattering (see Redheffer (1962), Verghese *et al* (1980), Kailath (1991a)); this is especially nice since as we mentioned in the beginning it was in these fields that the Wiener-Hopf equation was first studied. Moreover in the next section, we shall go on to a different kind of extension of Wiener's results, involving finite-time discrete time series. That discussion will lead us to a concept called *displacement structure,* which also had its roots in studies of the Riccati equation, as introduced by the astrophysicists V.A. Ambartsumian and S. Chandrasekhar in the mid-forties in order to find alternatives to the Wiener-Hopf technique (see Chandrasekhar (1950), Kailath (1991a,b)). Moreover the new results enabled improved solutions of some problems in scattering theory, *e.g.*, see Sidhu and Casti (1975). There is indeed a remarkably wide set of ideas that can be intertwined in many ways; a few more will appear below.

7 BEYOND STATE-SPACE MODELS / DISPLACEMENT STRUCTURE

In one of several different variations of Wiener's problem, his colleague N. Levinson in 1947 studied a finite-time discrete prediction problem, where the Wiener-Hopf equation was replaced by a set of linear equations with a Toeplitz coefficient matrix. He proposed a fast recursive solution, now known as the *Levinson algorithm*, very widely used in geophysical data processing (beginning in the mid-fifties) and in speech processing (beginning in the mid-sixties). Kolmogorov's formulation of the prediction problem gives an interesting insight into this algorithm, and leads to connections with the work of Szegő and Geronimus on orthogonal polynomials, of Schur (1917) on H^∞ functions, and then to new results on Toeplitz-like matrices and more generally on matrices with displacement structure (see below).

The Kolmogorov Isomorphism: The identity

$$\langle y_k, y_l \rangle = E y_k y_l^* = r_{k-l} = \int_{-\pi}^{\pi} z^k z^{-l} \frac{dF(\theta)}{2\pi} = \langle z^k, z^l \rangle_2, \quad z = e^{i\theta}$$

allows one to form an isometric mapping between the Hilbert space of random variables spanned by $\{y_k\}$, and the Hilbert space of functions on the unit circle spanned by the $\{z^k\}$.

Then the finite-interval prediction problem: find $\{a_{k,j}\}$ to minimize

$$E||y_k + a_{k,1} y_{k-1} + \ldots + a_{k,m} y_{k-m}||^2$$

is equivalent to the polynomial approximation problem: find $\{a_{k,j}\}$ to minimize

$$\int_{-\pi}^{\pi} |z^k + a_{k,1} z^{k-1} + \ldots + a_{k,m} z^{k-m}|^2 \frac{dF(\theta)}{2\pi}, \quad z = e^{i\theta}.$$

It turns out that around 1920 Szegő had shown that the minimizing polynomials

$$a_m(z) = z^m + a_{m,1} z^{m-1} + \ldots + a_{m,m}$$

had the nice property that they were orthogonal to each other w.r.t. the measure $F(z)$. Szegő and others went on to make many studies of these orthogonal polynomials. Among other results, in 1939, Szegő and Geronimus independently discovered that these polynomials obeyed a two-term (rather than the usual 3-term) recursion:

$$a_{m+1}(z) = a_m(z) - k_{m+1} z a_m^{\#}(z), \quad a_m^{\#}(z) = \text{the reciprocal polynomial,}$$

where $k_{m+1} = -a_{m+1,m+1}$, the constant term in $a_{m+1}(z)$. This is in fact almost the same as the recursion discovered by Levinson in 1947, except that to obtain a true recursion one needs to be able to compute k_{m+1} in terms of $\{F(z), a_m(z)\}$. This could have been done by Szegő or Geronimus, had they been interested in actual computation; however they were more interested in the asymptotic properties of the polynomials (in fact, a famous Szegő formula is just the formula discovered by Kolmogorov and Wiener for the irreducible error in prediction). A survey of the connections between orthogonal polynomial theory and linear estimation, and their fascinating continuous-time analogs, can be found in Kailath *et al* (1978).

Later it was discovered that a more far-reaching connection could be made with some of the work of I. Schur, who was well ahead of his time with his interest in computation. In 1917, he wrote a remarkable paper giving a computationally efficient solution to the Caratheodory moment problem that, in effect, also gave a fast algorithm for factorizing Toeplitz matrices; Levinson's algorithm factorizes the *inverse* of a Toeplitz matrix. It turns out that Schur's algorithm offers an alternative to the Levinson algorithm: it is somewhat slower for serial computation, but can be much faster for (software or hardware) parallel implementation!

There are many aspects to these algorithms arising from pursuing the prediction problem. One of the most fascinating is the concept of *displacement structure.* One way of motivating it is by asking questions such as the following:

If there are fast algorithms for factoring Toeplitz matrices, what about factoring non-Toeplitz matrices that are known to have Toeplitz inverses? Similarly, should it be much harder to factor the non-Toeplitz matrix T_1T_2 or $T_1T_2^{-1}T_3$ than T_1 (or T_2 or T_3) alone?

The answer is that these problems in fact have the same order of complexity as purely Toeplitz problems do. The reason is that what allows fast algorithms for Toeplitz matrices is not their Toeplitzness, which is lost under inversion and under multiplication, but something called *displacement structure*: R has displacement structure if $R - FRA$, or more generally $\Omega R\Delta - FRA$ has low rank for appropriate (low complexity) matrices $\{\Omega, \Delta, F, A\}$. The interested reader can verify that when $F = A^* = Z$, the lower shift matrix with ones on the first subdiagonal and zeros elsewhere, a Hermitian Toeplitz matrix and its inverse have displace rank less than or equal to 2. It is not hard to show that products, inverses and Schur complements essentially inherit the displacement structure. This fact can be exploited to obtain a generalized Schur algorithm for the fast recursive factorization of such matrices. Moreover there is a very useful physical structure – a cascade network or generalized transmission line – that can be associated with the generalized Schur algorithm, a fact that has lots of implications and applications. We may mention, among others, problems in linear algebra, inverse scattering, coding theory, complex interpolation, matrix completion, etc. The generalized Schur algorithm and several applications are discussed in a recent survey, Kailath and Sayed (1995).

8 NONLINEAR ESTIMATION

In the late 1950's, Wiener gave a series of lectures on the problem of nonlinear least-mean squares estimation, which were transcribed into a monograph Wiener (1958). Wiener proposed to use a so-called 'Volterra series' characterization of nonlinear systems as a sum of linear + quadratic + ... systems. However this approach had several limitations, especially of computational complexity and the difficulty of approximation (how many or which terms should we keep for a particular nonlinear system?).

The success of the state-space models for the linear problems led to a significant effort to try to obtain similar results for the nonlinear case. Thus suppose we have a nonlinear system, in state-space form,

$$\begin{cases} \dot{x}(t) = f(x(t), t, u(t)), & t \geq 0 \\ y(t) = h(x(t), t) + v(t) = s(t) + v(t), & \text{say.} \end{cases}$$

The minimum mean-square estimator of $s(t)$ given $\{y(\tau), \tau < t\}$ is no longer linear, and its computation requires full statistical knowledge of the non-Gaussian processes $x(\cdot)$, $s(\cdot)$ and $y(\cdot)$. When $\{x(\cdot), s(\cdot), y(\cdot)\}$ are jointly Gaussian, one has the Kalman filter recursions; in the general case, one has an ascending chain of coupled nonlinear equations for which no really satisfactory practical algorithms, or approximations, have been found. Therefore the nonlinear problem is effectively still open.

However there have been several interesting theoretical results. One set arises from the introduction of ideas from martingale theory (with some of the results later pursued in finance theory and on Wall Street). Martingale theory first enters through the fact that the white Gaussian measurement noise, $v(\cdot)$, of the engineers is the formal derivative of the special process introduced by Wiener in his study of Brownian motion:

$$\int_0^t v(\tau) d\tau = W(t), \quad \text{the Wiener (-Lévy) process.}$$

The martingale properties of $W(\cdot)$ lead to a striking generalization of the innovations process first introduced in the linear theory. Let us recall from Sec. 5 that with (scalar) observations containing white noise,

$$y(t) = s(t) + v(t), \quad \langle v(t), v(\tau) \rangle = \delta(t - \tau)$$

the optimum linear filter for finding $\hat{s}(\cdot)$ has transfer function (note that now $R \equiv 1$ in (2.8))

$$K(\omega) = 1 - \Psi^{-1}(\omega).$$

This implies in the time domain that

$$\hat{s}(t) = y(t) - e(t)$$

or that the innovations can be expressed as

$$e(t) = y(t) - \hat{s}(t).$$

Now when we deal with nonlinear operations on white noise, the formal manipulations become harder to justify: linear operations on white noise give smoother processes, but what is the square of white noise? Therefore one now works with integrated processes,

$$Y(t) \triangleq \int_0^t y(\tau)d\tau = \int_0^t s(\tau)d\tau + \int_0^t v(\tau)d\tau = \int_0^t s(\tau)d\tau + W(t)$$

and uses the Ito theory of stochastic integrals, especially as developed by the Japanese and French schools (see *e.g.*, Meyer (1975)).

In this language, one can show (see Kailath (1971a), Meyer (1973)) the following: Let

$$Y(t) = \int_0^t s(\tau)d\tau + W(t),$$

with

$$\int_0^T E|s(t)|dt < \infty, \quad E[W(t) - W(\tau)z(\tau)] = 0, \quad t > \tau.$$

Then, the process $E(\cdot)$ defined as

$$E(t) = Y(t) - \int_0^t \hat{s}(\tau)d\tau$$

where

$$\hat{s}(t) \triangleq E[s(t)|\mathcal{F}\{y(\tau), \tau \leq t\}]$$

is also a Wiener process w.r.t. the (nested) family of sigma fields $\mathcal{F}\{Y(\tau),\ \tau \leq t\}$. The main idea of the proof is to show first that $E(\cdot)$ is a martingale with respect to these sigma fields, and then to show that $E(\cdot)$ and $W(\cdot)$ have the same "quadratic variation" (again a concept introduced by Wiener). Then a theorem of Levy's gives the result. This is a nice result, since the process $y(\cdot)$ might be much more complicated than $E(\cdot)$; it shows the power of the assumption of additive white noise. Now in the linear case, results from the theory of integral equations enable us to show that (Kailath (1968,1972))

$$\mathcal{F}\{E(\tau),\ \tau \leq t\} = \mathcal{F}\{Y(\tau),\ \tau \leq t\}, \quad 0 \leq t \leq T$$

so that the processes $\{Y(\cdot)\}$ and $\{E(\cdot)\}$ are replaceable each by the other, without any loss of information. As mentioned earlier, this was the idea behind the innovations approach to the Wiener filter (Bode-Shannon (1950), Zadeh-Ragazzini (1950)); in the nonstationary finite-time case, the above result allows for a similar approach to the Kalman filter and several related problems (Kailath (1970b), Davis (1977)).

Therefore an important question is under what conditions this equality of sigma fields holds in the general case. The problem turned out to be quite difficult (Benes(1976)) and only after attempts by several researchers, did Allinger and Mitter (1981) succeed in proving the equality for the case where $s(\cdot)$ and $W(\cdot)$ are independent of each other and $\int_0^T E|s(t)|^2 dt < \infty$.

However even without the equivalence, the process $E(\cdot)$ leads to several nice results. One is that even though the sigma fields generated by $E(\cdot)$ and $Y(\cdot)$ may not be equivalent, Fujisaki *et al* (1972) showed that any function measurable w.r.t. the Y sigma fields can be written as a stochastic integral w.r.t. the Wiener process $E(\cdot)$. This then allows for a simpler description of the nonlinear filtering equations; however, as mentioned before, they are not useful for actual computation.

Another application that exploits only the fact that $E(\cdot)$ is a Wiener process is a generalized Cameron-Martin formula for the Radon-Nikodym derivative of the measures P_y and P_w induced by the processes $Y(\cdot)$ and $W(\cdot)$ (Kailath (1971b)):

$$\frac{dP_y}{dP_w} = \exp\left\{\int_0^T \hat{s}(t)dY(t) - \frac{1}{2}\int_0^T |\hat{s}(t)|^2 dt\right\}.$$

This expression has useful implications for the problem of detecting the presence or absence of a random signal $s(\cdot)$ in the presence of noise. When the signal $s(\cdot)$ is deterministic (and therefore known apriori) $\hat{s}(\cdot) \equiv s(\cdot)$, this is a result of Cameron and Martin (1944), contemporaries of Wiener at MIT. It is an interesting and useful fact that for random $s(\cdot)$ the deterministic formula still applies with the unavailable random signal $s(\cdot)$ being replaced by the observable least-squares estimate $\hat{s}(\cdot)$. This interpretation allows a lot of the insights and results of estimation theory to be carried over to signal detection theory (Kailath (1969), Kailath (1970a), Kailath and Duttweiler (1972), Davis and Andreadakis (1977)). We may remark that the Cameron-Martin formula arose as a theory of 'linear changes of variables' in Wiener space (the space of sample functions of a Wiener process). The generalized Cameron-Martin formula follows from a nonlinear version of this theory introduced in a seminal paper of Girsanov (1960), which has since been much studied and extended (see, *e.g.*, Liptser and Shiryayev (1977)).

9 CONCLUDING REMARKS

This has been an account of some of the ways in which Norbert Wiener's work and ideas significantly influenced, and in fact really initiated the modern development of the mathematical side of electrical engineering. The key contributions were his emphasis on the statistical nature of the communication process and his introduction of the use of optimization criteria into system design. I should hasten to add that many other notable researchers (Shannon, Rice, Tukey, Bode, Bellman, Kalman, to name just a few) had major roles in the post-1942 development of these topics. Wiener's own specific mathematical contributions to mathematical engineering are too numerous to cover in a single article. Here I have tried to indicate, in a very sketchy way and with some focus on things I know best, some of the wide range of ideas and techniques stimulated by Wiener's work on prediction and filtering.

Acknowledgements

The work was supported in part by the Advanced Research Projects Agency of the Department of Defense grant No. F49620-93-1-0085, monitored by the Air Force Office of Scientific Research, and by the Army Research Office under grant No. DAAH04-93-G-0029.

REFERENCES

[1] D. F. Allinger and S. K. Mitter, *New results on the innovations problem for nonlinear filtering*, Stochastics, vol. 4 (1981), pp. 339-348.

[2] V. A. Ambartsumian, *Diffuse reflection of light by a foggy medium,* Dokl. Akad. Sci. SSSR, vol. 38 (1943), pp. 229-322.

[3] R. E. Bellman, *On a class of variational problems*, Quart. Appl. Math., vol. 14 (1957), pp. 353-359.

[4] V. E. Benes, *On Kailath's innovations conjecture*, Bell Syst. Tech. J., vol. 55 (1976), pp. 981-1001.

[5] S. Bennett, *Nobert Wiener and control of anti-aircraft guns,* IEEE Control Systems, vol. 14, no. 6 (1994), pp. 58-62.

[6] H. W. Bode and C. E. Shannon, *A simplified derivation of linear least squares smoothing and prediction theory,* Proc. IRE, vol. 38 (1950), pp. 417-425.

[7] S. P. Boyd and C. H. Barratt, *Linear Controller Design: Limits of Performance,* Prentice-Hall, N.J., 1991.

[8] R. H. Cameron and W. T. Martin, *Transformation of Wiener integrals under translations*, Ann. Math., vol. 45 (1948), pp. 368-396.

[9] S. Chandrasekhar, *Radiative Transfer,* Oxford University Press, N.Y., 1950; Dover, N.Y., 1960.

UNIVERSITY OF HERTFORDSHIRE LRC

[10] M. H. A. Davis, *Linear Estimation and Stochastic Control,* Halsted Press, N.Y., 1977.

[11] M. H. A. Davis and E. Andreadakis, *Exact and approximate filtering in signal detection: an example,* IEEE Trans. on Inform. Thy., IT-23 (1977), pp. 768-772.

[12] J. L. Doob, *The elementary Gaussian processes*, Ann. Math. Stat., vol. 15 (1944), pp. 229-282.

[13] ——, *Time series and harmonic analysis*, Proc. Berkeley Symp. on Math. Stoc. and Prob., Calif. Univ., Calif. Press, 1949, pp. 303-343.

[14] ——, *Stochastic Processes*, Wiley, N.Y., 1953.

[15] A. Einstein, *Méthode pour la détermination de valeurs statistiques d'observations concernant des grandeurs soumises à des fluctuations irrégulières*, Archives des sciences Physiques et Naturelles, vol. 37 (1914), pp. 254-256, (A translation and commentary appear in IEEE Magazine on Acoustics, Speech and Signal Processing, Oct. 1987).

[16] M. Fujisaki, G. Kallianpur and H. Kunita, *Stochastic differential equations for the nonlinear filtering problem*, Osaka, J. Math., vol. 9 (1972), pp. 19-40.

[17] I. V. Girsanov, *On transforming a certain class of stochastic processes by absolutely continuous changes of measure*, Th. Prob. App., vol. 5 (1960), pp. 285-301.

[18] M. Green and D. J. N. Limebeer, *Linear Robust Control*, Prentice-Hall, N.J., 1995.

[19] B. Hassibi, A. Sayed and T. Kailath, *Linear estimation in Krein spaces: Parts I and II*, IEEE Trans. Autom. Contr., vol. AC-41 (1996), pp. 18-49.

[20] T. Kailath, *An innovations approach to least-squares estimation, Part I: Linear filtering in additive white noise,* IEEE Trans. Autom. Contr., vol. AC-13 (1968), pp. 646-655.

[21] ——, *A general likelihood ratio formula for random signals in noise,* IEEE Trans. Inform. Thy., IT-15, no. 3 (1969), pp. 350-361: ibid, IT-16 (1970), pp. 393-395.

[22] ——, *The innovations approach to detection and estimation theory,* Proc. IEEE, vol. 58 (1970a), pp. 680-695.

[23] ——, *Likelihood ratios for Gaussian processes,* IEEE Trans. on Inform. Thy., IT-16, no. 3 (1970b), pp. 276-288.

[24] ——, *Some extensions of the innovations theorem,* Bell Syst. Tech. J., vol. 50 (1971a), pp. 1487-1494.

[25] ——, *The structure of Radon-Nikodym derivatives with respect to Wiener and related measures*, Ann. Math. Stat., vol. 42 (1971b), pp. 1054-1067.

[26] ——, *A note on least-squares estimation by the innovations method,* J. SIAM Control, vol. 10 (1972), pp. 477-486.

[27] ——, *From Kalman filtering to innovations, martingales, scattering and other nice things*, in Mathematical System Theory, ed. A.C.Antoulas, Springer-Verlag, N.Y., (1991a), pp. 55-88.

[28] ——, *Remarks on the origins of the displacement rank concept*, Appl. Math. Comp., vol. 45 (1991b), pp. 193-206.

[29] T. Kailath, L. Ljüng and M. Morf, *Generalized Krein-Levinson equations for efficient calculation of Fredholm resolvents of nondisplacement kernels,* in Topics in Functional Analysis, Essays dedicated to M. G. Krein, I.C. Gohberg and M. Kac, eds., Academic Press, N.Y., 1978, pp. 169-184.

[30] T. Kailath and D. L. Duttweiler, *An RKHS approach to detection and estimation problems: Part III - Generalized innovations and a likelihood-ratio formula,* IEEE Trans. Info. Thy., IT-18 (1972), pp. 730-745.

[31] T. Kailath and A. Sayed, *Displacement structure: theory and applications*, SIAM Review, 37, no. 3 (1995), pp. 297-386.

[32] T. Kailath, A. Vieira and M. Morf, *Inverses of Toeplitz operators, innovations and orthogonal polynomials,* SIAM Rev., 20, no. 1 (1978), pp. 106-119.

[33] R. E. Kalman, *A new approach to linear filtering and prediction problems,* J. Basic Eng., vol. 82 (1960a), pp. 34-45.

[34] ——, *Contributions to the theory of optimal control*, Bol. Soc. Mat. Mex., vol. 5 (1960b), pp. 102-119.

[35] R. E. Kalman and R. S. Bucy, *New results in linear filtering and prediction theory,* Trans. ASME, Ser. D., J. Basic Eng., vol. 83 (1961), pp. 95-107.

[36] A. N. Kolmogorov, *Sur l'interpolation et extrapolation des suites stationnaires*, C. R. Acad. Sci., vol. 208 (1939), pp. 2043-2045.

[37] ——, *Stationary sequences in Hilbert spaces*, (in Russian), Bulletin Moscow State Univ., vol. 2 (1941), pp. 1-40. (English translation in Linear Least-Squares Estimation (1977), edited by T. Kailath, Academic Press, N.Y.)

[38] M. G. Krein, *Integral equations on a half-axis with kernel depending on the difference of the arguments,* Usp. Math. Nauk., vol.13 (1958), pp. 3-120, (Amer. Math. Socy. Transl).

[39] P. Lancaster and L. Rodman, *Solutions of the continuous and discrete time algebraic Riccati equations: A Review,* in The Riccati Equation, S. Bittanti, A.J. Laub and J.C. Willems eds., Springer-Verlag, 1991, pp. 11-51,

[40] N. Levinson, *The Wiener RMS (root-mean-square) error criterion in filter design and prediction,* J. Math. Phys., vol. 25 (1947), pp. 261-278.

[41] R. S. Liptser and A. N. Shiryayev, *Statistics of Random Processes, I and II*, Springer-Verlag, N.Y., 1978, (Russian original, 1974).

[42] L. Ljung, T. Kailath and B. Friedlander, *Scattering theory and linear least-squares estimation, Part I – Continuous-time problems,* Proc. IEEE, vol. 64 (1976), pp. 131-139.

[43] P. Masani and R. S. Phillips, *Anti-aircraft fire-control and the emergence of cybernetics,* Norbert Wiener, Collected Works, vol. IV (1985), ed. P. Masani, MIT Press, pp. 141-179.

[44] P. A. Meyer, *Sur un probléme de filtration*, Séminaire de Probabilités, VII, Lecture Notes in Mathematics, vol. 321 (1973), Springer-Verlag, N.Y., pp. 223-247.

[45] ——, *Un cours sur les intégrales stochastiques, Seminaire de Probabilités X,* Lecture Notes in Math., vol. 511 (1975), Springer-Verlag, N.Y., pp. 245-400.

[46] D. A. Mindell, *Automation's finest hour: Bell labs and automatic control in World War II,* IEEE Control Systems, vol. 15, no. 6 (1995), pp. 72-80.

[47] R.V.Patel, A.J.Laub and P.M.Van Dooren, eds, *Numerical linear algebra techniques for systems and control*, IEEE Press, N.J., 1994.

[48] J. Pincus, *Commentary on "Über eine klasse integralgleichungen",* Norbert Wiener, Collected Works, vol. III (1981), ed. P. Masani, MIT Press, pp. 44-53.

[49] R. M. Redheffer, *On the relation of transmission-line theory to scattering and transfer,* J. Math. Phys., vol. 41 (1962), pp. 1-41.

[50] A. Sayed, B. Hassibi and T. Kailath, *Inertia Conditions for the minimization of quadratic forms in indefinite metric spaces,* in *Operator Theory: Advances and Applications,* Birkhäuser, 1996.

[51] C. E. Shannon, *A mathematical theory of communication*, Bell Syst. Tech. J., vol. 27 (1948), pp. 379-423, 623-656.

[52] ——, *Letter to Vannevar Bush, in Collected Papers of C. E. Shannon, eds. N. J. A. Sloane and A. D. Wyner*, IEEE Press, NJ, 1994, pp. 455-456.

[53] I. Schur, *Über potenzreihen die im innern des Einheitskreises Beschränkt Sind*, J. für die Reine und Angewandte Mathematik, vol. 147 (1917), pp. 205-232, (English translation in *Operator Theory: Advances and Applications*, vol. 18 (1986), pp. 31–88, edited by I. Gohberg, Birkhäuser, Boston).

[54] G. Sidhu and J. Casti, *X and Y functions for planetary atmospheres with a Lambert law reflecting surface,* Astrophys. J., vol. 196 (1975), pp.607-612.

[55] H. L. Trentelman and J. C. Willems, *The dissipation inequality and the algebraic Riccati equation,* in *The Riccati Equation* (S. Bittanti, A. J. Laub, and J. C. Willems, eds.), pp. 197–242, Springer–Verlag, 1991.

[56] M. C. Yovits and J. L. Jackson, *Linear filter optimization with game theory considerations,* in IRE Nat. Conv. Rec., pt.4, (1955), pp. 193-199.

[57] G. Verghese, B. Friedlander and T. Kailath, *Scattering theory and linear least-squares estimation: Part III - The estimates*, IEEE Trans. Autom. Contr., vol. AC-25 (1980), pp. 794-802.

[58] N. Wiener, *Extrapolation, Interpolation and Smoothing of Stationary Time Series, with Engineering Applications,* N.Y.: Technology Press and Wiley, 1949 (Originally issued in February 1942, as a classified Nat. Defense Res. Council Rep.). (Reissued by MIT Press under the title *Time Series*).

[59] ——, *Nonlinear problems in random theory,* MIT Press, 1958.

[60] N. Wiener and E. Hopf, *Über eine klasse singulärer integralgleichungen,* S.-B. Preuss, Akad. Wiss. Berlin, Phys.-Math, Kl. 30/32, (1931), pp. 696-706.

[61] L. A. Zadeh, *From circuit theory to system theory*, Proc. IEEE, vol.50 (1962), pp. 856-865.

[62] L. A. Zadeh and J. R. Ragazzini, *An extension of Wiener's theory of prediction,* J. Appl. Phys., vol. 21 (1950), pp. 645-655.

PART I

COMMUNICATION AND INFORMATION SYSTEMS

3

RESIDUAL NOISE AFTER INTERFERENCE CANCELLATION ON FADING MULTIPATH CHANNELS

R. G. Gallager

Laboratory for Information and Decision Systems
Department of Electrical Engineering and Computer Science
Massachusetts Institute of Technology
Cambridge, MA 02139

Dedicated to Tom Kailath, one of my oldest and best friends, on the celebration of his 60th birthday. Much of the best early work on the fading multipath channels of interest in this paper was done by Tom [Ka59] while he was still a graduate student at MIT.

ABSTRACT

A popular information theoretic technique for multiaccess communication on a white Gaussian noise channel is to decode the users one by one. After each user is decoded, its encoded waveform is subtracted from the received signal, thus cancelling the interference from that user for the task of decoding subsequent users. This technique is not directly applicable to the fading multipath channels common to wireless communication. The problem is that what should be subtracted from the received signal is the response of the fading channel to the user's encoded waveform. Since the channel is unknown, the best that can be done is to subtract the convolution of the encoded waveform with an estimate of the channel. This leaves a residual noise term which is the convolution of the encoded waveform with the difference between the true channel and the estimated channel. The point of this paper is to show that this residual noise term is negligibly small for typical wireless situations.

1 INTRODUCTION

Cellular communication, personal communication systems (PCS), and packet radio systems all involve multiaccess communication, i.e., multiple transmitters sending data to the same receiver. Present day systems use a number of different techniques such as time division multiple access (TDMA), frequency division multiple access (FDMA), code division multiple access (CDMA), and collision resolution to control interference between the transmitters. The channels between these transmitters and receivers are typically fading multipath channels. For these channels, the response to each transmitted signal is a time-varying linear combination of delayed replicas of that transmitted signal, and the overall received signal is the sum of the responses to the various transmitters plus additive noise.

From an information theoretic standpoint, the above systems can be abstracted as multiaccess communication over time-varying multipath channels. In this abstraction, we restrict attention to a single receiver with multiple transmitters. The transmitters are limited to a common frequency band of width W and are each power constrained. The noise is modeled as white Gaussian, and the time-varying path strengths and delays are modeled as random processes. We can then attempt to find the multiaccess capacity region and achievable coding error probabilities as a function of rate and code block length.

For additive white Gaussian noise channels without time-varying multipath, the multiaccess capacity region is well known [CT91] and can be interpreted in terms of successive interference cancellation. That is, a set of m transmitter rates is within the capacity region if the receiver can decode the transmitters one by one. The first transmitter's code word is decoded with the other transmitters' code words treated as additional noise. The waveform for that code word is then subtracted from the received waveform, thus cancelling the interference of that waveform from successive decodings. The code words of subsequent transmitters are decoded and cancelled in the same way. The capacity region turns out to be the convex hull of the sets of rates decodable by this interference cancellation approach. Points in this convex hull can be achieved by time sharing between points achievable by interference cancellation. Time sharing has a number of system disadvantages for wireless systems [GJ91], but it turns out that arbitrary achievable points can also be achieved by interference cancellation directly. This can be done if some transmitters are conceptually split into two users, the available power of the transmitter and the required rate being split between the two users [RU96].

The situation for fading multipath channels is considerably more complex. As we shall see below, a fading multipath channel can be represented as an unknown time-varying linear filter. Thus the waveform from each of the m transmitters goes through an unknown time-varying linear filter. The received waveform is the sum of the outputs from these m filters along with white Gaussian noise. Suppose that we successfully decode a code word from one of the transmitters and attempt to do interference cancellation. We would like to subtract the convolution of the code word waveform and the time-varying linear filter from the received waveform, but since the time-varying filter is unknown, we instead use the convolution of the code word waveform and an estimate of the time-varying linear filter. Thus the interference cancellation is imperfect, and some residual noise, consisting of the code word waveform convolved with the difference between the true channel filter and the estimated channel filter, is left to interfere with subsequent decoding of other transmitters.

Our objective in this paper is to show that this residual noise is negligible for typical wireless situations. We will assume that each transmitter uses CDMA waveforms over a broad enough bandwidth that the interference from these transmitted signals can be modeled as white Gaussian noise over the bandwidth of interest. Because of this, we will study the residual noise in terms of a single transmitter, a single unknown filter, and additive Gaussian noise. We focus on the problem of estimating the unknown filter, and show that a Rake receiver [PG60] is an appropriate mechanism for both estimating the channel and detecting the transmitted signal. In the interest of simplicity, we shall make a number of simplifying assumptions as we proceed. Many of these assumptions can be avoided, but the results become less insightful.

Since we analyze only the issue of residual noise, we neglect many important problems associated with multiaccess fading multipath channels. One of these is the capacity region under the assumption that transmitters and receivers all know the fading multipath channels [CV93, Go94, KH95,TH96]. These analyses consider transmitters that dynamically change their power, spectral density, and/or rate as the channels change. Some results on the capacity when transmitters do not know the channels are contained in [Me95]. There is clearly a need for more work on the multiaccess capacity region of fading multipath channels under various feedback situations, but it seems clear that channel measurement must play a central role in this. We now proceed to analyze channel measurements and residual noise.

2 BASEBAND EQUIVALENTS

Consider M-ary signalling with the M signals $u_1(t), ..., u_M(t)$. Let T be the intersymbol duration, so that each T seconds, one of the signals $\{u_m(t), 1 \leq m \leq M\}$ is transmitted. We assume that $u_m(t)$ is essentially non-zero only for $-T \leq t \leq 0$ so that successive signals do not overlap. After passing through the multipath channel there will be some overlap which we discuss later. The signals all have bandwidth W, centered around some carrier frequency $f_0 \gg W$. Let $U_m(f) = \int u_m(t)e^{-j2\pi ft}dt$ be the Fourier transform of $u_m(t)$ for each m, and define the baseband equivalent waveforms $x_m(t)$ in terms of their Fourier transforms $X_m(f)$ where $X_m(f) = U_m(f + f_0)$ for $f > -f_0$ and $X_m(f) = 0$ otherwise (see Fig. 3.1).

This way of going from passband to baseband is not entirely conventional. In particular,

$$u_m(t) = 2\mathrm{Re}[\mathrm{x_m(t)e^{j2\pi f_0 t}}] = 2\mathrm{Re}[\mathrm{x_m(t)}]\cos(2\pi \mathrm{f_0 t}) - 2\mathrm{Im}[\mathrm{x_m(t)}]\sin(2\pi \mathrm{f_0 t}) \tag{3.1}$$

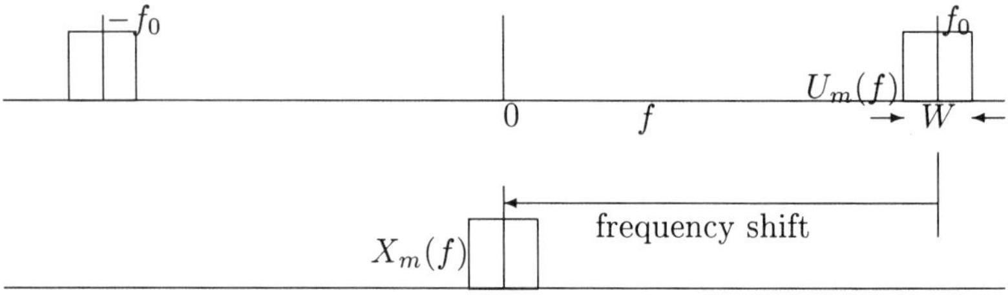

Figure 3.1 Baseband equivalent; $U_m(f)$ and $X_m(f)$ are generally complex.

Also, if $\mathcal{E}$ is the energy in the signal $u_m(t)$, then

$$\mathcal{E} = \int_{-\infty}^{\infty} [u_m(t)]^2 \, dt = 2 \int_{-\infty}^{\infty} |x_m(t)|^2 dt \; . \tag{3.2}$$

The desirable feature of this scaling is that if $u(t)$ is passed through a linear filter of impulse response $g(t)$ to get output $v(t)$, and if $x(t)$, $h(t)$, and $y(t)$ are the corresponding baseband equivalents of $u(t)$, $g(t)$, and $v(t)$, then

$$v(t) = u(t) * g(t) \qquad ; \qquad V(f) = G(f)V(f)$$

$$y(t) = h(t) * x(t) \qquad ; \qquad Y(f) = H(f)X(f) \tag{3.3}$$

Since the arguments to follow depend critically on being able to view signals and filters interchangeably, we have defined baseband waveforms so that (3.3) is satisfied, and by necessity this forces the peculiar energy scaling in (3.2).

Assume that the real and imaginary parts of $x_m(t); 1 \leq m \leq M$ are pseudo noise signals (as used in CDMA). These signals have the property that $|U_m(f)|$ is essentially constant over the signalling bandwidth $|f - f_0| \leq W/2$. It follows that $|X_m(f)|$ is essentially constant over $|f| \leq W/2$ and 0 elsewhere. For simplicity, we henceforth assume that $|X_m(f)|$ is exactly constant over $|f| \leq W/2$ and zero elsewhere. Applying Parseval's equation to (3.2), $\int_{-\infty}^{\infty} |X_m(f)|^2 df = \mathcal{E}/2$, so

$$|X_m(f)|^2 = \frac{\mathcal{E}}{2W} \quad \text{for} |f| \leq W/2 \quad ; \quad |X_m(f)|^2 = 0 \quad \text{for} |f| > W/2 \tag{3.4}$$

Since $|X_m(f)|^2$ and $R_m(t) = \int x_m^*(\tau) x_m(t+\tau) d\tau$ are Fourier transforms, it follows from (3.4) that $R_m(t) = (\mathcal{E}/2)\sin(\pi W t)/(\pi W t)$. Thus, if we view $x_m(t)$ in terms of its samples $x_{m,i}$ at rate W, we have $\sum_i x_{m,i}^* x_{m,i+j} = (\mathcal{E}W/2)\delta(j)$.

It is not possible to find waveforms $x(t)$ that are both time limited to the signal interval T and low pass limited to the band $W/2$. CDMA systems, however, have a relatively large time bandwidth product, $WT \gg 1$ (which is why they are called spread spectrum systems), and for this reason, waveforms can be found that are both approximately time limited and frequency limited. Finding such waveforms with desirable cross correlation properties is a large and very well studied field, but studying this would draw us away from our main purpose. Thus in what follows, we simply assume (3.4) to be valid, and recognize that the approximation can be quite good for $WT \gg 1$.

3 THE EFFECT OF MULTIPATH

Let $\tau_i(t)$ be the propagation delay of the ith propagation path at time t, and let $a_i(t)$ be the strength of that path, at least within the frequency range of interest around f_0. Both a_i and τ_i change slowly with time. The impulse response of the channel, i.e., the output at time t due to an impulse τ seconds earlier is then $g(\tau,t) = \sum_i a_i(t)\delta(\tau - \tau_i(t))$. Thus the response to a signal u(t) is

$$v(t) = \int u(t-\tau)g(\tau,t)d\tau = \sum_i u(t-\tau_i(t))a_i(t) \tag{3.5}$$

Defining $G(f,t) = \int g(\tau,t)e^{-j2\pi f\tau}d\tau$, we have $G(f,t) = \sum_i a_i(t)e^{-j2\pi f\tau_i(t)}$, again within the frequency range of interest. Define $H(f,t) = G(f+f_0,t)$ for the baseband region of interest. Then

$$H(f,t) = \sum_i a_i(t)e^{-j2\pi f_0\tau_i(t)}e^{-j2\pi f\tau_i(t)} = \sum_i \alpha_i(t)e^{-j2\pi f\tau_i(t)} \tag{3.6}$$

$$\text{where} \quad \alpha_i(t) = a_i(t)e^{-j2\pi f_0\tau_i(t)}$$

Letting $x(t)$ and $y(t)$ be the baseband equivalents of $u(t)$ and $v(t)$ respectively, and letting $X(f)$ and $Y(f)$ be the corresponding Fourier transforms, it can be shown after a little manipulation that

$$y(t) = \int_{-\infty}^{\infty} X(f)H(f,t)e^{j2\pi ft}df \tag{3.7}$$

This shows that $H(f,t)$ for $|f| > W/2$ has no effect on the output. Thus we arbitrarily define $H(f,t)$ to be 0 for $|f| > W/2$. Inverse Fourier transforming (3.6) with this modification, the baseband equivalent filter is

$$h(\tau,t) = \sum_i \alpha_i(t)\frac{\sin[\pi W(t-\tau_i(t))]}{\pi(t-\tau_i(t))} \tag{3.8}$$

Also, transforming the right hand side of (3.7), we get

$$y(t) = \int_{-\infty}^{\infty} x(t-\tau)h(\tau,t)d\tau \tag{3.9}$$

Note that $h(\tau,t)$ has one filtered impulse for each path, and that the sinc pulse representing the filtered impulse has a peak that increases with W and a width that decreases, thus keeping unit area. The multipath structure does not change as the bandwidth of the input is changed, but the filter $h(\tau,t)$ does change, since $h(\tau,t)$ represents only the effect of the channel over the given bandwidth. This is an important point, since the effect of the channel is typically very complex, but we need measure its effect only on the signals in the given band. Since we want to measure the channel over the bandwidth $W/2$, we want to characterize it in the simplest way over that band.

Define L as the multipath spread of the channel; this is the difference in propagation delay between the longest and shortest path. For W large, $h(\tau,t)$ is approximately 0 for $\tau < 0$ and for $\tau > L$ (in communication, one usually adjusts the time reference at the receiver to be delayed from that at the transmitter by the shortest (or sometimes the strongest) propagation delay). For smaller W, it can be seen that $h(\tau,t)$ is non-zero over an interval L' consisting of L plus several times $1/W$. For cellular mobile communication, L is typically between a few microseconds and 30 microseconds, and for PCS, L is typically much smaller, on the order of 100 nsec. If $L = 10$ μsec, and $W = 10^6$ H, then $h(\tau,t)$ could be represented (through the sampling theorem) by slightly more than 10 samples in τ; each sample is complex, so measuring h at any given time corresponds to measuring slightly more than 20 real numbers.

Define B as the Doppler spread of the channel; this is the difference between the largest and the smallest Doppler shift. Typical values in a mobile system are around 100 H. B determines how quickly $h(\tau,t)$ can change with t. The phase in the path strength $\alpha_i(t)$ can change significantly over the interval $1/B$, so that measurements of the channel become outdated over intervals of duration $1/B$. We will assume in what follows that the signalling interval T is very much smaller than $1/B$, and thus we assume that $h(\tau,t)$ is constant as a function of t over a signal interval T. Thus $h(\tau,t)$ is modeled as a linear time invariant filter over individual signal intervals, allowing one to play all the games of elementary linear systems. One must recognize, of course, that $h(\tau,t)$ changes significantly over many signalling intervals, so that one cannot simply measure h once and for all.

4 ESTIMATING $H(\tau,T)$

First ignore noise, assume that $x_m(\tau)$ is transmitted, and consider passing the channel output, $x_m(\tau) * h(\tau,t)$ through a filter matched to x_m (i.e., a filter with impulse response $x_m^*(-\tau)$) (see Fig. 3.2).

Taking Fourier transforms, we have

$$R_m(f) = X_m(f)H(f,t)X_m^*(f) = |X_m(f)|^2 H(f,t) = \frac{\mathcal{E}}{2W} H(f,t) \tag{3.10}$$

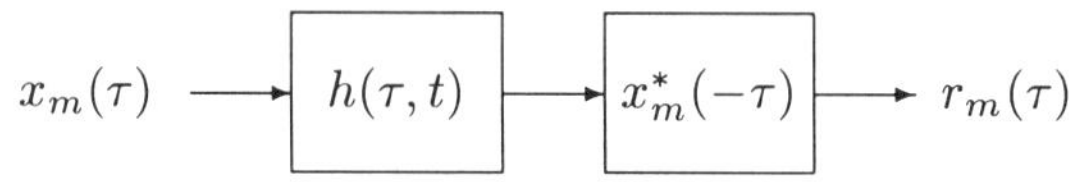

Figure 3.2

where we have used (3.4). Taking the inverse Fourier transform, we see that $r_m(\tau) = \frac{\mathcal{E}}{2W} h(\tau, t)$. Since we are looking at an input in the interval $[-T, 0]$, and we are assuming that $h(\tau, t)$ does not change in t over intervals of duration T, the parameter t can be taken to be 0. This suggests that the output should be attenuated by $2W/\mathcal{E}$ in order to obtain an estimate of $h(\tau, t)$ at t=0.

We now put the white noise back in the picture and look at the output of the attenuated matched filter including noise (see Fig. 3.3). Assume the noise has spectral density $N_0/2$. Filtering the noise to $|f - f_0| \leq W/2$, and defining the baseband equivalent noise, as the upper sideband shifted down by f_0, the baseband equivalent noise process is complex Gaussian and has the spectral density $N_0/2$ for $|f| \leq W/2$. It follows that the noise power of the baseband waveform is $N_0 W/2$, which is half the noise power of the band pass waveform. Thus we have scaled the noise in the same way as the signal. Physically, when one demodulates a passband waveform into quadrature baseband components, one can scale those baseband waveforms arbitrarily, but the signal and noise must be scaled the same way.

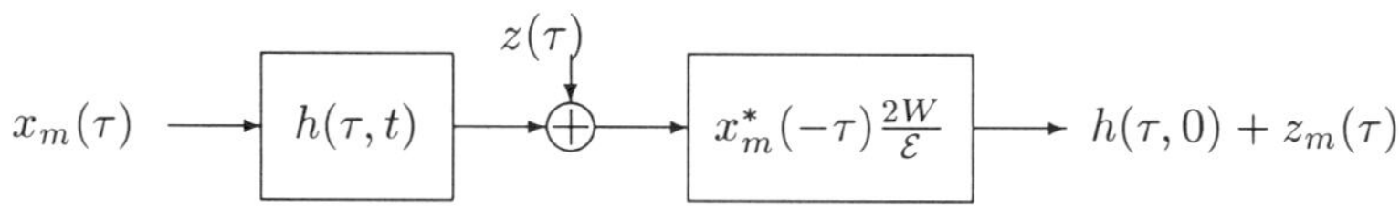

Figure 3.3

We have seen that the component of the output due to the signal $x_m(\tau)$ is $h(\tau, 0)$. To analyze a sample $z_m(\tau)$ of the output noise process, note that the Fourier transform of the attenuated matched filter is $X_m^*(f)2W/\mathcal{E}$. The spectral density of the filter, i.e., the magnitude squared of the Fourier transform, is $\mathcal{E}/(2W))[2W/\mathcal{E}]^2 = 2W/\mathcal{E}$. Since the input process has the spectral density $N_0/2$, the output $\{z_m(\tau)\}$ comes from a process with the spectral density $N_0 W/\mathcal{E}$. Since this output noise process has bandwidth $W/2$, $z_m(\tau)$, for any given τ is a sample value of a random variable of variance $N_0 W^2/\mathcal{E}$, so

$$\text{Var}[z_m(t)] = N_0 W^2/\mathcal{E} \tag{3.11}$$

Now consider a rake receiver (see Fig. 3.4). If $h(\tau, t)$ is known, then the optimal detector for the M-ary signal set $x_m(\tau)$, $1 \leq m \leq M$, through the filter $h(\tau, t)$, is simply a set of matched filters matched to the convolution of x_m and h, i.e., $x_m^*(-\tau) * h^*(-\tau, t)$. The decision on m in Fig. 3.4 replaces $h^*(-\tau, t)$ with $\hat{h}^*(-\tau, t)$, which is reasonable if the estimate is good. Assume that $h(\tau, t)$ is well estimated, and that a correct decision is made on the input m. Given this decision, the output of the filter matched to the signal x_m yields a new estimate of h plus additive Gaussian noise. The device to estimate h then uses the decision on m to accept the output from the mth matched filter to update the old estimate of h.

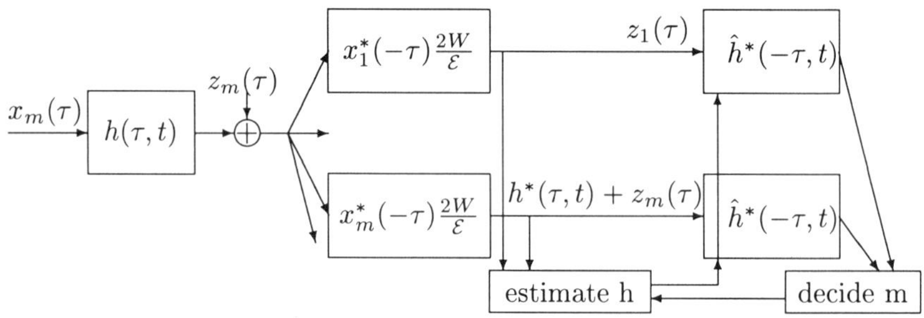

Figure 3.4 Rake receiver.

To avoid worrying about the optimal estimate of h, we can get an approximate answer by assuming that the estimate $\hat{h}$ of h is simply the linear average of the previous n measurements. Here one measurement is made each T seconds and the ith such measurement, made at time $-iT$, comes from matched filter m_i, where signal m_i was sent at that time. Since the channel filter remains almost constant for a time on the order of $1/B$, we take $n = 1/(BT)$. Let $z'(\tau)$ be the error in $\hat{h}(\tau, 0)$, i.e., $\hat{h}(\tau, t) = h(\tau, t) + z'(\tau)$. Since taking an average over n measurements with IID noise reduces the noise variance by a factor of n, it can be shown that

$$\mathrm{Var}(z'(\tau)) = \frac{N_0 W^2}{n\mathcal{E}} = \frac{BTN_0W^2}{\mathcal{E}} \quad ; \quad 0 \leq \tau \leq L' \tag{3.12}$$

We assume that the multipath spread L' (including the limitation to bandwidth $W/2$) is known, and thus that $\hat{h}(\tau, t)$ is taken to be 0 for $\tau < 0$ and $\tau > L'$. Since all of the noise processes being averaged are white over $|f| \leq W/2$, $\{z'(\tau)\}$ is a sample function of a process that is white over $|f| \leq W/2$ and is non-zero only over the interval $[0, L']$.

5 RESIDUAL NOISE

Finally we have the problem of determining the residual noise if the effect of the detected signal is subtracted from the received waveform (again assuming the correct signal was detected). The effect of the signal $x_m(\tau)$ on the received signal is $x_m(\tau) * h(\tau, t)$. The quantity subtracted from the received signal in cancelling the interference from this user is $x_m(\tau) * \hat{h}(\tau, t)$, and thus the residual noise after interference cancellation is $\phi(\tau) = x_m(\tau) * z'(\tau)$. Taking Fourier transforms, $\Phi(f) = X_m(f)Z'(f)$ and

$$|\Phi(f)|^2 = |X_m(f)|^2|Z'(f)|^2 = \frac{\mathcal{E}}{2W}|Z'(f)|^2 \tag{3.13}$$

Thus,

$$\int \frac{\mathcal{E}}{2W}|Z'(f)|^2 df = \int |\phi(\tau)|^2 d\tau = \int \frac{\mathcal{E}}{2W}|z'(\tau)|^2 d\tau \tag{3.14}$$

Taking the expected value of the final terms in (3.14),

$$\int \mathrm{Var}|\phi(\tau)|d\tau = \int \frac{\mathcal{E}}{2W}\mathrm{Var}|z'(\tau)|d\tau = \frac{BTN_0WL'}{2} \tag{3.15}$$

This is the baseband expected energy of the residual noise in the band $|f| \leq W/2$ and over the interval $0 \leq t \leq T$. Since $z'(\tau)$ is white over the band $|f| \leq W/2$, $\phi(\tau)$ is also white over $|f| \leq W/2$. Thus the spectral density of this noise power (averaged over the time interval $(0, T)$) is $BL'N_0/2$. Since the spreading product BL' is small for most wireless situations, this indicates that the residual noise is small relative to the ordinary additive noise of spectral density $N_0/2$. When multiaccess communication is taken into account, the noise that effects the filter measurement becomes not only the white noise but also the other users signals, which have been passed through their own multipath filters before contributing to the measurement of the filter in question.

6 DISCUSSION

The analysis here indicates that the residual noise is a factor BL' of the background noise and the interference from other users, and rather surprisingly, it does not depend on either the signal power or the signal bandwidth. The assumption, however, was that the signals could be successfully detected, and, of course, successful detection does depend on signal power and bandwidth. In fact, it can be shown that pseudo-noise signals of any given power and duration cannot be spread arbitrarily in bandwidth and still be detected on a fading multipath channel. This effect appears only indirectly here, where in it can be seen that the estimate of the channel becomes poor as W becomes large. This increasingly poor estimate does not increase the residual noise spectral density because the signal spectral density is decreasing with W, but it does increase the difficulty of detection.

It is not difficult to actually analyze the estimation error in estimating h. For any fixed τ, if we know the correlation function in t for the random process with sample functions $h(\tau, t)$, then we can use discrete Kalman filtering to find the minimum mean square error linear estimate. One finds, on doing this, that our assumption of $(BT)^{-1}$ estimates with IID noise is very optimistic, but it should be clear that the result only changes by a scale factor. A more serious issue is that with coding, there is significant delay before symbols can be correctly decoded, and this increases the delay in estimating the channel. One can imagine an iterative approach where symbols are detected without delay, this is used to update the channel estimate, and then a better channel estimate is made after decoding. One could also use a Viterbi decoder where different channel estimates are carried along with different potential paths, but this is not very attractive.

Perhaps the best way to look at this is that whatever method is used in decoding a user, that decoded data can be used to estimate the channel and cancel interference if decoding is correct. This can introduce large delays overall, because each user is delayed until the interference cancellation is done for the earlier users to be decoded, but interference cancellation is possible whenever decoding is possible.

It appears to be possible to carry through the analysis without assuming a flat spectral density for the input waveforms, but this seems to be an exercise for people who like complex calculations. Finally, we have ignored inter-symbol interference for the transmitter being detected. This seems to be reasonable if L' is small relative to the signal interval T. It is also reasonable if the spreading factor WT is large, since then the neighboring symbols appear like noise spread over the band, and act much like the interference from other users.

In summary, we must ask whether interference cancellation might someday be practical for wireless communication. We have shown that residual noise is not a major problem there, but delay might be an insurmountable problem, and interence from users in other cells might be sufficiently large that it doesn't pay to cancel interference within a cell. Thus the question is still open.

Acknowledgements

This research was supported in part by ARO under grant DAAH04-95-1-0103, in part by NSF under grant 9314341-NCR, and in part by ARO under grant DAAL-03-92-G-115. A rough draft of the basic results appeared in July 1994 as a LIDS report, LIDS-P-2254, M.I.T.

REFERENCES

[CV93] R.S.Cheng, S.Verdu, "Gaussian Multiaccess Channels with Capacity Region and Multiuser Water-filling," IEEE Trans. IT, Vol 39, No. 3, May 1993.

[CT91] T.Cover and J.Thomas, *Elements of Information Theory*, Wiley, 1991.

[GJ91] K.S. Gilhausen, I.M.Jacobs, R.Padovani, A.J.Viterbi, L.A.Weaver, and C.E.Wheatley, "On the Capacity of a cellular CDMA system," IEEE Trans. Vehic. Tech., Vol 40, No.2, May 1991.

[Go94] A.Goldsmith, "Design and Performance of High-speed Communication Systems over time-varying radio channels," Ph.D. thesis, University of California, Berkeley, 1994.

[Ka59] T. Kailath, "Sampling Models for Linear Time-Variant Filters," TR 352, MIT Research Laboratory of Electronics, May 25, 1959.

[KH95] R.Knopp, P.H.Humblet, "Channel Control and Multiple-Access, ICC, Seattle, Wash.., June 1995

[Me95] M.Medard, "The Capacity of Time-Varying Multiple User Channels in Wireless Communication," Ph.D. thesis, MIT, 1996.

[PG60] R.Price and P.E.Green, "Signal Processing in Radar Astronomy - Communication via Fluctuating Multipath Media," TR 234, Lincoln Laboratory, MIT, 6 Oct. 1960.

[RU96] B.Rimoldi and R.Urbanke, "A Rate Splitting approach to the Gaussian Multiple-Access Channel," IEEE Trans. IT Vol 42, No. 2, March 1996.

[TH96] D.Tse, S.Hanly, "Capacity Region of the Multi-access Fading Channel under Dynamic resource allocation and Polymatroid optimization," IEEE Information Theory Workshop, Haifa Israel, June 1996.

4

GENERALIZED DECISION-FEEDBACK EQUALIZATION FOR PACKET TRANSMISSION WITH ISI AND GAUSSIAN NOISE

J.M. Cioffi* and G.D. Forney, Jr.**

**Information Systems Lab.*
Stanford, CA 94305
email: cioffi@isl.stanford.edu

***Motorola*
Mansfield, MA
email: LUSE27@email.mot.com

Dedicated with respect and deepest regards to Professor Tom Kailath on his sixtieth birthday.

ABSTRACT

A general theory for transmission of finite-length packets over channels with intersymbol interference and additive Gaussian noise is developed. The theory is based on general principles of maximum-likelihood (ML) and linear minimum-mean-squared error (MMSE) estimation, innovations and modal representations of random vectors via Cholesky factorizations, eigendecompositions, and information theory. Using these principles, equivalent forward and backward channel models with desirable properties are developed. Fundamental relations between these theories are presented; for example, the mutual information $I(\boldsymbol{X};\boldsymbol{Y})$ between the input $\boldsymbol{X}$ and output $\boldsymbol{Y}$, when $\boldsymbol{X}$ is a Gaussian vector, is equal to $\log\{\|R_{\boldsymbol{x}'\boldsymbol{x}'}\|/\|R_{\boldsymbol{e}'\boldsymbol{e}'}\|\}$, where $\|R_{\boldsymbol{x}'\boldsymbol{x}'}\|$ and $\|R_{\boldsymbol{e}'\boldsymbol{e}'}\|$ are the effective determinants of the covariance matrices of the effective input and of the input linear MMSE estimation error, respectively. A Generalized Decision Feedback Equalization (GDFE) receiver structure is developed and is shown to be canonical for arbitrary linear Gaussian channels- i.e., a reliably transmitted data rate of $I(\boldsymbol{X};\boldsymbol{Y})$ can be approached arbitrarily closely with this receiver structure on any linear Gaussian channel with any input covariance matrix $R_{\boldsymbol{xx}}$. For optimal $R_{\boldsymbol{xx}}$, the performance of this receiver is in aggregate the same as the well-known vector coding (VC) structure, but in detail the structure is quite different from VC or other previously proposed block DFE receiver structures.

1 INTRODUCTION

In [1], canonical minimum-mean-squared-error decision-feedback equalization (MMSE-DFE) receiver structures for infinite-length sequence transmission have been developed. That paper illustrated an intimate relationship between MMSE-DFE equalization performance and the mutual information $I(\boldsymbol{X};\boldsymbol{Y})$ in bits per complex symbol between channel input sequence $\boldsymbol{X}$ and output sequence $\boldsymbol{Y}$, given by the formula

$$I(\boldsymbol{X};\boldsymbol{Y}) = \log_2 \mathrm{SNR}_{\text{MMSE-DFE}} , \tag{4.1}$$

where $\mathrm{SNR}_{\text{MMSE-DFE}}$ is the signal-to-noise ratio at the decision point of an MMSE-DFE receiver. From (4.1), it follows that the capacity-achieving transmit power spectrum is the same spectrum that optimizes $\mathrm{SNR}_{\text{MMSE-DFE}}$. Thus, the performance of a MMSE-DFE transmission system[1], with optimized-spectrum transmit signals and powerful coding, can approach the channel capacity of an arbitrary stationary linear-ISI Gaussian sequence channel as closely as capacity can be approached on an ideal Gaussian channel with that same coding – a situation called "canonical" in [1].

In many applications, however, the number of input symbols and output samples is finite; e.g., in point-to-point packet transmission when finite complexity or delay constraints dictate a block structure, or in multi-user packet transmission.

In these applications an appropriate channel model is a finite-dimensional matrix model $\boldsymbol{Y} = H\boldsymbol{X} + \boldsymbol{N}$, where $\boldsymbol{X}$ is a random input m-tuple, H is an $n \times m$ channel-response matrix, and $\boldsymbol{N}$ is an additive Gaussian noise n-tuple. (All quantities are complex.) This paper shows that on such channels the mutual information $I(\boldsymbol{X};\boldsymbol{Y})$ in bits per block is

$$I(\boldsymbol{X};\boldsymbol{Y}) = \log_2 |\boldsymbol{SNR}_{\mathrm{GDFE}}| , \tag{4.2}$$

where $\boldsymbol{SNR}_{\mathrm{GDFE}}$ is an appropriately defined matrix. Furthermore, it shows that a certain Generalized DFE (GDFE) receiver structure is canonical for such channels.

This paper continues to call a receiver **canonical** if in combination with the <u>same</u> sufficiently powerful coding that approaches capacity on the ISI-free channel, this canonical receiver can achieve arbitrarily low error rates for data rates approaching the value of the mutual information $I(\boldsymbol{X};\boldsymbol{Y})$ between channel input and output on the ISI-channel. The mutual information that measures a

[1]This MMSE-DFE actually can become several parallel MMSE-DFE's, one for each disconnected band of frequencies in the capacity-achieving power spectrum.

canonical receiver is computed under the assumption that the input statistics are Gaussian. It should be emphasized that a canonical receiver is not necessarily an optimum receiver, and indeed with no coding or with only moderately powerful coding it may be distinctly inferior to an optimum receiver. The new MMSE-DFE receiver structure of this paper, like that of [1], is constructed using principles of optimum estimation theory, not optimum detection theory, and therefore may be suboptimum when the input sequence is a discrete digital sequence, as it always is in practice. As in [1], the point is that a receiver does not need to do optimum detection to approach channel capacity, when it is used in conjunction with sufficiently powerful codes.

1.1 Parallel Channels - a simple illustration of canonical transmission

Suppose H is a square nonsingular $n \times n$ diagonal matrix and $R_{\boldsymbol{nn}} = N_0 I$, then the channel is equivalent to n independent "parallel" subchannels, each with input/output relation $Y_i = H_i X_i + N_i$. The signal to noise ratio on the ith subchannel is $\mathrm{SNR}_i = S_{x,i}|H_i|^2/N_0$ with $S_{x,i}$ the mean-square value for the ith element of the input vector $\boldsymbol{X}$. For each of these parallel subchannels, the mutual information is $log_2(1 + \mathrm{SNR}_i)$ bits per subchannel and for the set of channels, the mutual information is easily determined as [2]

$$I(\boldsymbol{X};\boldsymbol{Y}) = \log_2 \prod_{i=1}^{n} (1 + \mathrm{SNR}_i) \ . \tag{4.3}$$

Each of the subchannels can be independently coded with a powerful code for the ideal additive white Gaussian noise channel so that the data rate achieved is arbitrarily close to the mutual information. The set of such codes and channels then has an aggregate data rate that is the mutual information for the aggregate channel. Figure 4.1 illustrates a set of parallel channels.

The energy (values of $S_{x,i}$ allocated to each subchannel can be determined by a "water-filling" solution [2] and capacity for this block-diagonal-H channel can then be achieved with the same powerful codes that would be used on an ISI-free white-Gaussian noise channel.

While the parallel channels example is trivial, it is also very important in the study of canonical transmission because all the structures that this paper derives for more general H eventually reduce to a set of parallel channels for which the mutual information is the same as the original channel and the same powerful codes that would be used on an AWGN channel can be applied to achieve the highest possible data rates. This paper often uses the example of a one-dimensional channel to illustrate various properties, which can tacitly be inferred to be equivalent to the set of parallel channels.

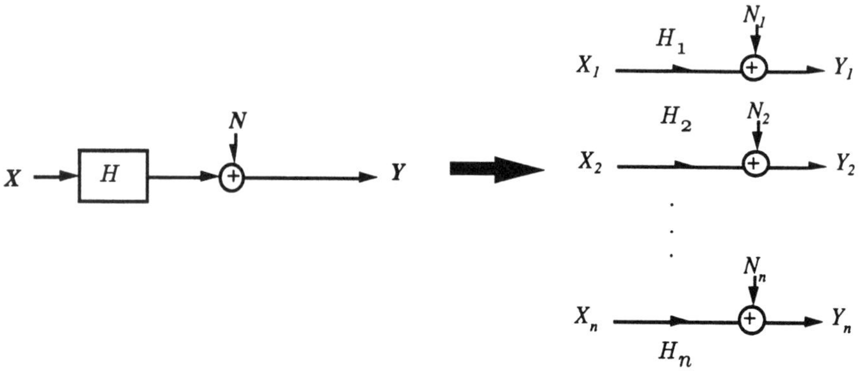

Figure 4.1 Parallel independent channels and equivalent channel.

1.2 More general canonical transmission

Like the well-known vector coding (VC) structure (which is shown to be a special case of the GDFE) [3] -[11], the canonical GDFE structure developed in this paper (which is not the same as the DFE receiver structures of [12] -[17]) effectively decomposes a matrix (block) channel into a number r_y of decoupled one-dimensional Gaussian subchannels with signal-to-noise ratios SNR_j, $1 \leq j \leq r_y$, such that the sum of the component mutual informations $I_j = \log_2(1+\mathrm{SNR}_j)$ is equal to $I(\boldsymbol{X};\boldsymbol{Y})$. It then follows from the channel coding theorem that for any rate $R_j < I_j$ there exists a discrete (non-Gaussian) code of rate R_j that is capable of achieving arbitrarily low error rates on the jth subchannel. By using such a code on each subchannel, an aggregate rate arbitrarily close to $I(\boldsymbol{X};\boldsymbol{Y})$ bits per block can be transmitted with arbitrarily low probability of error.

More generally than the VC special case of the GDFE, the subchannels in the GDFE receiver are not completely independent, but rather are decoupled by use of the "ideal DFE assumption," which is that the inputs to "past" subchannels are available to the receiver when decoding the current subchannel.

It is shown that the GDFE receiver is canonical even in the general case in which the input covariance matrix $R_{\boldsymbol{xx}}$ does not commute with the channel covariance matrix $H^*R_{\boldsymbol{nn}}^{-1}H$, in which case the vector coding special case is not defined. However, the optimum $R_{\boldsymbol{xx}}$, which is the same for all cases of the GDFE, always commutes with $H^*R_{\boldsymbol{nn}}^{-1}H$.

The set of $\{\mathrm{SNR}_j\}$ also differ. In the limit of large blocks (long packet lengths) and stationary channels, one special case known as the "packet" GDFE receiver approaches the MMSE-DFE receiver structure (or structures for disconnected transmission bands) of [1], Cholesky factorization becomes spectral factorization, and all SNR_j tend to become equal, provided that all subchannels are used. With the vector coding special case, the $\{\mathrm{SNR}_j\}$ are distributed in water-pouring fashion as a function of frequency and vector coding tends to what is known as multitone transmission [4]. However, the products of the $(1 + \mathrm{SNR}_j)$ for the set of subchannels in both cases are the same and equal to $2^{I(\boldsymbol{X};\boldsymbol{Y})}$, as is always the case with any GDFE.

After introducing the general linear Gaussian block channel model, Section 2 discusses modal representations of random vectors based on eigendecompositions and innovations representations (or "Cholesky" factorizations), which are the basic tools used to develop our canonical receivers. It then reviews general principles of linear MMSE estimation. Finally, it discusses additional information-theoretic properties that hold when $\boldsymbol{X}$ is Gaussian.

Section 3 begins by reducing the general channel model without loss of optimality to equivalent forward and backward matrix channel models that have many nice properties: unnecessary dimensions are eliminated, all matrices are square and nonsingular, and the channel-response matrix is equal to the noise covariance matrix. The operation of elimination of unnecessary dimensions is crucial in canonical receivers and asymptotically corresponds as the packet length increases to "symbol rate" optimization and carrier (center) frequency optimization in each used band for the MMSE-DFE. Elimination of unnecessary dimensions also corresponds to selecting good frequency bands for transmission in a vector coding (or multitone) transmission system as packet length increases. The optimum ML and linear MMSE estimators are developed from these models. When the input $\boldsymbol{X}$ is Gaussian, some interesting connections are developed between mutual information and optimal estimation. For example,

$$I(\boldsymbol{X};\boldsymbol{Y}) = \log \|R_{\boldsymbol{x}'\boldsymbol{x}'}\| / \|R_{\boldsymbol{e}'\boldsymbol{e}'}\| \ , \tag{4.4}$$

where $\|R_{\boldsymbol{x}'\boldsymbol{x}'}\|$ and $\|R_{\boldsymbol{e}'\boldsymbol{e}'}\|$ are the effective determinants of the covariance matrices of the effective input and of the error in the linear MMSE estimate of the input, respectively. Also,

$$I(\boldsymbol{X};\boldsymbol{Y}) = \log |\boldsymbol{SNR}_{\mathrm{GDFE}}| = \log |I + \boldsymbol{SNR}_{\mathrm{ML}}| \ , \tag{4.5}$$

where $\boldsymbol{SNR}_{\mathrm{GDFE}}$ and $\boldsymbol{SNR}_{\mathrm{ML}}$ are matrix generalizations of the usual one-dimensional SNRs for optimum linear MMSE and ML estimation, respectively.

Using an equivalent backward channel model and the "ideal DFE assumption," Section 4 then develops the GDFE receiver structure and shows that it is canonical.

Section 5 addresses the problem of choosing of the input covariance matrix $R_{\boldsymbol{xx}}$ for the GDFE to maximize $I(\boldsymbol{X};\boldsymbol{Y})$, which as is well known is solved by discrete water-pouring. The optimum $R_{\boldsymbol{xx}}$ is shown to commute with the channel covariance matrix $H^* R_{\boldsymbol{nn}}^{-1} H$. Vector coding is well defined in this situation, is also canonical, and uses the same $R_{\boldsymbol{xx}}$ and is a special case of the GDFE where the feedback section disappears.

Section 6 considers the passage to the limit of large blocks (long packets) for stationary Gaussian ISI channels and illustrates that the results of this paper converge to the results in [1] in the limit of infinite-length packets. Further Section 6 illustrates expanded interpretations of the results in [1] where while the MMSE-DFE converges to a stationary structure, there could be several such structures covering only those frequency bands that would also be used by water-pouring transmit optimization – this clearly shows that conventional MMSE-DFE structures such as those considered by Price [18], Salz [19] and others [20] are too generally claimed to be optimum as proposed. However, the necessary modifications (often not understood nor used) to restore optimality are illustrated generally by this paper and in the limit in Section 6.

2 THE BLOCK OR "PACKET" GAUSSIAN ISI CHANNEL

A block (or packet) transmission channel has a finite number of input samples and output samples. Such a channel model is appropriate when a finite-length information packet is transmitted, and detection is based on a finite number of samples of the received signal. Usually, the term packet refers to the situation where the samples are successively indexed in time within a block.

This section begins with a general block Gaussian ISI channel model. Two representations of random vectors are then discussed; in particular, modal representations based on eigendecompositions, and innovations representations based on Cholesky factorizations. These two types of representations are the basis of the canonical receivers to be discussed in this paper. This section progress to discussions of MMSE linear estimation, innovations recursions, and Gaussian random vectors.

2.1 Channel model

On a block Gaussian ISI channel, the received vector of sequence samples $\boldsymbol{Y}$ may be expressed by the matrix equation

$$\boldsymbol{Y} = H\boldsymbol{X} + \boldsymbol{N} \ , \tag{4.6}$$

where $\boldsymbol{X} = \{X_j, 1 \leq j \leq m\}$ is a complex random input m-vector, $\boldsymbol{Y} = \{Y_k, 1 \leq k \leq n\}$ is a complex random output n-vector, H is an $n \times m$ complex channel-response matrix, and $\boldsymbol{N}$ is a complex random Gaussian noise n-vector independent of $\boldsymbol{X}$. If $n = m$, the channel is square. All vectors are written as column vectors.

All random vectors, whether discrete, continuous or Gaussian, will be characterized solely by their second-order statistics. The mean of all unconditioned random variables is assumed to be zero, since a nonzero mean costs energy but carries no information. A random vector such as $\boldsymbol{X}$ is then characterized by its covariance matrix

$$R_{\boldsymbol{xx}} = E[\boldsymbol{X}\boldsymbol{X}^*] , \tag{4.7}$$

where the asterisk denotes conjugate transpose. The rank of $\boldsymbol{X}$ is the rank r_x of its covariance matrix $R_{\boldsymbol{xx}}$, which is the dimension of the complex vector space $\mathcal{S}_{\boldsymbol{X}}$ in which $\boldsymbol{X}$ takes its values. If $R_{\boldsymbol{xx}}$ is nonsingular, then $\boldsymbol{X}$ has full rank and $r_x = m$, otherwise $r_x < m$.

No restrictions are placed on the input covariance matrix $R_{\boldsymbol{xx}}$ or on the noise covariance matrix $R_{\boldsymbol{nn}}$, except that $\boldsymbol{N}$ is assumed to have full rank, $r_n = n$, so as to avoid noiseless channels of infinite capacity. Similarly, the channel-response matrix H is an arbitrary $n \times m$ complex matrix. The signal component of the output, namely the n-tuple

$$\hat{\boldsymbol{Y}}(\boldsymbol{X}) = H\boldsymbol{X} , \tag{4.8}$$

then has covariance matrix $HR_{\boldsymbol{xx}}H^*$. The notation $\hat{\boldsymbol{Y}}(\boldsymbol{X})$ indicates that $\hat{\boldsymbol{Y}}(\boldsymbol{X})$ is the conditional mean of $\boldsymbol{Y}$ given $\boldsymbol{X}$ (see Section 2.3). The vector space $\mathcal{S}_{\hat{\boldsymbol{Y}}}$ of $\hat{\boldsymbol{Y}}(\boldsymbol{X})$ is the image of the input space $\mathcal{S}_{\boldsymbol{X}}$ under the linear transformation H, and therefore the rank $r_{\hat{y}}$ of $\hat{\boldsymbol{Y}}(\boldsymbol{X})$ is not greater than r_x, with equality if and only if the map H acting on $\mathcal{S}_{\boldsymbol{X}}$ is one-to-one. Since $\boldsymbol{X}$ and $\boldsymbol{N}$ are independent, the output covariance matrix is

$$R_{\boldsymbol{yy}} = HR_{\boldsymbol{xx}}H^* + R_{\boldsymbol{nn}} . \tag{4.9}$$

Since $\boldsymbol{N}$ has full rank and $HR_{\boldsymbol{xx}}H^*$ is non-negative definite, $\boldsymbol{Y}$ has full rank, $r_y = n$ – however, $r_{\hat{y}} \leq \min(n, r_x)$.

2.2 Random vectors and covariance matrix factorizations

This section develops two characteristic representations of random vectors on which our canonical receiver structures will be based. A few preliminary remarks on the geometry of signal spaces may be helpful.

Geometries of vector spaces

There are two kinds of geometry that characterize a random vector such as $\boldsymbol{X}$, and two corresponding inner products:

1. First, there is the **ordinary Euclidean geometry** of the complex vector space $\mathcal{S}_{\boldsymbol{X}}$ in which $\boldsymbol{X}$ takes values. In $\mathcal{S}_{\boldsymbol{X}}$, the inner product of two ordinary ("deterministic") complex column vectors $\boldsymbol{x}$ and $\boldsymbol{y}$ is the ordinary Hermitian dot product:

$$\boldsymbol{x}^*\boldsymbol{y} = \sum_i x_i^* y_i , \tag{4.10}$$

where, as always in this paper, the asterisk denotes conjugate transpose. In ordinary Euclidean geometry the squared norm of a vector $\boldsymbol{x}$ is the usual Euclidean squared norm $\|\boldsymbol{x}\|^2$, namely the sum of the squared magnitudes $|x_i|^2$ of the components x_i, and two vectors $\boldsymbol{x}$ and $\boldsymbol{y}$ are orthogonal if their dot product $\boldsymbol{x}^*\boldsymbol{y}$ is zero.

2. Second, there is the **geometry of Hilbert spaces** of complex random variables, in which the inner product of two complex random variables X and Y is defined by their Hermitian cross-correlation

$$< X, Y >= E[XY^*] \ . \tag{4.11}$$

In Hilbert-space geometry the squared norm of a zero-mean random variable X is its variance $E[|X|^2]$, and two random variables X and Y are orthogonal if they are uncorrelated, $E[XY^*] = 0$.

The set $\{X_i\}$ of components of a random vector $\boldsymbol{X}$ generate a Hilbert space $V(\boldsymbol{X})$ consisting of all complex linear combinations

$$\sum_i a_i^* X_i = \boldsymbol{a}^* \boldsymbol{X} \tag{4.12}$$

of elements of $\boldsymbol{X}$. The inner product of two elements $\boldsymbol{a}^*\boldsymbol{X}$, $\boldsymbol{b}^*\boldsymbol{X} \in V(\boldsymbol{X})$ is

$$< \boldsymbol{a}^*\boldsymbol{X}, \ \boldsymbol{b}^*\boldsymbol{X} >= E[\boldsymbol{a}^*\boldsymbol{X}\boldsymbol{X}^*\boldsymbol{b}] = \boldsymbol{a}^* R_{\boldsymbol{x}\boldsymbol{x}} \boldsymbol{b} \ . \tag{4.13}$$

Thus the geometry of $V(\boldsymbol{X})$, which is characterized by the set of inner products between any two of its vectors, is entirely determined by the covariance matrix $R_{\boldsymbol{x}\boldsymbol{x}}$, which is the matrix of inner products (Gram matrix) of elements of $\boldsymbol{X}$.

Charateristic Representations

Characteristic representations will enable the design of canonical receivers:

Definition 1 (Characteristic representation of a random vector) *A* **characteristic representation** *of a random m-tuple* $\boldsymbol{X}$ *is a linear combination*

$$\boldsymbol{X} = F\boldsymbol{V} = \sum_j V_j \boldsymbol{f}_j \ , \tag{4.14}$$

where $\{V_j\}$ is a set of uncorrelated random variables–i.e., the covariance matrix $R_{\boldsymbol{v}\boldsymbol{v}}$ is diagonal- and F is a square matrix with determinant $|F| = 1$.

The covariance matrix of $\boldsymbol{X}$ is then

$$R_{\boldsymbol{x}\boldsymbol{x}} = E[F\boldsymbol{V}\boldsymbol{V}^*F^*] = FR_{\boldsymbol{v}\boldsymbol{v}}F^* \ . \tag{4.15}$$

Thus characteristic representations of the form $\boldsymbol{X} = F\boldsymbol{V}$ are closely related to covariance matrix factorizations of the form $R_{\boldsymbol{xx}} = FR_{\boldsymbol{vv}}F^*$, where $|F| = 1$ and $R_{\boldsymbol{vv}}$ is diagonal. Indeed, given such a factorization, define $\boldsymbol{V} = F^{-1}\boldsymbol{X}$; then $\boldsymbol{V}$ has the diagonal covariance matrix $R_{\boldsymbol{vv}}$ that occurs in the factorization and $\boldsymbol{X} = F\boldsymbol{V}$.

Since F is nonsingular, it is rank-preserving; i.e., the rank of $\boldsymbol{V}$ is equal to the rank of $\boldsymbol{X}$, $r_v = r_x$, which implies that precisely r_x of the random variables V_j are not identically zero. Since every element $\boldsymbol{a}^*\boldsymbol{X}$ of $V(\boldsymbol{X})$ is a linear combination of these rx nonzero random variables V_j via

$$\boldsymbol{a}^*\boldsymbol{X} = \boldsymbol{a}^*F\boldsymbol{V} \ , \tag{4.16}$$

it follows that $V(\boldsymbol{X}) = V(\boldsymbol{V})$ and that these r_x nonzero random variables V_j form an orthogonal basis for $V(\boldsymbol{X})$, whose dimension is thus also equal to r_x. The r_x corresponding complex vectors $\boldsymbol{f}_j$ generate the deterministic r_x-dimensional Euclidean space $\mathcal{S}_{\boldsymbol{X}}$, although they are not necessarily orthogonal in $\mathcal{S}_{\boldsymbol{X}}$.

Finally, the unimodular condition $|F| = 1$ implies that F and its inverse F^{-1} are volume-preserving transformations, provided that $\boldsymbol{X}$ has full rank. In other words, F is determinant-preserving:

$$|R_{\boldsymbol{xx}}| = |F||R_{\boldsymbol{vv}}||F^*| = |R_{\boldsymbol{vv}}| \ . \tag{4.17}$$

However, if $\boldsymbol{X}$ does not have full rank, then F is not necessarily a volume-preserving transformation from the r_x-dimensional subspace $\mathcal{S}_{\boldsymbol{V}}$ that supports the r_x nonzero random variables $\{V_j\}$ to the r_x-dimensional subspace $\mathcal{S}_{\boldsymbol{X}}$ that supports the random vector $\boldsymbol{X}$. There are two types of characteristic representations of interest:

Modal representations

A covariance matrix $R_{\boldsymbol{xx}}$ is square, Hermitian-symmetric, and nonnegative definite. Such a matrix has a (nonunique, in general) eigendecomposition

$$R_{\boldsymbol{xx}} = U\Lambda_x^2U^* = (U\Lambda_x)(U\Lambda_x)^* \ , \tag{4.18}$$

where U is a unitary matrix ($UU^* = U^*U = I$, so $U^{-1} = U^*$ and $|U| = 1$) and Λ_x^2 is a nonnegative real diagonal matrix whose diagonal elements are the eigenvalues of $R_{\boldsymbol{xx}}$. The set of eigenvalues is invariant in any eigendecomposition. The last expression shows that $U\Lambda_x$ may be regarded as a square root of $R_{\boldsymbol{xx}}$.

Correspondingly, if the modal variables $\boldsymbol{M}$ are defined by

$$\boldsymbol{M} = U^{-1}\boldsymbol{X} = U^*\boldsymbol{X} \ , \tag{4.19}$$

then $R_{\boldsymbol{mm}} = U^*R_{\boldsymbol{xx}}U = \Lambda_x^2$ and $\boldsymbol{X} = U\boldsymbol{M}$, where $|U| = 1$. Thus any eigendecomposition of $R_{\boldsymbol{xx}}$ leads to a characteristic representation of $\boldsymbol{X}$, called a **modal representation**.

Since the columns u_j of a unitary matrix U are orthonormal, a modal representation

$$\boldsymbol{X} = U\boldsymbol{M} = \sum_j M_j \boldsymbol{u}_j \ , \tag{4.20}$$

has the desirable property that the r_x vectors $\boldsymbol{u}_j$ corresponding to the r_x nonzero modal variables M_j form an orthonormal basis for $\mathcal{S}_{\boldsymbol{X}}$; i.e., both kinds of orthogonality occur in a modal decomposition.

Consequently, a unitary transformation is length-preserving; that is,

$$\|U\boldsymbol{m}\|^2 = \boldsymbol{m}^* U^* U \boldsymbol{m} = \boldsymbol{m}^* \boldsymbol{m} = \|\boldsymbol{m}\|^2 \ . \tag{4.21}$$

A fortiori, U is volume-preserving regardless of whether $\boldsymbol{X}$ has full rank.

Example 4.1 (Modal Representation Example) Let $\boldsymbol{X}$ be a random vector $[X_1, X_2]^*$ with covariance matrix

$$R_{\boldsymbol{xx}} = \begin{bmatrix} a & b \\ b & a \end{bmatrix} , \tag{4.22}$$

where a and b are real and $0 \le |b| \le a$. Then, $|R_{\boldsymbol{xx}}| = a^2 - b^2$, the eigenvalues of $R_{\boldsymbol{xx}}$ are $a+b$ and $a-b$, its eigenvectors are $\frac{1}{\sqrt{2}}[11]^*$ and $\frac{1}{\sqrt{2}}[-11]^*$, and its rank r_x is 2 unless $|b| = a$, when $r_x = 1$. An eigendecomposition of $R_{\boldsymbol{xx}}$ is thus

$$R_{\boldsymbol{xx}} = \begin{bmatrix} 2^{-1/2} & -2^{-1/2} \\ 2^{-1/2} & 2^{-1/2} \end{bmatrix} \begin{bmatrix} a+b & 0 \\ 0 & a-b \end{bmatrix} \begin{bmatrix} 2^{-1/2} & 2^{-1/2} \\ -2^{-1/2} & 2^{-1/2} \end{bmatrix} \tag{4.23}$$

and a modal representation of $\boldsymbol{X}$ is

$$\begin{bmatrix} X_1 \\ X_2 \end{bmatrix} = \begin{bmatrix} 2^{-1/2} & -2^{-1/2} \\ 2^{-1/2} & 2^{-1/2} \end{bmatrix} \begin{bmatrix} M_1 \\ M_2 \end{bmatrix} , \tag{4.24}$$

where $M_1 = \frac{1}{\sqrt{2}}(X_1 + X_2)$ has variance $a+b$, $M_2 = \frac{1}{\sqrt{2}}(X_2 - X_1)$ has variance $a-b$ and M_1 and M_2 are uncorrelated. If $b = a$, then $X_1 = X_2$ and $M_1 = \sqrt{2}X_1$, $M_2 = 0$, whereas if $b = -a$, then $X_1 = -X_2$ and $M_1 = 0$, $M_2 = \sqrt{2}X_1$. Note that if $b = 0$ then $R_{\boldsymbol{xx}} = U(aI)U^*$ for any 2×2 unitary matrix U, so there is a family of eigendecompositions of which the one given above is only one member.

Innovations representations

Alternatively, a covariance matrix $R_{\boldsymbol{xx}}$ has a unique factorization of the form

$$R_{\boldsymbol{xx}} = LD_x^2L^* = (LD_x)(LD_x)^* \ , \tag{4.25}$$

where L is a lower triangular matrix that is monic (i.e., which has ones on the diagonal, so $|L| = 1$) and D_x^2 is diagonal. This factorization is called the

Cholesky factorization of $R_{\boldsymbol{XX}}$, and the diagonal elements of D_x^2 (which must be real and nonnegative, with r_x of them nonzero) are called the **Cholesky factors** of $R_{\boldsymbol{XX}}$. The matrix LD_x is another square root of $R_{\boldsymbol{XX}}$.

Correspondingly, the innovations variables $\boldsymbol{W}$ are

$$\boldsymbol{W} = L^{-1}\boldsymbol{X} \ , \tag{4.26}$$

and $R_{\boldsymbol{WW}} = L^{-1}R_{\boldsymbol{XX}}L^{-*} = D_x^2$ and $\boldsymbol{X} = L\boldsymbol{W}$, where $|L| = 1$. (Here L^{-*} denotes $(L^{-1})^* = (L^*)^{-1}$.) Thus, the Cholesky factorization of $R_{\boldsymbol{XX}}$ leads to a unique characteristic representation of $\boldsymbol{X}$, called the **innovations representation**.

Since L is lower triangular, the innovations representation

$$\boldsymbol{X} = L\boldsymbol{W} = \sum_j W_j \boldsymbol{l}_j \ , \tag{4.27}$$

has the desirable property that, for any k, the first k components of $\boldsymbol{X}$ depend only on the first k components of $\boldsymbol{W}$ (and, since L^{-1} is also lower triangular, vice versa). From a dynamical point of view, an innovations representation thus has a kind of causality property, which is important when the sequential ordering of the components of $\boldsymbol{X}$ is important. Also, in matrix terms, this property implies that a Cholesky factorization has a nesting property that leads to recursive implementations. Again, the r_x columns $\boldsymbol{l}_j$ corresponding to the r_x nonzero innovations variables W_j span $\mathcal{S}_{\boldsymbol{X}}$, although they are not in general orthogonal.

The Cholesky factorization of $R_{\boldsymbol{XX}}$ and corresponding innovations representation of $\boldsymbol{X}$ depend very much on the ordering of the components of $\boldsymbol{X}$. If $\boldsymbol{X}'$ is a permutation of $\boldsymbol{X}$, then the innovations representation of $\boldsymbol{X}'$ and its Cholesky factors will be different (although because of the invariance of the effective determinant, the product of the nonzero Cholesky factors will be unchanged). In particular, if $\boldsymbol{X}'$ is the reversal of $\boldsymbol{X}$, then the Cholesky factorization of $R_{\boldsymbol{X}'\boldsymbol{X}'}$ can be permuted to give an upper-diagonal-lower factorization of $R_{\boldsymbol{XX}}$ of the form

$$R_{\boldsymbol{XX}} = (L')^*(D_x')^2 L' \ , \tag{4.28}$$

where $(L')^*$ is upper triangular, and a corresponding reverse innovations representation of $\boldsymbol{X}$ is then obtained:

$$\boldsymbol{X} = (L')^* \boldsymbol{W}' \ . \tag{4.29}$$

Example 4.2 (Innovations Representation Example) Again let $\boldsymbol{X}$ be a random vector $[X_1, X_2]^*$ with covariance matrix

$$R_{\boldsymbol{XX}} = \begin{bmatrix} a & b \\ b & a \end{bmatrix} \ , \tag{4.30}$$

with a, b real and $0 \leq |b| \leq a$. Then the unique Cholesky decomposition of $R_{\boldsymbol{xx}}$ is

$$R_{\boldsymbol{xx}} = \begin{bmatrix} 1 & 0 \\ b/a & 1 \end{bmatrix} \begin{bmatrix} a & 0 \\ 0 & (a^2 - b^2)/a \end{bmatrix} \begin{bmatrix} 1 & b/a \\ 0 & 1 \end{bmatrix} , \tag{4.31}$$

and an innovations representation of $\boldsymbol{X}$ is

$$\begin{bmatrix} X_1 \\ X_2 \end{bmatrix} = \begin{bmatrix} 1 & 0 \\ b/a & 1 \end{bmatrix} \begin{bmatrix} W_1 \\ W_2 \end{bmatrix} , \tag{4.32}$$

where $W_1 = X_1$ has variance a, $W_2 = X_2 - (b/a)X_1$ has variance $(a^2 - b^2)/a$, and W_1 and W_2 are uncorrelated. If $b = a$, then $X_1 = X_2$ and $W_1 = X_1$, $W_2 = 0$, whereas if $b = -a$, then $X_1 = -X_2$ and again $W_1 = X_1$, $W_2 = 0$. Note that when $\boldsymbol{X}$ does not have full rank, this map between the one-dimensional spaces $\mathcal{S}_{\boldsymbol{W}}$ and $\mathcal{S}_{\boldsymbol{X}}$ is not volume- (length-) preserving. But note that even when $b = 0$, the Cholesky decomposition is unique.

2.3 MMSE linear estimation

Suppose that the Hilbert space $V(\boldsymbol{X})$ generated by the elements of $\boldsymbol{X}$ is a subspace of a larger Hilbert space $V(\boldsymbol{X})^+$, and that the complex scalar Y is a random variable in $V(\boldsymbol{X})^+$. Then by the projection theorem, the closest variable to Y in $V(\boldsymbol{X})$ is the projection of Y onto $V(\boldsymbol{X})$, denoted by $Y_{|\boldsymbol{x}}$.

By the orthogonality principle, the projection $Y_{|\boldsymbol{x}}$ is the unique element of $V(\boldsymbol{X})$ such that the estimation error

$$E = Y - Y_{|\boldsymbol{x}} \tag{4.33}$$

is orthogonal to (uncorrelated with) all elements of $V(\boldsymbol{X})$, or equivalently to all elements X_i of $\boldsymbol{X}$. Since $Y_{|\boldsymbol{x}}$ is some linear combination of elements of $\boldsymbol{X}$, $Y_{|\boldsymbol{x}} = \boldsymbol{a}^*\boldsymbol{X}$, this implies that for all i

$$< X_i, E >=< X_i, Y > - < X_i, \boldsymbol{a}^*\boldsymbol{X} >=< X_i, Y > - < X_i, \boldsymbol{X} > \boldsymbol{a} = 0 . \tag{4.34}$$

Equation (4.34) for all i may be written as a matrix equation

$$< \boldsymbol{X}, E >=< \boldsymbol{X}, Y > - < \boldsymbol{X}, \boldsymbol{X} > \boldsymbol{a} = \boldsymbol{r}_{\boldsymbol{x}Y} - R_{\boldsymbol{xx}}\boldsymbol{a} = 0 , \tag{4.35}$$

where $\boldsymbol{r}_{\boldsymbol{x}Y} =< \boldsymbol{X}, Y >$ is the column vector with components $< X_i, Y >= E[X_i Y^*]$, and $R_{\boldsymbol{xx}}$ is the covariance matrix $< \boldsymbol{X}, \boldsymbol{X} >= E[\boldsymbol{X}\boldsymbol{X}^*]$. When $R_{\boldsymbol{xx}}$ is nonsingular, this determines a unique solution for $\boldsymbol{a}$:

$$\boldsymbol{a} = R_{\boldsymbol{xx}}^{-1}\boldsymbol{r}_{\boldsymbol{x}Y} . \tag{4.36}$$

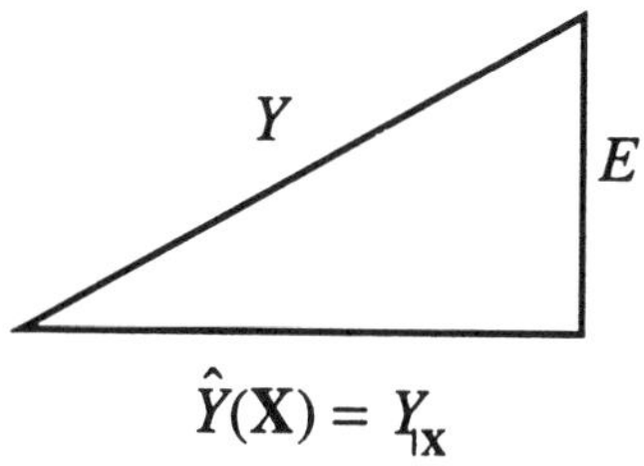

Figure 4.2 Orthogonality of MMSE linear estimate and error.

More generally, $\boldsymbol{a}$ may be uniquely determined by using an orthogonal basis for $V(\boldsymbol{X})$ with r_x elements, as discussed in Section 2.2.

To say that $Y_{|\boldsymbol{x}}$ is the closest variable to Y in $V(\boldsymbol{X})$ is to say that the variance of the difference variable $E = Y - Y_{|\boldsymbol{x}}$ is a minimum over all linear combinations of elements of $\boldsymbol{X}$. Therefore $Y_{|\boldsymbol{x}}$ is called the minimum-mean-squared-error (MMSE) linear estimate of Y given $\boldsymbol{X}$, and is alternatively denoted by

$$\hat{Y}(\boldsymbol{X}) = Y_{|\boldsymbol{x}} \; . \tag{4.37}$$

From the above development, any random variable Y may be written uniquely as

$$Y = \hat{Y}(\boldsymbol{X}) + E \; , \tag{4.38}$$

where $\hat{Y}(\boldsymbol{X})$ is in $V(\boldsymbol{X})$ and E is orthogonal to all variables in $V(\boldsymbol{X})$. This is illustrated by the right triangle of Figure 4.2. By the Pythagorean theorem for Hilbert spaces, the variance of Y is the sum of the variances of $\hat{Y}(\boldsymbol{X})$ and E. The estimation error variable E is zero if and only if $Y \in V(\boldsymbol{X})$. Since the mean of E is zero and E is orthogonal to $\boldsymbol{X}$, $\hat{Y}(\boldsymbol{X})$ is the conditional mean of Y given $\boldsymbol{X}$.

The above development generalizes straightforwardly to a set $\boldsymbol{Y} = \{Y_j\}$ of random variables Y_j. The MMSE linear estimate of $\boldsymbol{Y}$ given $\boldsymbol{X}$ is the vector

$$\hat{\boldsymbol{Y}}(\boldsymbol{X}) = \boldsymbol{Y}_{|\boldsymbol{x}} \tag{4.39}$$

of MMSE linear estimates $\hat{Y}_j(\boldsymbol{X}) = \boldsymbol{a}_j^* \boldsymbol{X}$, so $\hat{\boldsymbol{Y}}(\boldsymbol{X}) = A^* \boldsymbol{X}$ for some matrix A. The components E_j of the estimation error vector

$$\boldsymbol{E} = \boldsymbol{Y} - \hat{\boldsymbol{Y}}(\boldsymbol{X}) \tag{4.40}$$

are each orthogonal to all components of $\boldsymbol{X}$, and thus $\boldsymbol{E}$ satisfies the matrix equation

$$< \boldsymbol{X}, \boldsymbol{E} > = < \boldsymbol{X}, \boldsymbol{Y} > - < \boldsymbol{X}, \boldsymbol{X} > A = R_{\boldsymbol{xy}} - R_{\boldsymbol{xx}} A = 0 \; , \tag{4.41}$$

which yields the solution $A = R_{\boldsymbol{xx}}^{-1} R_{\boldsymbol{xy}}$ when $R_{\boldsymbol{xx}}$ is nonsingular. [2]

The orthogonality illustrated in Figure 4.2 continues to hold, since $\hat{\boldsymbol{Y}}(\boldsymbol{X})$ is a vector of elements of $V(\boldsymbol{X})$, while $\boldsymbol{E}$ is is vector of elements that are orthogonal to $V(\boldsymbol{X})$. However, the "Pythagorean theorem" now becomes

$$R_{\boldsymbol{yy}} = A^* R_{\boldsymbol{xx}} A + R_{\boldsymbol{ee}} \; ; \tag{4.42}$$

i.e., the covariance matrix of the diagonal is the sum of the covariance matrices of the two sides of the right triangle.

The covariance matrix $R_{\boldsymbol{ee}}$ of the minimum mean square linear estimation error $\boldsymbol{E}$ is minimum in every sense. Let $\hat{\boldsymbol{Y}}'(\boldsymbol{X}) = B^*\boldsymbol{X}$ be an arbitrary linear estimate of $\boldsymbol{Y}$ given $\boldsymbol{X}$, and let $\boldsymbol{E}' = \boldsymbol{Y} - \hat{\boldsymbol{Y}}'(\boldsymbol{X})$ be the corresponding error vector. Then since $\boldsymbol{Y} = A^*\boldsymbol{X} + \boldsymbol{E}$, it follows that $\boldsymbol{E}'$ has the orthogonal decomposition

$$\boldsymbol{E}' = (A^* - B^*)\boldsymbol{X} + \boldsymbol{E} = C^*\boldsymbol{X} + \boldsymbol{E} \; , \tag{4.43}$$

where $C^*\boldsymbol{X}$ is in $V(\boldsymbol{X})$ and $\boldsymbol{E}$ is orthogonal to $V(\boldsymbol{X})$. Consequently

$$R_{\boldsymbol{e}'\boldsymbol{e}'} = C^* R_{\boldsymbol{xx}} C + R_{\boldsymbol{ee}} \; , \tag{4.44}$$

where both $C^* R_{\boldsymbol{xx}} C$ and $R_{\boldsymbol{ee}}$ are nonnegative definite covariance matrices. It follows that $R_{\boldsymbol{ee}}$ is "less than" $R_{\boldsymbol{e}'\boldsymbol{e}'}$ in every sense; its determinant is less, its trace is less, its eigenvalues are less, its Cholesky factors are less, and so forth. For any vector $\boldsymbol{a}$, the variance of the linear combination $\boldsymbol{a}^*\boldsymbol{E}'$ is not less than that of $\boldsymbol{a}^*\boldsymbol{E}$, since

$$E[\boldsymbol{a}^*\boldsymbol{E}'(\boldsymbol{E}')^*\boldsymbol{a}] = \boldsymbol{a}^* R_{\boldsymbol{e}'\boldsymbol{e}'} \boldsymbol{a} \geq \boldsymbol{a}^* R_{\boldsymbol{ee}} \boldsymbol{a} \; , \tag{4.45}$$

by the nonnegative definiteness of $C^* R_{\boldsymbol{xx}} C$. Indeed, the nonnegative definiteness of a Hermitian-symmetric square matrix A is sometimes denoted by $A \geq 0$; in this notation, one may write

$$R_{\boldsymbol{e}'\boldsymbol{e}'} - R_{\boldsymbol{ee}} \geq 0 \; , \text{ or} \tag{4.46}$$

$$R_{\boldsymbol{e}'\boldsymbol{e}'} \geq R_{\boldsymbol{ee}} \; . \tag{4.47}$$

It follows that for any optimality criterion based on error variances, the vector MMSE linear estimate is optimum among all linear estimators.

2.4 Innovations representations via recursive MMSE prediction

The innovations representation of a random vector $\boldsymbol{X}$ may be developed by sequential MMSE linear prediction. Let $\boldsymbol{X}(j-1)$ denote the "past" relative to

[2]The inverse may be replaced by any one of many generalized inverses when $R_{\boldsymbol{xx}}$ is singular, see [21].

a component X_j of $\boldsymbol{X}$; i.e.,

$$\boldsymbol{X}(j-1) = \{X_k | k < j\}. \tag{4.48}$$

The MMSE linear prediction of X_j given $\boldsymbol{X}(j-1)$ is then the projection $X_j|_{\boldsymbol{X}(j-1)}$, and the jth innovations variable W_j may then be defined as the prediction error

$$W_j = X_j - X_j|_{\boldsymbol{X}(j-1)} \; . \tag{4.49}$$

By the orthogonality principle, W_j is orthogonal to the past space $V(\boldsymbol{X}(j-1))$; however, $V(\boldsymbol{X}(j-1))$ and W_j together span $V(\boldsymbol{X}(j))$. It follows that $V(\boldsymbol{W}(j)) = V(\boldsymbol{X}(j))$, and thus that the elements of $\boldsymbol{W}$ are orthogonal (uncorrelated). An innovations variable is zero if and only if it is in the past space $V(\boldsymbol{X}(j-1))$. Since $X_j|_{\boldsymbol{X}(j-1)}$ may be expressed as a linear combination of the elements either of $\boldsymbol{X}(j-1)$ or of $\boldsymbol{W}(j-1)$, the prediction error equations may be expressed in matrix form as either

$$\boldsymbol{W} = L^{-1}\boldsymbol{X} \; , \tag{4.50}$$

or

$$\boldsymbol{X} = L\boldsymbol{W} \; , \tag{4.51}$$

where L and L^{-1} are both lower triangular and monic. Then

$$R_{\boldsymbol{xx}} = L R_{\boldsymbol{ww}} L^* \; , \tag{4.52}$$

is the Cholesky factorization of $R_{\boldsymbol{xx}}$ since such a factorization is unique.

2.5 Gaussian random vectors

Heretofore random vectors $\boldsymbol{X}$ have not been assumed to be Gaussian. However, Gaussian random vectors have particularly nice properties. In particular, information-theoretic quantities are simple functions of the second-order statistics (covariance matrices) of Gaussian random vectors. Therefore in a model in which only second-order statistics are given, it is often helpful to analyze the case in which all variables are Gaussian; this usually simplifies the analysis and yields structures and bounds that are useful for the general case.

The probability distribution of a zero-mean complex Gaussian random vector $\boldsymbol{X}$ is completely determined by its covariance matrix $R_{\boldsymbol{xx}}$. If $R_{\boldsymbol{xx}}$ is nonsingular, then

$$p_{\boldsymbol{x}}(\boldsymbol{X}) = \pi^{-r_x} |R_{\boldsymbol{xx}}|^{-1} e^{-\boldsymbol{X}^* R_{\boldsymbol{xx}}^{-1} \boldsymbol{X}} \; . \tag{4.53}$$

The separability property of this distribution implies that uncorrelated Gaussian random variables are independent.

More generally, as shown in Section 2.2, given $R_{\boldsymbol{xx}}$, a Gaussian vector $\boldsymbol{X}$ may be expressed as a linear combination $\boldsymbol{X} = F\boldsymbol{V}$ of r_x nonzero uncorrelated and thus independent Gaussian random variables V_j. If F is unitary, then this map from $\mathcal{S}_{\boldsymbol{V}}$ to $\mathcal{S}_{\boldsymbol{X}}$ is volume-preserving.

If $\boldsymbol{Y}$ and $\boldsymbol{X}$ are jointly Gaussian, then it is straightforward to show that the MMSE linear estimate $\hat{\boldsymbol{Y}}(\boldsymbol{X})$ is actually the unconstrained MMSE estimate of $\boldsymbol{Y}$ given $\boldsymbol{X}$, since $\boldsymbol{Y}$ may be written as

$$\boldsymbol{Y} = \hat{\boldsymbol{Y}}(\boldsymbol{X}) + \boldsymbol{E} , \tag{4.54}$$

where $\boldsymbol{E}$ is a Gaussian random vector that is independent of $\boldsymbol{X}$.

As shown in [2], the differential entropy of a complex Gaussian vector $\boldsymbol{X}$ of rank r_x with nonsingular covariance matrix $R_{\boldsymbol{xx}}$ is

$$h(\boldsymbol{X}) = r_x \log_2 \pi e |R_{\boldsymbol{xx}}|^{1/r_x} . \tag{4.55}$$

More generally, since the differential entropy is invariant under volume-preserving transformations, and a modal representation $\boldsymbol{X} = U\boldsymbol{M}$ is volume-preserving regardless of whether $\boldsymbol{X}$ has full rank, the differential entropy $h(\boldsymbol{X})$ is equal to $h(\boldsymbol{M})$, where $\boldsymbol{M}$ is a set of independent complex Gaussian variables M_j with variances λ_j^2 equal to the eigenvalues of $R_{\boldsymbol{xx}}$. Thus

$$h(\boldsymbol{X}) = h(\boldsymbol{M}) = \sum_{j \in J} \log_2 \pi e \lambda_j^2 , \tag{4.56}$$

where the sum is only over the set $J = \{j \mid \lambda_j^2 > 0\}$ of r_x indices corresponding to the r_x nonzero eigenvalues of $R_{\boldsymbol{xx}}$. In other words,

$$h(\boldsymbol{M}) = r_x \log_2 \pi e \|R_{\boldsymbol{mm}}\|^{1/r_x} , \tag{4.57}$$

where $\|R_{\boldsymbol{mm}}\|$ is the **effective determinant** of the diagonal covariance matrix $R_{\boldsymbol{mm}}$:

> **Definition 2 (Effective Determinant)** *The effective determinant of a matrix is the product of its nonzero eigenvalues,*
>
> $$\|R_{\boldsymbol{mm}}\| = \prod_{j \in J} \lambda_j^2 . \tag{4.58}$$

Note that $\|R_{\boldsymbol{mm}}\|^{1/r_x}$ is the geometric mean of the nonzero eigenvalues of $R_{\boldsymbol{xx}}$.

Since $\|R_{\boldsymbol{mm}}\|$ is the product of the nonzero eigenvalues of $R_{\boldsymbol{xx}}$, $\|R_{\boldsymbol{mm}}\|$ is invariant in any modal representation of $\boldsymbol{X}$. Therefore $\|R_{\boldsymbol{mm}}\| = \|R_{\boldsymbol{xx}}\|$, and the differential entropy of $\boldsymbol{X}$ is equal to

$$h(\boldsymbol{X}) = r_x \log_2 \pi e \|R_{\boldsymbol{xx}}\|^{1/r_x} . \tag{4.59}$$

Example 4.3 (Two-Dimensional Example continued) Again let $\boldsymbol{X}$ be a random vector $[X_1^*, X_2^*]^*$ with covariance matrix

$$R_{\boldsymbol{xx}} = \begin{bmatrix} a & b \\ b & a \end{bmatrix} , \tag{4.60}$$

with a, b real and $0 \leq |b| \leq a$. The eigenvalues of $R_{\boldsymbol{xx}}$ are $(a+b, a-b)$, and the rank r_x is 2 unless $|b| = a$. The effective determinant of $R_{\boldsymbol{xx}}$ is thus equal to

$$\|R_{\boldsymbol{xx}}\| = \begin{cases} |R_{\boldsymbol{xx}}| = a^2 - b^2, & \text{if } |b| < a \\ 2a, & \text{if } |b| = a \; . \end{cases} \tag{4.61}$$

Note that the effective determinant is equal to the product of the Cholesky factors of $R_{\boldsymbol{xx}}$ when $\boldsymbol{X}$ has full rank, but not when $r_x = 1$. Note also that there is a discontinuity in the differential entropy $h(\boldsymbol{X})$ as $|b| \to a$. This discontinuity often occurs when $R_{\boldsymbol{xx}}$ is optimized as in Section 6. These discontinuities leads to "symbol-rate" and "center-frequency" optimization for each used frequency band in the stationary case.

The differential entropy of any random vector $\boldsymbol{X}$ with covariance matrix $R_{\boldsymbol{xx}}$ is upperbounded by the differential entropy of a Gaussian vector with the same covariance matrix:

$$h(\boldsymbol{X}) \leq r_x \log_2 \pi e \|R_{\boldsymbol{xx}}\|^{1/r_x} , \tag{4.62}$$

with equality if and only if $\boldsymbol{X}$ is Gaussian. The maximum entropy inference principle therefore suggests that if only the second-order statistics of $\boldsymbol{X}$ are known, then $\boldsymbol{X}$ should be presumed to be Gaussian. The effective determinant $\|R_{\boldsymbol{xx}}\|$ determines the differential entropy $h(\boldsymbol{X})$ of this presumed Gaussian density.

Since the mutual information between the input $\boldsymbol{X}$ and output $\boldsymbol{Y} = H\boldsymbol{X} + \boldsymbol{N}$ of a Gaussian ISI channel may be written as

$$I(\boldsymbol{X};\boldsymbol{Y}) = H(\boldsymbol{Y}) - H(\boldsymbol{Y}/\boldsymbol{X}) , \tag{4.63}$$

and the conditional differential entropy $H(\boldsymbol{Y}/\boldsymbol{X})$ is equal to $h(\boldsymbol{N})$, it follows that the mutual information is maximized for a given $R_{\boldsymbol{yy}}$ when $\boldsymbol{Y}$ is Gaussian, which in turn occurs when $\boldsymbol{X}$ is Gaussian.

These information-theoretic relations can be used to develop many determinantal inequalities, as shown by Cover and Thomas [2]. For example, Hadamard's inequality, which will be needed below, states that if R is a covariance matrix (a square Hermitian-symmetric nonnegative-definite matrix), then

$$|R| \leq \prod_j R_{jj} , \tag{4.64}$$

with equality if and only if R is diagonal. For, suppose that $\boldsymbol{X}$ is a Gaussian random vector with covariance matrix R; then Hadamard's inequality follows from the information-theoretic inequality

$$h(\boldsymbol{X}) = \sum_j h(X_j|X_1, ..., X_{j-1}) \leq \sum_j h(X_j) \ , \tag{4.65}$$

where equality holds if and only if the components X_j of $\boldsymbol{X}$ are independent.

3 EQUIVALENT CHANNEL MODELS, LINEAR ESTIMATION, AND MUTUAL INFORMATION

In this section, given a linear Gaussian channel model $\boldsymbol{Y} = H\boldsymbol{X}+\boldsymbol{N}$, equivalent forward and backward channel models that eliminate singularities and have many other nice properties are developed. Using these equivalent models, a number of relations are obtained between ML estimation, MMSE estimation, and mutual information (when $\boldsymbol{X}$ is Gaussian). In Section 4, the equivalent backward channel model will be used to develop the canonical GDFE receiver structure.

3.1 Forward and backward channel models

Given two random vectors $\boldsymbol{X}$ and $\boldsymbol{Y}$, either may be expressed uniquely as the sum of its MMSE linear estimate given the other and an orthogonal error vector:

$$\boldsymbol{Y} = \hat{\boldsymbol{Y}}(\boldsymbol{X}) + \boldsymbol{F} = A^*\boldsymbol{X} + \boldsymbol{F}; \tag{4.66}$$

$$\boldsymbol{X} = \hat{\boldsymbol{X}}(\boldsymbol{Y}) + \boldsymbol{G} = B^*\boldsymbol{Y} + \boldsymbol{G} \ , \tag{4.67}$$

where A and B are matrices to be determined, and $\boldsymbol{F}$ and $\boldsymbol{G}$ are orthogonal to $V(\boldsymbol{X})$ and $V(\boldsymbol{Y})$, respectively. The estimation error vector $\boldsymbol{F}$ is the innovations vector of $\boldsymbol{Y}$ given $\boldsymbol{X}$, while $\boldsymbol{G}$ is the innovations vector of $\boldsymbol{X}$ given $\boldsymbol{Y}$.

Suppose that the forward channel model

$$\boldsymbol{Y} = H\boldsymbol{X} + \boldsymbol{N} \tag{4.68}$$

is given, where $\boldsymbol{N}$ is independent of $\boldsymbol{X}$, so $\boldsymbol{N}$ is orthogonal to $V(\boldsymbol{X})$. Then since the decomposition $\boldsymbol{Y} = A^*\boldsymbol{X} + F$ is unique, $H\boldsymbol{X}$ must be the MMSE linear estimate $\hat{\boldsymbol{Y}}(\boldsymbol{X})$ of $\boldsymbol{Y}$ given $\boldsymbol{X}$, and $\boldsymbol{N}$ must be the estimation error or innovations of $\boldsymbol{Y}$ given $\boldsymbol{X}$.

The alternative representation above is then called the backward channel model, which may be written with the notation

$$\boldsymbol{X} = \hat{\boldsymbol{X}}(\boldsymbol{Y}) + \boldsymbol{E} = C\boldsymbol{Y} + \boldsymbol{E} \ , \tag{4.69}$$

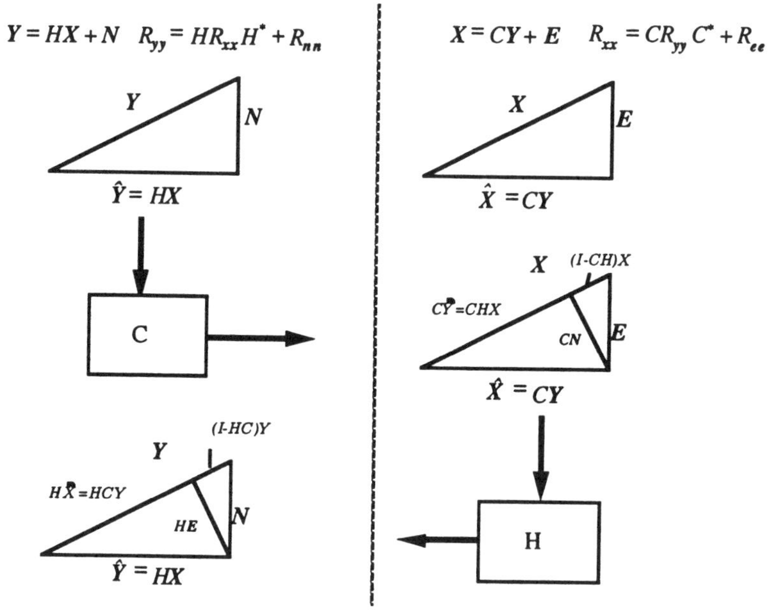

Figure 4.3 Pythagorean relations for forward and backward channel models.

where CY denotes the MMSE linear estimate $\hat{X}(Y)$ of X given Y, and $E = X - \hat{X}(Y)$ is the estimation error or innovations vector of X given Y. Thus, $B = C$ and $G = E$.

Figure 4.3 shows how the two Pythagorean representations of the forward and backward channel models may be combined, in two different ways. Thus in the forward channel

$$\hat{Y}(X) = HX = HCY + HE \tag{4.70}$$

is the sum of the orthogonal vectors $HCY \in V(Y)$ and $HE \in V(Y)^{\perp}$, and

$$N = (I - HC)Y - HE \, . \tag{4.71}$$

is also the sum of two orthogonal vectors. Similarly, in the backward channel there are orthogonal decompositions

$$\hat{X}(Y) = CY = CHX + CN \tag{4.72}$$

$$E = (I - CH)X - CN \, , \tag{4.73}$$

where CHX, $(I - CH)X \in V(X)$ and $CN \in V(X)^{\perp}$. All right triangles are geometrically similar; the "angle" between the two spaces $V(X)$ and $V(Y)$ is determined by the cross-correlation matrix $R_{xy} =< X, Y >$.

3.2 Canonical forward and backward channel models

The principles of Section 2 and of optimum estimation theory are now used to reduce the general channel models of the previous section to canonical forms in which extraneous dimensions are eliminated, and which have other nice properties.

> **Definition 3 (Canonical Channel Model)** *A channel model $\boldsymbol{Y} = H\boldsymbol{X} + \boldsymbol{N}$ is canonical if H is square, $R_{\boldsymbol{xx}}$ and $R_{\boldsymbol{nn}}$ are nonsingular, and furthermore $R_{\boldsymbol{nn}} = H$, which implies that H is a positive definite Hermitian-symmetric matrix.*

Our first observation is that any part of the input $\boldsymbol{X}$ that lies in the right null space (kernel) of H may be disregarded. In general, the matrix H defines a linear transformation H: $\mathcal{C}^m \to \mathcal{C}^n$ from the input space $\mathcal{C}^m$ of all possible complex m-vectors to the output space $\mathcal{C}^n$. The right null space of H is the kernel $K \subseteq \mathcal{C}^m$ of this transformation.

Any $\boldsymbol{x} \in \mathcal{C}^m$ may be written uniquely as

$$\boldsymbol{x} = \boldsymbol{x}_{|K} + \boldsymbol{x}_{|K^\perp} , \tag{4.74}$$

where $\boldsymbol{x}_{|K}$ is the projection of $\boldsymbol{x}$ onto K and $\boldsymbol{x}_{|K^\perp} = \boldsymbol{x} - \boldsymbol{x}_{|K}$ is the projection of $\boldsymbol{x}$ onto the orthogonal space $K^\perp$ to K. The signal component of the channel output then depends only on $\boldsymbol{x}_{|K^\perp}$, since

$$H\boldsymbol{x} = H\boldsymbol{x}_{|K^\perp} , \tag{4.75}$$

independent of $\boldsymbol{x}_{|K}$, since $H\boldsymbol{x}_{|K} = 0$. Thus, the input is effectively $\boldsymbol{x}_{|K^\perp}$, and $\boldsymbol{x}_{|K}$ does not affect the channel output. The projection $\boldsymbol{x}_{|K}$ will be called the **undetectable part** of the input $\boldsymbol{x}$, and $\boldsymbol{x}_{|K^\perp}$ will be called the **effective input**.

If the input is a random vector $\boldsymbol{X}$ with covariance matrix $R_{\boldsymbol{xx}}$ and signal space $\mathcal{S}_{\boldsymbol{X}} \subseteq \mathcal{C}^m$, then $\boldsymbol{X}$ may similarly be decomposed uniquely into

$$\boldsymbol{X} = \boldsymbol{X}_{|K} + \boldsymbol{X}' , \tag{4.76}$$

where $\boldsymbol{X}_{|K}$ is an undetectable input random vector defined on the space $K \cap \mathcal{S}_{\boldsymbol{X}}$, while $\boldsymbol{X}' = \boldsymbol{X}_{|K^\perp}$ is an effective input random vector defined on the effective input space $\mathcal{S}_{\boldsymbol{X}'} = K^\perp \cap \mathcal{S}_{\boldsymbol{X}}$. The probability density of the effective input $\boldsymbol{X}'$ and its covariance matrix $R_{\boldsymbol{x}'\boldsymbol{x}'}$ are induced from those of $\boldsymbol{X}$ by this definition.

The output signal then depends only on $\boldsymbol{X}'$:

$$H\boldsymbol{X} = H\boldsymbol{X}' . \tag{4.77}$$

The linear transformation H: $\mathcal{S}_{\boldsymbol{X}'} \to \mathcal{S}_{\hat{\boldsymbol{Y}}}$ is one-to-one over these spaces (but is not necessarily one-to-one on the larger spaces $\mathcal{C}^m \to \mathcal{C}^n$), and the signal space $\mathcal{S}_{\hat{\boldsymbol{Y}}}$ is the image of $\mathcal{S}_{\boldsymbol{X}'}$ under the transformation H. It follows that $\mathcal{S}_{\boldsymbol{X}'}$ and $\mathcal{S}_{\hat{\boldsymbol{Y}}}$ both have the same dimension, which will be called the effective rank of the channel and denoted as $r_{\hat{y}}$. This rank often is less than the input or output dimensionality of the original channel matrix H, so that $r_{\hat{y}} = r_{x'} \leq \min(n, r_x)$ and $r_{x'} \leq r_x \leq m$. Strict inequalities in fact often apply for optimized covariance $R_{\boldsymbol{xx}}$ as shown in later sections. Thus, both $\boldsymbol{X}'$ and $\hat{\boldsymbol{Y}}(\boldsymbol{X}) = H\boldsymbol{X} = H\boldsymbol{X}'$ have a dimensionality associated with the with $V(\boldsymbol{X}')$, that is rank $r_{\hat{y}} = r_{x'}$. Clearly only the $r_{\hat{y}}$-dimensional effective input $\boldsymbol{X}' = \boldsymbol{X}_{|K^\perp}$ can convey information through the channel, and any power applied to the $(n - r_{\hat{y}})$-dimensional undetectable part $\boldsymbol{X}_{|K}$ is wasted.

Since $R_{\boldsymbol{x}'\boldsymbol{x}'}$ has rank $r_{\hat{y}}$, the effective input $\boldsymbol{X}'$ may be represented as

$$\boldsymbol{X}' = U\boldsymbol{M}' , \tag{4.78}$$

where U is an $n \times n$ "unitary" matrix and is therefore an volume-preserving transformation, regardless of whether $\boldsymbol{X}'$ has full rank ($r_{\hat{y}} = r_{x'} = m$) or not ($r_{\hat{y}} < r_x \leq m$), and $\boldsymbol{M}'$ is a set of random variables with covariance matrix $R_{\boldsymbol{m}'\boldsymbol{m}'}$. It may be desirable for the elements of $\boldsymbol{M}'$ to be uncorrelated, in which case (4.78) becomes the modal representation of Section 2.2. The rank and effective determinant of $R_{\boldsymbol{m}'\boldsymbol{m}'}$ are then the same as those of $R_{\boldsymbol{x}'\boldsymbol{x}'}$:

$$r_{m'} = r_{x'} = r_{\hat{y}} \; ; \tag{4.79}$$

$$\|R_{\boldsymbol{m}'\boldsymbol{m}'}\| = \|R_{\boldsymbol{x}'\boldsymbol{x}'}\| \; . \tag{4.80}$$

The identically zero components of $\boldsymbol{M}'$ and the associated columns of U may be eliminated to obtain an equivalent one-to-one volume-preserving transformation from $\boldsymbol{M} \in \mathcal{C}^{r_{\hat{y}}}$ to $\boldsymbol{X}' \in \mathcal{S}_{\boldsymbol{X}'}$:

$$\boldsymbol{X}' = U'\boldsymbol{M} \; . \tag{4.81}$$

Then $\boldsymbol{M}$ has full rank $r_{\hat{y}}$, and the determinant of $R_{\boldsymbol{mm}}$ is equal to the effective determinant of $R_{\boldsymbol{m}'\boldsymbol{m}'}$:

$$m = r_{m'} = r_{x'} = r_{\hat{y}}; \tag{4.82}$$

$$|R_{\boldsymbol{mm}}| = \|R_{\boldsymbol{m}'\boldsymbol{m}'}\| = \|R_{\boldsymbol{x}'\boldsymbol{x}'}\| \; . \tag{4.83}$$

Although the matrix U' is not square in general, the map U' remains a one-to-one volume-preserving transformation from $\mathcal{C}^{r_y}$ to $\mathcal{S}_{\boldsymbol{X}'}$. It is clear that estimation of $\boldsymbol{M}$ is equivalent to estimation of $\boldsymbol{X}'$. Because $\boldsymbol{M}$ is full rank, then any characteristic representation of Section 2.2 in the form

$$\boldsymbol{M} = FV \tag{4.84}$$

will have a volume preserving F of rank $r_{\hat{y}} = m = r_v$. A convenient form of the GDFE for H corresponding to stationary scalar channels will use the innovations decomposition in Section 4 while vector coding in Section 5 will use the modal decomposition.

The forward channel model may now be written as

$$\boldsymbol{Y} = HU'\boldsymbol{M} + \boldsymbol{N} = G\boldsymbol{M} + \boldsymbol{N} \ , \tag{4.85}$$

where $\boldsymbol{M}$ is a complex random $r_{\hat{y}}$-vector with a covariance matrix $R_{\boldsymbol{mm}}$ that is positive definite and thus invertible, and $\boldsymbol{N}$ is a Gaussian noise vector independent of $\boldsymbol{M}$ with nonsingular covariance matrix $R_{\boldsymbol{nn}}$.

Finally, a series of information-lossless linear transformations may be applied to the channel output $\boldsymbol{Y}$ to obtain the final form of a canonical model. First, let S be any square root of $R_{\boldsymbol{nn}}$; i.e., let S be an invertible square matrix such that $R_{\boldsymbol{nn}} = SS^*$. Then the invertible noise-whitening matrix S^{-1} applied to $\boldsymbol{Y}$ yields the equivalent model

$$\boldsymbol{Y}' = S^{-1}\boldsymbol{Y} = S^{-1}G\boldsymbol{M} + S^{-1}\boldsymbol{N} = G'\boldsymbol{M} + \boldsymbol{N}' \ , \tag{4.86}$$

where $\boldsymbol{N}' = S^{-1}\boldsymbol{N}$ is a Gaussian noise vector with an identity covariance matrix,

$$R_{\boldsymbol{n}'\boldsymbol{n}'} = S^{-1}R_{\boldsymbol{nn}}S^{-*} = I \ , \tag{4.87}$$

and $G' = S^{-1}G = S^{-1}HU'$. The principle of the sufficiency of matched filtering may be applied: in the presence of white Gaussian noise, the outputs of a bank of matched filters matched to the responses of each input M_j (the columns of G') form a set of sufficient statistics for the detection of $\boldsymbol{M}$. In matrix notation, this set of outputs is the $r_m = r_{\hat{y}}$-dimensional vector

$$\boldsymbol{Z} = (G')^*\boldsymbol{Y}' = (G')^*G'\boldsymbol{M} + (G') * \boldsymbol{N}' = R_f\boldsymbol{M} + \boldsymbol{N}'' \ , \tag{4.88}$$

where the $r_m \times r_m$ full-rank positive-definite matrix R_f is

$$R_f = (G')^*G' = (U')^*H^*S^{-*}S^{-1}HU' = (U')^*H^*R_{\boldsymbol{nn}}^{-1}HU' \tag{4.89}$$

a Hermitian-symmetric matrix, $\boldsymbol{M}$ is a set of $r_{\hat{y}} = r_m$ uncorrelated random variables with nonsingular covariance matrix $R_{\boldsymbol{mm}}$, and $\boldsymbol{N}''$ is an independent Gaussian noise vector with covariance matrix

$$R_{\boldsymbol{n}''\boldsymbol{n}''} = (G')^*R_{\boldsymbol{n}'\boldsymbol{n}'}G' = (G')^*G' = R_f \ . \tag{4.90}$$

Thus the noise covariance matrix is equal to the equivalent channel-response matrix R_f. This yields our desired canonical forward channel model. This construction is summarized in Figure 4.4.

In summary:

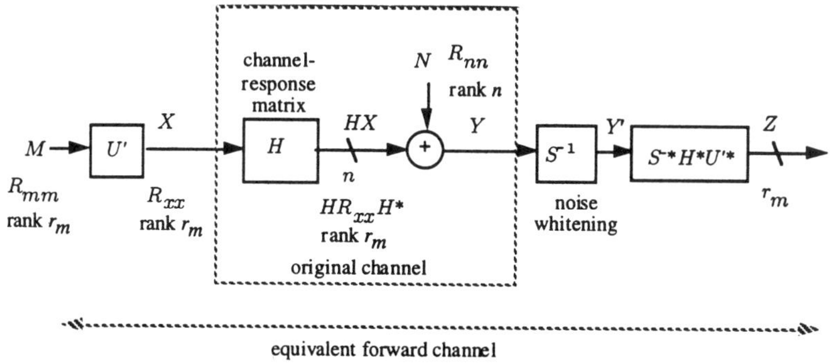

Figure 4.4 Construction of canonical forward channel model.

Theorem 4.1 (Equivalency of the Canonical Forward Channel Model) *Let $\boldsymbol{Y} = H\boldsymbol{X} + \boldsymbol{N}$, where $H\boldsymbol{X}$ has rank $r_m = r_{\hat{y}}$ and $\boldsymbol{N}$ is a full-rank ($r_n = n$) Gaussian random vector independent of $\boldsymbol{X}$. Without loss of optimality, an equivalent forward channel model is the canonical model*

$$\boldsymbol{Z} = R_f \boldsymbol{M} + \boldsymbol{N}'' , \tag{4.91}$$

where the channel-response matrix R_f is a square $r_m \times r_m$ full-rank covariance matrix, the input $\boldsymbol{M}$ is a full-rank $r_{\hat{y}}$-vector with nonsingular covariance matrix $R_{\boldsymbol{mm}}$, and the noise $\boldsymbol{N}''$ is a full-rank Gaussian $r_{\hat{y}}$-vector independent of $\boldsymbol{M}$ whose covariance matrix $R_{\boldsymbol{n}''\boldsymbol{n}''}$ is equal to R_f. There is a one-to-one volume-preserving map from $\boldsymbol{M}$ to the effective part $\boldsymbol{X}'$ of the input $\boldsymbol{X}$, and

$$r_{x'} = r_m = r_{\hat{y}} \leq r_y; \|R_{\boldsymbol{x}'\boldsymbol{x}'}\| = |R_{\boldsymbol{mm}}| \; . \tag{4.92}$$

The output $\boldsymbol{Z}$ is a sufficient statistic for detection of $\boldsymbol{M}$ or of $\boldsymbol{X}'$, and consequently the mutual information between input and output is the same in both models:

$$I(\boldsymbol{M};\boldsymbol{Z}) = I(\boldsymbol{X}';\boldsymbol{Y}) = I(\boldsymbol{X};\boldsymbol{Y}) \text{ bits/block.} \tag{4.93}$$

Since all $r_{\hat{y}} \times r_{\hat{y}}$ matrices are nonsingular, it is possible to solve explicitly for the corresponding backward channel model. The MMSE linear estimate of $\boldsymbol{M}$ given $\boldsymbol{Z}$ is $\hat{\boldsymbol{M}}(\boldsymbol{Z}) = R_b \boldsymbol{Z}$, where the matrix R_b is determined by

$$R_b^* = R_{\boldsymbol{zz}}^{-1} R_{\boldsymbol{zm}} \; . \tag{4.94}$$

Since

$$R_{\boldsymbol{zm}} = E[\boldsymbol{Z}\boldsymbol{M}^*] = R_f R_{\boldsymbol{mm}}; \tag{4.95}$$

$$R_{\boldsymbol{zz}} = E[\boldsymbol{Z}\boldsymbol{Z}^*] = R_f R_{\boldsymbol{mm}} R_f + R_f = R_f R_{\boldsymbol{mm}}(R_{\boldsymbol{mm}}^{-1} + R_f) \; , \tag{4.96}$$

R_b is determined by the following fundamental formula:

$$R_b^{-1} = R_{mm}^{-1} + R_f \ . \qquad (4.97)$$

This formula shows that R_b is Hermitian-symmetric, $R_b^* = R_b$. Also, it shows that as the input covariance R_{mm} becomes large, R_b tends to the inverse of R_f, meaning the noise/errors $\boldsymbol{E}$ and $\boldsymbol{N}''$ can be ignored.

The covariance matrix R_{zz} is most easily determined from the following relationships between the four matrices R_f, R_b, R_{mm} and R_{zz}:

$$R_{zz} = R_f R_{mm} R_b^{-1} = R_b^{-1} R_{mm} R_f \ . \qquad (4.98)$$

Since the covariance matrix R_{ee} of the estimation error vector $\boldsymbol{E} = \boldsymbol{M} - R_b\boldsymbol{Z}$ satisfies

$$R_{mm} = R_b R_{zz} R_b + R_{ee} = R_{mm} R_f R_b + R_{ee} \ , \qquad (4.99)$$

it follows that in the equivalent backward channel model the noise covariance matrix is again equal to the backward channel-response matrix:

$$R_{ee} = R_{mm} - R_{mm} R_f R_b = R_{mm}(R_b^{-1} - R_f)R_b = R_b \ . \qquad (4.100)$$

In summary:

Theorem 4.2 (Equivalency of the Backward Canonical Model) *Under the same conditions as in Theorem 4.1, there is an equivalent canonical backward channel model*

$$\boldsymbol{M} = R_b\boldsymbol{Z} + \boldsymbol{E} \ , \qquad (4.101)$$

where R_b is a square nonsingular Hermitian-symmetric channel-response matrix

$$R_b = (R_{mm}^{-1} + R_f)^{-1} \ , \qquad (4.102)$$

$\boldsymbol{Z}$ is a random "backward input" $r_{\hat{y}}$-vector with nonsingular covariance matrix

$$R_{zz} = R_f R_{mm} R_b^{-1} = R_b^{-1} R_{mm} R_f \ , \qquad (4.103)$$

and $\boldsymbol{E}$ is a random error $r_m = r_{\hat{y}}$-vector uncorrelated with $\boldsymbol{M}$ whose covariance matrix R_{ee} is equal to the channel-response matrix R_b.

Example 4.4 (Parallel Channels Example) Let the forward channel correspond to the previous "parallel channels" model of Section 1.1 and so any of the subchannels (with normalization of gain to one) is an ideal one-dimensional complex Gaussian channel $Y = X + N$ with signal and noise variances S_x and S_n, respectively. The corresponding equivalent canonical forward channel model for any such subchannel is

$$Z = S_n^{-1} Y = S_n^{-1} X + S_n^{-1} N = R_f M + N'' \ , \qquad (4.104)$$

where $R_f = S_n^{-1}$, $M = X$ with $R_{mm} = S_x$, and $N'' = S_n^{-1}N$ with $S_{n''n''} = S_n^{-1} = R_f$. The corresponding equivalent canonical backward channel model is

$$X = M = R_b Z + E \ , \tag{4.105}$$

where

$$R_b = (R_{mm}^{-1} + R_f)^{-1} = (S_x^{-1} + S_n^{-1})^{-1} = S_x S_n/(S_x + S_n) \ ; \tag{4.106}$$

$$R_{zz} = R_f R_{mm} R_b^{-1} = (S_x + S_n)/S_n^2 \ ; \tag{4.107}$$

$$R_{ee} = R_b = S_x S_n/(S_x + S_n) \ . \tag{4.108}$$

If any of the one-dimensional channels (or subchannel) had $S_x = 0$, then the reduction procedure from $\boldsymbol{X}$ to $\boldsymbol{M}$ would have resulted in this channel being eliminated from the set in the canonical forward and backward realizations, which would then consist of r_m subchannels corresponding to those with nonzero input energy. A subchannel with $h_i = 0$ would also suggest that the corresponding $S_{x,i}$ be set to zero and eliminated - that is that dimension is in the kernel $K \cap \mathcal{S}_{\boldsymbol{X}}$ of H and so is eliminated. Thus the models in (4.105) and (4.106) correspond to only the used subchannels from the parallel set.

Thus the equivalent canonical backward channel model is similar to the forward model in that it is square, nonsingular, and has noise covariance equal to the channel matrix. It differs in that the elements of $\boldsymbol{Z}$ are not in general uncorrelated, and the "noise" $\boldsymbol{E}$ is not in general Gaussian; furthermore, $\boldsymbol{E}$ is merely uncorrelated with $\boldsymbol{Z}$ rather than independent of $\boldsymbol{Z}$. This is not surprising, since nothing in the derivation of the backward model depends on R_{mm} being diagonal, $\boldsymbol{N}''$ being Gaussian, or $\boldsymbol{N}''$ being independent of (rather than merely uncorrelated with) $\boldsymbol{M}$.

By rederiving the forward model from the backward model, or by direct substitution, one may obtain the symmetrical relations

$$R_f^{-1} = R_{zz}^{-1} + R_b \ ; \tag{4.109}$$

$$R_{mm} = R_b R_{zz} R_f^{-1} = R_f^{-1} R_{zz} R_b \ . \tag{4.110}$$

Many other matrix relations follow easily. In particular:

$$R_f R_b = I - R_f R_{zz}^{-1} = I - R_{mm}^{-1} R_b \ ; \tag{4.111}$$

$$R_b R_f = I - R_{zz}^{-1} R_f = I - R_b R_{mm}^{-1} \ ; \tag{4.112}$$

$$R_b^{-1} R_f^{-1} = I + R_b^{-1} R_{zz}^{-1} = I + R_{mm}^{-1} R_f^{-1} \ ; \tag{4.113}$$

$$R_f^{-1} R_b^{-1} = I + R_f^{-1} R_{mm}^{-1} = I + R_{zz}^{-1} R_b^{-1} \ . \tag{4.114}$$

Example 4.5 (Parallel Channels continued) For the ideal one-dimensional channel (or any of the used subchannels in the parallel set), these relations become

$$R_f R_b = S_x/(S_x + S_n) \; ; \tag{4.115}$$

$$R_f^{-1} R_b^{-1} = 1 + \frac{S_n}{S_x} \; . \tag{4.116}$$

From these equations, one may obtain the determinantal relations

$$\begin{aligned} |I - R_f R_b| &= |I - R_b R_f| = |R_f|/|R_{\boldsymbol{zz}}| \\ &= |R_b|/|R_{\boldsymbol{mm}}| \; ; \end{aligned} \tag{4.117}$$

$$\begin{aligned} |R_f R_b|^{-1} &= |R_b R_f|^{-1} = |I + R_b^{-1} R_{\boldsymbol{zz}}^{-1}| \end{aligned} \tag{4.118}$$

$$\begin{aligned} &= |I + R_{\boldsymbol{mm}}^{-1} R_f^{-1}| \\ &= |I + R_f^{-1} R_{\boldsymbol{mm}}^{-1}| = |I + R_{\boldsymbol{zz}}^{-1} R_b^{-1}| \; . \end{aligned} \tag{4.119}$$

Using these relations, one may verify that all the right triangles shown in Figure 4.4 are in fact similar, if the squared length of each side is identified with the determinant of its covariance matrix; the ratio of the squared lengths of the long side to the hypotenuse is always $|R_f R_b|$, and the ratio of the squared lengths of the short side to the hypotenuse is always $|I - R_f R_b|$. (Note again that here the "Pythagorean theorem" involves the sums of covariance matrices, not of their determinants.)

The forward and backward equivalent canonical channel models are two completely equivalent ways of specifying the joint probability distribution $p_{\mathbf{M/Z}}(\boldsymbol{m}, \boldsymbol{z})$. The forward model corresponds to specifying first $\boldsymbol{M}$, then $\boldsymbol{Z}$ given $\boldsymbol{M}$; i.e., to specifying $p_{\mathbf{M/Z}}(\boldsymbol{m}, \boldsymbol{z})$ as the product $p_{\mathbf{M}}(\boldsymbol{m}) p_{\mathbf{Z/M}}(\boldsymbol{z}|\boldsymbol{m})$. The backward model corresponds to specifying first $\boldsymbol{Z}$, then $\boldsymbol{M}$ given $\boldsymbol{Z}$; i.e., to specifying $p_{\mathbf{M/Z}}(\boldsymbol{m}, \boldsymbol{z})$ as the product $p_{\mathbf{Z}}(\boldsymbol{z}) p_{\mathbf{M/Z}}(\boldsymbol{m}|\boldsymbol{z})$.

In the forward channel, the conditional probability $p_{\mathbf{Z/M}}(\boldsymbol{z}|\boldsymbol{m})$ is specified by an independent Gaussian noise variable $\boldsymbol{N}''$ via $p_{\mathbf{Z/M}}(\boldsymbol{z}|\boldsymbol{m}) = p_{\mathbf{N}''}(\boldsymbol{z} - R_f \boldsymbol{m})$. If $\boldsymbol{M}$ and therefore $\boldsymbol{Z}$ and $\boldsymbol{E}$ are Gaussian, then a similar separation formula holds in the backward channel.

3.3 ML and MMSE estimation

ML Estimation

Given an output $\boldsymbol{y}$, a maximum-likelihood (ML) estimator chooses the input $\boldsymbol{x} \in C^m$ that maximizes the likelihood $p_{\mathbf{Y/X}}(\boldsymbol{y}|\boldsymbol{x}) = p_{\mathbf{N}}(\boldsymbol{y} - H\boldsymbol{x})$. Since $H\boldsymbol{x} = H\boldsymbol{x}'$, where $\boldsymbol{x}'$ is the effective part of the input, all that an ML estimator can actually do is estimate the effective input $\boldsymbol{x}' \in \mathcal{S}_{\boldsymbol{X}'} \subset C^m$, or the corresponding $r_m = r_{\hat{y}}$-vector $\boldsymbol{m}$ such that $\boldsymbol{x} = U'\boldsymbol{m}$.

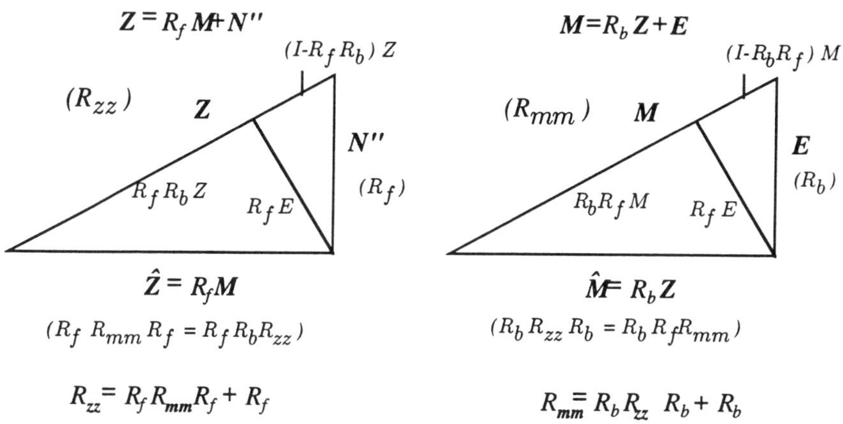

Figure 4.5 Similar right triangles.

Theorem 4.3 (The ML Estimator and Zero-Forcing Equalizer) *The ML estimates of $\boldsymbol{X}'$ or $\boldsymbol{M}$ from $\boldsymbol{Y}$ or $\boldsymbol{Z}$ are*

$$\hat{\boldsymbol{M}}(\boldsymbol{Z}) = R_f^{-1}\boldsymbol{Z} \; ; \tag{4.120}$$

$$\hat{\boldsymbol{X}}'(\boldsymbol{Z}) = U'\hat{\boldsymbol{M}}(\boldsymbol{Z}) = U'R_f^{-1}\boldsymbol{Z} \; . \tag{4.121}$$

Proof: Since $\boldsymbol{Z} = R_f\boldsymbol{M}+\boldsymbol{N}''$ is a sufficient statistic for estimation of $\boldsymbol{X}'$ or $\boldsymbol{M}$ from $\boldsymbol{Y}$, there is a one-to-one map between $S_{x'}$ or $S_m = C^{r_m}$ and $S_z = C^{r_m}$. Since $p_{\mathbf{N}''}(z - R_f\boldsymbol{m})$ is maximized for the $\boldsymbol{m}$ such that $z = R_f\boldsymbol{m}$ when $\boldsymbol{N}''$ is Gaussian, then the theorem follows. **QED.**

In other words, the block ML estimator simply computes the unique (effective) input that would give the observed matched-filter output vector $\boldsymbol{Z}$ in the absence of noise. For this reason the ML estimator is sometimes called a **zero-forcing equalizer.**

The ML estimation error is

$$\hat{\boldsymbol{E}} = \boldsymbol{M} - \hat{\boldsymbol{M}}(\boldsymbol{Z}) = \boldsymbol{M} - R_f^{-1}(R_f\boldsymbol{M} + \boldsymbol{N}'') = R_f^{-1}\boldsymbol{N}'' \; , \tag{4.122}$$

a Gaussian random vector with covariance matrix $R_f^{-1}R_{\boldsymbol{n}''\boldsymbol{n}''}R_f^{-1} = R_f^{-1}$.

ML Detection

It is important to distinguish an ML estimator from an ML detector, the latter of which is optimum for discrete uniform input distributions on $\boldsymbol{M}$. The ML estimator is only "optimum" when the input distribution for $\boldsymbol{M}$ is continuous uniform, which never occurs in practice. However, for some choices of receiver, an ML estimator followed by an ML detector designed only on knowledge of coding applied to $\boldsymbol{M}$ (and therefore not using any knowledge of the channel) to be optimum. The vector coding methods of Section 5 illustrate this property.

MMSE Estimation and MMSE Equalization

As observed in Section 2.3, for any optimality criterion based on error variances, the vector MMSE linear estimate is optimum among all linear estimators. Therefore, without more precisely specifying the optimality criterion, the linear MMSE estimator of $\boldsymbol{X}$ given $\boldsymbol{Y}$ (or the MMSE equalizer) may be defined as the vector MMSE linear estimate $\hat{\boldsymbol{X}}(\boldsymbol{Y}) = C\boldsymbol{Y}$. The linear estimation error vector $\boldsymbol{E} = \boldsymbol{X} - \hat{\boldsymbol{X}}(\boldsymbol{Y})$ is then minimized in every sense among linear estimators.

Theorem 4.4 (MMSE Estimator and MMSE Equalizer) *The MMSE estimator is given by*

$$\hat{\boldsymbol{M}}(\boldsymbol{Z}) = R_b\boldsymbol{Z} \; ; \tag{4.123}$$

$$\hat{\boldsymbol{X}}'(\boldsymbol{Z}) = U'\hat{\boldsymbol{M}}(\boldsymbol{Z}) \tag{4.124}$$
$$= U'R_b\boldsymbol{Z} \ . \tag{4.125}$$

In other words, the block MMSE estimator simply computes the unique (effective) input that minimizes the error vector covariance and does not ignore noise. For this reason the MMSE estimator is sometimes called a **linear MMSE equalizer.**

Proof: Follows directly from the equivalent backward channel model. **QED.**

The linear MMSE estimation error $\boldsymbol{E} = \boldsymbol{M} - \hat{\boldsymbol{M}}(\boldsymbol{Z})$ has covariance matrix $R_{\boldsymbol{ee}} = R_b$, which is "less than" R_f^{-1} since

$$R_f^{-1} - R_b = R_{\boldsymbol{zz}}^{-1} \tag{4.126}$$

is a positive definite matrix (sometimes written $R_{\boldsymbol{zz}}^{-1} > 0$, or $R_f^{-1} > R_b$). However, as the signal-to-noise ratio becomes large, R_b approaches R_f^{-1}.

The estimation error for $\boldsymbol{X}'$ is simply

$$\boldsymbol{E}_{\boldsymbol{x}'} = \boldsymbol{X}' - \hat{\boldsymbol{X}}'(\boldsymbol{Z}) = U'\boldsymbol{M} - U'\hat{\boldsymbol{M}}(\boldsymbol{Z}) = U'\boldsymbol{E} \ . \tag{4.127}$$

Because U' is a volume-preserving transformation, the effective determinant of the covariance matrix of $\boldsymbol{E}_{\boldsymbol{x}'}$ is equal to $|R_{\boldsymbol{ee}}| = |R_b|$.

Example 4.6 (Parallel Channels continued) For the ideal one-dimensional channel (or one of several subchannels in a parallel set) $Y = X + N$ with $R_{\boldsymbol{xx}} = S_x$ and $R_{\boldsymbol{nn}} = S_n$, or the equivalent channel $Z = S_n^{-1}M + N''$ with $R_{\boldsymbol{mm}} = S_x$ and $R_{\boldsymbol{n}''\boldsymbol{n}''} = S_n^{-1}$, the ML estimate of $M = X$ is $S_nZ = Y$, which has error variance S_n. The MMSE estimate of $M = X$ is $(S_xS_n/(S_x+S_n))Z = (S_x/(S_x+S_n))Y$, which has error variance $R_{\boldsymbol{ee}} = S_xS_n/(S_x + S_n) < S_n$. For deleted subchannels, no estimate occurs.

MAP Detection and Estimation

If the input $\boldsymbol{X}$ is Gaussian, then $\boldsymbol{E}$ is Gaussian and the linear MMSE estimator is the unconstrained MMSE estimator. Furthermore, since it maximizes the a posteriori probability density $p_{\mathbf{M}/\mathbf{Z}}(\boldsymbol{m}|\boldsymbol{z}) = p_{\mathbf{E}}(\boldsymbol{v} - R_b\boldsymbol{m})$, it may alternatively be called the maximum a posteriori (MAP) estimator. It is important to distinguish the MAP detector from the MAP estimator, just as it is important to distinguish the ML detector from the ML estimator. The detector is optimum when the input distribution is, as always the case in practice, discrete. The estimator is only defined for continuous distributions, and particularly for the this case, a continuous Gaussian distribution. Nonetheless, a MMSE estimator on a channel with a discrete input distribution for $\boldsymbol{M}$, followed by a detector that has a structure based only on $\boldsymbol{M}$ and not on the channel can be canonical, with specific structures illustrating this property in Sections 4 and 5.

Estimator and Detector Bias

The bias of an estimator $c\boldsymbol{X}$ of a random vector $\boldsymbol{X}$ is the difference between $\boldsymbol{X}$ and the expected value $E[c\boldsymbol{X}|\boldsymbol{X}]$. The ML estimator is unbiased, since

$$\bar{\boldsymbol{M}}(\boldsymbol{Z}) = \boldsymbol{M} + R_f^{-1}\boldsymbol{N}'' , \tag{4.128}$$

so $E[\bar{\boldsymbol{M}}|\boldsymbol{M}] = \boldsymbol{M}$. Indeed, it is clear that the ML estimator is the unique unbiased linear estimator of $\boldsymbol{M}$ given $\boldsymbol{Z}$, since if $\bar{\boldsymbol{M}} = C\boldsymbol{Z}$, then $E[\bar{\boldsymbol{M}}|\boldsymbol{M}] = CR_f\boldsymbol{M}$, which is equal to $\boldsymbol{M}$ everywhere if and only if $C = R_f^{-1}$.
The linear MMSE estimator is biased:

$$E[\hat{\boldsymbol{M}}|\boldsymbol{M}] = R_bR_f\boldsymbol{M} = (I - R_bR_{\boldsymbol{mm}}^{-1})\boldsymbol{M} = \boldsymbol{M} - R_bR_{\boldsymbol{mm}}^{-1}\boldsymbol{M} . \tag{4.129}$$

The bias is $R_bR_{\boldsymbol{mm}}^{-1}\boldsymbol{M}$, which tends to zero as the signal-to-noise ratio becomes large.

3.4 Mutual information

If the input $\boldsymbol{X}$ is Gaussian, then the mutual information $I(\boldsymbol{X};\boldsymbol{Y}) = I(\boldsymbol{M};\boldsymbol{Z})$ between input and output may be expressed in either of two ways:

$$I(\boldsymbol{M};\boldsymbol{Z}) = h(\boldsymbol{Z}) - h(\boldsymbol{Z}|\boldsymbol{M}) = h(\boldsymbol{Z}) - h(\boldsymbol{N}'') = \log|R_{\boldsymbol{zz}}|/|R_f| \; ; \tag{4.130}$$
$$I(\boldsymbol{M};\boldsymbol{Z}) = h(\boldsymbol{M}) - h(\boldsymbol{M}|\boldsymbol{Z}) = h(\boldsymbol{M}) - h(\boldsymbol{E}) = \log|R_{\boldsymbol{mm}}|/|R_b| \tag{4.131}$$

These relations recall the determinantal relations derived earlier,

$$|R_f|/|R_{\boldsymbol{zz}}| = |R_b|/|R_{\boldsymbol{mm}}| = |I - R_fR_b| = |I - R_bR_f| , \tag{4.132}$$

from which it follows that

$$I(\boldsymbol{M};\boldsymbol{Z}) = -\log|I - R_fR_b| = -\log|I - R_bR_f| . \tag{4.133}$$

The determinant $|R_{\boldsymbol{mm}}|$ is equal to the effective determinant $\|R_{\boldsymbol{x}'\boldsymbol{x}'}\|$ of the effective input $\boldsymbol{X}'$ in its r_y-dimensional space $\mathcal{S}_{\boldsymbol{X}'}$. The determinant $|R_b|$ is equal to the effective determinant $\|R_{\boldsymbol{e}'\boldsymbol{e}'}\|$ of the error of the MMSE estimator $U'\boldsymbol{M}$ of $\boldsymbol{X}'$. Therefore there is an interesting connection between MMSE estimation and mutual information, as follows:

Theorem 4.5 (Sufficiency of Canonical Transmission with the Forward and Backward Channel Models) *Given a channel model $\boldsymbol{Y} = H\boldsymbol{X} + \boldsymbol{N}$ where $\boldsymbol{N}$ is full-rank and Gaussian, let*

$\|R_{\boldsymbol{x}'\boldsymbol{x}'}\|$ be the effective determinant of the effective input $\boldsymbol{X}'$, and let $\|R_{\boldsymbol{e}'\boldsymbol{e}'}\|$ be the effective determinant of the linear MMSE estimate of $\boldsymbol{X}'$ given $\boldsymbol{Y}$. Then the mutual information $I(\boldsymbol{X};\boldsymbol{Y})$ when $\boldsymbol{X}$ is Gaussian is given by

$$I(\boldsymbol{X};\boldsymbol{Y}) = \log \|R_{\boldsymbol{x}'\boldsymbol{x}'}\| / \|R_{\boldsymbol{e}'\boldsymbol{e}'}\| \ . \tag{4.134}$$

The above theorem implies the potential existence of canonical transmission systems that use only the forward or backward canonical models, thus ignoring all eliminated dimensions and inputs. As noted earlier, the mutual information $I(\boldsymbol{X};\boldsymbol{Y})$ when $\boldsymbol{X}$ is Gaussian is an upper bound to the mutual information when $\boldsymbol{X}$ is an arbitrary random vector with the same second-order statistics.

Example 4.7 (Parallel Channels (cont.)) On the ideal one-dimensional Gaussian channel, the input variance is $R_{\boldsymbol{xx}} = S_x$ and the MMSE error variance is $R_{\boldsymbol{ee}} = S_x S_n/(S_x + S_n)$, so $I(\boldsymbol{X};\boldsymbol{Y}) = \log R_{\boldsymbol{xx}}/R_{\boldsymbol{ee}} = log(S_x + S_n)/S_n$.

Matrix SNRs

Mutual information results suggest some matrix SNR definitions that allow generalization of many of the results in [1].

Definition 4 (MMSE-SNR Matrix) *Define the square matrix $\boldsymbol{SNR}_{\text{GDFE}}$ by*

$$\boldsymbol{SNR}_{\text{GDFE}} = R_{\boldsymbol{mm}} R_b^{-1} \ ; \tag{4.135}$$

then

$$I(\boldsymbol{M};\boldsymbol{Z}) = I(\boldsymbol{X};\boldsymbol{Y}) = \log |\boldsymbol{SNR}_{\text{GDFE}}| \ . \tag{4.136}$$

(Alternatively, the same result is obtained using the conjugate transpose $\mathbf{SNR}^* = \mathbf{SNR} = R_b^{-1} R_{\boldsymbol{mm}}$*.)*

$\boldsymbol{SNR}_{\text{GDFE}}$ is the matrix generalization of $\text{SNR}_{\text{MMSE-DFE}}$ for the infinite-length MMSE-DFE. The SNR is well-understood through the ratio of the transmitted message energy (covariance) $R_{\boldsymbol{mm}}$ to the minimized square-error power (covariance) $R_{\boldsymbol{ee}} = R_b$. Similarly, recalling that the ML error variance is R_f^{-1}, define

$$\boldsymbol{SNR}_{\text{ML}} = R_{\boldsymbol{mm}} R_f \ , \tag{4.137}$$

(or alternatively $\boldsymbol{SNR}^*_{\text{ML}} = \boldsymbol{SNR}_{\text{ML}} = R_f R_{\boldsymbol{mm}}$). Then, since $R_b^{-1} = R_{\boldsymbol{mm}}^{-1} + R_f$, it follows that

$$\boldsymbol{SNR}_{\text{GDFE}} = I + \boldsymbol{SNR}_{\text{ML}} \ . \tag{4.138}$$

Equation (4.138) is the matrix equivalent of the expression $\text{SNR}_{\text{MMSE-DFE}} = \text{SNR}_{\text{MMSE-DFE,U}} + 1$ in [1].

In summary:

Theorem 4.6 *Given canonical forward and backward channel models $\boldsymbol{Z} = R_f\boldsymbol{M} + \boldsymbol{N}''$ and $\boldsymbol{M} = R_b\boldsymbol{Z} + \boldsymbol{E}$, define $\boldsymbol{SNR}_{\text{GDFE}} = R_{\boldsymbol{mm}}R_b^{-1} = R_f^{-1}R_{\boldsymbol{zz}}$, and define $\boldsymbol{SNR}_{\text{ML}} = R_{\boldsymbol{mm}}R_f = R_bR_{\boldsymbol{zz}}$. Then the mutual information $I(\boldsymbol{M};\boldsymbol{Z})$ when $\boldsymbol{M}$ is Gaussian is given by*

$$I(\boldsymbol{M};\boldsymbol{Z}) = \log|\boldsymbol{SNR}_{\text{GDFE}}| = \log|I + \boldsymbol{SNR}_{\text{ML}}| \ . \tag{4.139}$$

Bias Results

Since the ML estimator is the unique unbiased linear estimator, this result may also be interpreted as a relation between mutual information, linear MMSE estimation, and unbiased linear estimation.

Since the MMSE estimator is $R_b\boldsymbol{Z}$ and the unbiased ML estimator is $R_f^{-1}\boldsymbol{Z}$, an MMSE estimate may be converted to an unbiased ML estimate by multiplication by $\boldsymbol{SNR}_{\text{ML}}^{-1}\boldsymbol{SNR}_{\text{GDFE}}$, or by $\boldsymbol{SNR}_{\text{GDFE}}\boldsymbol{SNR}_{\text{ML}}^{-1}$. The bias of the MMSE estimate is equal to

$$R_bR_{\boldsymbol{mm}}^{-1}\boldsymbol{M} = \boldsymbol{SNR}_{\text{GDFE}}^{-1}\boldsymbol{M} \ . \tag{4.140}$$

Example 4.8 (Example 2 (cont.)) On the ideal one dimensional Gaussian channel, $\boldsymbol{SNR}_{\text{GDFE}} = R_{\boldsymbol{xx}}/R_{\boldsymbol{ee}} = (S_x + S_n)/S_n$ and $\boldsymbol{SNR}_{\text{ML}} = R_{\boldsymbol{xx}}/R_{\boldsymbol{nn}} = S_x/S_n = \boldsymbol{SNR}_{\text{GDFE}} - 1$. The biased MMSE estimate $(S_x/(S_x + S_n))Y$ may be converted to the unique unbiased linear estimate Y by multiplication by $\text{SNR}_{\text{MMSE-DFE}}/\text{SNR}_{\text{ML}} = (S_x + S_n)/S_x$.

Notice that $\boldsymbol{SNR}_{\text{GDFE}}$ and $\boldsymbol{SNR}_{\text{ML}}$ are diagonalized by the same unitary transformations U and therefore commute; for if

$$\boldsymbol{SNR}_{\text{ML}} = U\Lambda_{ML}^2U*, \tag{4.141}$$

then

$$\boldsymbol{SNR}_{\text{GDFE}} = I + \boldsymbol{SNR}_{\text{ML}} = U(I + \Lambda_{ML}^2)U* = U\Lambda_{MMSE}^2U* \ . \tag{4.142}$$

This implies that the eigenvalues of $\boldsymbol{SNR}_{\text{GDFE}}$ are equal componentwise to 1 plus the eigenvalues of $\boldsymbol{SNR}_{\text{ML}}$:

$$\lambda_{MMSE,j}^2 = 1 + \lambda_{ML,j}^2 \ , \tag{4.143}$$

regardless of whether the matrices R_f, $R_{\boldsymbol{mm}}$, R_b and $R_{\boldsymbol{zz}}$ commute. Thus each of the individual modes in the block channel has a relationship between MMSE SNR and unbiased SNR that parallels the relationship established in [1], namely $\text{SNR}_{\text{MMSE-DFE}} = \text{SNR}_{\text{MMSE-DFE,U}} + 1$.

4 THE GENERALIZED DFE RECEIVER STRUCTURE

This section introduces and develops the canonical GDFE receiver structure for a general block Gaussian ISI channel $\boldsymbol{Y} = H\boldsymbol{X} + \boldsymbol{N}$ with arbitrary H, $R_{\boldsymbol{xx}}$ and (full-rank) $R_{\boldsymbol{nn}}$. This structure is an apparently novel structure that is a finite-length generalization of the usual infinite-length MMSE decision-feedback equalization (MMSE-DFE).

The starting point is an equivalent canonical backward channel model

$$\boldsymbol{M} = R_b\boldsymbol{Z} + \boldsymbol{E} \ . \tag{4.144}$$

Since R_b is a nonsingular covariance matrix, it has a unique Cholesky factorization

$$R_b = L_b D_b^2 L_b^* \ , \tag{4.145}$$

where L_b is a monic lower triangular matrix and D_b^2 is a nonsingular positive-definite diagonal matrix.

Premultiplication by the lower triangular matrix L_b^{-1} yields the equivalent channel model

$$\boldsymbol{M}' = L_b^{-1}\boldsymbol{M} = D_b^2 L_b^* \boldsymbol{Z} + L_b^{-1|}\boldsymbol{E} = \boldsymbol{Z}' + \boldsymbol{E}' \ , \tag{4.146}$$

where $\boldsymbol{Z}' = D_b^2 L_b^* \boldsymbol{Z}$ may be viewed as the result of passing $\boldsymbol{Z}$ through an upper-triangular "feedforward filter" $D_b^2 L_b^*$, and the noise $\boldsymbol{E}' = L_b^{-1}\boldsymbol{E}$ has a diagonal covariance matrix

$$R_{\boldsymbol{e}'\boldsymbol{e}'} = L_b^{-1} R_{\boldsymbol{ee}} L_b^{-*} = D_b^2 \ ; \tag{4.147}$$

i.e., its components are uncorrelated.

The usual assumptions of decision-feedback equalization are now invoked:

- symbol-by-symbol decisions may be made on the components M_j of $\boldsymbol{M}$;
- in the detection of M_j, it may be assumed that all previous decisions are correct (the "ideal DFE assumption").

Now since L_b^{-1} is lower triangular, M_j is a linear combination of Z_j', E_j' and previous components $\boldsymbol{M}(j-1) = [M_1, ..., M_{j-1}]$. The MMSE symbol estimate of M_j given Z_j' and $\boldsymbol{M}(j-1)$ is therefore equal to Z_j' minus the linear combination of the past components $[M_1, ..., M_{j-1}]$ that is specified by the jth row of L_b^{-1} (the "feedback filter" at time j). The error in this estimate is E_j'. The signal-to-noise ratio for the jth symbol is thus

$$\text{SNR}_j = E[|M_j|^2]/E[|E_j'|^2] = \lambda_{m,j}^2 / d_{b,j}^2 \ , \tag{4.148}$$

where $\lambda_{m,j}^2$ is the jth diagonal element of the diagonal matrix $R_{\boldsymbol{mm}}$, and $d_{b,j}^2$ is the jth Cholesky factor of R_b.

Theorem 4.7 (GDFE is Canonical) *If R_{mm} is diagonal, or equivalently, the input vector $\boldsymbol{M}$ has uncorrelated elements, then the GDFE is canonical.*

Proof: The product of the symbol SNRs is $|\boldsymbol{SNR}_{\text{GDFE}}|$, since

$$\prod_j \text{SNR}_j = |R_{mm}|/|D_b^2| = |R_{mm}|/|R_b| = |\boldsymbol{SNR}_{\text{GDFE}}| \; . \quad (4.149)$$

This expression is the key to showing that this receiver structure is canonical. To complete the proof, assume that $\boldsymbol{X}$ and thus all random vectors are Gaussian. Then

$$I(\boldsymbol{M};\boldsymbol{Z}) = \log |\boldsymbol{SNR}_{\text{GDFE}}| \; . \quad (4.150)$$

Furthermore, from the chain rule of information theory,

$$I(\boldsymbol{M};\boldsymbol{Z}) = \prod_j I(M_j; Z_j|\boldsymbol{M}(j-1)) \; . \quad (4.151)$$

The mutual information in the jth symbol transmission may be expressed as

$$\begin{aligned} I(M_j; Z_j|\boldsymbol{M}(j-1)) &= h(M_j|\boldsymbol{M}(j-1)) - h(M_j|Z_j', \boldsymbol{M}(j-1)) \\ &= h(M_j) - h(E_j') && (4.152) \\ &= \log \frac{\lambda_{m,j}^2}{d_{b,j}^2} && (4.153) \\ &= \log \text{SNR}_j \; , && (4.154) \end{aligned}$$

since M_j is independent of $\boldsymbol{M}(j-1)$ when $\boldsymbol{M}$ is Gaussian and E_j' is the estimation error (innovations) for M_j given $[Z_j', \boldsymbol{M}(j-1)]$. Thus

$$I(\boldsymbol{M};\boldsymbol{Z}) = \prod_j \log \text{SNR}_j \; . \quad (4.155)$$

Now use a long code of rate arbitrarily close to $\log \text{SNR}_j$ on each subchannel that has an arbitrarily low error probability. Decode the subchannels in order so that the "past" decisions $M_1, ..., M_{j-1}$ are available when decoding M_j (which justifies the ideal DFE assumption). Then one can send at an aggregate rate arbitrarily close to $I(\boldsymbol{M};\boldsymbol{Z}) = \log |\boldsymbol{SNR}_{\text{GDFE}}|$ per block with arbitrarily low probability of error. Hence this block MMSE-DFE receiver structure is canonical.**QED**.

In practice, as in vector coding systems, one can code "across subchannels" to avoid excessive decoding delay and buffering. The ideal DFE assumption then fails, but this problem may be elegantly handled by a kind of "transmitter precoding" similar to the precoding techniques that have been developed for single-channel transmission systems.

4.1 The Packet GDFE - stationary special case

The GDFE is general but does not converge to the usual MMSE-DFE for infinite-length packets on a stationary channel without the additional transmitter alterations in this subsection. In general, these alterations add additional complexity for no improvement in performance, other than they allow a recursive implementation of the transmit filter via Cholesky factorization.

The input vector $\boldsymbol{M}$ can be decomposed according to its innovation representation as in Section 2.2 as

$$\boldsymbol{M} = L\boldsymbol{W} \ , \tag{4.156}$$

where $R_{\boldsymbol{mm}} = LR_{\boldsymbol{ww}}L^*$ and L is lower-triangular and nonsingular, and where $R_{\boldsymbol{ww}}$ is diagonal. The elements of $\boldsymbol{W}$ are the innovations of $\boldsymbol{M}$. For the Packet GDFE, the elements of $\boldsymbol{W}$ are considered the coded input sequence and should be the values estimated by the GDFE receiver. The alteration necessary to the receiver is simply to replace the feedback section by the rows of $\tilde{L} = L^{-1}L_b^{-1}$ instead of the rows of L_b^{-1}. This new matrix feedback section is still lower triangular and previous decisions on elements of $\boldsymbol{W}$ can be used to aid future decisions just as the elements of $\boldsymbol{M}$ were used in the diagonal-$R_{\boldsymbol{mm}}$ case.

In the transmitter, the input becomes

$$\boldsymbol{X} = U'\boldsymbol{M} = U'L\boldsymbol{V} \tag{4.157}$$

and $\|R_{\boldsymbol{xx}}\| = |R_{\boldsymbol{mm}}| = |R_{\boldsymbol{vv}}|$ so that $I(\boldsymbol{X};\boldsymbol{Y}) = I(\boldsymbol{M};\boldsymbol{Z}) = I(\tilde{\boldsymbol{X}};\boldsymbol{Y})$.

Lemma 4.1 (Packet GDFE is Canonical) *The packet GDFE, which estimates* $\boldsymbol{W}$ *directly in the feedback section and uses the additional transmit filter of L is canonical.*

Proof: The proof is identical to the proof for the GDFE with diagonal $\boldsymbol{M}$ with $\boldsymbol{W}$ replacing $\boldsymbol{M}$. **QED.**

The transmit signal decomposition has an interesting interpretation:

- The lower-triangular filter L relates the innovations or underlying transmitted signal to the filtered channel output for whatever transmit covariance $R_{\boldsymbol{xx}}$ is used. When $R_{\boldsymbol{xx}}$ is stationary, the rows of L will converge for long packet length to the filters that relate the innovations to the channel input. Different parts of L may converge to different filters, corresponding to the different frequency bands used.
- The filter U' is not triangular and is necessary when dimensions have been reduced from the original channel H. This filter combines different sets of L into a single transmit signal - the transmit signal thus contains potentially nonoverlapping frequency bands in the the stationary case and

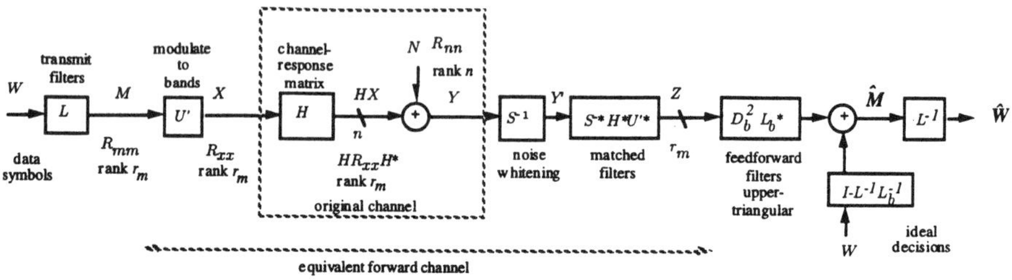

Figure 4.6 Packet GDFE.

> U' is an orthogonal matrix that is volume preserving and corresponds to essentially modulation of the various baseband signals generated by the triangular L into the different frequency bands.

Figure 4.6 illustrates the various parts of the GDFE, including the special packet case. When U' corresponds to a modal decomposition, $L = I$ and $\boldsymbol{M} = \boldsymbol{V}$, and the input cannot be realized by triangular filtering.

5 TRANSMITTER OPTIMIZATION AND VECTOR CODING

To this point the input covariance matrix $R_{\boldsymbol{xx}}$ has been assumed to be given. In this section $R_{\boldsymbol{xx}}$ will be chosen subject to a power constraint to maximize the mutual information $I(\boldsymbol{X};\boldsymbol{Y})$ assuming Gaussian input statistics, or equivalently to maximize the determinant $|\boldsymbol{SNR}_{\text{GDFE}}|$.

An optimized $R_{\boldsymbol{xx}}$ has a natural diagonal representation that suggests the well-known method of vector coding, which yields an alternative canonical receiver structure for the optimum $R_{\boldsymbol{xx}}$. When $R_{\boldsymbol{xx}}$ is optimized, or equivalently $R_{\boldsymbol{mm}}$ is optimized, then VC and the GDFE have the same canonical performance.

5.1 Input covariance optimization

The channel model is the general linear Gaussian channel $\boldsymbol{Y} = H\boldsymbol{X} + \boldsymbol{N}$ as before, with H and $R_{\boldsymbol{nn}}$ given, and $R_{\boldsymbol{xx}}$ to be optimized subject to a constraint on the average input energy $E[\boldsymbol{X}\boldsymbol{X}^*]$, which is the trace of $R_{\boldsymbol{xx}}$- i.e., the sum $\sum_j R_{\boldsymbol{xx}}(jj)$ of the variances $R_{\boldsymbol{xx}}(jj)$ of the components X_j of $\boldsymbol{X}$.

As discussed earlier, a general input vector $\boldsymbol{X}$ may be decomposed into an undetectable part $\boldsymbol{X}_{|K}$ in the right null space K of H and an effective input $\boldsymbol{X}_{|K^\perp}$ in the orthogonal subspace $K^\perp$. It is clear that no energy should be wasted on $X_{|K}$ and therefore an optimized $\boldsymbol{X}'$ should be constrained to lie in $K^\perp$. The dimension $r_{\hat{y}} = r_m$ of $K^\perp$ is equal to the dimension of the range space of H, since H is a one-to-one map from $K^\perp$ to its range space.

It is convenient to choose an orthonormal basis for $K^\perp$ consisting of $r_{x'}$ orthonormal vectors $\boldsymbol{U} = \{\boldsymbol{u}_j, 1 \le j \le r_{x'}\}$. Then any $\boldsymbol{x} \in K^\perp$ may be written as a linear combination of the basis vectors, $\boldsymbol{x} = U\boldsymbol{m}$, and furthermore because of orthonormality, $\|\boldsymbol{x}\|^2 = \|\boldsymbol{m}\|^2$. Then an average energy constraint on $\boldsymbol{X}$ translates directly into an equivalent average energy constraint on the random r_x-vector $\boldsymbol{M}$. The VC design then chooses $R_{\boldsymbol{mm}}$ in the channel model $\boldsymbol{Y} = HU\boldsymbol{M} + \boldsymbol{N}$ to maximize $I(\boldsymbol{X};\boldsymbol{Y}) = I(\boldsymbol{M};\boldsymbol{Y})$ subject to an average energy constraint on $\boldsymbol{M}$.

Again noise whitening and matched filtering may be used without loss of optimality to reduce to obtain an equivalent canonical channel model

$$\boldsymbol{Z} = R_f\boldsymbol{M} + \boldsymbol{N}'' \,. \tag{4.158}$$

where $R_f = U^*H^*R_{\boldsymbol{nn}}^{-1}HU$ is a square full-rank covariance matrix that is both the channel-response matrix and the covariance matrix of $\boldsymbol{N}''$. Again $I(\boldsymbol{M};\boldsymbol{Z}) = I(\boldsymbol{X};\boldsymbol{Y})$. This equivalent canonical forward channel model is similar to that derived in Section 3, except that $R_{\boldsymbol{mm}}$ is yet to be determined and is not necessarily diagonal.

Now it is desired to maximize $I(\boldsymbol{M};\boldsymbol{Z}) = \log|I + \boldsymbol{SNR}_{\mathrm{ML}}| = \log|I + R_{\boldsymbol{mm}}R_f|$:

Theorem 4.8 (Optimum transmit vectors) *The optimum $R_{\boldsymbol{mm}}$ must have the same eigenvectors as R_f: i.e., $R_{\boldsymbol{mm}}$ must commute with R_f.*

Proof: Let V be a unitary transformation that diagonalizes R_f; i.e., $V^*R_fV = \Lambda_f^2$, where $\Lambda_f^2 = \mathrm{diag}\{\lambda_{f,j}^2\}$ is a diagonal matrix whose diagonal components are the eigenvalues $\lambda_{f,j}^2$ of R_f. Let $R_{m,j}^2$ be the diagonal elements of $V^*R_{\boldsymbol{mm}}V$. Then the diagonal elements of $I + V^*R_{\boldsymbol{mm}}VV^*R_fV = I + (V^*R_{\boldsymbol{mm}}V)\Lambda_f^2$ are $\{1 + R_{m,j}^2\lambda_{f,j}^2\}$, so by Hadamard's inequality,

$$|I + R_{\boldsymbol{mm}}R_f| = |I + V^*R_{\boldsymbol{mm}}VV^*R_fV| \le \prod_j (1 + R_{m,j}^2\lambda_{f,j}^2) \,, \tag{4.159}$$

with equality if and only if $V^*R_{\boldsymbol{mm}}V$ is diagonal. Since V is a unitary transformation the trace (sum of the diagonal components) of $V^*R_{\boldsymbol{mm}}V$ is the same as the trace of $R_{\boldsymbol{mm}}$; therefore setting the off-diagonal components of $V^* R_{mm} V$ to zero will not change the average energy of $\boldsymbol{M}$, but will necessarily decrease $I(\boldsymbol{M};\boldsymbol{Y}) = I(\boldsymbol{M};\boldsymbol{Z})$ unless $V^* R_{mm} V$is already diagonal. Thus the optimum $V^* R_{mm} V$ is diagonal. **QED**.

Given that $V^* R_{mm} V$ is a diagonal matrix $\Lambda_m^2 = diag\{\lambda_{m,j}^2\}$, the optimum variances $\lambda_{m,j}^2$ may then be determined by discrete water-pouring in the usual manner, with the result that

$$\lambda_{m,j}^2 = K - \frac{1}{\lambda_{f,j}^2}, \text{if } K \geq \frac{1}{\lambda_{f,j}^2}; \tag{4.160}$$

$$\lambda_{m,j}^2 = 0 \text{ , otherwise,} \tag{4.161}$$

where K is a constant chosen so that the average energy constraint on $\prod_j \lambda_{m,j}^2$ is met. Thus this water-pouring optimization may cause some of the subchannels to be unused and thus reduce the effective rank of the channel below $r_{x'}$ to a new value of r_m that would then force $r_{x'}$ to be smaller and equal to r_m through the original definitions of these ranks, which depend on the choice of R_{xx}, or equivalently, R_{mm}.

The optimum R_{xx} is then determined from the optimum Λ_m^2 via

$$R_{xx} = U R_{mm} U^* = U V \Lambda_M^2 V^* U^* \ . \tag{4.162}$$

A canonical GDFE receiver may then be constructed from this optimum R_{xx} and may be used to approach the maximized $I(\boldsymbol{M};\boldsymbol{Y})$, namely the channel capacity of the given linear Gaussian packet channel.

Finally, since

$$R_{xx} = U R_{mm} U^* \ ; \tag{4.163}$$

$$H^* R_{nn}^{-1} H = U R_f U^* \ , \tag{4.164}$$

it follows readily from the orthonormality of U that if R_{mm} and R_f commute, then R_{xx} and $H^* R_{nn}^{-1} H$ commute.

5.2 Commuting channels

The above argument shows that an optimum R_{mm} commutes with R_f. A canonical channel model $\boldsymbol{Z} = R_f \boldsymbol{M} + \boldsymbol{N}$ will be called commuting if R_{mm} and R_f commute:

$$R_{mm} R_f = R_f R_{mm} \ . \tag{4.165}$$

Equivalently, since $(R_{mm} R_f)^* = R_f R_{mm}$, a canonical channel is commuting if $R_{mm} R_f$ is Hermitian-symmetric. Since

$$\boldsymbol{SNR}_{\text{ML}} = R_{mm} R_f \ ; \tag{4.166}$$

$$\boldsymbol{SNR}_{\text{GDFE}} = I + \boldsymbol{SNR}_{\text{ML}}, \tag{4.167}$$

a channel is commuting if either $\mathrm{SNR}_{\mathrm{ML}}$ or $\mathrm{SNR}_{\text{MMSE-DFE}}$ is Hermitian-symmetric. A one-dimensional channel is necessarily commuting.

The inverses $R_{\boldsymbol{mm}}^{-1}$ and R_f^{-1} of commuting covariance matrices commute with with $R_{\boldsymbol{mm}}$ and R_f and with each other. Moreover, from the defining equations for R_b and $R_{\boldsymbol{zz}}$,

$$R_b = (R_{\boldsymbol{mm}}^{-1} + R_f)^{-1} \; ; \qquad (4.168)$$

$$R_{\boldsymbol{zz}} = R_f R_{\boldsymbol{mm}} R_b^{-1} = R_b{-1} R_{\boldsymbol{mm}} R_f \; , \qquad (4.169)$$

it follows that R_b and $R_{\boldsymbol{zz}}$ and their inverses also commute with each other and with $R_{\boldsymbol{mm}}$ and R_f. Thus the corresponding backward canonical channel model is commuting as well.

5.3 Vector coding

If $\boldsymbol{Z} = R_f \boldsymbol{M} + \boldsymbol{N}$ is a commuting equivalent forward channel model of rank r_y, then $R_{\boldsymbol{mm}} = VL\Lambda_m^2 V^*$ and $R_f = V\Lambda_f^2 V^*$ for some unitary matrix V, so

$$\boldsymbol{Z} = V\Lambda_f^2 V^* \boldsymbol{M} + \boldsymbol{N} \; ; \qquad (4.170)$$

$$\boldsymbol{Z}' = V^* \boldsymbol{Z} = \Lambda_f^2 V^* \boldsymbol{M} + V^* \boldsymbol{N} = \Lambda_f^2 \boldsymbol{M}' + \boldsymbol{N}' \; , \qquad (4.171)$$

where $\boldsymbol{Z}' = V^*\boldsymbol{Z}$, $\boldsymbol{M}' = V^*\boldsymbol{M}$, and $\boldsymbol{N}' = V^*\boldsymbol{N}$; i.e., the random vectors are represented in the basis determined by the unitary transformation V. Now

$$R_{\boldsymbol{m}'\boldsymbol{m}'} = V^* R_{\boldsymbol{mm}} V = \Lambda_m^2 \; ; \qquad (4.172)$$

$$R_{\boldsymbol{n}'\boldsymbol{n}'} = V^* R_f V = \Lambda_f^2 \; . \qquad (4.173)$$

The channel therefore naturally decomposes into r_y decoupled one-dimensional subchannels of the form

$$\boldsymbol{Z}_j' = \lambda_{f,j}^2 M_j' + N_j', 1 \le j \le r_y \; , \qquad (4.174)$$

where the variance of M_j' is $\lambda_{m,j}^2$ and N_j' is an independent Gaussian variable of variance $\lambda_{f,j}^2$. This is just a standard one-dimensional Gaussian channel model of the type of Example 2, with $S_{x,j} = \lambda_{f,j}^4 \lambda_{m,j}^2, S_{n,j} = \lambda_{f,j}^2$, and therefore

$$\frac{S_{x,j}}{S_{n,j}} = \lambda_{f,j}^2 \lambda_{m,j}^2 \; . \qquad (4.175)$$

The mutual information over such a channel is

$$I(M_j'; Z_j') = \log\left(1 + \frac{S_{x,j}}{S_{n,j}}\right) = \log(1 + \lambda_{f,j}^2 \lambda_{m,j}^2) \; . \qquad (4.176)$$

The aggregate mutual information of all parallel subchannels is

$$\begin{aligned} I(\boldsymbol{M}';\boldsymbol{Z}') &= \prod_j \log(1+\lambda_{f,j}^2\lambda_{m,j}^2) = \log|I+\lambda_f^2\Lambda_m^2| \\ &= \log|I+R_f R_{\boldsymbol{mm}}| = \log|\boldsymbol{SNR}_{\text{GDFE}}| \ . \end{aligned} \tag{4.177}$$

It follows that this structure, called vector coding, is canonical for any commuting channel. In particular, it is canonical for any channel for which $R_{\boldsymbol{mm}}$ or $R_{\boldsymbol{xx}}$ has been optimized.

Lemma 4.2 (Optimality and Canonical properties of VC) *Vector Coding is both optimal and canonical for a commuting channel.*

Proof: It follows directly from (4.177) that VC is canonical. VC is also a ML estimator for which each subchannel can use an ML detector for the applied code. If the input $\boldsymbol{X}$ is uniform discrete over the $r_{\hat{y}}$-dimensional subspace, then this ML detector minimizes error probability. **QED.**

If the channel is not commuting, however, then it cannot be decomposed into completely decoupled one-dimensional subchannels in this way; i.e., vector coding is not well defined for noncommuting channels. Thus in certain cases where $R_{\boldsymbol{xx}}$ is predetermined and cannot be optimized, the GDFE structure may be the only canonical receiver structure available.

DMT - Discrete Multitone

DMT or Discrete Multitone is a special case of VC when the channel correlation matrix matrix $HR_{\boldsymbol{nn}}^{-1}H^*$ is circulant. This circulant property is forced by the use of a cyclic prefix in each transmitted packet, which is simply a repeat of the last few samples at the beginning and end of a packet. The eigenvectors needed for a commutative channel and for the optimized input are essentially the vectors associated with a Discrete Fourier Transform, thus allowing very efficient optimal and canonical implementations through the use of Fast Fourier Transform methods.

6 LIMITING RESULTS WITH INCREASING PACKET LENGTH

The results in this paper all converge to generalizations of the known results in [1] for infinite-length (continuous non-packet) transmission on a stationary dispersive channel with additive Gaussian noise. This convergence requires that the individual elements of the vectors $\boldsymbol{X}$ and $\boldsymbol{N}$ are successive samples from stationary random processes and that H for any values of $m \geq n$ has each successive row moved one position to the right with respect to the previous row, but the row elements are otherwise the same. That is, H is "Toeplitz" as $n \to \infty$.

Perhaps not well established in [1] is the situation in which these well-known results exist, namely that the input process $\boldsymbol{X}$ must have nonsingular covariance as $n \to \infty$, which requires a resampling or "optimization of symbol and center frequencies" as a function of the channel, which tacitly may involve multiple disjoint frequency bands and multiple MMSE-DFE's. The GDFE more accurately describes these multiple MMSE-DFE's in the limit, each of which exhibits the properties discussed in [1].

6.1 Channel Models

The D-transform of a discrete time sequence or random process X_k (the samples of $\boldsymbol{X}$ as $m \to \infty$) is $X(D) \triangleq \sum_k X_k D^k$. Convolution of sequences in discrete time corresponds to multiplication of their D-transforms. The matrix channel with Toeplitz H corresponds to convolution of $X(D)$ with $h(D)$ (the D-transform of the first row of H). Thus, the dual channel model becomes:

$$\boldsymbol{Y} = H\boldsymbol{X} + \boldsymbol{N} \Longrightarrow Y(D) = h(D)X(D) + N(D) \tag{4.178}$$

$$\boldsymbol{X} = C\boldsymbol{Y} + \boldsymbol{E} \Longrightarrow X(D) = c(D)Y(D) + E(D) \ . \tag{4.179}$$

Multiplication of a vector by H^* corresponds to convolution with $h^*(D^{-*})$. Thus, a matched filter output is

$$Z(D) = h^*(D^{-*})Y(D) \ . \tag{4.180}$$

For stationary sequences, the autocorrelation function $r_{xx,k} = E[X_l X_{l-k}^*]$ has a D-transform

$$R_{xx}(D) = \sum_k r_{xx,k} D^k \ . \tag{4.181}$$

Pythagorean relationships are

$$R_{yy}(D) = h(D)R_{xx}(D)h^*(D^{-*}) + R_{nn}(D) \tag{4.182}$$

$$R_{xx}(D) = c(D)R_{yy}(D)c^*(D^{-*}) + R_{ee}(D) \ . \tag{4.183}$$

Also,

$$R_{ee}(D) = R_{ex}(D) = R_{xx}(D) - c(D)R_{xy}(D) \ . \tag{4.184}$$

6.2 Limiting Entropy, Mutual Information, and SNR

The innovations are stationary when $X(D)$ is stationary and critical to the generalization of entropy. A particularly crucial problem in establishing limiting results, and required by a stationary process, is the singularity of the process $X(D)$. Subsection 6.2 reviews results when $X(D)$ is nonsingular and Subsection 6.2 extends and generalizes these results in a heuristic way to the nonsingular case.

Nonsingular input sequences

A stationary random sequence $x(D)$ satisfies the Paley-Wiener Criterion:

$$S_w = |\frac{1}{2\pi}\int_{-\pi}^{\pi} \log R_{xx}(e^{-\jmath\theta})d\theta| < \infty \ , \tag{4.185}$$

which means it is also nonsingular. In practice, satisfaction of the PW criterion means the power spectral density $R_{xx}(e^{-\jmath\theta})$ cannot be infinite nor zero at more than a few discrete frequencies, a requirement often not met if the input sequence $X(D)$ tries to zero energy in certain regions of the band that water-filling arguments might dictate should be zeroed. For a nonsingular sequence the vector $\boldsymbol{X}$ will have a nonsingular $R_{\boldsymbol{xx}}$ for all packet lengths as $m \to \infty$. The types of singular processes of interest in Subsection 6.2 are actually very close to stationary in that within certain frequency bands (or at the right sampling rates and center/carrier frequencies), PW is individually satisfied for each of several disjoint bands.

When $X(D)$ is stationary and therefore nonsingular, the relation

$$\boldsymbol{W} = L^{-1}\boldsymbol{X} \tag{4.186}$$

directly corresponds to the chain rule for entropy ([2]) when $\boldsymbol{X}$ is Gaussian:

$$H(\boldsymbol{X}) = H(X_1) + H(X_2/X_1) + ... + H(X_m/\{X_1, X_2, ...X_{m-1}\}) \ . \tag{4.187}$$

That is W_k is the MMSE sample corresponding to the estimate of X_k, given all previous values of $X(D)$,

$$H(\boldsymbol{X}) = H(W_1) + H(W_2) + ...H(W_m) \ . \tag{4.188}$$

L^{-1} is a linear prediction filter operating on $\boldsymbol{X}$ to produce $\boldsymbol{W}$. Clearly, since X_k is stationary, this filter is constant, meaning that L^{-1} tends towards a Toeplitz matrix when m gets large, and $R_{\boldsymbol{ww}}$ tends towards a constant diagonal matrix with linear minimum mean square error S_w along the diagonal.

In this case, Cholesky factorization corresponds to

$$R_{\boldsymbol{xx}} = LR_{\boldsymbol{ww}}L^* \Longrightarrow R_{xx}(D) = l(D)S_w l^*(D^{-*}) \ , \tag{4.189}$$

where $l(D)$ is monic ($l_0 = 1$), causal ($l_k = 0 \ \forall \ k < 0$) and minimum-phase (all roots and poles outside the unit circle), and

$$S_w = \frac{1}{2\pi}\int_{-\pi}^{\pi} \log R_{xx}(e^{-\jmath\theta})d\theta \ , \tag{4.190}$$

implying that $R_{xx}(D)$ satisfies the discrete-time Paley-Wiener criterion [22]. The linear prediction filter is $1/l(D)$, and the innovations sequence is $w(D) = x(D)/l(D)$.

The entropy for a stationary process is defined as

$$H(X) = \lim_{m\to\infty} \frac{mS_w}{m} = \lim_{m\to\infty} \frac{H(\boldsymbol{X})}{m} \ . \tag{4.191}$$

For the Gaussian random process $X(D)$, this value is clearly

$$H(X) = log(\pi e S_w) \ , \tag{4.192}$$

and because $R_{\boldsymbol{ww}}$ is a constant diagonal, the prediction error sequence or innovations $W(D)$ is white. Similarly, if $\boldsymbol{X}$ and $\boldsymbol{Y}$ are jointly stationary and Gaussian, the limit is found using Toeplitz distribution results [23],

$$H(X/Y) = \frac{1}{2\pi}\int_{-\pi/2}^{\pi/2} \log_2 R_{ee}(e^{-\jmath\theta})d\theta \ . \tag{4.193}$$

Essentially, the conditional entropy is equal to the entropy of the error sequence associated with estimation of the random variable based on the given random variable. This stationary Gaussian error sequence itself also has a innovations representation and the conditional entropy is thus also equal to the entropy of this innovations sequence. Thus,

$$\begin{aligned} H(X/Y) &= H(X_k/(Y,[E_{k-1},E_{k-2}...])) = H(X_k/(Y,[X_{k-1},X_{k-2}...])) \\ &= \log_2(\pi e \frac{N_0}{D_b}) \ , \end{aligned} \tag{4.194}$$

where the rightmost relation is obtained by recognizing that the MMSE estimation associated with $H(X_k/(Y,[X_{k-1},X_{k-2}...]))$ is that of the MMSE-DFE. Further, D_b must also converge to a constant since the matrix R_b^{-1} is Toeplitz when $R_{\boldsymbol{ww}}$ is constant, which it must be when the system is stationary and infinite length. The error sequence for the MMSE-DFE is white because this sequence is the innovations sequence for the linear prediction of the error sequence corresponding to the linear MMSE of $X(D)$ given $Y(D)$.

The value of D_b is determined from the spectral factorization $R_b^{-1}(D) = l_b(D)D_b l_b^*(D^{-*})$ (where $l(D)$ is causal, monic, and minimum-phase, $D_b > 0$ is real, and $l^*(D)$ is anticausal, monic, and maximum-phase - see [1])

$$D_b = \frac{1}{2\pi}\int_{-\pi/2}^{\pi/2} \log R_b^{-1}(e^{-\jmath\theta})d\theta \ , \tag{4.195}$$

where,

$$R_b^{-1}(D) = R_{mm}^{-1}(D) + R_f(D) = R_m^{-1}(D) + h(D)R_{nn}^{-1}(D)h^*(D^{-*});. \tag{4.196}$$

The factorization of (4.196) is called the "key equation" in [1].

The single-band MMSE-DFE: The mutual information for jointly stationary and Gaussian $X(D)$ and $Y(D)$ also has a limiting definition

$$I(X;Y) = \lim_{m\to\infty} \bar{I}(\boldsymbol{X};\boldsymbol{Y}) \ . \tag{4.197}$$

The formula $I(X;Y) = H(X) - H(X/Y) = H(Y) - H(Y/X)$ leads to

$$I(X;Y) = \frac{1}{2\pi}\int_{-\pi/2}^{\pi/2} \log_2(\text{SNR}(\theta))d\theta \ , \tag{4.198}$$

where

$$\text{SNR}(\theta) = \frac{S_w}{R_{ee}(e^{-\jmath\theta})} = \frac{S_w|l(e^{-\jmath\theta})h(e^{-\jmath\theta})|^2 + R_{nn}(e^{-\jmath\theta})}{R_{nn}(e^{-\jmath\theta})} \ . \tag{4.199}$$

The MMSE-DFE is biased, but simple scaling can remove the bias and the relation

$$I(X;Y) = \log_2(1 + \text{SNR}_{\text{MMSE-DFE,U}}) = \log_2\left(S_w D_b\right) \ , \tag{4.200}$$

shows the MMSE-DFE to be canonical for a given fixed choice of input spectrum given by $R_{xx}(D) = l(D)S_w l^*(D^{-*})$. Thus, a maximum likelihood detector is not necessary because best performance can be attained by applying a good code with as small a gap from mutual information (on an AWGN) to an intersymbol interference channel that uses a MMSE-DFE and still maintain that same small gap from mutual information.

The Case of Singular Input

Technically, a singular input sequence is not stationary because it does not satisfy the Paley-Wiener Criterion. However, it is often possible in practice to resample a sequence at a lower rate, and possibly with a carrier offset in bandpass processes, so that an equivalent complex baseband random process is stationary. Such stationary processes can be added together, again with carrier offset, to create a nonstationary process (in this case, cyclostationary with period equal to the greatest common multiple of carrier periods or in the finite-length case to the packet period).

In effect, each of the frequency bands used now has a stationary process within it and all the results of Section 6.2 apply individually to each band. The data rate is of course the sum of the data rates. The SNR is the geometric average where each band's SNR is weighted by its ratio of bandwidth to the total used by all bands. The union of all these disjoint bands is denoted by Ω and a modification of the PW criterion holds such that

$$\frac{1}{2\pi}\int_{\Omega} \log R_{xx}(e^{-\jmath\theta})d\theta| < \infty \ . \tag{4.201}$$

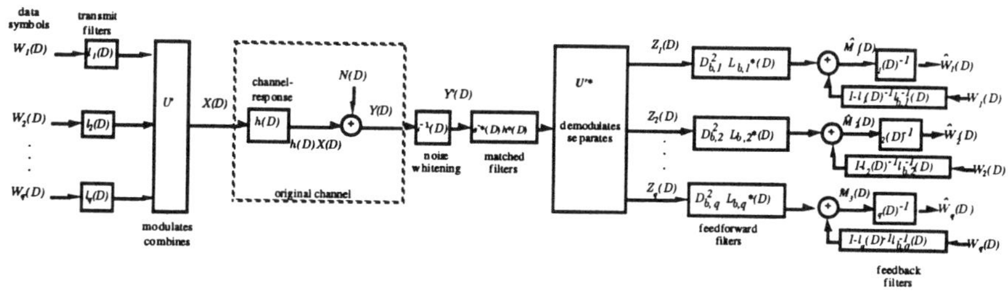

Figure 4.7 Singular GDFE in the limit.

With the GDFE, this situation is illustrated much more clearly than in [1]. The transmit filter U' of the canonical channel models combines the various bands via interpolation and translation. Translation in frequency is a unitary matrix operation. Recall U' was $m \times r_m$ "unitary" matrix, thus allowing for interpolation of the input to effectively a higher sampling rate for the combined signals The matrix L does not converge to a single filter, but rather essentially becomes triangular with disjoint blocks, each of which internally exhibits the convergence of the rows to the innovations filter for the corresponding band. U' then combines these signals into an aggregate (cyclostationary) packet transmit signal.

This situation is depicted in Figure 4.7.

6.3 Vector Coding to Multitone

The VC case, as in Section 5, corresponds to the forward canonical model

$$\boldsymbol{Z} = R_f \boldsymbol{M} + \boldsymbol{N}'' \ . \tag{4.202}$$

for which the GDFE is both canonical and ML if the input $\boldsymbol{M}$ is already, or is decomposed by, a modal decomposition

$$\boldsymbol{M} = V\boldsymbol{M}' \tag{4.203}$$

where both $\boldsymbol{M}$ and $\boldsymbol{M}'$ have full rank $r_m \leq m$. Singularity is trivially handled by U' in the VC case as it corresponds to ignoring subchannels for which $\lambda_{f,i} = 0$ or for which $\lambda_{m,i} = 0$.

In the limit as packet length goes to zero, Toeplitz distribution arguments ([24]) lead to the limit

$$\begin{aligned}\lim_{m\to\infty} \log_2 |\boldsymbol{SNR}_{\mathrm{ML}}|^{1/m} &= I(X;Y) \\ &= \lim_{m\to\infty} \frac{1}{m} \sum_{i=1}^{m} \log_2\left(1+\lambda_{m,i}|\lambda_{f,i}|^2\right) \\ &= \frac{1}{2\pi}\int_{\Omega} \log_2\left(1+\frac{S_x(\theta)|H(e^{-\jmath\theta})|^2}{R_{nn}(e^{-\jmath\theta})}\right) d\theta \quad (4.204)\end{aligned}$$

The vector-coding system becomes equivalent to a "multitone" transmission system as packet length goes to infinity. Thus, the GDFE and VC converge to the highest performance levels possible, namely a data rate possibility of $I(X;Y)$ if good known codes for the AWGN are applied. Both must use the same frequency bands and the channel is always commutative at infinite length.

6.4 Infinite-length Transmit Optimization

The well-known water-filling energy distribution [24] [1] satisfies

$$S_x(\theta) + \frac{R_{nn}(e^{-\jmath\theta})}{|H(e^{-\jmath\theta})|^2} = \kappa \; . \quad (4.205)$$

The solution must exhibit $S_x(\theta) \geq 0$. There is a band Ω^* such that for all $\theta \in \Omega^*$, $S_x(\theta) > 0$. When $|\Omega^*| = 2\pi$, an innovations representation of the thus stationary input can be found through the canonical factorization

$$R_{xx}(D) = l(D) S_w l^*(D^{-*}) \; . \quad (4.206)$$

Then, $l(D)$ is the stationary MMSE-DFE transmit filter that acts on the input data innovations $w(D)$ to produce the proper water-fill spectrum of the channel input sequence $x(D)$. When $S_x(\theta) = 0$ over a measurable band, then separate MMSE-DFE's should be applied to each of the measurable frequency bands for which $S_x > 0$ for all but a countable number of discrete points. The bit rate for each connected subregion of Ω^*, and the GDFE will converge to a constant on all dimensions used by water-filling that correspond to a connected subregion. Each band may have a different symbol rate (equal to the measure of the corresponding connected region of used frequencies) and possibly a carrier-frequency (corresponding to the center frequency of each such band). In effect, one independently designs a MMSE-DFE and takes limits for each of the connected sub-bands of Ω^*. The limiting case of the GDFE is the infinite-length canonical transmission structure called the MMSE-DFE in [1] in each of the optimum bands of Ω^*, which is used by either the VC GDFE for which the feedback section is trivially zero or for the packet GDFE for which feedback sections are nontrivial.

7 SUMMARY AND CONCLUSION

The concept of canonical transmission has been refined to characterize systems that may not be optimum detectors, but for which nevertheless the highest possible data rates may be transmitted with the careful application of the same good codes that near capacity on the ideal additive white Gaussian noise channel. The GDFE structure is a generalization of decision feedback that allows for any characteristic representation of an input and derives from canonical forward and backward channel models that remove unnecessary dimensions and force nonsingular transmission over only those dimensions that can carry information. Various forms of the GDFE, corresponding for instance to an innovations representation of the input, i.e., the "packet GDFE," or to a modal representation of the input, otherwise known as Vector Coding. The VC case is indeed very special, because it is both canonical and optimal and the feedback section of the GDFE trivially disappears, avoiding the need for precoding methods. The VC case, however, must use only special inputs that commute with the forward channel characterization matrix R_f while the GDFE exists in general when this condition is not met. Other characteristic representations could also be used to form other types of GDFE's.

The GDFE is always canonical. The GDFE, however, is not equivalent to the fixed DFE's in common use in data transmission, the latter of which are decidely suboptimum and not canonical unless special conditions hold that are often not met. For this reason, the GDFE is the preferred method for high-performance design of transmission on channels with ISI and additive Gaussian noise. Various methods can be used to simplify a GDFE, most notably the elimination of the feedback section with the Vector-Coding GDFE, which can be further simplified through the use of fast Fourier Transform methods in the implementation known as DMT.

Other areas of simplification of the GDFE remain open to study in addition to the study of specific performance differences on various channels, which can run from very small to very large. The existence of a packet channel model $\boldsymbol{Y} = H\boldsymbol{X} + \boldsymbol{N}$ has been postulated and indeed is a research topic in itself as to appropriate ways to synthesize a channel design such that this relationship holds exactly or approximately.

Acknowledgement: The authors would like to thank Dr. R. Gallager of MIT and Dr. C. Rohrs of Tellabs Research for their contributions to some of the basic concepts in this paper. The first author would like to acknowledge a series of conversations with Dr. Paul Algoet in 1990 that precipitated a few of the concepts in this paper.

Both authors would like to dedicate this paper to Dr. Thomas Kailath of Stanford University on his 60th Birthday. Through Prof. Kailath's efforts, many of the concepts used in this paper have been popularized and developed over his long and distinguished career. The first author also adds his appreciation for the profound impact that TK has had on his career and for the patient and continued support through many stages of progress in the understanding of all aspects of information systems.

REFERENCES

[1] J.M. Cioffi, G.P. Dudevoir, M. V. Eyuboglu, and G.D. Forney. "MMSE Decision-Feedback Equalizers and Coding – Parts I and II". *IEEE transactions on Communications*, 43(10):2582–2604, October 1995.

[2] T.M. Cover and J.A. Thomas. *"Elements of Information Theory"*. Wiley & Sons, New York, 1991.

[3] M.L. Doelz, E.T. Heald, and D.L. Martin. "Binary Data Transmission Techniques for LInear Systems". *Proceedings of the IRE*, 45:656–661, May 1957.

[4] J.A.C. Bingham. "Multicarrier Modulation for Data Transmission: An Idea Whose Time has Come". *IEEE Communications Magazine*, 28(4):5–14, April 1990.

[5] A. Ruiz, J.M. Cioffi, and S. Kasturia. "Discrete Multiple Tone Modulation with Coset Coding for the Spectrally Shaped Channel". *IEEE Transactions on Communications*, 40(5), May 1992.

[6] J.C. Rault, D. Castelain, and B. Le Floch. "The Coded Orthogonal Frequency Division Multiplexing (COFDM) Technique, and Its Application to Digital Radio Broadcasting toward Mobile Receivers". In *Proceedings of Globecom 1989*, Dallas, TX, November 1989.

[7] American National Standards Institute (ANSI). "Metallic Interfaces for Asymmetric Digital Subscriber Lines (ADSL)". In *ANSI Standard T1.413*, Washington, D.C., December 1995.

[8] G. Plenge. "DAB - A new sound broadcasting system – Status of the development – Routes to its introduction". *European Broadcasting Union Review*, (246), April 1991.

[9] Per Applequist. "HD-DEVINCE, a Scandinavian terrestrial HDTV project". In *Proceedings 1993 National Association of Broadcasters*, Las Vegas, NV, April 1993.

[10] S. Kasturia, J. Aslanis, and J.M. Cioffi. "Vector Coding for Partial-Response Channels". *IEEE Transactions on Information Theory*, 36(4):741–762, July 1990.

[11] L.C. Barbosa. "Maximum Likelihood Sequence Estimators: A Geometric View". *IEEE Transactions on Information Theory*, 35(2):419–427, March 1989.

[12] N. Al-Dhahir and J.M. Cioffi. "MMSE Decision-Feedback Equalizers: Finite-Length Results". to appear, IEEE Transactions on Information Theory, 1995.

[13] N. Al-Dhahir and J.M. Cioffi. "Optimal Finite-Complexity Transmit Filters for Packet-Based Data Transmission on Dispersive Channels with Application to the FIR MMSE-DFE". to appear, IEEE transactions on Information Theory, May 1993.

[14] G.K. Kaleh. "Channel Equalization for Block Transmission Systems". *IEEE Journal on Selected Areas in Communication*, 13(1):110–121, January 1995.

[15] N. Zervos, S. Pasupathy, and A. Venetsanopoulos. "The Unified Decision Theory of Non-Linear Equalization". In *Proceedings, IEEE 1984 Globecom*, pages 683–687, Atlanta, December 1984.

[16] M. Austin. "Decision Feedback Equalization for Digital Communication over Dispersive Channels". *M.I.T. Research Lab Electronics Technical Report. 461*, August 1967.

[17] C.A. Belfiore and J.H. Park Jr. "Decision Feedback Equalization". *Proceedings IEEE*, 67(8):1143–1156, August 1979.

[18] R. Price. "Nonlinearly Feedback-Equalized PAM vs Capacity for Noisy Filter Channels". In *International Conference on Communications*, June 1972. 22-12 - 22-17.

[19] J. Salz. "Optimum Mean-Square Decision Feedback Equalization". *Bell System Technical Journal*, 52(8):1341, October 1973.

[20] E.A. Lee and D. G. Messerschmitt. *Digital Communications.* Kluwer, Boston, 1988.

[21] C.L. Lawson and R. J. Hanson. *Solving Least Squares Problems.* Prentice-Hall, Englewood Cliffs, NJ, 1974.

[22] T. Kailath. *Lectures on Least-Squares Estimation.* Springer-Verlag, New York, 1976.

[23] R.E. Blahut. *Principles and Practice of Information Theory.* Addison-Wesley, Menlo Park, CA, 1987.

[24] R.G. Gallager. *Information Theory and Reliable Communication.* Wiley, New York, NY, 1968.

[25] R. Wesel and J.M. Cioffi. "Fundamentals of Coding for Broadcast OFDM". In *Proceedings of the 29th Asilomar Conference on Signals, Systems, and Computers*, Pacific Grove, CA, October 1995.

5

PACKET RADIO MULTIPLE ACCESS

Norman Abramson

ALOHA Networks, Inc.
521 Lake Street
San Francisco, CA 94118
e-mail: alohanet@delphi.com

Dedicated with ALOHA to Tom Kailath on the occasion of his 60th birthday.

1 INTRODUCTION

Wireless random access channels can provide communications from many terminals to a single hub station. Examples of such channels occur in cellular networks, personal communication networks (PCN's), packet radio networks, local area networks (LAN's) and very small aperture terminal (VSAT) satellite networks.

Two random access techniques have been used in wireless networks when the number of terminals is large and rapidly varying. The first of these techniques involves the use of ALOHA channels while the second uses Code Division Multiple Access (CDMA). Conventional ALOHA networks have the advantage of simplicity but cannot easily be used in wideband networks. CDMA is much more complicated to implement but allows the use of wide bandwidth spread spectrum operation and low transmitter power.

In this paper we explain the close relationship between these two multiple access architectures. And we describe a new advanced multiple access technique, Spread ALOHA, which combines the proven simplicity and operational flexibility of ALOHA with the high bandwidth and low power of spread spectrum CDMA.

2 ALOHA

2.1 ALOHA Throughput

In an ALOHA multiple access channel packets are buffered at each terminal and transmitted over a common channel to the hub station. No control is imposed on the channel in order to synchronize transmission from the various users, and therefore the start times of packets from different users in the network can be modeled as a Poisson point process. Under these circumstances packets from different users will be transmitted with a high probability of success if there is a moderate amount of traffic on the network. As the traffic on the network increases the probability of a collision between packets from different users increases. The pessimistic assumption that a collision results in the loss of two packets is usually made in the analysis of an ALOHA channel. Using this assumption we can derive a relation between the total traffic on the network and the throughput of the network. If we let G be the total traffic expressed as a fraction of the maximum possible data rate and S be the channel throughput expressed as the same fraction, then the throughput will vary as a function of the offered traffic.

In figure 5.1 we show G as the fraction of time covered by the line segments for all transmitted packets, while S is the fraction of time covered by the line segments for all correctly received packets.

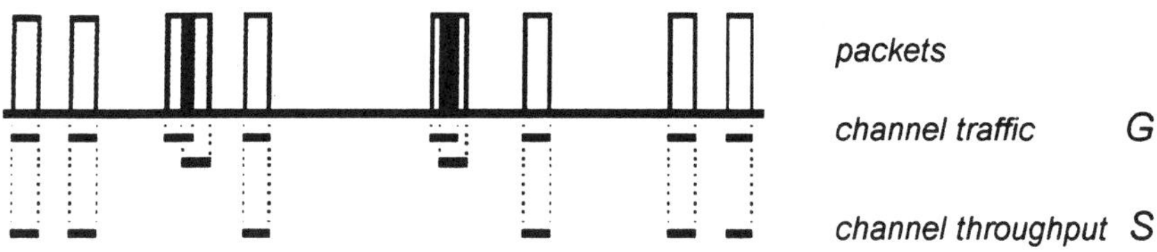

Figure 5.1 Definition of the ALOHA Traffic (G) and Throughput (S).

The relationship between the throughput, S, and the traffic, G, for an ALOHA channel is given by [1].

$$S = Ge^{-2G} \tag{5.1}$$

The relationship is shown in figure 5.2. From figure 5.2 we see that the ALOHA throughput reaches a maximum value of 0.186 when the value of the traffic is 0.5.

2.2 ALOHA Efficiency

The throughput of the channel however is not necessarily the most appropriate figure of merit for a multiple access channel. The throughput of a channel is simply the fraction of time during which the channel can be used to transmit data. In some cases, such as average power limited satellite channels or battery operated transmitters, the average data rate of the channel for a fixed transmitter power and a fixed bandwidth is a more appropriate figure of merit.

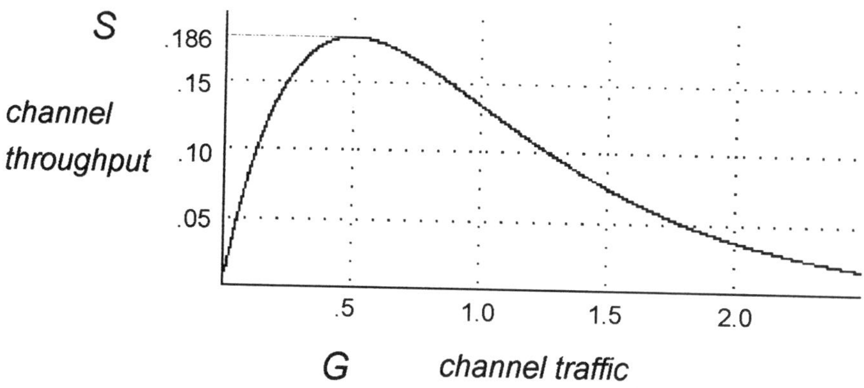

Figure 5.2 ALOHA Throughput (S) vs. Traffic(G).

We can define a different figure of merit for multiple access channels, called the *efficiency* of the channel, which takes into account the system resources of average power and bandwidth. When these channel resources are taken into account the picture of ALOHA efficiency which emerges is much different from that of ALOHA throughput.

The transmission of packets in a typical ALOHA channel is shown in figure 5.3.

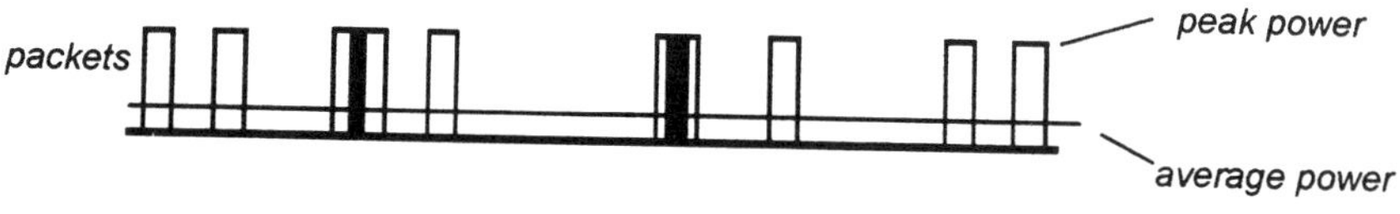

Figure 5.3 Packets in an ALOHA Channel

From figure 5.3 we see that the average power in the channel can be much less than the peak power (or the average power during a packet). If the average power in the channel is P, then the peak power is P/G, where G is the channel traffic. This higher power level during the transmission of a packet can compensate, in part, for the fact that the throughput of the channel is less than one.

We restrict our interest in these arguments to values of channel traffic, G, less than one, so that the terminology "average power" and "peak power" makes sense. Although the theoretical results we derive apply equally well for larger values of G, such values of channel traffic, G, is usually in the range of 0.10 or less; in these situations there can be more than 10 db. difference between the peak and average powers.

The capacity of the additive white Gaussian noise channel is given by the well known Shannon equation

$$C = W \log \left(1 + \frac{P}{N}\right) \text{ bits/second} \tag{5.2}$$

where W is the bandwidth of the channel in Hertz, and P/N is the average signal to noise power ratio of the channel. For an ALOHA channel transmitting with a throughput of S we calculate the multiple access channel capacity C_a by multiplying the capacity expression in equation (5.2) by S and replacing P in equation (5.2) by P/G, the average power during the transmission of a packet.

$$\begin{aligned} C_a &= SW \log \left(1 + \frac{P}{GN}\right) \\ &= Ge^{-2G} W \log \left(1 + \frac{P}{GN}\right) \end{aligned} \tag{5.3}$$

We can define r, the efficiency of the ALOHA multiple access channel as the ratio of the ALOHA channel capacity to the capacity of the continuous channel using the same average power and the same total bandwidth.

$$r = \frac{C_a}{C} = \frac{Ge^{-2G} \log \left(1 + \frac{P}{GN}\right)}{\log \left(1 + \frac{P}{N}\right)} \tag{5.4}$$

The efficiency is plotted as a function of the channel traffic, G, for various values of the average channel signal to noise ratio, P/N, in figure 5.4.

In the original ALOHANET where the individual transmitters were not average power limited the throughput was an appropriate figure of merit for the channel. In a multiple access channel which is average power limited, such as a battery operated terminal or a satellite transponder, the efficiency of the channel protocol is a more appropriate figure of merit.

From equation 5.4 and figure 5.4, we see that the channel efficiency of an ALOHA channel approaches one for the important case of small values of throughput and small values of the signal to noise power ratio. In other words, under these conditions it is not possible to find a multiple access protocol which has a higher capacity for a given value of average power and a given bandwidth.

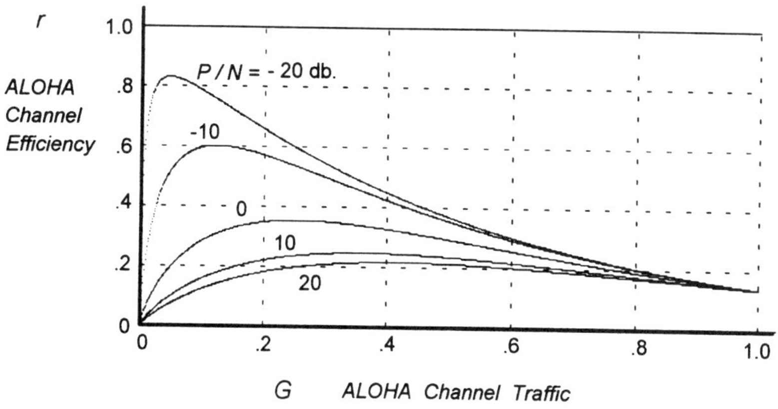

Figure 5.4 Efficiency of an ALOHA Multiple Access Channel

2.3 Narrowband Limitations of ALOHA

In theory conventional ALOHA multiple access can be scaled in frequency so that the results of section 2.1 and 2.2 apply without regard to the bandwidth available for the ALOHA channel. In practice however there is a problem in attempting to use high bandwidth ALOHA channels.

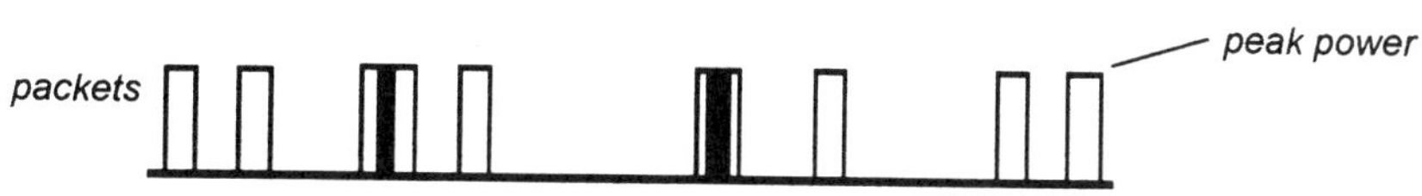

Figure 5.5 Peak Power in an ALOHA Channel

As the channel bandwidth increases the duration of each packet decreases. This decrease of course will decrease the number of collisions in the channel and allow the channel to be used at a higher peak data rate. However in order to maintain the same bit error probability as the packet duration decreases it is necessary to increase the signal power to maintain a constant energy per bit. This requirement on the transmitter power is the same problem which constrains the data rate of any TDMA system, for ALOHA may be looked on as a form of random TDMA. Because of the constraint on peak transmitter power the average data rate of most ALOHA channels is limited to a few tens of kilobits per second. (For the case of the CSMA/CD protocol used in Ethernet, based on ALOHA, this limitation does not apply because transmitter power is not a problem in a cable based channel.)

Thus while ALOHA is a proven technology based upon a simple to implement contention principle, the implementation of ALOHA at high data rates has always been a problem. For many of the applications of interest today this narrow band limitation is the primary obstacle to the use of an ALOHA based multiple access protocol.

3 CODE DIVISION MULTIPLE ACCESS (CDMA)

3.1 A CDMA representation

The transmission of a signal from a single user in a multiple access CDMA channel can be described in terms of a complex signal representation as shown in figure 5.6.

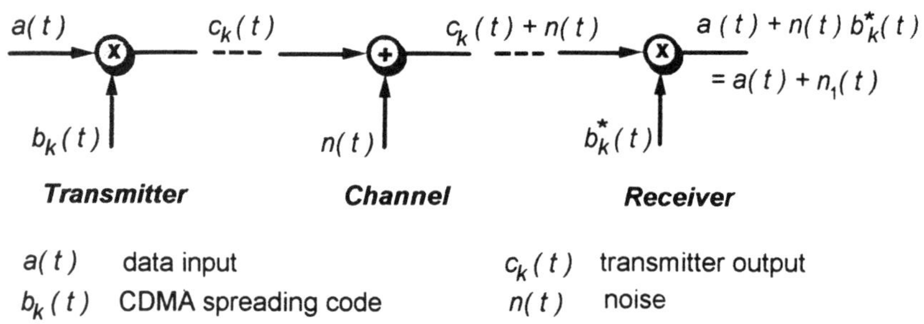

Figure 5.6 Single User Transmission in a CDMA Channel

The CDMA process is represented [2] as the multiplication of input signal $a(t)$ by a different coding signal, $b_k(t)$, for each user. The resulting transmitted signal is

$$c_k(t) = a(t)b_k(t) \tag{5.5}$$

If we require that $b_k(t)$ the CDMA code spreading signal for this user satisfies

$$|b_k(t)|^2 = 1 \tag{5.6}$$

then the fundamental operation of the Spread Spectrum receiver is to multiply the received signal plus noise, $c_k(t) + n(t)$, by the complex conjugate of the CDMA code spreading signal, $b_k^*(t)$, in order to recover the original information in $a(t)$.

$$[c_k(t) + n(t)]\, b_k^*(t) = a(t) + n(t)b_k^*(t) \tag{5.7}$$

This representation is general enough to include conventional forms of modulation, when $b_k(t)$ is a narrowband. But in the case of CDMA, the modulating carrier, $b_k(t)$, is a different wideband carrier for each possible terminal in the network. This multiplicity of possible transmitter carriers is reflected in a multiplicity of receivers required at a CDMA based station in order to demodulate the received signal. Clearly if the total number of potential data terminals in the network is much larger than the number of terminals active at any given time this requirement can introduce considerable complexity into the design of a CDMA multiple access system.

3.2 Commercial CDMA Networks

Only two significant commercial CDMA networks have been built so far. The design of each of these two networks addressed the problem posed by the requirement of a separate receiver for each possible transmitter in a CDMA network in a different way. The first of these networks was the C-200 CDMA network put into operation by Equatorial Communications in 1984. The Equatorial network provided two way data communications from micro earth terminals to a hub station.

Figure 5.7 C-200 CDMA Network Configuration

In the C-200 CDMA network each station in the network was assigned a fixed spreading code. A separate filter, called an "ear card", matched to each spreading code was installed at the hub for each station. The network worked well when it was first set up with a small number of stations. But as the number of stations in the network grew the logistics of installing and maintaining a large number of ear cards in the hub station made the operation of the network more and more difficult. The network is no longer in operation.

The second significant commercial CDMA network to go into operation is the IS-95 implementation of CDMA for cellular telephone networks. In an IS-95 network the problems associated with installing a separate "ear card" in the hub for each possible transmitter are handled by software rather than by hardware as in the C-200 network.

In an IS-95 network an ear card for each station is replaced by a bank of 64 programmable ear cards or programmable correlators. The number of possible stations in the network is much larger than 64 although 64 is an upper bound to the number of stations which can be authorized to transmit at any one time. In practice the number of channels which can be assigned to transmitters at any one time is somewhere between 20 and 40. The stations are assigned a fixed spreading code as in the C-200 network. Rather each station is assigned a random code by the hub after the station initiates a request for a channel. A simplified representation of this class of CDMA architecture is shown in figure 5.8.

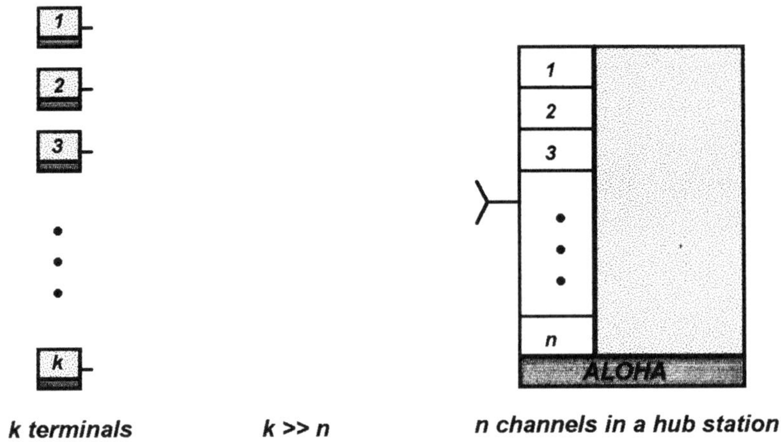

Figure 5.8 Simplified Representation of a CDMA Network Architecture

This architecture solves the problem of linear growth of the network as the number of possible stations which can transmit increases, but two other problems are introduced. First there is a time delay introduced by the process of setting up a channel, including selecting and synchronizing the spreading code at the hub and at the station (said to be in excess of 3 seconds). Second, a separate channel has to be installed in the network so that a station can request a spreading code when it wishes to establish a channel. In the IS-95 architecture this request channel is an ALOHA channel.

3.3 CDMA Network Limitations

The network architecture we have described transforms the random access packet nature of the CDMA channel into a demand assigned circuit switched network with dedicated channels assigned to a given station during its active period. In satellite networks this architecture is usually referred to as a DAMA (Demand Assigned Multiple Access) system although this designation is not common in the cellular industry where IS-95 is employed. For voice traffic in cellular networks the setup delay and overhead, and the circuit switched nature of the network is not a major problem. However, these considerations are important in digital networks which are designed to handle transaction traffic and file traffic as well as digitized voice.

Thus while a CDMA Spread Spectrum network of the sort used in IS-95 has the bandwidth to provide the high speed data services of interest for today's data applications, the IS-95 architecture we have described cannot effectively service such applications. While there is little double the frequency reuse capabilities of CDMA provide an advantage over TDMA for a network of voice traffic, it is hard to see how the properties of a CDMA network architecture can be made compatible with a network serving transaction traffic, intermittent file traffic or a combination of such traffic with digital voice traffic.

4 SPREAD ALOHA

4.1 Integrated Services Data for Wireless Networks

From section 2.3 we see that conventional ALOHA can provide true packet switched operation for wireless networks but only at a relatively low bandwidth. From sections 3.2 and 3.3 we see that spread spectrum CDMA has the higher bandwidth required for today's applications but that a true packet switched CDMA network is not practical for large numbers of transmitters while a circuit switched form of CDMA is not practical for a diverse mix of user applications. In this section we describe how a second generation version of ALOHA can combine the proven simplicity and operational flexibility of ALOHA with the high bandwidth and low power of spread spectrum CDMA to provide true packet switched operation in a wideband wireless medium.

4.2 Combining ALOHA and Spread Spectrum

Conventional spread spectrum CDMA signal detection of the sort described in figure 5.8 involves a bank of correlators or matched filters, b_j, where each filter is matched to the distinct spreading sequence of a different transmitter.

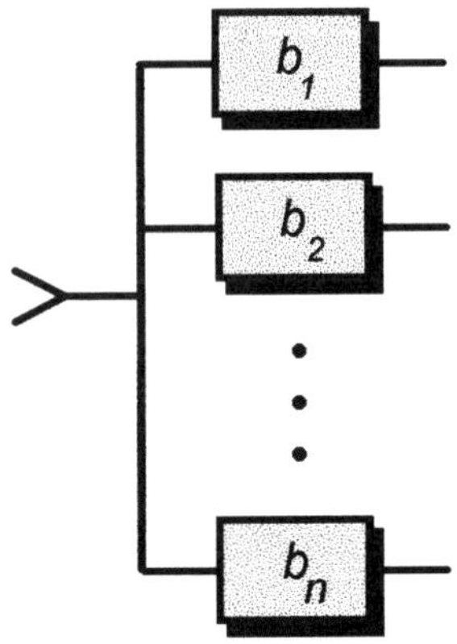

Figure 5.9 Conventional CDMA Signal Detection.

Each branch of figure 5.9 can be implemented as either a correlator or a matched filter, matched to the spreading sequence for each transmitter. The spreading gain, g, of the CDMA channel can be defined as the number of Nyquist samples per individual bit in the system. If we choose the matched filter implementation, the output of a typical branch will appear as shown in figure 5.10, with the output for each bit in a packet appearing at regularly spaced intervals separated by g Nyquist samples, or chips.

Figure 5.10 Output of a CDMA Matched Filter.

The bit sequence for the output shown in figure 5.10 is *10100* ... If two packets from different transmitters are present in the channel the output of the matched filter matched to each of the two signal will appear similar to figure 5.10 since the different spreading sequences will tend to suppress each signal in the other matched filter as long as the total number of packets in the channel at the same time is not too large. A number of results [[3],[4],[5]] indicate that not too large in this context means that the number of simultaneous transmitters should not exceed 0.10 g to 0.20 g, where g is the spread spectrum spreading gain used in the channel.

Note however that the use of different codes for different transmitters is not essential to the operation of a multiple access capability in the channel. Indeed if we use the same spreading code for all of the transmitters in the channel we can still separate packet transmissions due to different transmitters as long as the chip phase difference between two transmitters is large enough. For example, take the case of exactly two packet transmitters using the same spreading code, but offset by p Nyquist samples. Then the output of a single matched filter, matched to the spreading sequence of both of the transmitters is shown in figure 5.11.

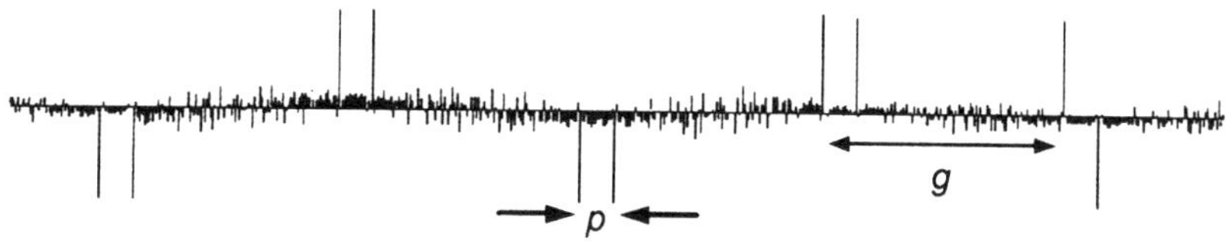

Figure 5.11 Output of a Spread ALOHA Matched Filter.

We call this form of multiple access, *Spread ALOHA*. In figure 5.11 the bit sequence for the first packet shown is *01011* ... while the sequence for the second is *01010* ...

Spread ALOHA can be thought of as a version of Code Division Multiple Access (CDMA) without Code Division. Indeed from figure 5.11 we see that different codes are not necessary to distinguish transmitted signals from different transmitters in a multiple access channel. Rather the multiple access capability of such a channel derives from the high bandwidth spread spectrum nature of the signal rather than from any form of code division. In the case of Spread ALOHA channel the throughput limit and the efficiency limit of the channel are the same as that of a conventional ALOHA channel and equations (5.1) and (5.4) apply.

REFERENCES

[1] Norman Abramson, "The Throughput of Packet Broadcasting Channels", *IEEE Transactions on Communications*, **25**(1):117-128, January 1977.

[2] Marvin K. Simon, Jim K. Omura, Robert A. Scholtz and Barry K. Levitt, *Spread Spectrum Communications, Volumes 1, 2 and 3*, Computer Science Press, Rockville, Maryland, 1985.

[3] George Turin, "The Effects of Multipath and Fading on the Performance of Direct-Sequence CDMA Systems", *IEEE Journal of Selected Areas in Communications*, **2**(4):597-603, July, 1984.

[4] Pickholtz, et al., "Theory of Spread Spectrum Communications — A Tutorial", *IEEE Transactions on Communications*, **30**(5):855-884, May, 1982.

[5] Joseph Y. N. Hui, "Throughput Analysis for Code Division Multiple Accessing of the Spread Spectrum Channel", *IEEE Journal of Selected Areas in Communication*, **2**(4):482-486, July, 1984.

6

THE EVOLUTION OF MOBILE COMMUNICATIONS

Arogyaswami Paulraj

Information Systems Laboratory
Stanford University
Stanford, CA 94305

Dedicated with affection, admiration and appreciation, on the occasion of his sixtieth birthday, to Professor Thomas Kailath - a devoted scholar, an inspired teacher and a generous mentor.

ABSTRACT

One early application of the wireless telegraph developed by Marconi was ship-to-shore communications. This maritime version of mobile communications finally evolved into the ubiquitous cellular networks of today through a series of technology breakthroughs. This paper traces the evolution of this fascinating technology.

1 INTRODUCTION

It is a pleasure to contribute this short history of mobile communications to the proceedings of the "International Conference on Communications, Control and Signal Processing" held in June, 1995 on to celebrate Prof. Thomas Kailath's 60th birthday. It has been a pleasure and a privilege to have known Dr. Kailath for nearly 25 years, initially as his student and later as a colleague. His great accomplishments are an inspiration to us all. Prof. Kailath is well known for his many contributions. Most admirable, perhaps, has been his recent efforts in using advanced mathematical techniques in lithography and process control for semiconductor manufacturing. Dr. Kailath has successfully applied new mathematical approaches into areas traditionally dominated by empirical methods. I wish him many more decades of fruitful research.

1995 is the 100th anniversary of the invention of the radio by that extraordinary pioneer - Guglielmo Marconi. Marconi's thoroughness in conduct of experiments, his emphasis on a deeper understanding of the underlying science, his enthusiasm, his commitment to making things work, and his entrepreneurial spirit has given us an inestimable heritage. It is therefore appropriate to trace Marconi's early work in radio before we embark on the more recent history of mobile communications.

2 MARCONI AND THE BIRTH OF RADIO

Early origins of radio date back to 1861 when, J.M.C. Maxwell while at King's College in London, proposed a mathematical theory of electromagnetic waves. A practical demonstration of the existence of such waves was done by H. Hertz in 1887 at Univ. of Karlsruhe using stationary or standing waves. Improvements in the generation and reception of electromagnetic waves was pursued by many workers in Europe. In 1890, Branly in Paris developed a "coherer" that could detect the presence of electromagnetic waves using iron filings in a glass bottle. The coherer was further refined by A. Righi at Univ. of Bologna and O. Lodge in England. Other contributions came from Popov in Russia who probably devised the first radio antenna in his attempts to detect electromagnetic radiation from lightings.

Figure 6.1 Marconi and his patented wireless telegraph, London 1986

Marconi, at the young age of 21, inspired by the lectures on radio waves by Prof. Righi at Univ. of Bologna, built and demonstrated the first radio telegraph in summer of 1895. He used Hertz's spark transmitter, Lodge's coherer and added antennas to assemble his instrument. In the fall of 1895, he was able to communicate from a transmitter at his home in Villa Grifoni in Bologna to a receiver behind a nearby hill and thus show that "wireless" telegraphy was indeed a reality. He tried to interest the Italian Government in his invention but without success. His mother, who was Irish, took him to England to pursue his invention, in a more conducive environment. Marconi sailed for England in February 1896 where he was introduced to Sir William Preece, then Chief Engineer of Post and Telegraph. Preece soon became a strong supporter. With Preece's help Marconi began successful experiments, first in the Salisbury plains and later across the Bristol channel. In 1897 the Italian Government invited Marconi back home for a demonstration, and he successfully transmitted signals across water over a distance of 18 km. Returning to England later that year, he founded the Wireless Telegraph Trading Signal Company to build radio receivers for the Royal Navy who had become an early and enthusiastic customer. Credit for the Royal Navy's interest goes to Admiral Battenburg and Captain Johnson. Johnson who had been also pursuing wireless experiments of his own, abandoned them after it became clear that Marconi had a superior technology [1].

In 1898, Marconi improved the telegraph by adding a four-circuit tuning device, allowing simultaneous use of two radio circuits. That year, he bridged the English channel between Wimeraux and Dover, a distance of 52 km. He began to build a successful company, now named Marconi Wireless Telegraph Co. and turned his sights to bridging the Atlantic. He achieved this in 1901 between Poldhu in Cornwall, England and St.John's in New Foundland. His attempts at commercial service was realized a year later with the North American station at Glace Bay in Nova Scotia. His technical developments around this time included the magnetic detector, which was an improvement over the less efficient coherer, the rotary spark, and the use of directive antennas at Glace bay to increase the signal level and to reduce interference into the duplex receiver circuits. Later Marconi adapted many new technologies into his increasingly sophisticated radio equipment, including the diode valve developed by Fleming, the crystal detector, CW transmission developed Poulsen, Fessenden and Alexanderson in the US, and the triode valve or audion developed by Lee De Forrest also in the US.

Marconi, who came to be known as the father of wireless communications, spent his later years as an elder statesman and an inventor, and in 1909 he shared the Nobel Prize for physics. He died in 1937 and was accorded the unique tribute of a two minute silence by all radio stations across the world.

3 EARLY MOBILE RADIO

Civilian use of wireless technology began with the installation of the first 2 MHz land mobile radio telephone system in 1921 by the Detroit Police Department for police car dispatch. The advantages of mobile communications were quickly realized, but its wider use was limited by the lack of channels in low frequency band. Progressively, higher frequency bands were used, opening up

Figure 6.2 Marconi at Glace Bay, 1905

more channels. A key advance came in 1933, when Major Armstrong invented Frequency Modulation (FM), which made possible high quality two-way radio communications.

In 1946 a *Personal Correspondence System* introduced by Bell Systems began service [2] and operated at 150 MHz with speech channels 120 kHz apart. As demand for public wireless services began to grow, the *Improved Mobile Telephone Service* (IMTS) using FM technology was developed by AT&T. These were the first mobile radio systems to connect with the public telephone network using a fixed number of radio channels in a single geographic area. The years 1964 and 1969 saw the introduction of two other systems: the MJ and MK. These systems used an offset frequency plan, but supported very few channels [2].

Extending such technology to a large number of users with full duplex channels needed excessive bandwidth. A solution was found in the cellular concept conceived by D.H. Ring at Bell Laboratories in 1947. Cellularization meant dividing the service area into several smaller cells, and using a subset of the total available radio channels in each cell. Radio channels could therefore be reused in cells that were far enough apart to avoid co-channel interference. For a fixed number of available channels, this implied that the system capacity could be increased by reducing the size of cells, thereby increasing the number of times a radio channel could be reused. Cellular systems needed a *handoff* mechanism which allows a mobile leaving a cell to switch to a new channel available in the next cell. This required techniques for tracking mobile units as they moved from cell to cell.

The implementation of the cellular concept was a regulatory challenge because spectrum in the desirable frequency bands was not available. The regulatory issue was solved when the Federal Communications Commission (FCC) set aside frequencies in the 800 MHz band in 1970. The technical problems of handoff were solved when AT&T proposed the first high capacity cellular telephone system called the *Advanced Mobile Phone Service* (AMPS) in 1970, which implemented handoffs to switch an active call between cells automatically. The design of cellular networks began to mature as the role of propagation, terrain, fading, and shadowing became better understood. An improved version of AMPS is still the primary standard for analog cellular telephones.

4 EVOLUTION IN EUROPE

In Europe, the Scandanavian countries first began to develop radio based communications to cope with a sparsely distributed population. The first mobile communication started in 1978 [3] and began to grow rapidly after the introduction of the *Nordic Mobile Telephony* (NMT) system in 1981. The NMT 450 operated in the 450-470 MHz band. When this channel became crowded, NMT 900 was introduced in the 900 MHz band. NMT had reached a penetration of 6% of the population by 1991 [3] [4].

Parallel to the NMT systems in Scandanavia, a *Total Access Communications System* (TACS) began to be developed in the United Kingdom while the Germans developed a C-system with many advancements over the NMT/AMPS systems. The C-system used digital signaling, but retained analog FM voice modulation and also used advanced power control in the mobile and the base stations to vary cell boundaries to adapt to fluctuating traffic patterns.

4.1 GSM

As analog systems evolved in Europe, each country developed its own set of standards which were incompatible with each other. This was undesirable not only because the mobile equipment was limited to operation within national boundaries, but also since there was a limited market for each type of equipment, making for poor economies of scale.

Europe realized this and in 1982 the Conference of European Posts and Telegraphs (CEPT) formed a study group called the Groupe Spéciale Mobile (GSM) to study and develop a pan-European public land mobile system. In 1989, the GSM standards responsibility was transferred to the European Telecommunication Standards Institute (ETSI), and phase I of the GSM specifications were published in 1990. Commercial service was started in mid-1991, and by 1993 there were 36 GSM networks in 22 countries, with 25 countries considering or having already selected GSM. By the beginning of 1995, there were 5.8 million GSM subscribers worldwide. The acronym GSM now (aptly) stands for *Global System for Mobile* [5].

4.2 GSM Technology

GSM uses TDMA technology with 200 kHz channels and a bit rate of 270 kbps. Eight slots are used per frame with a slot duration of 57 μsecs (see Figure 6.3). GSM uses Guassian Minimum Shift Keying (GMSK) modulation with a Bandwidth-Time product of 0.3. The voice channel uses a 14.1 kbps hybrid coder (RPE-LTP). Frequency hopping and a reuse factor of four provide for good voice quality. Spectrum efficiency in GSM is moderate and is about 2.5 times better than the analog AMPS system [6].

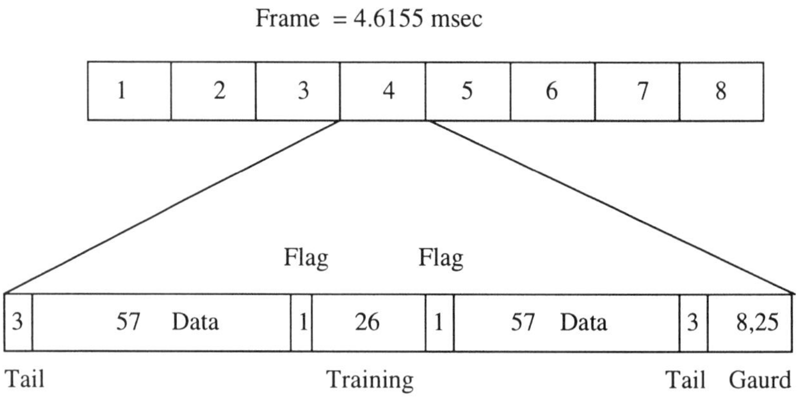

Figure 6.3 GSM slot structure

From the beginning, the GSM planners wanted ISDN compatibility in services offered and in control signaling. The basic teleservice supported is voice telephony. Overlayed is an emergency service (similar to 911), a Group 3 fax, and a unique feature called Short Message Service (SMS). SMS is a bi-directional service for sending short (up to 160 bytes) alphanumeric messages in a store-and-forward method with an acknowledgement of receipt provided to the sender. SMS can also be used in a cell broadcast mode for sending messages such as traffic updates. Supplementary services are provided on top of teleservices or bearer services and include features such as caller identification, call forwarding, call waiting, multi-party conversations, and barring of outgoing (international) calls.

5 EVOLUTION IN THE U.S.

The first commercial service of the AMPS system began in 1983. The U.S. cellular industry began in earnest in 1986 when the FCC divided the country into metropolitan statistical areas (MSAs) and rural service areas (RSAs) and began conducting lotteries for cellular licenses. At present there are 734 cellular service areas, including 428 RSAs and 306 MSAs, with two cellular operators licensed in each RSA or MSA. There are more than 400 carriers operating approximately 1,500 cellular systems. Each carrier operates on 25 MHz of radio spectrum in the band between 824-849 MHz and 869-894 MHz, close to the ultra-high frequency (UHF) TV band. The AMPS system now has nearly 16,000 base stations and serves over 20 million customers. AMPS was a first-generation cellular standard.

5.1 IS-54 - TDMA Digital Standard

In 1990 when the AMPS networks began to reach capacity limits in certain markets, the Telecommunications Industry Association approved a new interim standard, IS-54, to meet the increased demands and ensure compatibility with existing networks. This standard uses narrowband time-division multiple access (TDMA) digital radio technology. TDMA systems carry three digital voice channels on a 30 kHz analog AMPS channel. IS-54 uses a fairly efficient $\pi/4$ DQPSK modulation scheme with a time slot of 6.66 ms (see Figure 6.4). The gross bit rate is 41.6 kbps on the channel. Handoffs in IS-54 are performed with an improved, mobile-assisted technology. IS-54 is considered to be a second-generation cellular standard. Seven percent of the cells in the United States currently deploy IS-54 technology, and this figure is expected to increase to 30% by the year 2000 [7].

IS-54 has been recently revised to IS-136 which incorporates a several improvements including SMS, sleep mode (for battery life extension), and private systems. Future revisions of IS-136 will support over-the-air channel activation.

G	R	Data	Synch	Data	SACCH	CDVCC	Data
6	6	16	28	122	12	12	122

324 bits in 6.66 msec

Figure 6.4 IS-54 slot structure

5.2 CDMA

While capacity needs increased and the demand for improved voice quality grew, new approaches to cellular technology were needed. A standard based on code-division multiple access (CDMA) technology, was developed by Qualcomm Inc., and the new standard (IS-95) was approved in July, 1993. IS-95 uses a 1.25 MHz bandwidth that can be simultaneously used by several subscribers, and its capacity is expected to be ten times higher than AMPS. Unlike AMPS and IS-54, CDMA is a wideband, spread spectrum technology that "spreads" multiple conversations over a wide segment of the cellular broadcast spectrum. Each telephone or data call is assigned a unique code that permits it to be distinguished from the multitude of other calls simultaneously transmitted over the same spectrum. It eliminates frequency planning, which is required in TDMA. IS-95 is considered to be a third-generation cellular standard.

CDMA's origins go back several decades to military communications where its superiority to tolerate jamming made it very popular. In CDMA each user's data is *spread* using a noise-like random sequence. This spread signal occupies a much larger bandwidth than the original signal; hence, these are also known as spread spectrum techniques. Several users may transmit in the same frequency band, but each uses a different spreading code. A particular user's data stream can be extracted from the received signal by correlating it with the user code. This correlation boosts the desired user's signal relative to the other users by a factor known as the coding gain. See Figure 6.5.

In the forward link, the base communicates to multiple users using spreading codes that are mutually orthogonal. The mobile receives the signals from the base station and extracts its own signal by correlating it with the known user code. The signal transmitted from the base may arrive via different paths due to scattering from buildings and hills. This multipath propagation can be used by CDMA networks to some advantage. Paths separated by one microsecond or more appear to be distinct signals and a so-called "RAKE" combiner collects energy from these distinct paths. This exploitation of multipath is a key advantage of CDMA. In non-CDMA systems, multipath often is seen as a problem, rather than as an advantage.

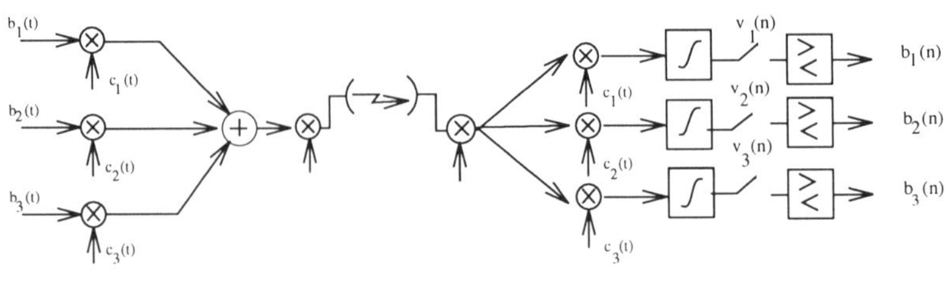

Figure 6.5 Users communicate using different spreading codes

In the reverse link, the voice stream from the mobile user is first digitally coded, then spread by the user's code, and finally transmitted. The base receives signals from a large number of users and extracts individual signals by correlating them with the appropriate code. Of particular importance is power control wherein the signal power of each user must arrive at the base station at the same power level. Failing this, the number of users supported by the system will reduce dramatically [8].

Power Control

The transmission path from the users to the base can show very wide variations in signal attenuation, and moreover, these losses can fluctuate rapidly (more than 100 times a second). Therefore, special techniques are needed to ensure that the transmitted power of mobile users are controlled so that these signals arrive at the base at almost equal power level. This is made possible through several control loops that include open loop and closed loop control techniques.

In open loop control, the mobile estimates the average attenuation in the forward link and adjusts its transmit power in the reverse link accordingly. Open loop control compensates for certain slowly varying components of the path attenuation. The more rapidly changing components caused by mobile motion and multipath are compensated by a closed loop control where the base station checks the received power from the user and sends commands on the forward link to make necessary adjustments.

Soft Handoff

In cellular systems, a call in progress is kept alive as the mobile unit moves out of the current cell and into another cell using handoff. In IS-95 a *soft handoff* technique is used which allows both the original cell and the new cell to temporarily serve the mobile during the transition period. Soft handoff minimizes probability of dropped calls—a bane in most cellular systems—and makes the handoff transparent to the user. In soft handoff the mobile continuously keeps track of the power of the certain signals called pilots from the base stations in its vicinity and then starts communications with a base when its pilot exceeds a set threshold.

CDMA is yet to begin service in the 800 MHz band. Customer service is expected to start in the fall of 1995 in Los Angeles.

6 PCS

In 1992 the FCC allocated additional spectrum in the 1800 MHz band (see Figure 6.6) for Personal Communications Services (PCS), a term encompassing

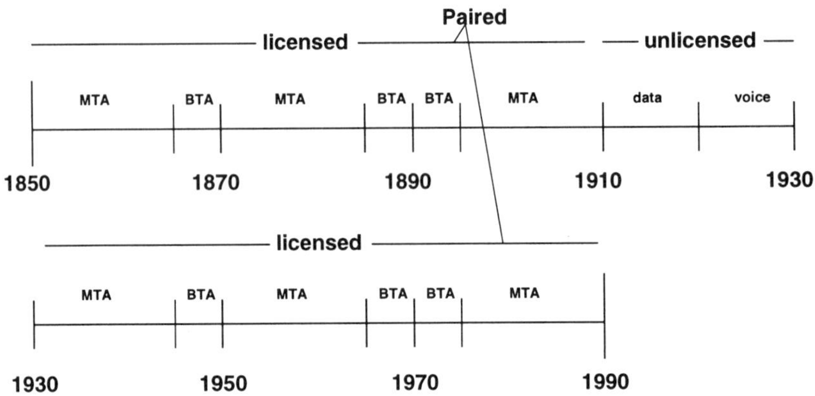

Figure 6.6 PCS Frequency Allocation in the U.S.

a wide range of wireless mobile technologies including cellular, paging, cordless voice, mobile data, and specialized mobile radio (SMR). The opening of this new band is expected to unleash a whole range of new services [9].

The FCC completed the auction of 99 broadband PCS licenses in 51 Major Trading Areas (MTAs) in the summer of 1995. The total bid amount for a total of 99 licenses was over $7 billion. Including the narrowband PCS licenses, the total auction revenue exceeded $8 billion.

PCS standards work has been progressing well and most interim technical standard bulletins have been issued for the PCS band high tier services. TDMA IS-136 and CDMA IS-95 Rev A issued in summer 1995 and have also become ANSI standards. The roaming standard IS-41 has also been revised for the PCS band. Other high tier standards approved for the PCS band include the IS-991 (Omnipoint system) and PCS1800 (the GSM standard modified for the 1800 MHz band). Standards activity for the low tier standards is still in progress.

7 MOBILE COMM. CLASSIFICATION

Though we have largely focused on two-way voice communications, mobile communications has evolved into a variety of services spanning data and voice. Figure 6.7 shows a classification of mobile communications based on transport technology and mobility. With evolution of PCS we should see increasing convergence of such technologies where a single instrument offers a variety of services in multiple environments.

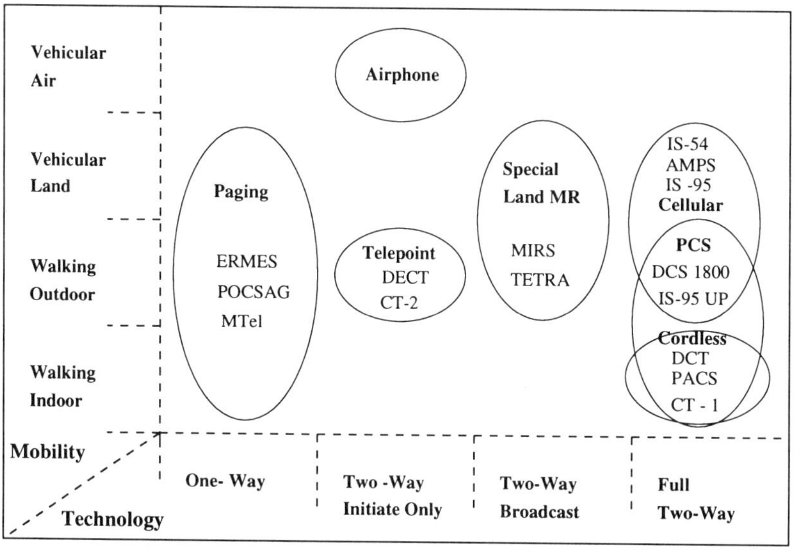

Figure 6.7 Classification of Mobile Communications

8 MOBILE COMM. - MARKET GROWTH

Figure 6.8 shows projections for growth in mobile communications revenue. Clearly, wireless is likely to remain a major growth area in telecommunications for another decade. PCS is slated for the major growth in the developed world, while wireless local loop is likely to see explosive expansion in the developing countries [10] [11].

At the end of 1994, Scandanavian countries had the highest cellular phone penetration of 15% compared to the U.S. level of 9%. The world total of cellular subscribers today is estimated to be 60 million, a 55% increase over the past year. The United States has about 25 million cellular subscribers today. The cellular plus PCS market in the US is expected to grow to 60 million subscribers in the next 6 to 8 years.

9 SUMMARY

Mobile communications is at the threshold of tremendous growth worldwide. The technology offers many advantages to an increasingly mobile population. New technologies are needed to support a variety of services in an integrated and cost-effective manner. Multi-mode instruments that can offer voice, data, and even video services to subscribers who may be sitting or walking indoors, moving around in public places like shopping malls, traveling in high speed cars

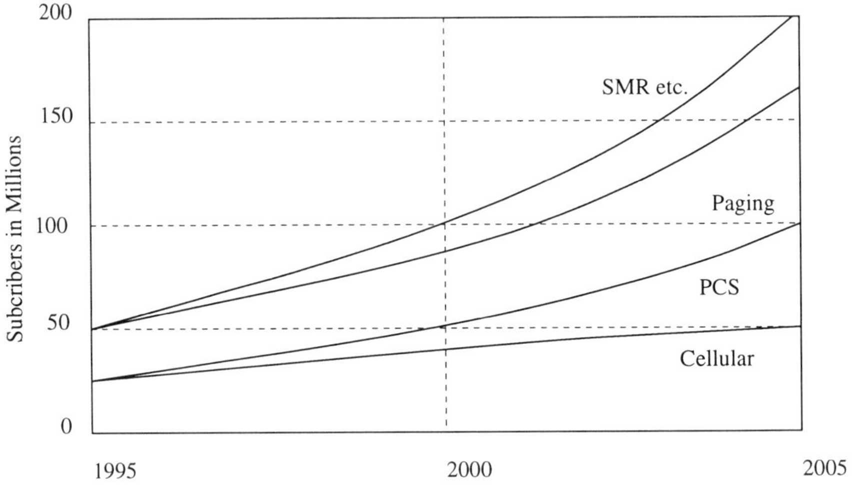

Figure 6.8 Current and projected revenues from various forms of mobile communications

or aircraft, or even camping out in wilderness areas, pose a great challenge. Equally important is international standardization so that mobile users can move seamlessly across countries. While this dream is within reach in Europe, worldwide mobility is still some years away. The role of CCIR through the Future Personal Land Mobile Telecommunication Service (FPLMTS) program offers a vision for such a future, a vision worthy of Marconi's great and precious legacy.

REFERENCES

[1] Dunlap, O. E., *Marconi: The Man and his wireless*, New York: Macmillan, 1938.

[2] Young, W. R., "Advanced mobile phone serive: Introduction, background and objectives," *The Bell System Technical Journal*, January 1979.

[3] Hellstrom, K. and A. Lundqvist, "Trends in mobile communications," *Ericsson Review*, vol. 3, 1991.

[4] Hild, S., "A brief history of mobile communications," Tech. Rep. 372, Univ. of Cambridge, January 1995.

[5] Hodges, M. R. L., "The GSM radio interface," *British Telecom Technological Journal*, vol. 8(2), January 1990.

[6] Lee, W. C., *Mobile Communications Design Fundamentals*, New York: John Wiley, 1993.

[7] Chang, J. J. C., R. A. Miska, and R. A. Shober, "Wireless systems and technologies: An overview," *AT&T Technical Journal*, July/August 1993.

[8] Anonymous, "An overview of the application of code division multiple access (cdma) to digital cellular systems and personal cellular networks," Tech. Rep. EX60-10010, Qualcomm, Inc., May 1992.

[9] Anonymous, "Notice of enquiry on pcs," Tech. Rep. Docket 90-314, Federal Communications Commission, October 1991.

[10] Paulraj, A., "Wireless Local Loop for Developing Countries: A Technology Perspective," in *Annual Review of International Engineering Consortium*, Chicago: Intl. Engg. Consortium, 1995.

[11] Saunders, R. J. J. Warford, and B. Willenius, "Telecommunications and economic development," tech. rep., World Bank, 1994.

7

BLIND EQUALIZATION AND SMART ANTENNAS

Guanghan Xu

Engineering Science Bldg. Rm. 343
Dept. of Electrical & Computer Engineering
The University of Texas at Austin
Austin, TX 78712-1084
e-mail: xu@globe.ece.utexas.edu

Dedicated to Professor Kailath with respect and admiration on the occasion of his sixtieth birthday.

ABSTRACT

With the exponential increasing demand for wireless communications, the capacity of current cellular systems will soon become incapable of handling the growing traffic. Since radio frequencies are diminishing natural resources, there seems to be a fundamental barrier to further capacity increase. Smart antenna systems were proposed to open up another resource, *i.e.*, space, for a significant increase of channel capacity and for solving or mitigating the aforementioned problems. Smart antenna systems for military applications have been extensively researched by Kailath's group at Stanford over the last 10 years. Working with Paulraj and Kailath, the author started to convert the military technology to more cost-sensitive commerical technology. This paper briefly summarizes the research on civilian smart antenna systems the author initiated as a postdoctoral scholar at Stanford and further developed at The University of Texas at Austin.

1 INTRODUCTION

Wireless[1] communications, compared to wireline applications, have some unique problems such as capacity limitation due to finite spectrum allocation, multipath fading due to environment variation, power constraints due to limited battery life, *etc.* The search for effective technologies to overcome these difficulties has been active for years, motivated by the rapidly increasing de-

[1]This work was sponsored in part by National Science Foundation CAREER Program under Grant MIP-9502695, the Joint Services Electronics Program under Contract F49620-95-C-0045, Office of Naval Research under Grant N00014-95-1-0638, Southwestern Bell Technology Resources, Inc., Motorola, Inc., and Texas Instruments.

mand for wireless services. Most techniques, *e.g.*, channel coding for combating multipath fading, were developed to address individual issues, and have seemingly little correlation to other problems in this discipline. Recently, Smart Antenna System (SAS) , which exploit the spatial diversity among different users have been found very useful in solving or alleviating several key problems in wireless systems. With proper signal processing techniques, a smart antenna system can increase coverage and capacity, mitigate multipath fading, simplify handoff, and enhance security and privacy [1, 2, 3, 4]

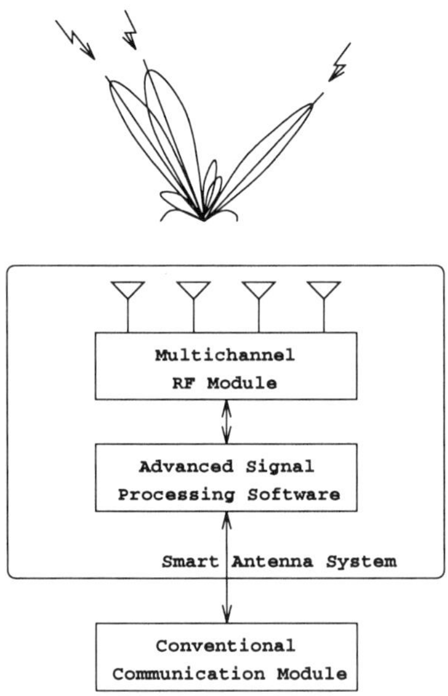

Figure 7.1 A smart antenna system

Figure 7.1 shows a typical smart antenna architecture. To exploit the spatial diversity, multiple antennas are placed at the base station. Following these RF channels is an advanced signal processing module that capitalizes on the spatial dimension by adaptively suppressing the interference and optimally combining signals from the desired users, thereby enhancing both the performance and the capacity of wireless systems. Furthermore, directional transmission and reception can provide better coverage and power efficiency, leading to significant increase in battery life and simplification in the MTSO (Mobile Telephone Switching Office) administration. Moreover, signal pre-processing techniques can be used to concentrate the complexity in the centralized base-station, so that the subscriber units can be simplified.

2 SMART ANTENNA CONCEPTS AND PROBLEM STATEMENT

Antenna array processing techniques were originally developed for target tracking and anti-jamming military communications. After a decade of extensive research (see *e.g.*, [5]) in this area, coupled with the rapid advance of microelectronics technology, the so-called high resolution antenna array technology can now achieve superior performance with affordable cost. The main goal of our research was to apply this high-performance military technology to improve the quality of wireless communications, especially that of the cellular telephony. Figure 7.2 shows a typical antenna array installed at a base station.

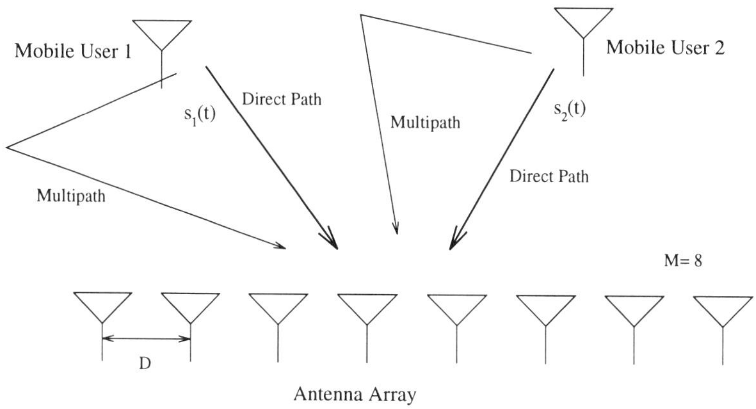

Figure 7.2 An 8-element uniform linear antenna array and two co-channel sources

The signals received by an antenna array contain both direct path and multipath signals, which are most likely from different directions of arrival (DOAs). Let us assume the array response vector to a transmitted signal $s_1(t)$ from a direction of arrival θ is $\mathbf{a}(\theta) = [1, a_1(\theta), \ldots, a_M(\theta)]$, where $a_i(\theta)$ is a complex number denoting the amplitude gain and phase shift of the signal at the $(i+1)^{th}$ antenna relative to that at the first antenna. For a uniform linear array with separation D as shown in Figure 7.2, $\mathbf{a}(\theta) = [1, e^{j2\pi f_c \sin\theta D/c}, \ldots, e^{j2\pi f_c \sin\theta(M-1)D/c}]^T$, where f, c, ^{T}, and * denote the carrier frequency, speed of light, transpose, and Hermitian, respectively. Clearly, the so-called array response vectors depend strongly on the direction or angle of arrival θ but are not functions of time. In a typical wireless scenario, an omni-directional antenna array not only receives signals from direct paths, but also receives multipath interference with different DOAs. Assuming that the multipath delays are much smaller than the symbol period of $s_1(t)$, the signal vector or snapshot, corresponding to one source, received by the antenna array can be written as:

$$\mathbf{x}(t) = \underbrace{\mathbf{a}(\theta_1)s_1(t)}_{\text{direct path}} + \underbrace{\sum_{l=2}^{N_m} \alpha_l \mathbf{a}(\theta_l)s_1(t)}_{\text{multipath}} = \mathbf{a}_1 s_1(t), \tag{7.1}$$

where $N_m - 1$ is the total number of multipath signals, α_l the phase and amplitude difference between the lth multipath and the direct path, and $\mathbf{a}_1 = \sum_{l=1}^{N_m} \alpha_l \mathbf{a}(\theta_l)$, which is referred to as the *spatial signature* associated with the first source. If there are d multiple co-channel signals present, then the data vector

$$\mathbf{x}(t) = \sum_{k=1}^{d} \mathbf{a}_k s_k(t) + \mathbf{n}(t), \tag{7.2}$$

where $\mathbf{a}_k$ is the kth spatial signature and $s_k(t)$ is the kth signal waveform and $\mathbf{n}(t)$ denotes noise. Clearly, the spatial signatures corresponding to different locations are different in most cases.

In the above, we assume that the multipath delays are much smaller than the symbol period, *i.e.*, there is no intersymbol interference (ISI). This may not be true in reality and a more general data model is

$$\mathbf{x}(t) = \sum_{k=1}^{d} \sum_{n=0}^{L-1} \mathbf{h}_k(n) s_k(t - nT_s) + \mathbf{n}(t), \tag{7.3}$$

where T_s is the sampling interval, $t = 0, T_s, 2T_s, \ldots$, and $\{\mathbf{h}_k(n)\}$ is the vector channel and L is the channel order.

One objective of the smart antenna system is to find all or some of the $\{s_k(\cdot)\}_{k=1}^{d}$, based on the received data vectors $\mathbf{x}(\cdot)$. This step is usually referred to as *smart uplinks*. Another objective of the smart antenna system is to find *appropriate* vector FIR filters or pre-equalizers $\{\mathbf{w}(\cdot)\}$ to form the *appropriate* transmit vector $\mathbf{y}(t) \triangleq \sum_{k=1}^{d} \sum_n \mathbf{w}_k(n) s_k(t - nT_s)$ such that the signal stength is selectively distributed. More specifically, we design $\mathbf{w}_i(\cdot)$ to minimize the co-channel interference (CCI) $\sum_{k \neq i} \mathbf{w}_i(n) \otimes \mathbf{h}_k(n) \otimes s_k(n)$ and ISI $1 - \mathbf{w}_i(n) \otimes \mathbf{h}_i(n)$. This step is called *smart downlink*. In the following, we shall summarize our research results on both smart uplinks and downlinks.

3 SMART UPLINKS

Our results on smart uplinks originated from our research on blind channel equalization, which was motivated by Kailath while the author was a student and later a research associate at Stanford.

3.1 Single-User Blind Equalization

Blind channel equalization (BCE) [6], *i.e.*, determining and equalizing the channel response based solely on the channel output without the use of a training sequence, has received considerable attention recently in communications and signal processing. As shown in Figure 7.3, we need to identify the channel $h(\cdot)$ based on the output $x(\cdot)$ and the statistics of $s(\cdot)$. In a reason soon to be noted, earlier approaches to blind identification were based on using higher-order (than second) statistics of the output $x(\cdot)$; see [7, 8, 9, 10, 11, 12] and the references therein. These methods, although useful in many applications, are computationally expensive, and so are not suitable for mobile communications due to the rapid variation of multipath channel characteristics. In 1991, when the author was still at Stanford, Kailath brought in as a post-doctoral scholar, Lang Tong, a brillant student of Prof. Ruey-Wen Liu at Notre Dame. The joint work [13, 14] of Tong, Kailath and myself led to the first blind equalization approach based on second-order statistics *only*. In the signal processing community, it is well-known that the second-order statistics of $x(\cdot)$ are not sufficient for blind channel identification due to the phase ambiguity. We observed that the *oversampled* digital communication signal is no longer stationary but *cyclostationary* [15] and that exploitation of cyclostationarity could resolve the phase ambiguity.

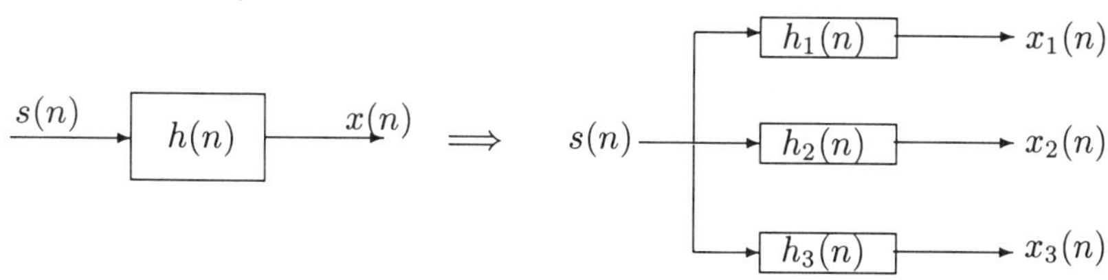

Figure 7.3 From a single-input and (oversampled) single-output (SISO) channel to a single-input and multiple-output (SIMO) channel

Now the question is whether all the channels are *blindly* identifiable. By going into the frequency domain, we found that the cause is closely related to Sylvester Resultants and Toeplitz approximation theory. Using a nice theorem proved by Kung, Kailath and Morf as early as 1976, we were able to develop and rigorously prove several sufficient and necessary identifiability conditions. Actually, the results are quite simple, *i.e.*, the channels are *blindly* identifiable iff $h(\cdot)$ does not have zeros located on a circle with spacing $2\pi/N$, where N is the oversampling rate.

In the course of searching for a better understanding, Kailath suggested to look into the standard observability condition in control theory and this led to the discovery of a new deterministic or least squares approach [16] that does not require the knowledge of the statistics of $s(\cdot)$. Slightly later, Moulines *et. al.* and Slock, another former Kailath student, reported another deterministic approach [17, 18] by exploiting the Toeplitz structure of the channel matrix formed by the elements of $h(\cdot)$. These approaches are especially attractive since it can identify the channel with a small number of samples, *e.g.,* 50 symbols. The basic idea of the deterministic approach is illustrated in Figure 7.3: a single-input, single-output (SISO) channel can be equivalently expressed as a single-input, multiple-output (SIMO) channel if the input is a digital communication signal, where the oversampling rate is 3. We can write $x_1(\cdot)$ and $x_2(\cdot)$ as

$$x_1(n) = h_1(n) \otimes s(n), \quad x_2(n) = h_2(n) \otimes s(n). \tag{7.4}$$

Then,

$$h_1(n) \otimes x_1(n) = h_2(n) \otimes (h_1(n) \otimes s(n)) = h_1(n) \otimes \underbrace{(h_2(n) \otimes s(n))}_{=x_2(n)} = h_1(n) \otimes x_2(n). \tag{7.5}$$

The above equation shows that the outputs of each channel pair are related by their channel responses. Clearly, if we have adequate data samples of the outputs, by (7.5), we can write out an overdetermined set of linear equations involving $h_1(\cdot)$ and $h_2(\cdot)$, which may be determined uniquely up to a scalar multiple. Note that this structure is not available in a single channel system. The identifiability condition, not surprisingly, is that all the channel responses $\{h_i(\cdot)\}_{i=1}^{M}$ should not share any common zeros and the signal $s(\cdot)$ must be *persistently excited.* More detailed descriptions of sufficient and necessary conditions for the deterministic approach can be found in [19, 16, 17, 20, 21, 22].

Later, we also found that temporal oversampling is equivalent to spatial oversampling (via an antenna array) [23] if we properly arrange the data from multiple receivers. In fact, this is somewhat obvious by treating $h_i(\cdot)$ as the channel response corresponding to the ith receiver. Due to the rich spatial diversity of wireless communication environments, a *spatially* oversampled channel is more easily identified or the identifiability conditions are more easily met. In this case, a smart antenna system at a base station can equalize a multipath channel for a single user in the cell without relying on any training sequences.

3.2 Multi-User Blind Channel Equalization

The above discussions are restricted to single-user blind equalization, i.e. to the elimination of ISI. Now, what if there are multiple co-channel signals? Can the above techniques be extended to eliminate both ISI and co-channel interference (CCI)? If this can be accomplished, then the smart uplinks are complete. Let us first approach this problem by looking at another approach to blind symbol estimation [24, 25] not requiring channel identification. Then, we will extend this approach to handle multi-user blind channel equalization.

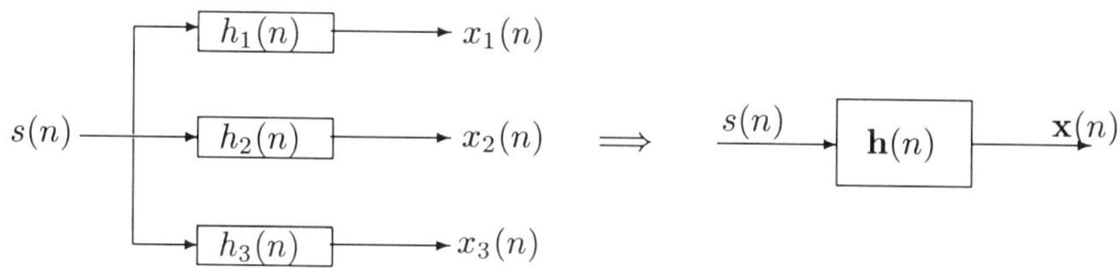

Figure 7.4 From a SIMO channel to a vector channel

If we define $\mathbf{h}(n) = [h_1(n), h_2(n), h_3(n)]^T$ and $\mathbf{x}(n) = [x_1(n), x_2(n), x_3(n)]^T$, then we will have a vector channel $\mathbf{h}(\cdot)$ as shown in Figure 7.4. Thus

$$\mathbf{x}(n) = \mathbf{h}(n) \otimes s(n). \tag{7.6}$$

Let us form a block Hankel matrix with $\mathbf{x}(\cdot)$, *i.e.*,

$$\mathbf{X}(K) = \underbrace{\begin{bmatrix} \mathbf{x}(L+1) & \mathbf{x}(L+2) & \cdots \mathbf{x}(N-K+1) \\ \mathbf{x}(L+2) & \mathbf{x}(L+3) & \cdots \mathbf{x}(N-K+2) \\ \vdots & \vdots & \cdots & \vdots \\ \mathbf{x}(L+K) & \mathbf{x}(L+K+1) & \cdots & \mathbf{x}(N) \end{bmatrix}}_{N-K-L+1}, \quad K = 1, 2, \cdots \tag{7.7}$$

where L is the channel order, N the number of output vectors $\mathbf{x}(\cdot)$ available, and K is defined as a *smoothing factor.* By (7.6), we can express $\mathbf{X}(K)$ in terms of $\mathbf{h}(\cdot)$ and $\mathbf{s}(t)(\cdot)$ as follows,

$$\mathbf{X}(K) = \underbrace{\begin{bmatrix} \mathbf{h}_L & \mathbf{h}_{L-1} & \cdots & \mathbf{h}_0 & \mathbf{0} & \cdots & \mathbf{0} \\ \mathbf{0} & \mathbf{h}_L & \mathbf{h}_{L-1} & \cdots & \mathbf{h}_0 & \cdots & \mathbf{0} \\ \vdots & \vdots & \ddots & \ddots & \ddots & \ddots & \vdots \\ \mathbf{0} & \cdots & \mathbf{0} & \mathbf{h}_L & \mathbf{h}_{L-1} & \cdots & \mathbf{h}_0 \end{bmatrix}}_{\mathbf{H}(K),\ K+L\ blocks} \underbrace{\begin{bmatrix} s(1) & s(2) & \cdots s(N-r+1) \\ s(2) & s(3) & \cdots s(N-r+2) \\ \vdots & \vdots & \cdots & \vdots \\ s(r) & s(r+1) & \cdots & s(N) \end{bmatrix}}_{\mathbf{S}(r),\ r=L+K}.$$

Now Hankel or Toeplitz matrices have the following interesting property.

Lemma 7.1 *The input sequence $\{s(1), \ldots, s(N)\}$ can be uniquely determined, up to a scalar multiplier, from the row span of $\mathbf{S}(r)$ if $\mathbf{S}(r)$ is of full row rank.*

The proof of this lemma can be found in [24, 25]. If we properly select K, then we can show that $\mathbf{H}(K)$ is of full column rank and therefore the row span of $\mathbf{X}(K)$ is the same as the that of $\mathbf{S}(L+K)$. By Lemma 7.1, we can determine the input symbols up to a scalar multiple, $\rho\mathbf{S}(1) = \rho[s(1), s(2), \ldots, s(N)]$, where ρ

is a complex scalar. The detailed algorithm is shown in [24, 25]. Here the key is again oversampling $M > 2$, so that, $\mathbf{H}(K)$ has more columns than rows and $\mathbf{S}(L+K)$ is the opposite.

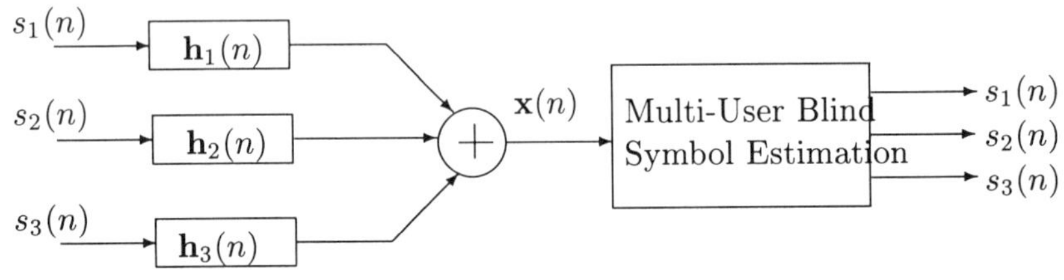

Figure 7.5 Multi-user blind symbol estimation

In the multi-user problem as illustrated in Figure 7.5, we cannot estimate $\mathbf{s}_1 = \mathbf{S}_1(1) = [s_1(1), \ldots, s_1(N)], \ldots, \mathbf{s}_d = \mathbf{S}_d(1) = [s_d(1), \ldots, s_d(N)]$. However, as shown in [25], we can estimate the row span of $\mathbf{S} = [\mathbf{s}_1^T, \mathbf{s}_2^T, \ldots, \mathbf{s}_d^T]^T$ or the matrix $\mathbf{WS}$, where $\mathbf{W}$ is a $d \times d$ nonsingular matrix. Without additional knowledge of $\{s_k(\cdot)\}$, the user symbols are not identifiable. In digital communication systems, $\{s_k(\cdot)\}$ usually belong to a finite alphabet and this knowledge can be exploited to identify $\mathbf{S}$ or $\{s_k(\cdot)\}$ from $\mathbf{WS}$ using, *e.g.*, the iterative linear squares prediction (ILSP) method of Talwar *et. al.* [26, 25]. Therefore, the complicated multi-user blind symbol estimation problem can be nicely solved and both ISI and CCI can be eliminated simultaneously. This completes the smart uplink problem.

4 SMART DOWNLINKS

In wireless communications, there are two types of duplex schemes: time-division-duplex (TDD) and frequency-division-duplex (FDD). In TDD systems, the uplink and downlink share the same carrier but occupy different time slots. In FDD systems, the carriers for the uplink and downlink are different. The duplex modes determine the smart antenna techniques for selective downlink transmission that transmit multiple signals to their designated locations without mutual interference.

In TDD systems, reciprocity holds and the uplink and downlink channels are identical. Therefore, the channels $\{\mathbf{h}_i(\cdot)\}$ can be estimated by the above blind channel identification; then we can design a pre-equalization filter $\{\mathbf{g}_i(\cdot)\}$ such that $\mathbf{h}_i(n) \otimes \mathbf{g}_j(n) \triangleq \sum_l \mathbf{h}_i^H(l)\mathbf{g}_j(n-l) = \delta_{ij}(n)$, where $\delta_{ij}(n) = 1$ for $n = 0$, $i = j$ and $\delta_{ij}(n) = 0$ otherwise. If the signal transmitted from the antenna array is $\mathbf{y}(n) = \sum_{k=1}^{d} \mathbf{g}_k(n) \otimes s_k(n)$, then the signal reaches the ith user is

$$\begin{aligned} t_i(n) = \mathbf{h}_i(n) \otimes \mathbf{y}(n) &= \mathbf{h}_i(n) \otimes \sum_{k=1}^{d} \mathbf{g}_k(n) \otimes s_k(n) \\ &= \sum_{k=1}^{d} (\mathbf{h}_i(n) \otimes \mathbf{g}_k(n)) \otimes s_k(n) = s_i(n) \end{aligned} \tag{7.8}$$

Therefore, the pre-equalized signal arrives at each user free of ISI and CCI. Although it is well-known that an FIR filter cannot equalize an FIR channel, we can eliminate ISI with an FIR filter because of the SIMO (or oversampling) structure mentioned earlier. The techniques of finding the pre-equalizers $\{\mathbf{g}_k(\cdot)\}$ can be found in [27].

In FDD systems, due to the different channel characteristics in uplinks and downlinks as a result of different carriers, it is almost impossible to exactly determine the downlink channel from the corresponding uplink channel that can be estimated. One technique is to allow some feedback from the mobile user to determine the downlink characteristics [28]. This, however, requires modification of current cellular standards. The only invariant parameters between the uplink and downlink are direction-of-arrivals (DOAs) of the direct path and multipath components. Traditional DOA estimation techniques (see *e.g.*, [29]) can only estimate up to $2M/3$ DOA's, where M is the number of antennas. These techniques only exploit the spatial property of the impinging signal and ignore its temporal property. Recently, Liu and Xu [30] proposed an integrated approach that exploits both the spatial and temporal properties (*e.g.*, finite alphabet) and can resolve up to $2M^2/3$ DOA's, a factor of M increase. The basic idea of the approach is to first find the spatial signature of each signal by applying the property restoral techniques, *e.g.*, ILSP [26, 25] and then find the DOAs from the estimated spatial signatures since spatial signatures contain the sufficient information of the DOA's.

5 SMART ANTENNA TESTBED AND EXPERIMENTAL RESULTS

Strongly encouraged by Kailath, the author started to establish a smart antenna testbed as soon as he joined UT-Austin. After much hard work, we finally completed an advanced and flexible smart antenna testbed. In the testbed, we have 12-channel transceivers, 8 of which are for base station smart antennas and 4 of which are for mobile units.

This testbed has been extensively used to conduct channel propagation studies and to validate the developed techniques. Due to space limitation, we can only show a few representative experimental results. In the first experiment, multiple transmitters were used to transmit two BPSK signals from mobile units.

The signal received by the base smart antennas was acquired and processed using the aforementioned blind equalization techniques. The first plot of Figure 7.6 shows the eye pattern of the signal received from one antenna. Obviously, there is both ISI and CCI. After processing the received data using the aforementioned technique, the BPSK patterns are clearly shown in the second and third plots of Figure 7.6. Downlink experiments were also performed for both TDD and FDD schemes. Here, we only show the results for the FDD case which is more difficult for smart antennas. The downlink experiment was carried out in a multipath rich indoor environment. The traditional DOA-based approach and the new integrated approach [30] were utilized to process the uplink data and calculate the downlink beamforming weights. Then two tones are beamformed separately and the sums are sent to the base smart antenna transmitters. If the smart downlink is successful, each mobile unit is supposed to receive only one tone while another tone should be suppressed. Figure 7.7 show the Fourier transforms of the signals received by the two mobile receivers using the traditional DOA-based techniques. Due to the rich multipath, both tones were picked by one mobile unit and the DOA-based smart downlink did not perform well in this scenario. Figure 7.8 shows the results obtained from the integrated approach. Clearly, the co-channel interference (CCI) was significantly suppressed with the integrated approach.

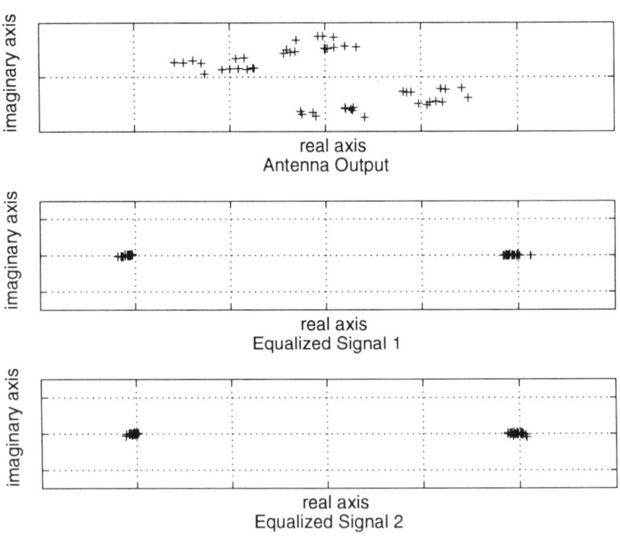

Figure 7.6 Comparison of eye patterns before and after smart uplink processing

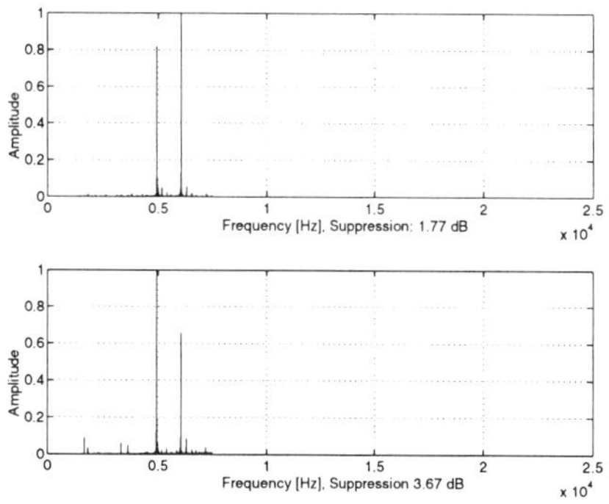

Figure 7.7 Results of the DOA-based downlink approach

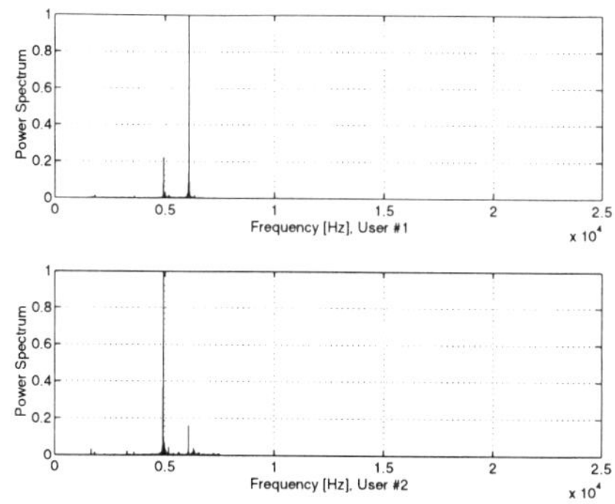

Figure 7.8 Results of our integrated downlink approach

6 CONCLUDING REMARKS

In this paper, we have summarized the smart antenna research that the author initiated with Paulraj and Kailath at Stanford. We have presented innovative signal processing techniques for smart uplinks and downlinks and validated their promising performance in field experiments using our advanced and flexible smart antenna testbed.

REFERENCES

[1] S. Andersson, M. Millnert, M. Viberg, and B. Wahlberg, "An Adaptive Array for Mobile Communication Systems", *IEEE Trans. on Veh. Tec.*, vol. 40, pp. 230–236, 1991.

[2] J.H. Winters, J. Salz, and R.D. Gitlin, "The Impact of Antenna Diversity on the Capacity of Wireless Communication Systems", *IEEE Trans. on Communications*, vol. 42, pp. 1740–1751, February/March/April 1994.

[3] G. Xu, H. Liu, W.J. Vogel, H.P. Lin, S.S. Jeng, and G.W. Torrence, "Experimental Studies of Space-Division-Multiple-Access Schemes for Spectral Efficient Wireless Communications", *in Proc. ICC'94*, New Orleans, LA, May 1994.

[4] P. Balaban and J. Salz, "Optimum Diversity Combining and Equalization in Digital Data Transmission with Applications to Cellular Mobile Radio - Part I: Theoretical Considerations", *IEEE Trans. on Communications*, vol. 40, pp. 885–894, May 1992.

[5] A. Paulraj, B.E. Ottersten, R.H. Roy, G. Xu A. Swindlehurst, and T. Kailath, "Subspace Methods for Direction Finding and Parameter Estimation", in N.K. Bose Ed. and C.R. Rao, editors, *Handbook of Statistics Volume 10 "Signal Processing and Its Applications"*. Elsevier Science Publishers B.V., 1992.

[6] Y. Sato, "A Method of self-recovering equalization for multilevel amplitude-modulation", *IEEE Trans. Commun.*, vol. 23, pp. 679–682, June 1975.

[7] A. Benveniste and M. Goursat, "Blind equalizers", *IEEE Trans. Commun.*, vol. 32, pp. 871–883, August 1984.

[8] Z. Ding, R.A. Kennedy, B.D.O. Anderson, and C.R. Johnson, "Ill-convergence of Godard blind equalizers in data communication systems", *IEEE Trans. Communications*, pp. 1313–1327, September 1991.

[9] J.K. Tugnait, "Identificaiton of Linear Stochastic System via Second and Fourth-order Cumulant Matching", *IEEE Trans. on Information Theory*, vol. 3, pp. 393–407, May 1987.

[10] O. Shalvi and E. Weinstein, "New Criteria for Blind Deconvolution of Nonminimum Phase Systems (Channels)", *IEEE Inform. Theory*, vol. 36, pp. 312–320, March 1990.

[11] D. Hatzinakos and C. Nikias, "Estimation of multipath channel response in frequency selective channels", *IEEE J. Select. Areas Commun.*, vol. S7, pp. 12–19, January 1989.

[12] A.P. Petropulu and C.L. Nikias, "Blind Deconvolution Using Signal Reconstruction from Partial Higher Order Cepstral Information", *IEEE Trans. on SP*, vol. 41, pp. 2088–2095, June 1993.

[13] L. Tong, G. Xu, and T. Kailath, "A new approach to blind identification and equalization of multipath channel", *in Proc. of the 25th Asilomar Conference on Signals, Systems, and Computers*, Pacific Grove, CA, November 1991.

[14] L. Tong, G. Xu, and T. Kailath, "Blind Identification and Equalization Based on Second-Order Statistics: A Time Domain Approach", *IEEE Trans. on Information Theory*, vol. 40, March 1994.

[15] W.A. Gardner, "Characterization of Cyclostationary Random Signal Processes", *IEEE Trans. on Information Theory*, vol. 21, pp. 4–14, 1975.

[16] G. Xu, H. Liu, L. Tong, and T. Kailath, "A Least-Squares Approach to Blind Channel Identification", *accepted by IEEE Trans. on Signal Processing.*

[17] E. Moulines, P. Duhamel, J. Cardoso, and S. Mayrargue, "Subspace methods for the blind identification of multichannel FIR filters", *in Proc. IEEE ICASSP'94*, pp. IV573–IV576, April 1994.

[18] D.T.M. Slock, "Blind fractionally-spaced equalization, perfect-reconstruction filter banks and multichannel linear prediction", *in Proc. IEEE ICASSP'94*, pp. IV585–IV588, April 1994.

[19] H. Liu, G. Xu, and L. Tong, "A Deterministic Approach to Blind Equalization", *in Proc. Asilomar Conference on Signals, Systems, and Computers*, Pacific Grove, CA, Nov. 1993.

[20] J.K. Tugnait, "On Blind Equalization of Multipath Channels Using Fractional Sampling and Second-Order Cyclostationary Statistics", *in Proc. of Globecom'93*, Houston, TX, Dec. 1993.

[21] Z. Ding and Y. Li, "Channel Identification Using Second Order Cyclic Statistics", *in Proc. Asilomar Conf. on Signal, Systems, and Computers*, Pacific Grove, CA, Oct. 1992.

[22] Y. Li and Z. Ding, "Blind Channel Identification Based on Second-Order Cyclostationary Statistics", *in Proc. ICASSP'93*, pp. 81–84, Minneapolis, MN, April 1993.

[23] L. Tong, G. Xu, and T. Kailath, "Fast Blind Equalization of Multipath Channels via Antenna Arrays", *in Proc. ICASSP 93 Conf.*, pp. 272–275, Mineapolis, MN, April 1993.

[24] H. Liu and G. Xu, "A Deterministic Approach to Blind Symbol Estimation", *IEEE Signal Processing Letters*, vol. 1, pp. 205–208, December 1994.

[25] H. Liu and G. Xu, "Closed-form Blind Symbol Estimation in Digital Communications", *IEEE Trans. on Signal Processing*, vol. 41, 12 1995.

[26] S. Talwar, M. Viberg, and A. Paulraj, "Blind estimation of multiple co-channel digital signals using an antenna array, *IEEE Signal Processing Letters*, vol. 1, pp. 29–31, February 1994.

[27] H. Liu and G. Xu, "Multiuser Blind Channel Estimation and Spatial Channel Pre-Equalization", *in Proc. ICASSP'95*, Detroit, MI, May 1995.

[28] D. Gerlach and A. Paulraj, "Base Station Transmitter Antenna Arrays with Mobile to Base Feedback", *in Proc. 27th Asilomar Conf. on Signals, Systems, and Comp.*, pp. 1432–1436, Pacific Grove, CA, Nov. 1993.

[29] T.J. Shan, A. Paulraj, and T. Kailath, "On Smoothed Rank Profile Test in Eigenstructure Approach to Direction-of-Arrival Estimation", *IEEE Trans. ASSP*, vol. 33, pp. 1377–1385, October 1987.

[30] H. Liu and G. Xu, "An Effective Transmission Beamforming Scheme for Frequency-Division-Duplex Digital Wireless Communication Systems", *in Proc. ICASSP'95*, Detroit, MI, April 1995.

8

COMPLEXITY MANAGEMENT: A MAJOR ISSUE FOR TELECOMMUNICATIONS

David G. Messerschmitt

Department of Electrical Engineering and Computer Sciences
University of California
Berkeley, CA 94720

Dedicated with great admiration to Professor Thomas Kailath upon the occasion of his sixtieth birthday.

ABSTRACT

As a result of years of geometrical advances in underlying electronics and photonics technology, traditional efficiency and performance considerations (which have been dominant activities in telecommunications research) will play a somewhat diminished role in the future. Simultaneously we are accumulating multiple standards, protocols, and transmission media, proliferating a variety of user-oriented applications, and seeing cost-effective software implementations and hardware systems with enormous complexity. These trends imply that an increasing barrier to progress in telecommunications is not cost or efficiency, but managing the tremendous complexity of heterogeneous networks, media, terminals, and applications in a multi-vendor environment. More generally, while complexity management has been a traditional issue in software engineering, and later in integrated circuit design, for the future it will be an increasingly important issue in large-scale system design. Our hypothesis is that complexity management will be an increasing factor in telecommunications research and development. This does not imply that interesting issues in signal processing and communications theory disappear; to the contrary, complexity management considerations raise a number of new issues and will doubtless revitalize these fields. We briefly describe complexity management methodologies that have arisen in the software domain, and speculate on the nature of complexity management in large system design. Is it largely an issue in the management of the development process, or is it amenable to systematic and rigorous approaches? To be more concrete, we give examples from the telecommunications realm drawing on our own work.

1 INTRODUCTION

Telecommunications has traditionally been driven by advances in underlying electronics and photonics technologies, and had focused on a small set of universal applications – namely voice telephony, video conferencing, and data transmission – with most of the effort devoted to lowering costs and increasing efficiency. The dominant research themes have been largely efficiency-driven, such as the classical problems of increasing the bit rate required to represent a given source. Our thesis is that this situation is rapidly changing, with complexity management considerations replacing efficiency as the dominant challenge in telecommunications research. We outline arguments in favor of this thesis, describe some general techniques available for complexity management, and give examples from telecommunications.

2 THE CHANGING ENVIRONMENT FOR RESEARCH

2.1 Advances in the underlying technology

Advances in electronics technology, roughly doubling in cost-performance every couple years, have greatly contributed to advances in communications efficiency by making sophisticated compression and coding techniques both feasible and affordable. However, in many transmission media, we appear to be approaching fundamental limits, even in the context of affordable implementations, as a result in the accumulated research in compression and coding. Simultaneously, photonics and storage technology have been advancing at the same rate as electronics, or even faster, making efficiency less crucial in backbone networks and storage environments. Indeed, traditional communication theory has made little headway in affecting the practice of fiber optics transmission.

In considering appropriate research themes, we should differentiate between bottlenecks that will disappear with advances in the underlying processing and transmission technologies, as contrasted with more fundamental problems that are not subject to mere technological solution. Premier examples of the latter include traffic capacity of wireless multiaccess channels, and achieving adequate interactive response in the face of latency (which is lower bounded by propagation delay, already significant in global networks). In contrast, continuing advances in fiber optics, optical networking, electronic switching, and magnetic and optical storage media are providing ample (and increasing) bandwidth resources in backbone networks and storage capacity. There is of course continuing interest in better utilizing existing facilities like wirepair and the voiceband telephone channel – where efficiency is critical – but here technology is approaching fundamental limits. Further, we seem poised to finally provide widespread broadband access through some combination of new fiber and existing coaxial facilities.

Advances in electronics technology not only provide ample performance, but they also have two other important impacts:

- The feasible number of devices per chip is increasing, from millions today to hundreds of millions within the next decade. A dominant design problem for such chips is architecture and complexity management; indeed the functional design of such chips is conceptually similar to developing a large software system.
- Software implementation of most functions becomes feasible. Today, audio-frequency functions are typically implemented in software (often on a specialized, but increasingly on a general purpose, processor). In the future, video signal processing will fall in the same category, and customized hardware design will be relegated to radio and microwave frequencies.

Both of these factors imply an increasing convergence between the technical problems faced in the telecommunications and applications software industries. This has long been the case in the signalling, control, and switching aspects of telecommunications as well, where the dominance of software is legendary.

2.2 Advanced applications

Most of the telecommunications applications today are functionally simple applications with universal interest, like voice telephony and point-to-point video teleconferencing. Indeed, the telecommunications industry perceives itself as selling and provisioning these universal applications, rather than underlying services like audio, video, and data transport[1]. However, as desktop and untethered programmable platforms become standard foundations for networked applications, the cost-effective software realization of functionally complex applications becomes feasible. In addition, we can expect dynamic deployment of software-defined applications over the network to dramatically speed up the rate of innovation in commercially available telecommunications applications; that is, the telecommunications marketplace will begin to look more and more like desktop computing[1]. Most new applications will be functionally complex, often involving multipoint participants and integrated multimedia like audio, video, and graphics and shared data.

[1]We distinguish between *applications*, which provide functionality to the end user, and *services*, like audio, video, and data transport, which are available to be incorporated in those applications.

2.3 Heterogeneity

In the distant past, telecommunications was characterized by a relatively straightforward homogeneous network providing a single application: voice telephony. As time has passed, this simple environment has evolved in different directions:

- There has been a proliferation of new applications (such as voiceband data, videophone, and facsimile) utilizing the telephony infrastructure.
- New standards have proliferated for any given application, such as voice or video encoding or voiceband data.
- New telecommunications media, such as coaxial cable, fiber optics, microwave radio, and recently mobile radio and wireless infrared have appeared.
- To date, most applications have been realized by dedicated special-purpose and non-programmable terminals, like telephones. This is rapidly changing, as programmable platforms such as desktop computer, notebook computer, and personal digital assistant are becoming increasingly viewed as communications devices.
- An impact of global telecommunications deregulation is a proliferation of service providers, often with more than one service provider involved in provisioning a single application, along with the involvement of new equipment vendors (like the computer companies).

From a technical perspective, these developments imply a rapidly increasing level of *heterogeneity* in applications, *heterogeneity* in transport systems, and *heterogeneity* in terminals. Meanwhile, application developers, and especially end users, would like to be isolated from the myriad of technologies and enterprises involved in their provisioning. They want applications to operate seamlessly across the telecommunications infrastructure, with both applications and networks appropriately scaling and configuring to whatever detailed technological components are involved. It is not acceptable to users to have their telecommunications applications restricted to only a portion of the network (and hence to a subset of the other users), or to equipment from a particular vendor. All parties involved – users, application providers, content providers, and equipment vendors – want a flexible and dynamic network that can scale and evolve to meet whatever demands are placed on it, and accommodate new advances without major dislocations or disinvestment. Users don't want a closed, proprietary network that limits the available applications to those envisioned by a limited set of vendors and service providers, but rather an environment in which a variety of competing vendors and service providers can flourish, and innovation can reign supreme. These properties that have led to the recent success of the Internet, for example.

These factors imply a telecommunications infrastructure in the future that is vastly more complex than any existing software system. (In fact, the infrastructure will incorporate *many* existing and new large software systems, among other elements, like unreliable physical channels.) Unlike typical large software systems, it will be designed not by a single organization, but literally hundreds of equipment and software vendors, and tens of standards bodies and provider organizations. If experience with the development of software systems is a guide – and it *should* be – success will hinge on the containment of the complexity inherent in this large heterogeneous system.

3 SHORT TUTORIAL ON COMPLEXITY MANAGEMENT

We have asserted that management of complexity is a central issue in the future of telecommunications. What does this mean? How do we accomplish it? Can it be systematized, or is it largely an organizational management issue?

While large telecommunications systems are different from software systems, experience with the management of complexity in the latter domain is relevant, and can serve as a useful starting point in our thinking. Thus, we briefly summarize some techniques for complexity management from that domain.

Three basic components of complexity management are shown in Figure8.1:

- *Architecture* is the prior plan of the system that partitions functionality among a set of interacting modules[2].
- *Theory* exploits the constrained behavior of architectural modules to establish certain predictable properties, or perhaps the absence of certain undesired behaviors (such as deadlock, instability, etc.)[3].
- *Tools* are software systems that keep track of the large design databases typical of complex systems, systematically synthesize more routine parts of the system, etc[4].

Architecture is by far the most critical element, and will now be discussed in greater detail, including some related elements like interfaces, abstraction, configurability and scalability, and reuse.

---

[2]It is interesting to compare with the more general dictionary definition of these terms. *Architecture* is defined as "the manner in which the components of a computer or computer system are organized and integrated." [2]

[3]*Theory* is defined as "the analysis of a set of facts in their relation to one another" [2].

[4]A *tool* is defined as "something (as an instrument of apparatus) used in performing an operation or necessary in the practice of a vocation or profession"[2].

Tools	**Automate and manage large design databases**
Theory	**Predictable behavioral limitations**
Architecture	**Modularity and functionality**

Figure 8.1 Three basic elements of complexity management.

3.1 Architecture

The architecture of the system is the basic plan that insures it performs the functions for which it is intended. The most important aspect of the architecture is the basic *modularity*, or partitioning of functionality into mutually interacting elements[5]. The interacting modules, which display limited and predictable behaviors, should be as *independent* from one another as possible. Further, they are designed with carefully constructed *interfaces*, with the internal implementation beyond what is displayed at those interfaces carefully hidden. By independence of modules, we mean that the functionality of one module should depend on the functionality of other modules only the extent appropriate and necessary, and not on incidental implementation details. This independence makes the system more open to change and evolution, since changes to the implementation of one module should not affect other modules.

A familiar example of an architecture is a computing system, shown in Figure 8.2. This oversimplified architecture divides a computer system into basic modules of arithmetic and logic unit (ALU), cache memory, main memory, secondary storage, and a bus that serves to connect the other modules. Each module has a clearly defined and limited function. The bus is key to insuring independence of the other modules, since it forces them to communicate in a standardized (module-independent) way. By being asynchronous (using handshaking techniques), valid operation can even be insured independent of speed.

[5]A *module* is defined as an "an independently-operable unit that is a part of the total structure"[2].

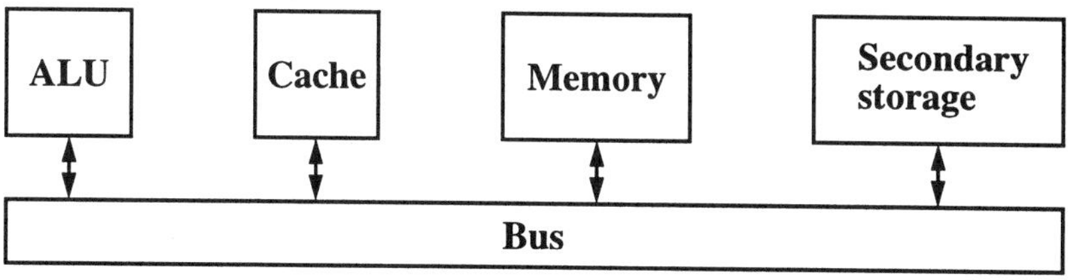

Figure 8.2 A simplified and familiar example of an architecture.

3.2 Abstraction

Abstraction is a key concept underlying modularity, as well as other aspects of the design process[6]. Abstraction refers to the conscious hiding of unnecessary implementation or functional details while making visible behavioral properties that are essential and important to other modules. Abstraction helps insure the independence of the architectural modules, as one module cannot be dependent on the hidden properties of another module. For example, in Figure 8.2, the visible properties of the memory may be sequenced read or write requests, information stored in fixed sizes units, an address associated with each unit and random access. Deliberately hidden at the interface are the technology (be it CMOS, bipolar, bubble memory, etc.), the speed of memory access (variations are accommodated by handshaking), and internal organization. The internal organization may be quite different than suggested by the interface abstraction, for example turning streamed access into a random access by internally performing multiple accesses (as would be the case in magnetic storage).

Abstraction can be used in other ways that are important to containing the complexity of a design. One is illustrated in Figure 8.3, where a layered logical (as opposed to physical or functional) representation of the same computer system is shown. The electronic *device* can be considered an abstraction of the underlying semiconductor physics, hiding unnecessary details like holes and electronics while preserving terminal properties like the voltage-current curve. *Logic* defines modules (consisting internally of transistors) with limited behaviors like "nor gates" and "inverters" that hide device properties and circuit details (like CMOS vs. bipolar technology). The logic layer also defines arguably one of the most important abstractions of modern technology, the "bit", which simplifies the logic outputs to have only two states (the reality is more complicated). *Reg-*

[6] *Abstract* is defined as "disassociated from any specific instance"[2].

isters define higher-level modules like "shift registers" and "two's complement adders" which form specific functions while hiding their internal logic implementation details (and indeed admitting many possible implementations). The register-transfer level defines the basic modules used in the architecture of the ALU, which externally hides such details through the definition of an instruction set. The *instruction set* presents a powerful abstraction to the software, since it separates the latter from all the internal details of the hardware. Similarly, the *operating system* layer defines a number of abstractions that separate the user-level process from the details of memory, storage, and communications resource management.

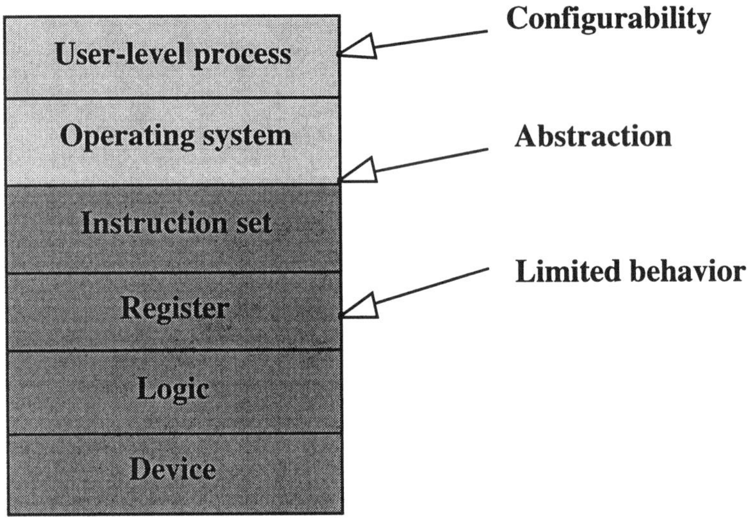

Figure 8.3 A familiar example of the use of abstraction in a computing system ALU.

3.3 Reusability: configurability, scalability, and adaptability

One basic approach of complexity management is reusability, which has several aspects:

- *Modularity* encourages reusability, since it defines a grouping of functionality that is carefully defined and documented. *Abstraction* at the interface enhances reusability since usage is separated from implementation specifics.

- *Configurability* insures that modules are not designed for a very specific environment or use, but rather internal parameters can be modified to accommodate new unanticipated uses.
- Closely associated with configurability is *scalability*, a property of the architecture that maximizes the flexibility to configure to different levels of performance (processing power, bandwidth, number of ports, etc.), as unconstrained as possible by technology. A scalable architecture is more broadly configurable.
- *Adaptability* requires configurability, and adds the capability to base the configuration on active observation of the module environment.

The simplest implication of reusability is the saving of design costs. In an operational system, many of the same characteristics that allow reusability can lead to a system that is self-organizing. Configurable and adaptable modules that adjust to their observed environment allow a distributed-control architecture that is in itself a powerful form of complexity management.

4 A TELECOMMUNICATIONS EXAMPLE

In addition to the generic issues of heterogeneity, future networks supporting multimedia datatypes will have a number of objectives, many of which interact in complicated ways. These include:

- Differing connection *topologies*, including point-to-point, multicast, and multisource. (For example, video conferencing requires both multicast and multisource topologies.)
- *Untethered* (no wires), *nomadic* (accessible from different locations), and *mobile* (accessible while moving) multimedia applications will be important for some users.
- *Privacy* by *end-to-end encryption* will be demanded by a subset of users, as they gain familiarity with privacy in their data applications.
- *Traffic efficiency*, which is particularly an issue on wireless access links. Wireless access traffic is interference-limited within a given volume of space and bandwidth, and will become an increasingly serious bottleneck as backbone networks become faster.
- High *subjective quality*, which is indirectly impacted by compression algorithms and by loss and corruption in the transport.

- *Low latency* is critical for some interactive applications. Since the propagation delay in terrestrial global networks is significant (hundreds of milliseconds), there is little headroom to increase latency through signal processing and queueing for these interactive applications.

Simultaneously satisfying these requirements requires a carefully crafted architecture. In part this is because these diverse requirements create dependencies among system elements that must be carefully managed for good modularity. We will now give illustrative examples from our own work.

Achieving high traffic capacity on wireless links requires joint source/channel coding (JSCC); that is, coordination of the source coding with resource costs in the transport. For example, the source coding will display a trade-off between bitrate and bit-error tolerance, while the transmission media associates a resource cost to the bitrate, delay, and reliability of each source stream. Much past source coding research has emphasized the minimization of bitrate – without regard for the reliability or delay requirements – but wireless links attach a high cost to the stringent reliability requirements that are often associated with aggressive compression. Maximizing traffic capacity requires adjusting the source bitrate, reliability and delay trade-offs, taking into account the resource cost of these quality-of-service (QOS) requirements in the transmission media. The source representation must be highly scalable to different bitrates and reliability, for example connections with and without wireless access. Further gains can be achieved by segmenting information from a single source into different QOS classes, and fine tuning the transmission resources (like power, coding redundancy, etc.) so that no source segment receives more transmission resources than required.

The concatenation of transmission media with quite different characteristics (for example a broadband backbone with wireless access will be common in the future) is a complication. It is easy in this context to seriously violate modularity, for example, by using transcoding from one source coding standard to another as shown in Figure 8.4a. Using a different compression standard on each link allows a customized per-link JSCC. (A past example of this is digital cellular telephony, which utilizes transcoding from 64 kb/s μ - 255 speech to 13 kb/s VCELP.) Transcoding has poor modularity because the source coding and the transport are designed tightly as a unit, and one cannot easily change without the other. Further, it is difficult to introduce new and improved compression standards if existing standards are already widely deployed within the network. There are problems related to the other requirements as well. The transcoder introduces considerable delay (a serious disadvantage in a global network) and the accumulation of quantization impairment, and is incompatible with end-to-end encryption.

Multicast connections (from a single source to two or more sinks) are a scalable solution to multiple sinks (as in group video conferencing for example), since the source generates a single stream irrespective of the number of sinks. However, multicast offers a more serious JSCC challenge, as illustrated in Figure 8.4b. Downstream from a multicast splitting point, there will in general be heterogeneous transport links that have to be accommodated *simultaneously*. Again, transcoding is a feasible solution, but one that has the same difficulties as in unicast.

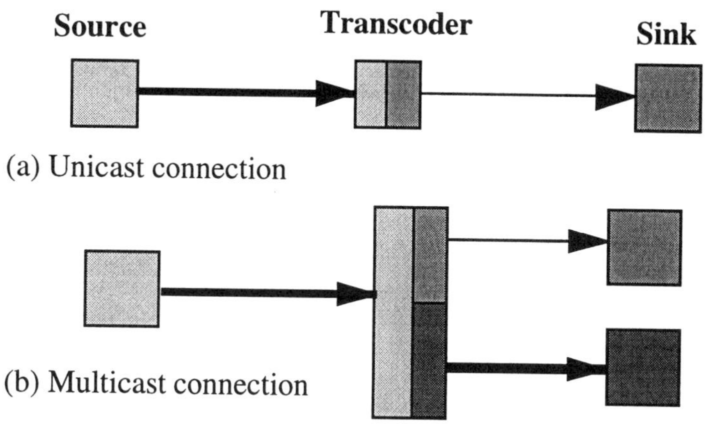

Figure 8.4 Transcoder architecture for JSCC in a heterogeneous transport environment.

Fortunately, JSCC in both unicast and multicast connections can be accommodated utilizing the alternative substream architecture [3] shown in Figure 8.5. The abstraction of the transport from the source perspective is a set of substreams, with different quality of service (QOS) requirements (delay and reliability) for each substream. Those QOS requirements can be negotiated between source and transport at setup. The source then configure itself for substreams with the expected QOS, trying to attain the highest subjective quality. To the transport, the source is abstracted as a set of substreams with specified QOS objectives. Internally, the transport disaggregates the negotiated QOS for each substream to configure the QOS of the individual links.

Multicast presents special scalability challenges. One the one hand, heterogeneous downstream links and terminals can be accommodated by choosing a subset of substreams for each downstream multicast subtree. On the other hand, sinks may be entering or leaving the multicast group at any time. It is not scalable to presume that the source can negotiate with an intermediate (and even varying) number of sinks and transport links, or that all sinks should be required to reconfigure whenever a sink enters or leaves the multicast group. Thus, a serious challenge is to configure the source coding and substreams to

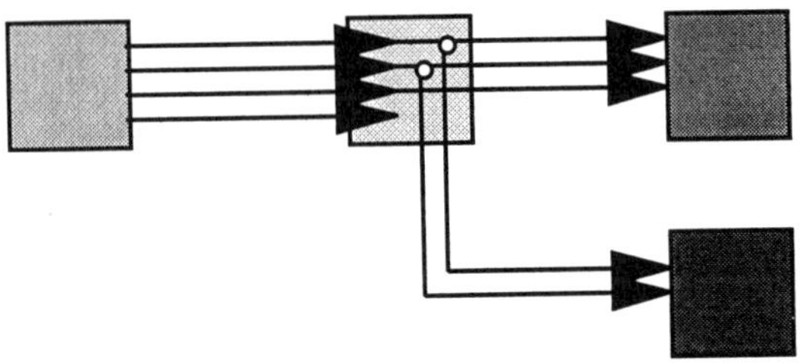

Figure 8.5 A substream architecture manages JSCC as well as heterogeneous transport and terminal environment in unicast and multicast connections.

simultaneously satisfy the differing needs of a generic and probably unknown set of downstream transport links and sinks.

The substream abstraction offers good modularity, since the source needs no knowledge of the details of how the transport achieves a given QOS. The only details of the transport visible to the source are the fundamental properties: delay and reliability. The transport has no knowledge of the service or compression standard being utilized, only the desired QOS (delay and reliability) and bitrate. Configurability allows reuse of the transport for new sources or compression standards in the future. Further, the architecture is consistent with end-to-end encryption, as long as each substream is independently encrypted.

Having defined an architecture, a number of interesting new research issues arise. Source coding must interface the variable-QOS substream transport abstraction, configuring to the available substream QOS while achieving the highest subjective quality [4]. In addition, source coding needs scalability to the varying bandwidth, processing, and display resolution of heterogeneous sinks. (In fact, if multicast connections are supported, the source representation has to embed these differing requirements in a common set of substreams.) Another interesting problem is JSCC in the delay dimension, which we have addressed for both video [5][6] and graphics [7]. Within the transport, exploiting the substream architecture for higher multiaccess wireless traffic capacity is particularly interesting, as it leads to the new problem of variable QOS in multiaccess environments. We have addressed this in CDMA, utilizing power control to provision variable reliability QOS [8][9], as well as packet scheduling to provision variable delay QOS [6].

5 CONCLUSIONS

As we move from homogeneous networks provisioning simple, universal, largely nonconfigurable telecommunications applications, to an environment that is heterogeneous in applications, transport media, and terminals, complexity management becomes a critical element of success. Carefully crafted architectures are needed to meet all the functional requirements, achieve adequate levels of performance, and offer a seamless environment for the provisioning of applications. Complexity management, far from displacing traditional signal processing, communications theory, and queuing considerations, raises many interesting new questions and serious challenges in all the detailed constituent traditional disciplines, such as compression, encryption, error-control, modulation, protocols, etc. While efficiency remains an important consideration in multiaccess wireless systems, many other considerations relating to the new functional, configurability, and scalability requirements are brought to the fore. These traditional disciplines will doubtless be revitalized by these new issues.

REFERENCES

[1] D.G. Messerschmitt, "The convergence of communications and computing: what are the implications today?", submitted to *IEEE Proceedings.* (Also available at http://www.eecs.berkeley.edu/~messer/PAPERS/PRESS/Convergence.html)

[2] *Merriam-Webster's Collegiate Dictionary*, Tenth Edition, Meriam-Webster, Inc., 1995.

[3] P. Haskell and D.G. Messerschmitt, "In favor of an enhanced network interface for multimedia services", to appear in *IEEE Multimedia Magazine.*

[4] L.C. Yun and D.G. Messerschmitt, "Digital video in a fading interference wireless environment", *IEEE Int. Conf. on Acoustics, Speech, and Signal Processing*, Atlanta, GA, May 1996.

[5] A. Lao, J. Reason, and D.G. Messerschmitt, "Layered asynchronous video for wireless services", *IEEE Workshop on Mobile Computing Systems and Applications*, Santa Cruz, CA, Dec. 1994.

[6] J.M. Reason, L.C. Yun, A.Y. Lao, D.G. Messerschmitt, "Asynchronous video: coordinated video coding and transport for heterogeneous networks with wireless access", *Mobile Computing*, H.F. Korth and T. Imielinski, Ed., Kluwer Academic Press, Boston, MA, 1995.

[7] R. Han and D.G. Messershmitt, "Asymptotically reliable transport of text/graphics over wireless channels", *Proc. Multimedia Computing and Networking*, San Jose, January 29-31, 1996.

[8] L.C. Yun and D.G. Messershmitt, "Power control and coding for variable QOS on a CDMA channel", *Proc. IEEE Military Communications Conference*, Oct. 1994.

[9] L.C. Yun and D.G. Messershmitt, "Variable quality of service in CDMA systems by statistical power control", *Proc. IEEE International Conference on Communications*, June 18-21, 1995, Seattle, WA.

PART II

MATHEMATICS AND COMPUTATION

9

HORIZONTAL, VERTICAL, ESPECIALLY DIAGONAL

Patrick Dewilde

Afdeling der Elektrotechniek
Delft University of Technology
Mekelweg 4, Postbus 5031
2600 GA Delft, The Netherlands

To my mentor, inspirer and friend, Tom Kailath
on his 60th birthday.

ABSTRACT

The paper retraces some of the steps that have led to two types of diagonal calculus: the displacement structure theory, and the realization theory for time-varying systems, and attemps at bringing the two together. It focuses attention on minimal computational complexity, which it studies using embedding or scattering theory. This approach produces on the way a simple proof and closed form solution of the (time-varying) 'bounded-real' (Kalman-Yacubovitch-Popov) lemma. The paper ends with a theorem that exhibits the differences and the connection between the two diagonal approaches.

1 INTRODUCTION

One of the most outstanding Flemish poets, Paul van Ostaaijen wrote a graphical poem in which he extolled the virtues of doing things 'especially' diagonal, as opposed to rigidly horizontal or vertical (my interpretation!). The diagonal dimension aptly describes how my interactions with Tom Kailath have been over the years, and in this paper I wish not only to retrace some of the steps, but also to place them in my present understanding. It so happened that Tom and I (with many students and collaborators) explored different though related diagonal dimensions (I'll soon explain), and although we knew of each other's work, the connections have remained somewhat mysterious. I hope that the present paper will contribute to making a synthesis of our different tracks. The two diagonal theories aim at reducing computational complexity, but of different problems and in different ways. Their conjunction should provide for a double reduction.

The classical horizontal, and its dual, vertical dimensions arose in linear algebra (undoubtedly a product of mechanics ...) when numbers were welded together to form vectors, and calculus for these new objects was developed. The epitome of this trend of thought was found in the Jacobi or Singular Value Decomposition (SVD) which describes the 'geometry' of a linear map. In this decomposition, a linear map which is represented by an $m \times n$ matrix of rank k, gets decomposed as a product of three matrices:

$$A = \begin{bmatrix} \vdots & & \vdots \\ u_1 & \dots & u_k \\ \vdots & & \vdots \end{bmatrix} \Sigma \begin{bmatrix} \cdots v_1^* \cdots \\ \cdots \\ \cdots v_k^* \cdots \end{bmatrix}$$

in which the set $\{u_1 \cdots u_k\}$ consists of orthonormal vectors spanning the range of A, Σ is a diagonal matrix of singular values, positive numbers $\sigma_1 \geq \sigma_2 \geq \cdots \geq \sigma_k$, and the set $\{v_1, \cdots, v_k\}$ consists of orthonormal vectors spanning the range of the adjoint A^*.

The diagonal dimension I wish to consider (not provided by Σ!) is one in which not a single map is involved, but a sequence of maps progressing in time. One first gets an intuition of a diagonal progression when one considers the flow graph of a computation. Fig. 1 shows the data flow graph for the QR-algorithm (sometimes called a 'Gentlemen-Kung' array, although such flow graphs have already a long history).

The dashed lines in fig. 1 show the progression in time - the signal wavefront. It was extensively studied by a number of Tom Kailath's students (S. Rao [17], S.Y. Kung [13] and H.V.Jagadish [8]). The evolution of time progresses diagonally through the algorithm and generates intermediate data which we call the 'state' of the computation. Since the operators used in the QR algorithm are orthogonal transformations, there is also conservation of energy in some sense, soon to be made more explicit. Such computations are reminiscent of the space-time relation of transmission line theory.

2 DISPLACEMENT CALCULUS

One of the great contributions of the 'Stanford School' under the direction of Tom Kailath, is the introduction of the notion of displacement structure, first in integral equations and then, in matrix calculus [9]. Starting with the thesis of Martin Morf [15], the calculus evolved as an underlying concept in the theory of square root and fast array algorithms, see the review [18]. Let T be a positive definite matrix (say a covariance matrix) and let us suppose that T

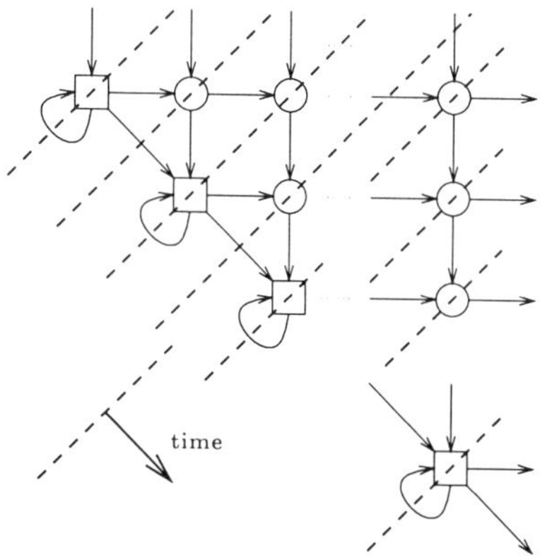

Figure 9.1 The flow graph for the QR algorithm, in which a matrix which is being fed in row-wise from the top, gets converted via 2 × 2 orthogonal rotations to an upper triangular form recursively.

has 'structure' in the following sense. Let

$$\sigma = \begin{bmatrix} 0 & & & 0 \\ 1 & \ddots & & \\ & \ddots & \ddots & \\ 0 & & 1 & 0 \end{bmatrix}$$

be a 'lower shift' matrix, and suppose that the matrix

$$T - \sigma T \sigma^* = X \qquad (9.1)$$

has low rank α - it is by definition the *displacement rank*, because the left hand side gives the difference between the matrix T and a diagonally down-shifted version of it. Suppose that X has the signature (it is a hermitian matrix)

$$J = \begin{bmatrix} 1_p & \\ & -1_q \end{bmatrix}$$

(in which '1_n' denotes an identity matrix of dimension n), then one can write

$$X = GJG^* \qquad (9.2)$$

for some $n \times (p+q)$ matrix which we further partition as:

$$G = \begin{bmatrix} \vdots & \vdots \\ g & h \\ \vdots & \vdots \end{bmatrix} = \begin{bmatrix} g_0 & h_0 \\ \vdots & \vdots \\ g_{n-1} & h_{n-1} \end{bmatrix} \tag{9.3}$$

It turns out that the Cholesky factorization of $T = LL^*$ can be found recursively in $O(\alpha n^2)$ operations instead of the usual $O(n^3)$. The recursive algorithm is an incarnation of the famous Schur algorithm [21, 5]. I summarize it briefly at the end of this section, see the theses of Hanoch Lev-Ari [14] and Ali Sayed [19], as well as the survey paper [18] for a complete account. The algorithm uses two basic operations:

- The embedding of a vector of Euclidean norm one in a unitary matrix: let u be a $1 : n$ vector of Euclidean norm one, decomposed as $u = e^{j\phi_1}[r_1 \; x_2]$, with $r_1 \geq 0$; define

$$\mathcal{U}(u) = \begin{bmatrix} e^{j\phi_1} r_1 & e^{j\phi_1} x_2 \\ -x_2^* & I - (1 - \sqrt{1 - x_2 x_2^*}) x_2^* x_2 \end{bmatrix} \tag{9.4}$$

  then $\mathcal{U}(u)$ is an 'efficient' unitary matrix with u as its first row (meaning that its product with a vector can be computed in $O(n)$ operations - (9.4) is of the same nature but numerically better than a Householder transformation).

- The embedded hyperbolic (Halmos) transformation: the basic hyperbolic transformation $H(\rho)$ with reflection coefficient ρ ($|\rho| < 1$) is given by:

$$H(\rho) = \frac{1}{\sqrt{1 - |\rho|^2}} \begin{bmatrix} 1 & \rho \\ \rho^* & 1 \end{bmatrix},$$

  embedded in a unit $(p+q)$ matrix it becomes a matrix $\mathcal{H}_{i,j}(\rho)$ of size $(p+q) \times (p+q)$ which is a unit matrix except on the intersections of the ith and $(p+j)^{\text{th}}$ rows and columns where it is equal to $H(\rho)$.

The Schur algorithm yields the following results:

- it produces the Cholesky factor L of the factorization $T = LL^*$;
- the corresponding 'chain scattering matrix' $\Theta(z)$ has a realization

$$\Theta(z) = \begin{bmatrix} D_{11} & D_{21} \\ D_{12} & D_{22} \end{bmatrix} + \begin{bmatrix} B_1 \\ B_2 \end{bmatrix} z(I - Az)^{-1}[C_1 \; C_2]$$

for which

$$\begin{bmatrix} A \\ \hline B_1 \\ B_2 \end{bmatrix} = \begin{bmatrix} \sigma^* \\ \hline g_0^* \cdots g_{n-1}^* \\ h_0^* \cdots h_{n-1}^* \end{bmatrix}$$

is a reachability pair. Moreover, Θ is J-lossless , which means that

1. $J - \Theta(z)^* J \Theta(z) \geq 0$ for $|z| \leq 1$,
2. $J = \Theta(z)^* J \Theta(z)$ for $|z| = 1$.

$\Theta(z)$ is of course a matrix of polynomials in z.

- connected to $\Theta(z)$ there is a $p \times q$ transfer operator $S(z)$ which is contractive in $\{z : |z| \leq 1\}$ and for which the interpolation property

$$h(z) = g(z)S(z) + O(z^n)$$

holds. A suitable $S(z)$ is given by:

$$S(z) = -\Theta_{12}(z)\Theta_{22}^{-1}(z),$$

(the inverse always exists!)

- etc....

There exist also time-continuous versions of the the Schur algorithm, see the work of Tom Kailath and Freddy Bruckstein [2, 4] [3], or our own contribution [16]. See also [20] [18], [19].

The moral of the story is that the alpha-stationary or displacement structure theory is closely related or equivalent to an interpolation theory for a linear time-invariant system. This observation holds also for many structured matrix inversion problems, see e.g. the recent work of Kailath and Olshevsky [11]. Interpolation problems of this sort have many other connections as well, a non-exhaustive list: H_∞-control, Kalman filtering, the LMS algorithm, Game theory, the Partial Lossless Inverse Scattering Problem, QR-factorization, Inner-Outer factorization etc...

The Schur Algorithm

Let (initialization):

$$G^{(0)}(z) = g_0^{(0)} + g_1^{(0)} z + \cdots + g_{n-1}^{(0)} z^{n-1} (= g(z))$$

$$H^{(0)}(z) = h_0^{(0)} + h_1^{(0)} z + \cdots + h_{n-1}^{(0)} z^{n-1} (= h(z))$$

The kth step:

Given:

$$G^{(k-1)}(z) = g_0^{(k-1)} + g_1^{(k-1)} z + \cdots$$

$$H^{(k-1)}(z) = h_0^{(k-1)} + h_1^{(k-1)} z + \cdots$$

Compute:

$$\theta_k = \left[\begin{array}{c|c} \mathcal{U}^*\left(\frac{g_0^{(k-1)}}{\|g_0^{(k-1)}\|}\right) & 0 \\ \hline 0 & \mathcal{U}^*\left(\frac{h_0^{(k-1)}}{\|h_0^{(k-1)}\|}\right) \end{array}\right] \mathcal{H}_{1,1}\left(-\frac{\| h_0^{(k-1)} \|}{\| g_0^{(k-1)} \|}\right)$$

$$[G^{(k)}(z)\ H^{(k)}(z)] = z^{-1}[G^{(k-1)}(z)\ H^{(k-1)}(z)].\theta_k. \left[\begin{array}{cccc|ccc} z & & & & & & \\ & 1 & & & & & \\ & & \ddots & & & & \\ & & & 1 & & & \\ \hline & & & & 1 & & \\ & & & & & \ddots & \\ & & & & & & 1 \end{array}\right]$$

which corresponds to the 'chain scattering matrix' filter given by (with $E(z) = \text{diag}\{z, 1, \cdots 1, 1, \cdots, 1\}$):

$$\Theta(z) = \theta_1 E(z) \theta_2 E(z) \cdots \theta_n,$$

and is represented by the signal flow diagram:

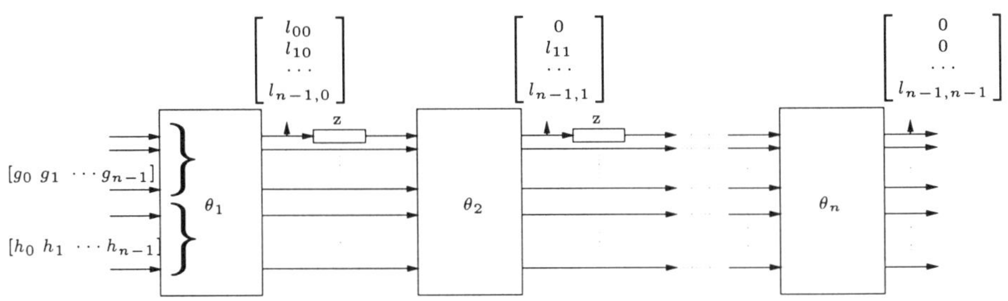

Figure 9.2 The Schur algorithm to compute the Cholesky decomposition of an alpha-stationary matrix.

3 THE OTHER DIAGONAL DIMENSION

However, there is another diagonal dimension! In the previous section we were looking at the change of the systems' characteristics over time, how its parameters vary. The other dimension has to do with the evolution of the data or signals processed by the system. One is time-update, the other state-update. In this section we start exploring this second dimension, and will harmonize the two in a later section.

An example

Suppose that T is a finite (positive definite real) Töplitz matrix:

$$T = \begin{bmatrix} 1 & c_1 & \cdots & & c_{n-1} \\ c_1 & \ddots & \ddots & & \vdots \\ \vdots & \ddots & \ddots & \ddots & \\ c_{n-1} & \cdots & & c_1 & 1 \end{bmatrix}.$$

The displacement rank is two here, since:

$$C - \sigma C \sigma^* = \begin{bmatrix} 1 & 0 \\ c_1 & c_1 \\ \vdots & \vdots \\ c_{n-1} & c_{n-1} \end{bmatrix} \begin{bmatrix} 1 & \\ & -1 \end{bmatrix} \begin{bmatrix} 1 & c_1 & \cdots & c_{n-1} \\ 0 & c_1 & \cdots & c_{n-1} \end{bmatrix}.$$

The dimension of the state space, however is variable, but at most equal to:

$$\delta = \max \operatorname{rank}_k \begin{bmatrix} c_1 & c_2 & \cdots & c_k \\ c_2 & \iddots & \iddots & \vdots \\ \vdots & \iddots & \iddots & \vdots \\ c_k & \cdots & \cdots & c_{n-1} \end{bmatrix}$$

which is also the minimum degree of any extension $\gamma(z) = 1 + 2c_1(z) + \cdots + 2c_{n-1}z^{n-1} + O(z^n)$, by Kronecker's theorem [12]. We shall see in the last section that this means that there exits computational schemes for the product Lx which utilize no more that δ intermediate quantities. This result can be strenghtened considerably to yield calculations which are also algebraically minimal, i.e. which use just as many parameters as the number of free algebraic variables. These ideas lead to a simple but major result of minimal realization theory for linear time-varying systems:

The minimum number of states at each stage of the computation determines the algebraic complexity of the problem.

We shall produce such minimal algebraic realizations using embedding theory.

The transfer operator of a linear time varying causal system can be represented by a block upper triangular operator:

$$[\cdots, u_{-1}, \boxed{u_0}, u_1, \cdots] \begin{bmatrix} \ddots & \ddots & \ddots & \ddots & \ddots & \ddots \\ 0 & T_{-1,-1} & T_{-1,0} & T_{-1,1} & T_{-1,2} & \ddots \\ & 0 & \boxed{T_{0,0}} & T_{0,1} & T_{0,2} & \ddots \\ & & 0 & T_{1,1} & T_{1,2} & \ddots \\ & 0 & & 0 & T_{2,2} & \ddots \\ & & & & & \ddots \end{bmatrix} =$$

$$= [\cdots, y_{-1}, \boxed{y_0}, y_1, y_2, \cdots].$$

All dimensions of input and output vectors may be variable (including empty). The 'Hankel map' H_k at time k maps an input with support on $t < k$ to the systems' output, restricted to $t \geq k$. It is given by the matrix or operator:

$$H_k = \begin{bmatrix} \vdots & \vdots & \\ T_{k-2,k} & T_{k-2,k+1} & \cdots \\ T_{k-1,k} & T_{k-1,k+1} & \cdots \end{bmatrix},$$

which we call the *Hankel operator at time k*, the global Hankel operator being the collection of H_k's over all k. The rank of H_k determines the minimal number of state variables needed at stage k of the computation. Let $\delta_k = \text{rank}(H_k)$, then a realization in terms of minimal state dimensions is given by fig. 9.3.

Interesting questions which arise at this point are:

(a) How is 'state complexity' related to 'computational complexity'?

(b) Complexity can be reduced further by allowing some tolerance on the accuracy of the computation: given a level of tolerance, what is then the minimal computational complexity?

(c) When does a low displacement rank system also have a low dimensional state representation?

It turns out that all these questions have (attractive) answers. In this presentation I shall not dwell on item (b), see the literature ([6, 23]) for a fairly exhaustive treatment, but I shall approach (a) from the point of view of embedding theory, the bounded real lemma (an important special case of the Kalman-Yacubovitch-Popov lemma), and present some preliminary thoughts on (c).

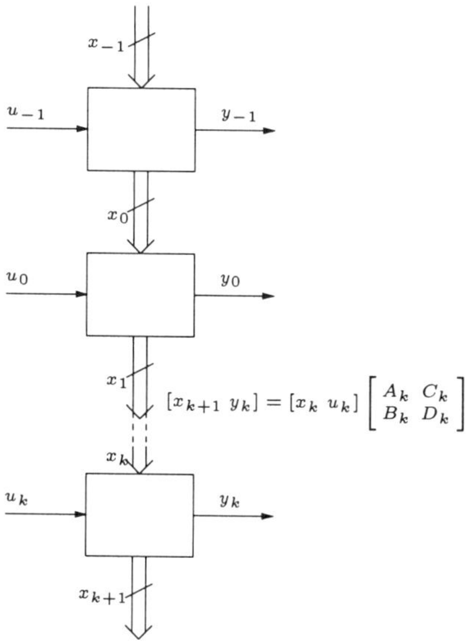

Figure 9.3 The signal flow graph for a state-minimal realization. At each stage the dimension of the incoming state space is equal to the rank of the corresponding 'Hankel' operator.

4 EMBEDDING THEORY

A very attractive road to minimal algebraic realizations - one which parallels the design theory of lossless filters - goes via unitary embedding theory for contractive (linear, time-varying) operators. The embedding theory yields in one breath a proof for the bounded real lemma, and a closed form solution to the related Riccati equation. I follow the treatment given by Alle-Jan van der Veen in his thesis ([22]).

We start out with a given contractive operator (matrix) S which we wish to embed in a unitary operator Σ of same degree (same local state complexity):

$$\Sigma = \begin{bmatrix} S & \Sigma_{12} \\ \Sigma_{21} & \Sigma_{22} \end{bmatrix}$$

We assume that S is given in state space form, i.e. that there exists a realization with finite local state dimension for S:

$$\begin{cases} x(k+1) = x(k)A(k) + u(k)B(k) \\ y(k) \quad\;\;\;\; = x(k)C(k) + u(k)D(k) \end{cases} \tag{9.5}$$

and we try to find matrices $B_2(k)$, $C_2(k)$, $D_{12}(k)$, $D_{21}(k)$ and $D_{22}(k)$ so that the augmented system:

$$\begin{cases} x(k+1) = x(k)A(k) + u(k)B(k) + u_2(k)B_2(k) \\ y(k) \quad\;\; = x(k)C(k) + u(k)D(k) + u_2(k)D_{21}(k) \\ y_2(k) \quad\; = x(k)C_2(k) + u(k)D_{12}(k) + u_2(k)D_{22}(k) \end{cases} \tag{9.6}$$

is a realization of a unitary system Σ. There is a more compact notation for state equations of time-varying systems, one which allows for matrix representations of complex objects such as the Hankel operator which I introduced earlier and which is a 'diagonal' collection of operators. The notation uses diagonals as single symbols, just as what one does when one represents a vector by a single symbol. Let us define:

$$A = \mathrm{diag}\{\cdots, A_{-1}, \boxed{A_0}, A_1, \cdots\},$$

$$B = \mathrm{diag}\{\cdots, B_{-1}, \boxed{B_0}, B_1, \cdots\}$$

etc... as diagonal, i.e. instantaneous, operators, as well as the unit shift

$$Z = \begin{bmatrix} \ddots & \ddots & & & \\ & 0 & I & 0 & \\ & & \boxed{0} & I & 0 \\ & & & 0 & I \\ & & & \ddots & \ddots & \ddots \end{bmatrix}$$

In this notation, the global state evolution becomes:

$$[xZ^{-1}\; y] = [x\; u] \begin{bmatrix} A & C \\ B & D \end{bmatrix}$$

and the corresponding transfer operator has the representation:

$$S = D + BZ(I - AZ)^{-1}C$$

provided the inverse is meaningful. This will e.g. be the case when the spectral radius $\rho(AZ) < 1$, see [1], a property which we just may assume here.

The embedding problem now consists in finding diagonal operators B_2, C_2, D_{ik} such that the realization

$$\begin{bmatrix} A & C & C_2 \\ B & D & D_{12} \\ B_2 & D_{21} & D_{22} \end{bmatrix}$$

corresponds to a unitary transfer operator. Some thought about this point leads to the search for a state transformation $R: \; x' = xR$ (in which R is a time-varying but invertible diagonal operator) which is such that the resulting transformed state realization is unitary. With $R^{(k)} \triangleq Z^{*k} R Z^k$ for a kth shift along the diagonal in the South-East direction, the transformed realization becomes:

$$\sigma = \begin{bmatrix} RA\left(R^{(-1)}\right)^{-1} & RC & RC_2 \\ B\left(R^{(-1)}\right)^{-1} & D & D_{12} \\ B_2\left(R^{(-1)}\right)^{-1} & D_{21} & D_{22} \end{bmatrix}$$

which will be unitary if (and only if) there exists a positive definite diagonal operator M which satisfies:

$$\begin{cases} A^*MA + B^*B + B_2^*B_2 &= M^{(-1)} \\ C^*MC + D^*D + D_{21}^*D_{21} &= I \\ C^*MA + D^*B + D_{21}^*B_2 &= 0 \end{cases} \tag{9.7}$$

M provides the needed state transformation as $M = R^*R$. The bounded real lemma states that S will be contractive if and only if there exists a positive definite M which satisfies the following two constraints: (1) $I - D^*D - C^*MC$ is non-singular and (2)

$$M^{(-1)} = A^*MA + B^*B + (A^*MC + B^*D)(I - D^*D - C^*MC)^{-1}(C^*MA + D^*B) \tag{9.8}$$

It should not be too difficult to see that solutions of (9.7) and (9.8) are equivalent modulo the invertibility condition.

Equation (9.8) is the famous Riccati equation related to the embedding problem. There is a huge literature describing its properties, convergence of recursions, see in particular [10] etc... Remarkably, however, the equation does have a closed form solution, which is simply an expression of 'conservation of energy'. I give a sketch of its derivation. Before doing so, let me remark that the time varying equation subsumes the time invariant or algebraic case, because then the diagonal shift $M^{(-1)} = M$ so that (9.8) becomes indeed algebraic.

To construct the solution we shall need the diagonal calculus defined earlier. Operators will now consist of entries which are themselves diagonals. Instead of working with a single input-output map $u \mapsto y$, we take such a map at each point k in time, i.e. we let the operator operate on a stack of inputs which we represent as a sequence of diagonals and map to a similar stack of outputs represented as a sequence of diagonals as well. Inputs (and outputs) form a 'tableau':

$$u \doteq \begin{bmatrix} \ddots & \ddots & \ddots & \ddots & \ddots \\ \cdots & u_{-1,-1} & u_{-1,0} & u_{-1,1} & \cdots \\ \cdots & u_{0,-1} & \boxed{u_{0,0}} & u_{0,1} & \cdots \\ \cdots & u_{-1,-1} & u_{-1,0} & u_{1,1} & \cdots \\ \ddots & \ddots & \ddots & \ddots & \ddots \end{bmatrix}$$

and we write this u as a sequence of diagonals:

$$U = \{\cdots, U_{\{-2\}}, U_{\{-1\}}, U_{\{0\}}, U_{\{1\}}, \cdots\}$$

in which the $U_{\{k\}}$'s are the diagonals of the tableau choosen so that $u = \sum U_{\{k\}} Z^k$ (energy is still measured as the overall sum squared, or in mathematical language, the Hilbert-Schmidt norm). The operator S will be contractive on sequences of diagonals just as before (the diagonalization allows us to consider all time points in one formula, alternatively one would have to introduce tensorcalculus to handle the sequences of matrices that the operators we are considering are built from).

Furthermore, we decompose U in two parts, strict past and future:

$$U = U_p + U_f$$

with

$$\begin{aligned} U_p &= \{U_{\{-1\}}, U_{\{-2\}}, \cdots\} \\ U_f &= \{U_{\{0\}}, U_{\{1\}}, U_{\{2\}}, \cdots\}. \end{aligned}$$

At each point in time k, U_p contains a past sequence.

We solve the Riccati equation by forcing the past of the embedded system to conserve energy, i.e. to be such that the map from past inputs to the product of past outputs and current state is unitary. The 'past' of a system is shown graphically in fig. 9.4. It consists of two operators: the reachability operator $\mathcal{R}$ which maps U_p to the state X (the state at time point i for input series i, $X = \text{diag}\{x_{i,i}\}$) and the past part of the transfer operator K which maps U_p to Y_p. Expressed in our diagonal notation, these two operators are given by:

$$\mathcal{R} = \begin{bmatrix} B^{(1)} \\ B^{(2)}A^{(1)} \\ B^{(3)}A^{(2)}A^{(1)} \\ \cdots \end{bmatrix}$$

$$K = \begin{bmatrix} T_0^{(1)} & 0 & \\ T_1^{(1)} & T_0^{(2)} & 0 \\ T_2^{(1)} & T_1^{(2)} & \ddots \\ \ddots & \ddots & \ddots \end{bmatrix} = \begin{bmatrix} D^{(1)} & & 0 & \\ B^{(2)}C^{(1)} & D^{(2)} & & 0 \\ B^{(3)}A^{(2)}C^{(1)} & B^{(3)}C^{(2)} & D^{(3)} & \\ \ddots & \ddots & \ddots & \ddots \end{bmatrix}.$$

The trick is now to find a state metric (state Gramian $xM^{-1}x^*$ with $M = R^*R$) which is such that the past of the embedded system produces a unitary map (at each and every time point), see fig. 9.5.

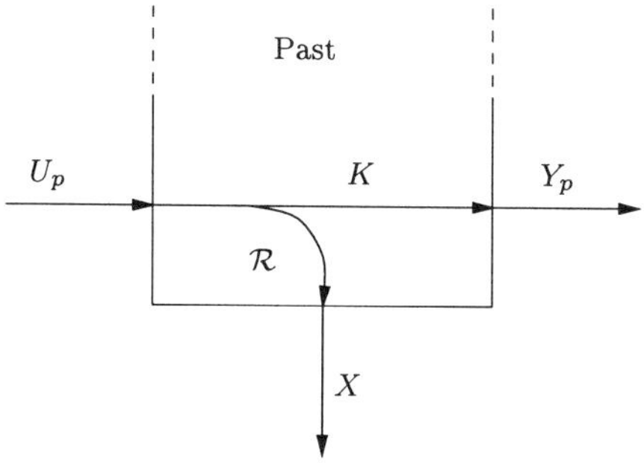

Figure 9.4 Global representation of the past of a system.

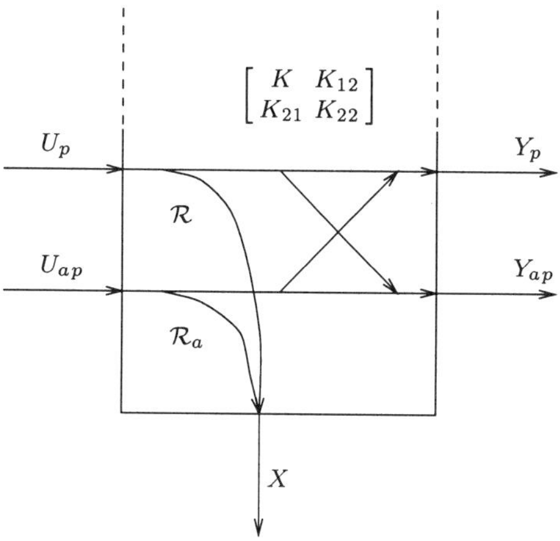

Figure 9.5 The embedded past of the system must yield a unitary map from past inputs to state $\times$ output.

The unitarity condition has as a consequence that:

$$\begin{bmatrix} \mathcal{R}^* & \mathcal{R}_a^* \\ K^* & K_{21}^* \\ K_{12}^* & K_{22}^* \end{bmatrix} \begin{bmatrix} \mathcal{R} & K & K_{12} \\ \mathcal{R}_a & K_{21} & K_{22} \end{bmatrix} = \begin{bmatrix} M & & \\ & I & \\ & & I \end{bmatrix} \tag{9.9}$$

or:

$$\begin{cases} M = \mathcal{R}^*\mathcal{R} + \mathcal{R}_a^*\mathcal{R}_a \\ 0 = K^*\mathcal{R} + K_{21}^*\mathcal{R}_a \\ I = K^*K + K_{21}^*K_{21} \end{cases}$$

K_{21} and $\mathcal{R}_a$ can be eliminated to yield a closed form expression for M as:

$$M = \mathcal{R}^*(I - KK^*)^{-1}\mathcal{R}$$

and a solution to the Riccati equation (9.8).

Is this solution 'computable'? It represents the solution in the sense of converging series. In his thesis [22], Alle-Jan van der Veen shows that the Riccati equation is equivalent to recursive applications of the Schur complementation theory on a well choosen set of operators, and leads again to an incarnation of the Schur algorithm.

5 ALGEBRAIC MINIMALITY

Can a transfer operator (linear time invariant or linear time varying) be realized with a number of 'elementary' operations equal to the number of free algebraic parameters? Do minimal state realizations play a role in this process? We shall see that the embedding theory is instrumental in answering these questions positively. It is not the only possible way, but an elegant one, since it produces realizations which are intrinsically numerically stable.

In the case of a real LTI system of degree δ, with input of dimension p and output of dimension q, the algebraic degree is given by:

$$(p+q)(\delta+\frac{1}{2}(p+q-1)),$$

as one can infer from a fractional expansion of T. In the LTV case the expressions are more complex due to the variable nature of the state dimension, but they are of the same nature.

Realization theory is dependent on the choice of the basic elementary operation. In the present case we choose the Jacobi or elementary orthogonal rotation. It is a function of a single rotation angle ϕ, and given by

$$\jmath(\phi)=\begin{bmatrix}\cos\phi & -\sin\phi \\ \sin\phi & \cos\phi\end{bmatrix}.$$

$\jmath(\phi)$ can be realized using a building block called the CORDIC, for which there exist nice realizations, see e.g. [7]. Its usage on an embedded, orthogonal realization starts out from one in which the A_k matrices are upper triangular (which can always be achieved using state transformations of the QR type - for details see [22]) and then reduces it further by annihilating off-diagonal elements. Due to the orthogonality, there occurs an automatic diagonalization because as soon as a column has only a single non-zero element (which has to be equal in magnitude to one), then the row corresponding to that element can also have only one non-zero entry. The process is illustrated in the following reduction:

$$\left[\begin{array}{c|cc} A & C_1 & C_2 \\ \hline B_1 & D_{11} & D_{12} \\ B_2 & D_{21} & D_{22}\end{array}\right] =$$

$$= \left[\begin{array}{ccc|cc} * & \cdots & * & * & * \\ 0 & \ddots & \vdots & \vdots & \vdots \\ 0 & 0 & * & * & * \\ \hline * & \cdots & * & * & * \\ & \cdots & * & * & * \end{array}\right] \Rightarrow \left[\begin{array}{ccc|cc} * & \cdots & \mathbf{0} & \mathbf{0} & \mathbf{0} \\ 0 & \ddots & \vdots & \vdots & \vdots \\ 0 & 0 & * & * & * \\ \hline 0 & \cdots & * & * & * \\ 0 & \cdots & * & * & * \end{array}\right] \Rightarrow \left[\begin{array}{cccccc|cc} * & 0 & 0 & \cdots & & 0 & 0 & 0 \\ 0 & * & \mathbf{0} & \cdots & & \mathbf{0} & \mathbf{0} & \mathbf{0} \\ 0 & 0 & * & \cdots & & * & * & * \\ & & & & \ddots & \vdots & \vdots & \vdots \\ & & & & & * & * & * \\ \hline 0 & 0 & * & \cdots & & * & * & * \\ 0 & 0 & * & \cdots & & * & * & * \end{array}\right]$$

etc... This procedure produces a realization for each stage which looks like fig. 9.6.

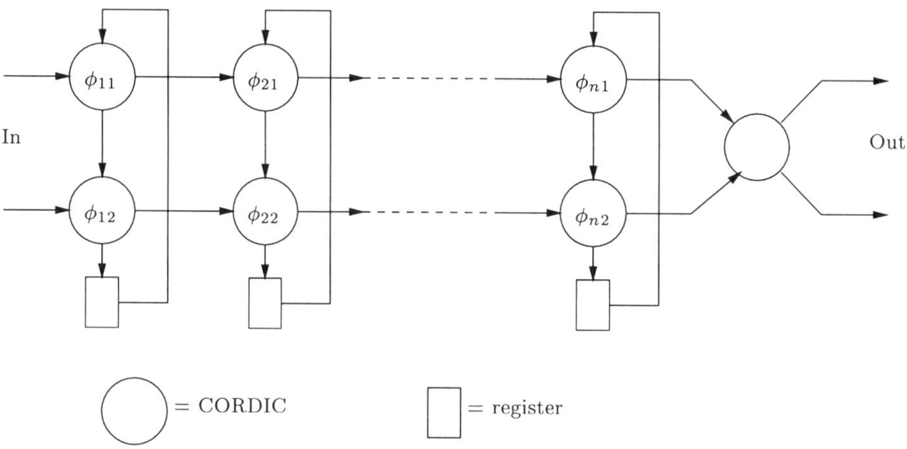

Figure 9.6 Algebraic minimal realization of a single input/single output map, obtained through embedding theory.

The realization which we obtain in this way is especially attractive, because it uses stable numerical operations throughout. It may even be converted into a realization which is insensitive for critical modes, the theory of which lies outside the scope of the present discussion.

6 WELDING THE DIAGONAL DIMENSIONS

In section 3, I showed how the displacement structure and the state complexity were complementary, they can vary independently of one another. Both are instrumental in defining the complexity of computations: the variation of parameters over time on the one hand and the number of algebraic operations on the other. Can the two be welded in one scheme? I dedicate the following theorem which answers that question in principle, to Tom Kailath on his 60th birthday.

Theorem 9.1 *Let T be a positive definite matrix and let α be its displacement rank, take*

$$T - \sigma T \sigma^* = GJG^* = \begin{bmatrix} g_0 \\ g_1 \\ \vdots \\ g_{n-1} \end{bmatrix} J [g_0^* \cdots g_{n-1}^*]$$

and let

$$\delta = \sup \operatorname{rank} \begin{bmatrix} g_1 & g_2 \cdots & g_k \\ g_2 & & g_{k+1} \\ & & \vdots \\ g_{n-k} & & g_{n-1} \end{bmatrix}.$$

Suppose further that T has the additive and Cholesky decompositions $T = \frac{1}{2}(C + C^*) = LL^*$ with C and L lower triangular matrices. Then there exist realizations for L and C of state degree at most δ.

Remark

The theorem can be strenghtened considerably by considering the local degree k at each point of the computation, but I do not want to get into details here.

Proof

Put $X = GJG^*$, then it is not hard to show that

$$T = X + \sigma X \sigma^* + \sigma^2 X \sigma^{*2} + \cdots + \sigma^{n-1} X \sigma^{*(n-1)}$$

Hence we find that

$$C = \begin{bmatrix} g_0 J g_0^* & 2g_0 J g_1^* & \cdots & 2g_0 J g_{n-1}^* \\ 0 & g_0 J g_0^* + g_1 J g_1^* & \cdots & 2(g_0 J g_{n-2}^* + g_1 J g_{n-1}^*) \\ 0 & 0 & \cdots & \\ 0 & 0 & 0 & \sum_{i=0}^{n-1} g_i J g_i^* \end{bmatrix}$$

and the kth Hankel matrix for C is given by:

$$H_k = 2 \begin{bmatrix} g_0 J g_k^* & g_0 J g_{k+1}^* & \cdots & g_0 J g_{n-1}^* \\ \vdots & \vdots & \vdots & \vdots \\ g_{k-1} J g_k^* + \cdots + g_0 J g_1^* & \cdots & & \end{bmatrix} =$$

$$= 2 \begin{bmatrix} g_0 J & & 0 \\ \vdots & \ddots & \\ g_{k-1} J & \cdots & g_0 J \end{bmatrix} \begin{bmatrix} g_k^* & g_{k+1}^* & \cdots & g_{k+n-1}^* \\ g_{k-1}^* & g_k^* & & \\ & & \cdots & \\ g_1^* & \cdots & & g_{n-1}^* \end{bmatrix}$$

and hence,

$$\operatorname{rank}(H_k) \leq \operatorname{rank} \begin{bmatrix} g_1 & g_2 & \cdots & g_k \\ g_2 & \cdots & & \\ & \cdots & & \\ & & \cdots & g_{n+k-1} \end{bmatrix}$$

and there is thus a realization for C of degree less than or equal to δ, by the Hankel theory of section 3. The realization for L can be derived from a realization for C using an embedding procedure like the procedure discussed in section 4 but now specialized to the present formulation, and hence it will also be of degree less than or equal to δ. □

7 ENVOY

I wish to conclude with the thesis that the diagonal calculus in its various guises, brings many interesting problems and ideas of signal processing and linear algebra into play. It unifies the two fields and provides for a common framework. From a mathematical point of view it succeeds in transfering important results and techniques from one field to the other, such as model reduction and norm approximation. I hope that the last section of my presentation has convinced the audience that there is much room for further research, and I look forward to cooperating with Tom on these and other matters for many years to come.

REFERENCES

[1] Daniel Alpay, Patrick Dewilde, and Harry Dym. Lossless inverse scattering and reproducing kernels for upper triangular operators. *Operator Theory, Advances and Applications*, 47:61–135, 1990.

[2] A.M. Bruckstein. *Scattering Models in Signal Processing.* PhD thesis, Stanford University, 1984.

[3] A.M. Bruckstein, B.C. Levy and T. Kailath, "Differential Methods in Inverse Scattering," *SIAM J. Appl. Math.*, **45**(2):312-335, April 1985.

[4] A.M. Bruckstein and T. Kailath. Inverse scattering for discrete transmission-line models. *SIAM Review*, (29(3)), September 1987.

[5] P. Dewilde, A. Vieira, and T. Kailath. On a generalized Szegö-levinson realization algorithm for optimal linear predictors based on a network theoretic approach. *IEEE Trans. on Circuits and Systems*, CAS-25(9):663–675, September 1978.

[6] P.M. Dewilde and A.J. van der Veen. On the Hankel-norm approximation of upper-triangular operators and matrices. *Integral Eq. Operator Th.*, 17(1):1–45, 1993.

[7] G.J. Hekstra and Ed.F.A. Deprettere. Floating point cordic. In *Proc. 11th Symposium on Computer Arithmetic.* IEEE, 1993.

[8] H.V. Jagadish. *Techniques for the Design of Parallel and Pipelined VLSI Systems for Numerical Computation.* PhD thesis, Stanford University, 1985.

[9] T. Kailath, S.-Y. Kung and M. Morf, "Displacement Ranks of Matrices and Linear Equations," *J. Math. Anal. and Appl.*, Vol. 68, No. 2, pp. 395-407, 1979. See also *Bull. Amer. Math. Soc.*, Vol. 1, pp. 769-773, Sept. 1979.

[10] T. Kailath and L. Ljung. Asymptotic behaviour of constant-coefficient Riccati differential equations. *IEEE Trans. Automatic Control*, **21**(3):385–388, June 1976.

[11] T. Kailath and V. Olshevsky. Displacement structure approach to chebyshev-vandermonde and related matrices. *Integral Equations and Operator Theory*, 1995.

[12] L. Kronecker. Algebraische reduktion der schaaren bilinearer formen. *S.B. Akad. Berlin*, pages 663–776, 1890.

[13] S.Y. Kung. *Array Processors.* Prentice Hall, 1988.

[14] H. Lev-Ari. *Non-stationary Lattice Filter Modeling.* PhD thesis, Stanford University, 1983.

[15] M. Morf. *Fast Algorithms for Multivariable Systems.* PhD thesis, Stanford University, 1974.

[16] J.T. Fokkema P. Dewilde and I. Widya. Inverse scattering and linear prediction, the time continuous case. In *Stochastic Systems: The Mathematics of Filtering and Identification and Applications*, pages 351–382. D. Reidel Publ. House, 1981.

[17] S. Rao. *Regular Iterative Algorithms and Their Implementations on Processor Arrays.* PhD thesis, Stanford University, 1985.

[18] A. Sayed and T. Kailath. Displacement structure: theory and applications. *SIAM review*, September 1995.

[19] A.H. Sayed. *Displacement Structure in Signal Processing and Mathematics.* PhD thesis, Stanford University, Stanford, CA, August 1992.

[20] A.H. Sayed, T. Kailath, H. Lev-Ari, and T. Constatinescu, "Recursive Solutions of Rational Interpolation Problems Via Fast Matrix Factorization," *Integral Equations and Operator Theory*, **20**:84-118, September 1994.

[21] I. Schur. Ueber potenzreihen, die im innern des einheitskreises beschränkt sind, i. *J. Reine Angew. Math.*, 147:205–232, 1917. Eng. transl. *Operator Theory: Adv. Appl.* vol. 18, pp. 31-59, Birkhäuser Verlag, 1986.

[22] A.J. van der Veen. *Time-Varying System Theory and Computational Modeling: Realization, Approximation, and Factorization.* PhD thesis, Delft University of Technology, Delft, The Netherlands, June 1993.

[23] A.J. van der Veen and P.M. Dewilde. On low-complexity approximation of matrices. *Linear Algebra and its Applications*, 205/206:1145–1201, July 1994.

10

FAST INVERSION OF VANDERMONDE AND VANDERMONDE-LIKE MATRICES

I.Gohberg* and V. Olshevsky**

** School of Mathematical Sciences*
Tel Aviv University
Ramat Aviv 69978, Israel

*** Information Systems Laboratory*
Stanford University
Stanford, CA 94305

Dedicated to Thomas Kailath on the occasion of his sixtieth birthday, with admiration and friendship.

ABSTRACT

In this paper we compare the numerical properties of the well-known Parker-Traub and Björck-Pereyra algorithms, which both use the special structure of a Vandermonde matrix to compute the entries of its inverse. The results of numerical experiments suggest that the Parker variant of $O(n^2)$ inversion algorithm allows one not only *fast* inversion of a Vandermonde matrix, but it also gives more accuracy in the computed solution than when using other structured algorithms, or standard (*non-fast*) numerically stable methods.

We show that the Parker-Traub algorithm is connected with the well-known concept of *displacement rank*, introduced by T.Kailath and his coauthors about two decades ago, and therefore this algorithm can be generalized to invert the more general class of *Vandermonde-like* matrices, naturally suggested by the idea of displacement.

1 INTRODUCTION

In this paper we consider the numerical inversion of Vandermonde matrices $V(x) = \left[x_i^{j-1}\right]_{1\leq i,j\leq n}$, where $x = (x_i)_{i=0}^{n-1} \in \mathbf{C}^n$. Such matrices are ill-conditioned, and standard numerically stable methods in general fail to compute the entries of their inverses accurately. The use of the structure of $V(x)$ may allow one to avoid the above difficulty, and to achieve high accuracy. Here we compare two well-known methods that exploit the special structure of Vandermonde matrices : the Traub and the Björck-Pereyra algorithms.

1.1 The Traub algorithm

The algorithm, proposed by Traub in [22], is *fast* in the sense that it computes all n^2 entries of $V(x)^{-1}$ in only $6n^2$ floating point operations (flops), which compares favorably with the complexity $O(n^3)$ flops of general purpose algorithms. At the same time this algorithm is generally regarded as being numerically unstable, and the first indication of this can be found in [22].

1.2 The Björck-Pereyra algorithm

In [1] Björck and Pereyra showed how to solve a Vandermonde linear system in only $5n^2/2$ flops. Clearly a Vandermonde matrix can be inverted by applying the Björck-Pereyra algorithm to solve n linear systems, using the columns of identity matrix for the right-hand sides. The latter $O(n^3)$ scheme is no longer fast. But, for the special case of positive and monotonically ordered points

$$0 < x_1 < x_2 < ... < x_n, \tag{10.1}$$

an error analysis of [10] implies the following pleasing bound :

$$|\hat{V}(x)^{-1} - V(x)^{-1}| \leq 5nu|V(x)^{-1}| + O(u^2). \tag{10.2}$$

Here $\hat{V}(x)^{-1}$ stands for the inverse matrix computed by the Björck-Pereyra algorithm, u is the machine precision, and the operation of taking the absolute value and comparison of matrices are understood in a componentwise sense.

The latter bound was used in [10] to suggest that the $O(n^2)$ Traub inversion algorithm is fast, but not accurate, whereas the $O(n^3)$ Björck-Pereyra algorithm (applied for inversion) is the other way around. To the best of our knowledge the possibility of simultaneously fast and accurate inversion of a Vandermonde matrix was not reported anywhere. The results of our numerical experiments indicate that the algorithm, described next, satisfies these requirements; thus one does not need to sacrifice the accuracy to achieve speed.

1.3 The Parker algorithm

As we discovered, prior to both the Traub and Björck-Pereyra algorithms, Parker suggested in [20] another fast $O(n^2)$ scheme for the inversion of Vandermonde matrices. However, the Parker algorithm differs from the Traub algorithm only in one nonessential detail, and hence it is also subject to Higham's comment in [10], implying that the result of the form (10.2) will not hold for the Parker variant as well. At the same time our numerical experiments show that the small difference between the Parker and Traub algorithms is crucial, from the numerical point of view. Moreover, even in the situation most favorable for the Björck-Pereyra algorithm, *viz.* when (10.1) holds, the numerical performance of the Parker algorithm turned out to be not worse than that of the Björck-Pereyra algorithm . In fact it is much better in the other cases, not captured by (10.1). This occurrence reminds us that just comparing error bounds alone cannot be a reliable basis for making practical recommendations.

1.4 Displacement structure

The concept of *displacement structure* was first introduced by T.Kailath and his coauthors in [16], and later it was much studied and generalized, see *e.g.* recent review [18] for historical remarks, list of applications, and further references. This approach provides a unified method for the design of fast algorithms for various classes of structured matrices. In particular we shall show that the Parker and Traub algorithms can be derived by using the fact that Vandermonde matrices have displacement structure. Moreover, we shall show that these algorithms can be generalized to invert the wider class of *Vandermonde-like matrices*, naturally suggested by the concept of displacement.

1.5 Contents

The Traub and the Björck-Pereyra algorithms are described in Sections 1 and 2, where the well-known results on their numerical properties are briefly discussed. Then in Section 3 we describe the Parker algorithm, and compare it with the Traub algorithm. The high accuracy of the Parker inversion algorithm is illustrated in Section 4 by computed examples. Finally, in Section 5 we reveal a connection between the Parker and Traub algorithms and the concept of displacement structure. This allows us to obtain in Section 6 a *generalized Parker-Traub algorithm* for inversion the more general Vandermonde-like matrices, naturally suggested by the concept of displacement.

1.6 Extended version

An extended version of this paper [9], containing more detailed results on connections with displacement structure, complete proofs, and numerical results for solving Vandermonde linear systems, will appear elsewhere.

2 THE TRAUB ALGORITHM

In [22] Traub derived a fast algorithm that is based on explicit formula for $V(x)^{-1}$. Let

$$P(x) = \prod_{k=1}^{n} (x - x_k) = x^n + \sum_{k=0}^{n-1} a_k \cdot x^k, \tag{10.3}$$

be the *master* polynomial, whose zeros are the nodes of $V(x)$. Following [22], consider the divided difference

$$P[t, x] = \frac{P(t) - P(x)}{t - x}, \tag{10.4}$$

and define the *associated polynomials* $\{q_k(x)\}_{0\leq k\leq n}$ with $q_n(x) = P(x)$ by

$$P[t,x] = q_{n-1}(x) + t \cdot q_{n-2}(x) + ...t^{n-2} \cdot q_1(x) + t^{n-1} \cdot q_0(x). \qquad (10.5)$$

From (10.3), (10.4) one sees that the bivariate function $P[t,x]$ has a Hankel structure :

$$P[t,x] = \sum_{i=1}^{n} a_i \cdot \frac{t^i - x^i}{t-x} = \sum_{i=1}^{n} a_i \cdot (t^{i-1} + t^{i-2}x + ... + tx^{i-2} + x^{i-1}),$$

which implies that the associated polynomials are given by

$$q_k(x) = x^k + a_{n-1} \cdot x^{k-1} + ... + a_{n-k+1} \cdot x + a_{n-k}. \qquad (10.6)$$

Equivalently, they satisfy the recurrence relations

$$q_0(x) = 1, \qquad q_k(x) = x \cdot q_{k-1}(x) + a_{n-k} \quad (k = 1, 2, ..., n). \qquad (10.7)$$

From $P[x,x] = P'(x)$ and (10.3), (10.4), (10.5) we obtain what Traub called the *basic orthonormality relation* :

$$\frac{P[x_j, x_k]}{P'(x_k)} = \sum_{i=0}^{n-1} x_j^i \cdot \frac{q_{n-1-i}(x_k)}{P'(x_k)} = \delta_{jk}.$$

The latter relation implies that the inverse of a Vandermonde matrix is given by

$$V^{-1}(x) = \begin{bmatrix} q_{n-1}(x_1) & q_{n-1}(x_2) & \cdots & q_{n-1}(x_n) \\ q_{n-2}(x_1) & q_{n-2}(x_2) & \cdots & q_{n-2}(x_n) \\ \vdots & \vdots & & \vdots \\ q_0(x_1) & q_0(x_2) & \cdots & q_0(x_n) \end{bmatrix} \cdot \operatorname{diag}([\frac{1}{P'(x_i)}]_{i=1}^{n}). \qquad (10.8)$$

Recall that (10.7) allows one to compute the entries of the first factor in (10.8). Differentiating (10.7), one sees that the entries $P'(x) = q_n'(x)$ of the second factor can be computed by

$$q_1'(x) = a_{n-1}, \qquad q_k'(x) = q_{k-1}(x) + x \cdot q_{k-1}'(x) \qquad (k = 2, 3, ..., n). \qquad (10.9)$$

Thus the inversion procedure can be summarized as follows.

Traub algorithm

1. Compute the coefficients of $P(x)$ in (10.3) via nested polynomial multiplication :

$$\begin{bmatrix} a_0^{(1)} \\ a_1^{(1)} \end{bmatrix} = \begin{bmatrix} -x_1 \\ 1 \end{bmatrix}, \qquad \begin{bmatrix} a_0^{(k)} \\ a_1^{(k)} \\ \vdots \\ a_k^{(k)} \end{bmatrix} = \begin{bmatrix} 0 \\ a_0^{(k-1)} \\ \vdots \\ a_{k-1}^{(k-1)} \end{bmatrix} - x_k \cdot \begin{bmatrix} a_0^{(k-1)} \\ \vdots \\ a_{k-1}^{(k-1)} \\ 0 \end{bmatrix}, \qquad (10.10)$$

with $a_j = a_j^{(n)}$.

2. For $j = 1, 2, ..., n$ do :
   (a) Compute $q_k(x_j)$ $(k = 0, 1, ..., n-1)$ using recursion (10.7).
   (b) Using these quantities compute $P'(x_j) = q_n'(x_j)$ using recursion (10.9).
   (c) Compute the the j-th column $\left[\frac{q_k(x_j)}{P'(x_j)} \right]_{0 \leq k \leq n-1}$ of $V^{-1}(x)$.

The Traub algorithms computes all n^2 entries of $V(x)$ in only $6n^2$ flops, which compares favorably with the complexity $O(n^3)$ flops of standard (structure-ignoring) methods.

3 THE BJÖRCK-PEREYRA ALGORITHM

Björck and Pereyra described in [1] a fast algorithm that solves a Vandermonde linear system in only $5n^2/2$ flops. Clearly it can be used for computing the entries of $V(x)^{-1}$ by solving n linear systems, using the columns of identity matrix for the right-hand sides. The latter $O(n^3)$ inversion scheme is no longer fast. But Björck and Pereyra observed in [1] that their algorithm frequently produces more accurate solutions than could be expected from the condition number of the coefficient matrix. In [10] Higham analyzed a specific class of Vandermonde matrices with positive and monotonically ordered points,

$$0 < x_1 < x_2 < ... < x_n, \tag{10.11}$$

and showed that in this case the Björck-Pereyra algorithm computes the inverse matrix $\hat{V}(x)^{-1}$ so that the error is as small as could possibly be expected :

$$|\hat{V}(x)^{-1} - V(x)^{-1}| \leq 5nu \cdot |V(x)^{-1}| + O(u^2). \tag{10.12}$$

Here the comparison and the operation of taking the absolute value of a matrix, are understood in a componentwise sense.

The latter bound was used in [10] to compare the above two algorithms, and it was pointed out there that (10.12) *" shows that contrary to what one might expect, $V(x)^{-1}$ can be computed with high relative accuracy"*. Then Higham turned to the Traub algorithms and wrote : *" However [22] does not contain a rounding error analysis, and it can be shown that Traub's $O(n^2)$ algorithm must involve subtraction of like-signed numbers, suggesting that a result of the form (10.12) will not hold."*

These arguments suggest that the $O(n^2)$ Traub inversion algorithm is fast, but not accurate; whereas the Björck-Pereyra algorithm is the other way around, i.e. it is accurate (for points as in (10.11)), but not fast. [1] To the best of our knowledge the possibility of simultaneously fast and accurate inversion of $V(x)$ was not reported anywhere. Numerical experiments suggest that the $O(n^2)$ algorithm described in the next section satisfies these requirements; thus it is possible to achieve a speed-up of computation without sacrificing the accuracy.

4 THE PARKER ALGORITHM

3.1. The Parker algorithm. We discovered that earlier than the above algorithms, Parker described in [20] another fast inversion procedure, which is based on the variant of the inversion formula described below. It is easy to see that the columns of

$$V^{-1}(x) = \left[l_{k,j} \right]_{1 \le k,j \le n} . \tag{10.13}$$

are formed from the coefficients of Lagrange polynomials $L_j(x) = \sum_{k=1}^{n} l_{k,j} \cdot x^{k-1}$, i.e. the unique polynomials of degree $n-1$ satisfying $L_j(x_j) = 1$, and $L_j(x_k) = 0$ $(k \neq j)$. The description (10.13) can be found in [15], and later it was often rederived by many authors, and in particular by Parker in [20], who, anticipating the Traub algorithm, observed that it suggests an efficient inversion procedure for $V(x)$. Indeed,

$$L_j(x) = \frac{P(x)}{(x - x_j) \cdot P'(x_j)}, \tag{10.14}$$

where $P(x)$ is the master polynomial, defined in (10.3), and therefore

$$P'(x_j) = (x_j - x_1) \cdot \ldots \cdot (x_j - x_{j-1}) \cdot (x_j - x_{j+1}) \cdot \ldots \cdot (x_j - x_n). \tag{10.15}$$

In fact Parker outlined the following scheme. First one computes the coefficients of $P(x) = \sum_{i=1}^{n} a_i x^{i-1}$, then divides it synthetically by $(x - x_j)$. It is easy to see that the synthetic division gives rise to the following recursion

$$q_{0,j} = 1, \qquad q_{k,j} = x_j \cdot q_{k-1,j} + a_{n-k} \tag{10.16}$$

for the coefficients of the quotient

$$\frac{P(x)}{(x - x_j)} = \sum_{k=0}^{n-1} q_{k,j} x^{n-k-1}. \tag{10.17}$$

To finally obtain the coefficients of $L_j(x)$ in (10.14), or equivalently the entries of the j-th column of $V^{-1}(x)$ in (10.13), it remains only to divide the polynomial

[1] It was demonstrated in [4] that if points x_k are of both signs (so that they cannot be reordered to meet the restriction (10.11)), then the $O(n^3)$ Björck-Pereyra inversion algorithm becomes less accurate. Motivated by this fact Gohberg and Koltracht suggested another another $O(n^3)$ algorithm, which showed better numerical properties, see *e.g.* [4].

in (10.17) by $P'(x_j)$, the latter is computed via (10.15). Thus the inversion algorithm can be summarized as follows.

Parker algorithm

1. Compute the coefficients of $P(x)$ in (10.3) by (10.10).
2. For $j = 1, 2, ..., n$ do
   (a) Compute $q_{k,j}$ $(k = 0, 1, ..., n-1)$ by recursion (10.16).
   (b) Compute $P'(x_j)$ by (10.15).
   (c) Compute the the j-th column $\left[\frac{q_{k,j}}{P'(x_j)}\right]_{0 \leq k \leq n-1}$ of $V^{-1}(x)$.

The computational complexity $6n^2$ flops of the above algorithm [2] was not counted by Parker, who however indicated that his algorithm is fast, by noting that "*pencil and paper calculation of the inverse of a matrix of order six takes about twenty minutes*".

Note that later this algorithm was rederived also in [23], [3] and in [13].

3.2. Comparison of the Traub and Parker algorithms. By direct comparing one sees that since recursions (10.7) and (10.16) coincide, the only difference between the Traub and the Parker $O(n^2)$ schemes is in the step 2.b, i.e. in computing the quantities $P'(x_j)$. Thus it is essentially the same algorithm, and hence all the remarks in Section 2 on the Traub algorithm are valid for the Parker algorithm as well. In particular the latter also involves the subtraction of like-signed numbers. Interestingly, if the condition (10.11) holds, then *each step* 2.a (since recursions (10.7) and (10.16) coincide, this step is the same for both algorithms) must contain such a subtraction. Indeed, in this case the coefficients of the polynomials $L_j(x)$ in (10.14) have a sign-oscillation property, and hence the same is true for the coefficients $q_{k,j}$ of the polynomials in (10.17). Therefore *each step* 2.a of both Parker and Traub algorithms must involve the subtraction of like-signed numbers, see *e.g.* (10.16). Therefore the Parker algorithm is also subject to the above Higham's comment, saying that a result of the form (10.12) will not hold for the Parker algorithm as well. At the same time the results of numerical experiments show that the Parker algorithm, combined with an appropriate reordering of the points x_k turns out to be very reliable in practice. Moreover, a careful search did not reveal any case where the Björck-Pereyra algorithm was more accurate than the Parker algorithm, whereas the latter demonstrated, in many examples, a strong numerical superiority over other algorithms compared. This occurrence reminds us that just comparing error bounds *alone* cannot be the basis for making reliable recommendations.

[2] By a proper implementation of the step 2.b, this complexity can be further reduced to $5.5n^2$ flops, see *e.g.* [7], where computing $P'(x_j)$ appears as a part of a fast $O(n^2)$ algorithm for inversion of Chebyshev-Vandermonde matrices.

5 NUMERICAL EXPERIMENTS

In this section we give fairly representative numerical examples, comparing the accuracy of the following algorithms : 1) $O(n^3)$ Gauss-Jordan elimination with complete pivoting (**GJECP**); 2) the $O(n^3)$ Björck-Pereyra algorithm (**B-P**), applied for inversion; 3) the $O(n^2)$ Traub inversion algorithm; 4) the $O(n^2)$ Parker inversion algorithm. For each algorithm we measured a relative error $e_s = \frac{\|\hat{V}(x)^{-1} - V(x)^{-1}\|}{\|V(x)^{-1}\|}$, where $\hat{V}(x)^{-1}$ was computed in 7-digit arithmetic (single precision). Here the "exact" solution $V(x)^{-1}$ was computed in 16-digit arithmetic (double precision) by the best, in this particular example, among compared algorithms.

4.1. Different orderings of the points. The GECP and the Traub algorithms did not show in computed examples any correlation with different orderings, whereas the accuracy of the Björck-Pereyra and of the Parker algorithm do depend on ordering of x_k. We considered three orderings of x_k, namely : 1) random ordering, 2) monotonic ordering $x_1 < x_2 < ... < x_n$, and 3) Leja ordering, where the points x_i are ordered so that $|x_1| = \max_{1 \le k \le n} |x_k|$, and

$$\prod_{j=1}^{k-1} |x_k - x_j| = \max_{k \le l \le n} \prod_{j=1}^{k-1} |x_l - x_j|, \qquad (2 \le k \le n). \tag{10.18}$$

The latter ordering is related to Leja work on interpolation, see e.g. [21] and references therein. In [11] Higham observed that (10.18) corresponds to the row permutation that would be obtained by applying *partial pivoting*, and showed how to reorder x_k so that (10.18) holds using only n^2 flops; thus incorporating the Leja ordering will not slow down any fast $O(n^2)$ algorithm. ,

4.2. Positive points. Here we present the results of numerical inversion of $V(x)$, where x_k are the equidistant in the interval (0,1) points.

Table 1. Equidistant in (0,1) points.

				Random ordering		Monotonic ordering		Leja ordering	
n	$\kappa_2(V)$	GJECP e_s	Traub e_s	B-P e_s	Parker e_s	B-P e_s	Parker e_s	B-P e_s	Parker e_s
5	2e+03	8e-07	9e-06	4e-08	1e-07	**9e-08**	1e-07	5e-08	**1e-07**
10	6e+07	3e-02	1e-01	2e-07	2e-07	**2e-07**	2e-07	3e-07	**2e-07**
20	4e+16	1e+00	1e+00	2e-06	4e-07	**5e-07**	4e-07	5e-07	**5e-07**
30	4e+18	1e+00	1e+00	6e-04	5e-07	**6e-07**	5e-07	3e-06	**6e-07**
40	8e+18	1e+00	1e+00	3e-04	8e-07	**7e-07**	7e-07	7e-06	**7e-07**
50	6e+18	1e+00	1e+00	Inf	Inf	Inf	Inf	Inf	Inf

The matrix in the Example 1 becomes extremely ill-conditioned already for $n = 10$, so it is not surprising that the Gauss-Jordan elimination failed. Furthermore, as Table 1 shows, the Traub variant of the inversion algorithm also breaks down numerically. Recall that since $x_k > 0$, the Björk-Pereyra algorithm combined with monotonic ordering (10.11) is guaranteed to provide in Example

1 an exceptionally high accuracy, see *e.g.* (10.12). At the same time, Table 1 demonstrates that even in this the most favorable for the Björck-Pereyra algorithm situation, the accuracy of the Parker algorithm is at least not worse. In fact it is much better in the other cases, as will be seen in the next example.

4.3. Points with both signs. Here we present the results for inversion of $V(x)$, where $x_i = \cos(\frac{(2i-1)\cdot\pi}{2\cdot n})$, i.e. they are of both signs. For this case there are no sharp stability results, like in (10.12) and moreover, the numerical data indicate that the Björck-Pereyra algorithm becomes less accurate. However it is not the case with the Parker algorithm, combined with Leja ordering, and it provides an exceptionally high accuracy, as shown in the next table.

Table 2. Chebyshev zeros in (-1,1).

				Random ordering		Monotonic ordering		Leja ordering	
		GJECP	Traub	B-P	Parker	B-P	Parker	B-P	Parker
n	$\kappa_2(V)$	e_s	e_s	e_s	e_s	e_s	e_s	e_s	e_s
5	2e+01	2e-07	3e-07	8e-08	8e-08	1e-07	5e-08	**2e-07**	**6e-08**
10	1e+03	2e-06	5e-06	3e-07	1e-07	5e-07	5e-07	**5e-07**	**1e-07**
20	9e+06	5e-02	2e-02	8e-06	2e-07	1e-05	1e-05	**2e-05**	**3e-07**
30	6e+10	1e+00	3e-01	9e-03	3e-07	2e-04	7e-05	**5e-05**	**3e-07**
40	4e+14	1e+00	5e-01	1e+00	4e-07	3e-03	6e-03	**2e-03**	**3e-07**
50	6e+17	1e+00	6e-01	1e-01	6e-07	5e-02	2e-02	**1e-02**	**6e-07**
60	7e+18	1e+00	7e-01	2e+01	6e-07	8e-01	1e+00	**5e-02**	**6e-07**

The above two examples are fairly representative, and a careful search did not reveal any deviations from the pattern shown in Tables 1 and 2.

Finally note that recently the Traub algorithm was generalized to invert *Chebyshev-Vandermonde* matrices in [7], and what we call *three-term* Vandermonde matrices in [2]. Now it is clear that both these algorithms are generalizations of the Parker rather than the Traub scheme. Favorable stability results were reported in both cases, whereas their prototype, i.e. the Traub algorithm, was generally regarded as being numerically unstable. We hope that the present paper explains this curious situation.

6 VANDERMONDE-LIKE MATRICES

5.1. Structure of the inverse of a Vandermonde matrix. Recall that the Parker-Traub algorithm is based on formula (10.8), which in view of (10.6) can be rewritten as follows

$$V^{-1}(x) = \begin{bmatrix} a_1 & a_2 \; a_3 & \cdots & a_{n-1} & 1 \\ a_2 & a_3 \; \cdots & a_{n-1} & 1 & 0 \\ a_3 & & \iddots \;\; \iddots & & \vdots \\ \vdots & \iddots \; \iddots & & & \vdots \\ a_{n-1} & \iddots & & & \vdots \\ 1 & 0 \; \cdots & \cdots & \cdots & 0 \end{bmatrix} \cdot V(x)^T \cdot \mathrm{diag}([\frac{1}{P'(x_i)}]_{i=1}^n). \quad (10.19)$$

More specifically, the Traub and the Parker algorithms exploit the following properties of a Vandermonde matrix to speed-up computing of its inverse.

(P1) $V(x)^{-1}$ has the special structure shown in (10.19), i.e. it is a product of a Hankel, transposed Vandermonde and diagonal matrices.

(P2) The entries of the matrices on the right-hand side of (10.19) can be computed in $O(n^2)$ flops.

(P3) The product of the Hankel and transposed Vandermonde matrices on the right-hand side of (10.19) can be computed in $O(n^2)$ flops, using the recursion in (10.7). Indeed, thsi can be easily seen by comparing (10.7), (10.8) and (10.19).

Moreover, the Parker algorithm achieves a high relative accuracy by the incorporation of Leja ordering, i.e. by exploiting the next property.

(P4) Partial pivoting technique can be incorporated without increasing the $O(n^2)$ complexity of the algorithm.

The Parker-Traub algorithm uses **(P1)-(P4)** to invert $V(x)$ in two basic steps :

(PT1) Compute the entries of the matrices on the right-hand side of (10.19).This is done in the steps 1 and 2.b of the Parker algorithm.

(PT2) Compute the product of the Hankel, transpose Vandermonde and a diagonal matrices on the right-hand side of (10.19).This is done in the steps 2.a and 2.c of the Parker algorithm.

It turns out that the properties **(P1) - (P4)** reflect the fact that $V(x)$ has *displacement structure.* Therefore the Parker-Traub algorithm in **(PT1) - (PT2)** can be extended to invert the more general class of *Vandermonde-like* matrices, naturally suggested by the idea of displacement, described next.

5.2. Displacement structure. The displacement structure theory was started by T.Kailath and his coauthors in [16], where it was first applied to the study of Toeplitz matrices. They considered a *displacement operator*

$\nabla_{\{Z_0,Z_0\}}(\cdot) : \mathcal{C}^{n\times n} \to \mathcal{C}^{n\times n}$ of the form

$$\nabla_{\{Z_0,Z_0\}}(R) = R - Z_0 \cdot R \cdot Z_0^T, \tag{10.20}$$

where Z_0 is the lower shift matrix, i.e. with ones on the first subdiagonal and zeros elsewhere; $Z_0 \cdot R \cdot Z_0^T$ then corresponds to shifting R downwards along main diagonal. The matrix $\nabla_{\{Z_0,Z_0\}}(R)$ was called $\{Z_0, Z_0\}$-*displacement* of R, and the number $\text{rank}\nabla_{\{Z_0,Z_0\}}(R)$ was called $\{Z_0, Z_0\}$-*displacement rank* of R. Kailath et al. observed that the shift-invariance property of Toeplitz matrices implies that their $\{Z_0, Z_0\}$-displacement rank do not exceed 2, and showed that many properties of ordinary Toeplitz matrices are naturally carried over to more general *Toeplitz-like* matrices, defined as those with low $\{Z_0, Z_0\}$-displacement rank. Later it was realized that the concept of displacement suggests a unified approach to the study of Hankel, Toeplitz-plus-Hankel, Vandermonde, Chebyshev-Vandermonde, Cauchy, Pick matrices, Bezoutians, and many other classes of structured matrices, see *e.g.* recent review [18] for historical remarks, and list of applications. In particular, Heinig and Rost observed in [12] that Vandermonde matrices are transformed to rank-one matrices by a suitable displacement operator, and used this fact to carry over many results to more general *Vandermonde-like* matrices, defined next.

5.3. Vandermonde-like displacement structure. For our purposes here it will be more convenient to exploit the displacement operator $\nabla_{\{D_x^{-1},Z_0^T\}}(\cdot) : \mathcal{C}^{n\times n} \to \mathcal{C}^{n\times n}$, defined by

$$\nabla_{\{D_x^{-1},Z_0^T\}}(R) = D_x^{-1} \cdot R - R \cdot Z_0^T, \tag{10.21}$$

which slightly differs from the ones used in [12], [8], [6]. Here $D_x = \text{diag}(x_1, x_2, ..., x_n)$ and Z_0 stands for the lower shift matrix. Matching the pattern of definitions in the previous subsection, the matrix $\nabla_{\{D_x^{-1},Z_0^T\}}(R)$ is called a $\{D_x^{-1}, Z_0^T\}$-*displacement* of R, and the rank of $\nabla_{\{D_x^{-1},Z_0^T\}}(R)$ is referred to as a $\{D_x^{-1}, Z_0^T\}$-*displacement rank* of R.

It is easy to check that the displacement operator (10.21) transforms a Vandermonde matrix $V(x)$ into rank-one matrix, thus justifying the name *Vandermonde-like* for matrices with low $\{D_x^{-1}, Z_0^T\}$-displacement rank.

5.4. Inversion formula for Vandermonde-like matrices. In [16] T.Kailath and his coauthors observed that the displacement structure of a matrix is inherited by its inverse, and used this fact to explain the form of the Gohberg-Semencul formula for inversion of Toeplitz matrices, and to obtain its generalization for more general Toeplitz-like matrices. Later a similar explanation was given in [19] for the form of the Gohberg-Olshevsky formulas for inversion of Chebyshev-Vandermonde matrices. Similarly, the the pattern of arguments of [16], [19] can be used to explain the form of the Traub inversion formula (10.19), and to obtain a more general formula for inversion of Vandermonde-like matrices, given in the next statement.

Theorem 10.1 *Let $\{D_x^{-1}, Z_0^T\}$-displacement rank of $R \in \mathcal{C}^{n\times n}$ be equal to α, and let R be specified by its $\{D_x^{-1}, Z_0^T\}$-displacement $G_1 = \begin{bmatrix} \varphi_1 \ \varphi_2 \cdots \varphi_\alpha \end{bmatrix} \in \mathbf{C}^{n\times\alpha}$, $B_1 = \begin{bmatrix} \psi_i \ \psi_2 \cdots \psi_\alpha \end{bmatrix}^T \in \mathbf{C}^{n\times\alpha}$ on the right-hand side of*

$$D_x^{-1} \cdot R - R \cdot Z_0^T = G \cdot B. \tag{10.22}$$

Then

$$R^{-1} = \sum_{m=1}^{\alpha} \begin{bmatrix} a_1^{(m)} & a_2^{(m)} & a_3^{(m)} & \cdots & a_n^{(m)} \\ a_2^{(m)} & a_3^{(m)} & \cdots & a_n^{(m)} & 0 \\ a_3^{(m)} & & \cdot^{\cdot^{\cdot}} & 0 & 0 \\ \vdots & \cdot^{\cdot^{\cdot}} & \cdot^{\cdot^{\cdot}} & & \vdots \\ a_n^{(m)} & 0 & \cdots & \cdots & 0 \end{bmatrix} \cdot V(x)^T \cdot \operatorname{diag}(\begin{bmatrix} c_i^{(m)} \end{bmatrix}_{1\le i\le n}), \tag{10.23}$$

where $a_m = \begin{bmatrix} a_i^{(m)} \end{bmatrix}_{1\le i\le n}$, $c_m = \begin{bmatrix} c_i^{(m)} \end{bmatrix}_{1\le i\le n}$ are determined from the 2α linear systems

$$R \cdot a_m = \varphi_m, \qquad R^T \cdot D_x^{-1} c_m = \psi_m, \qquad (m = 1, 2, ..., \alpha) \tag{10.24}$$

The proof of the above theorem can be found in [9], and it is based on the fact that the displacement structure of R is essentially inherited by its inverse :

$$R^{-1} \cdot D_x^{-1} - Z_0^T \cdot R^{-1} = \sum_{m=1}^{\alpha} (R^{-1}\varphi_m) \cdot (\psi_m R^{-1}), \tag{10.25}$$

and on an explicit formula to recover a Vandermonde-like matrix from its displacement.

A different inversion formula for Vandermonde-like matrices was obtained earlier in [12], Theorem II-2.5. Heinig and Rost did not use their formula to derive an inversion algorithm for Vandermonde-like matrix R. Instead, they gave in [12], Theorem II-2.6, a recursion for the rows of some auxiliary matrix P, then the entries of the inverse matrix are obtained by computing the matrix product $R^{-1} := V(x)^{-1} \cdot P$. The latter procedure of [H84] does not seem to be related to the Parker-Traub algorithm, and moreover it is not fast, requiring $O(n^3)$ flops. A fast algorithm for inversion of Vandermonde-like matrices is suggested in the next section.

7 GENERALIZED PARKER-TRAUB ALGORITHM

6.1. Properties (P1) - (P4). In this section we shall show that all the properties **(P1) - (P4)** hold not just for Vandermonde matrices, but for

Vandermonde-like matrices as well, and shall use this fact to derive a fast $O(n^2)$ *generalized Parker-Traub algorithms* for inversion of Vandermonde-like matrices.

Property (P1). First observe that the inversion formula in (10.23) represents R^{-1} as a sum of α terms, each of which is a product of a Hankel, transposed Vandermonde and a diagonal matrices; this extends the property **(P1)** to Vandermonde-like matrices.

Property (P2). The entries of the matrices on the right-hand side of (10.23) are obtained via solving 2α linear systems (10.24) with a Vandermonde-like coefficient matrix R. Such systems can be solved in $O(n^2)$ flops by using a suitable form of *generalized Schur algorithm* [14], [17], [18], [6]; or by many Levinson-type algorithms, see, *e.g.* [12], [5], to mention just a few. The use of any of these algorithms allows one fast computing the entries of the matrices on the right-hand side of (10.23), and thus extends the property **(P2)** to Vandermonde-like matrices. Among a variety of the above possibilities a variant of the generalized Schur algorithm from [6] is the most attractive for our purposes here, for the reason, specified next.

Property (P4). The algorithm in [6] exploits the fact that R satisfies the displacement equation of the form (10.22) with a *diagonal* matrix D_x, to incorporate partial pivoting technique into fast triangulazation of R. This allows us to remove the restriction on R to be *a strongly regular* matrix, required by the other algorithms, and moreover, to extend the property **(P4)** to Vandermonde-like matrices.

Property (P3). The expression on the right-hand side of (10.23) involves α products of a Hankel and a Vandermonde matrix, and each such product can be computed in $O(n^2)$ flops in a similar to (10.7) fashion.

Thus the basic steps of the algorithm are briefly outlined in the next subsection.

6.2. Generalized Parker-Traub algorithm. Given a displacement $\{G_1, B_1\}$ as on the right- hand side of (10.22), the following algorithm computes in $O(n^2)$ flops the entries of the Vandermonde-like matrix R^{-1}, and similarly to **(PT1)** - **(PT2)** above, it consists of the following two steps **(GPT1)** - **(GPT2)**.

GPT1. Computing the entries of the matrices on the right-hand side of (10.23).

GPT1.1. Triangular factorization of R. Compute in $O(n^2)$ flops the triangular factorization $R = PLU$ of the permuted version of R, using the variant of [6] of the generalized Schur algorithm, incorporating partial pivoting.

GPT1.2. Solving linear systems (10.24). This step uses the standard forward and back-substitution technique, and it requires performing $O(n^2)$ flops.

GPT2. Forming the inverse. Formula (10.23) represents R^{-1} as the sum of α products of a Hankel, transpose Vandermonde and diagonal matrices. These products can be computed via recursions similar to the one in (10.7). This step requires performing of $O(n^2)$ flops.

Generalized Parker-Traub algorithm

INPUT: The displacement $G_1 = \left[g_{i,j}^{(1)} \right]_{\substack{1\le i\le n \\ 1\le j\le\alpha}}$, $B_1 = \left[g_{i,j}^{(1)} \right]_{\substack{1\le i\le n \\ 1\le j\le\alpha}}$, as on the right hand side of (10.22).

OUTPUT: $R^{-1} = \left[h_{i,j} \right]$

COMPLEXITY: $O(n^2)$ flops.

GPT1 .

GPT1.1. Triangular factorization of R .

GPT1.1.1. Set $D^{(1)} = \mathrm{diag}(x_1, x_2, ..., x_n)$.

GPT1.1.2. For $k = 1, 2, , , , n-1$ do :

GPT1.1.2.1. Compute

$$\left[l_{i,k} \right]_{k\le i\le n} = \sum_{m=1}^{\alpha} b_{k,m}^{(k)} \cdot D^{(k)} \cdot \left[g_{m,k}^{(k)} \right]_{k\le m\le n}.$$

GPT1.1.2.2. Find $l_{q,k} = \max_{k\le i\le n} l_{i,k}$.
Swap x_k and x_q. Swap rows $l_{k,k}$ and $l_{q,k}$.
Swap rows $\left[g_{k,m}^{(k)} \right]_{1\le m\le\alpha}$ and $\left[g_{q,m}^{(k)} \right]_{1\le m\le\alpha}$.
Set P_k be the permutation of the k-th and the q-th entries.

GPT1.1.2.3. Compute

$$l_k = \left[l_{i,k} \right]_{k\le i\le n} = \frac{1}{l_{k,k}} \cdot \left[l_{i,k} \right]_{k\le i\le n},$$

$$u_k = \left[u_{k,j} \right]_{k\le j\le n} = x_k \cdot \sum_{m=1}^{\alpha} g_{k,m}^{(k)} \cdot v_m,$$

where $v_m = \left[v_{m,i} \right]_{k\le i\le n}$ $(m = 1, 2, ..., \alpha)$ are computed by

$$v_{m,k} = b_{m,k}^{(k)},$$

$$v_{m,t} = x_k \cdot v_{m,t-1} + b_{m,t}^{(k)} \qquad (t = k+1, k+2, ..., n).$$

GPT1.1.2.4. Update $G^{(k+1)}$, $B^{(k+1)}$ by

$$\begin{bmatrix} 0 \\ G^{(k+1)} \end{bmatrix} = G^{(k)} - l_k \cdot g^{(k)},$$

$$\begin{bmatrix} 0 \ B^{(k+1)} \end{bmatrix} = B^{(k)} - b^{(k)} \cdot \frac{u_k}{u_{k,k}},$$

where $g^{(k)} \in \mathbf{C}^{1\times\alpha}$ and $b^{(k)} \in \mathbf{C}^{\alpha\times 1}$ are the first row of $G^{(k)}$ and the first column of $B^{(k)}$, respectively.

GPT1.1.3. Set $l_{n,n} = 1$, $u_{n,n} = \frac{1}{x_n} \cdot \sum_{m=1}^{\alpha} g_{n,m}^{(n)} \cdot b_{m,n}^{(n)}$.

GPT1.1.4. Set $L = \begin{bmatrix} l_{ij} \end{bmatrix}$, $U = \begin{bmatrix} u_{ij} \end{bmatrix}$, $P = P_{n-1} \cdot P_{n-2} \cdot \ldots \cdot P_1$.

GPT1.2. Solving linear systems (10.24). Solve 2α linear systems (10.24) via forward and backsubstitution.

GPT2. Forming the inverse. For $j = 1, 2, ..., n$ do

GPT2.1. For $m = 1, 2, ..., \alpha$ set $q_m = a^{(m)}$.

GPT2.2. $h_{n,j} = \sum_{m=1}^{\alpha} \frac{q_m}{c_j^{(m)})}$.

GPT2.3. For $i = 1, 2, ..., n-1$ do

GPT2.3.1. $q_m = x_j \cdot q_m + a_{n-i}^{(m)}$.

GPT2.3.2. $h_{n-i,j} = \sum_{m=1}^{\alpha} \frac{q_m}{c_j^{(m)})}$.

Acknowledgements

It is our pleasure to thank Thomas Kailath for useful discussions.

REFERENCES

[1] A. Björck and V.Pereyra, " Solution of Vandermonde systems of linear equations, " Math. of Computation, **24** : 893 – 903 (1970).

[2] D.Calvetti and L.Reichel, " Fast inversion of Vandermonde-like matrices involving orthogonal polynomials," BIT, **33** (1993), 473 – 484 .

[3] G.Forney, " Concatenated codes, " The M.I.T. Press, 1966, Cambridge.

[4] I.Gohberg and I.Koltracht, " Mixed, componentwise and structured condition numbers, " SIAM J. of Matrix Anal. Appl., **14**(1993), 688–704.

[5] I.Gohberg, T.Kailath, I.Koltracht and P.Lancaster, " Linear Complexity parallel algorithms for linear systems of equations with recursive structure, " Linear Algebra Appl., **88/89** (1987), 271–315.

[6] I.Gohberg, T.Kailath and V.Olshevsky, " Fast Gaussian elimination with partial pivoting for matrices with displacement structure, " Math. of Computation, **64** (1995), 1557-1576.

[7] I.Gohberg, and V.Olshevsky, " Fast inversion of Chebyshev–Vandermonde matrices, " *Numerische Mathematik*, **67**, No. 1 : 71 – 92 (1994).

[8] I.Gohberg and V.Olshevsky, " Complexity of multiplication with vectors for structured matrices, " Linear Algebra Appl., **202** (1994), 163 – 192.

[9] I.Gohberg and V.Olshevsky, "Vandermonde and Vandermonde-like matrices," preprint, 1994.

[10] N.Higham, " Error analysis of the Björck–Pereyra algorithms for solving Vandermonde systems, " Numerische mathematic, **50** (1987), 613 – 632.

[11] N.Higham, " Stability analysis of algorithms for solving confluent Vandermonde-like systems," SIAM J. Matrix Anal. Appl., **11**(1) (1990), 23–41.

[12] G.Heinig and K.Rost, " Algebraic methods for Toeplitz-like matrices and operators, " Operator Theory, vol. **13**, Birkauser Verlag, Basel, 1984.

[13] I.Kaufman, "The inversion of the Vandermonde matrix and the transformation to the Jordan canonical form", IEEE Trans. on Automatic Control, **14**(1969), 774 - 777.

[14] H.Lev-Ari and T.Kailath, "Triangular factorization of structured matrices, " in I.Schur methods in operator theory and signal processing, Operator Theory: Advances and Applications, vol **18** (edited by I.Gohberg), Birkhauser Verlag, Basel, 1986.

[15] G. Kowalewski, " Interpolation und genäherte Quadratur, " Teubner, Berlin, 1932.

[16] T.Kailath, S.Kung and M.Morf, " Displacement ranks of matrices and linear equations, " J. Math. Anal. and Appl., **68** (1979), 395-407.

[17] T. Kailath and A.H.Sayed, " Fast algorithms for generalized displacement structures, " in Recent advances in mathematical theory of systems, control, networks and signal processing II, Proc. of the MTNS-91 (H.Kimura, S.Kodama, Eds.), Mita Press, Japan, 1992, 27 – 32.

[18] T. Kailath and A.H.Sayed, " Displacement structure : theory and applications, " SIAM Review, **37**(3):297-386, September 1995.

[19] T.Kailath and V.Olshevsky, " Displacement structure approach to Chebyshev–Vandermonde and related matrices, " *Integral Equations and Operator Theory*, **22**:65-92, May 1995.

[20] F.Parker, " Inverses of Vandermonde matrices, " Amer. Math. Monthly, **71** (1964), 410 - 411.

[21] L.Reichel, " Newton interpolation at Leja points, " BIT, **30**(1990), 23 – 41.

[22] J. Traub, " Associated polynomials and uniform methods for the solution of linear problems, " SIAM Review, **8**, No. 3 (1966), 277 – 301.

[23] H.J.Wertz, "On the numerical inversion of a recurrent problem : the Vandermonde matrix", IEEE Trans. on Automatic Control, AC-10, **4**(1965), 492.

11

IMPROVING THE ACCURACY OF THE GENERALIZED SCHUR ALGORITHM

S. Chandrasekaran and Ali H. Sayed

Dept. of Electrical and Computer Engineering
University of California
Santa Barbara, CA 93106-9560

Dedicated with admiration and gratitude to Prof. Thomas Kailath on the occasion of his 60th birthday.

ABSTRACT

We show how to stabilize the generalized Schur algorithm for the Cholesky factorization of positive-definite structured matrices R that satisfy $R - FRF^T = GJG^T$, where J is a signature matrix, F is a stable lower-triangular matrix, and G is a generator matrix. We use a perturbation analysis to indicate the best accuracy that can be expected from any finite precision algorithm that uses the generator matrix as the input data. We then show that the modified Schur algorithm proposed in this work essentially achieves this bound for a large class of structured matrices.

1 INTRODUCTION

One of the most frequent structures, at least in signal processing applications, is the Toeplitz structure with constant entries along the diagonals of the matrix [1]. A classical algorithm for the Cholesky factorization of the *inverses* of such matrices is the so-called Levinson-Durbin algorithm [1, 2], an error analysis of which has been provided by Cybenko [3]. He showed that, in the case of positive reflection coefficients, the residual error produced by the Levinson-Durbin procedure is comparable to the error produced by the Cholesky factorization [2, p.191].

A related analysis has been carried out by Sweet [4] for the Bareiss algorithm, which is closely related to an algorithm of Schur [1]. These are fast procedures for the Cholesky factorization of the Toeplitz matrix itself rather than its inverse. Sweet concluded that the Bareiss algorithm is asymptotically stable.

In recent work, Bojanczyk et al [5] further extended and strengthened the conclusions of Sweet [4] by employing elementary downdating techniques [6] that are also characteristic of array formulations of the Schur algorithm [7]. They considered the larger class of quasi-Toeplitz matrices, which includes the Toeplitz matrix as a special case, and provided an error analysis that establishes that the Schur algorithm for this class of matrices is asymptotically stable.

The interesting formulation of Bojanczyk et al. [5] has motivated us to take a closer look at the numerical stability of a generalized Schur algorithm [7] that applies to a wider class of structured matrices. This class is briefly introduced in the next section, where a lower-triangular matrix F is shown to be pivotal in characterizing the structure of the matrix. For example, in the Toeplitz or quasi-Toeplitz cases mentioned above, the matrix F is equal to the shift matrix Z (i.e., a Jordan block with zero eigenvalue and ones on the first subdiagonal). Multiplying a column vector u by Z simply corresponds to shifting down the entries of u by one position. In general, however, the matrix F can be any lower triangular matrix (e.g., diagonal, bidiagonal, strictly lower-triangular, etc.). This creates several complications that we address in order to guarantee a numerically reliable algorithm.

In the discussion that follows we use $\|\cdot\|$ to denote the 2−norm of its argument. We further assume, without loss of generality, that F is represented exactly in the computer. Also, the $\hat{\cdot}$ notation denotes computed quantities, while the $\bar{\cdot}$ notation denotes intermediate exact quantities. We assume $n\epsilon < 1$, where ϵ denotes the machine precision and n is the matrix size. We also use subscripted δ's to denote quantities smaller than machine precision in magnitude, and subscripted c's to denote low order polynomials in n. We further assume that in our floating point model, additions, subtractions, multiplications, divisions, and square-roots are done to full relative accuracy, i.e.,

$$fl(x \circ y) = (x \circ y)(1 + \delta),$$

where $\circ$ denotes $+, -, \times, \div$ and $|\delta| \leq \epsilon$. Likewise for the square-root operation.

2 DISPLACEMENT STRUCTURE

The notion of displacement structure has proven to be a very powerful and convenient tool in several problems in systems and mathematics. Since its introduction by Kailath et al. in the late seventies [8], it has provided a systematic way to elegantly describe and exploit different forms of matrix structures that often arise in applications. For reasons of brevity, however, we limit ourselves here to a brief review of the case of positive-definite symmetric structured matrices. For more details on the topic, the interested reader may consult [7]; a recent survey on the origins and theory of the displacement structure concept.

Consider an $n \times n$ symmetric positive-definite matrix R and an $n \times n$ lower-triangular real-valued matrix F. The displacement of R with respect to F is denoted by ∇_F and defined as

$$\nabla_F = R - FRF^T. \tag{11.1}$$

The matrix R is said to have low displacement rank with respect to F if the rank of ∇_F is considerably lower than n. In this case, R is said to have displacement structure with respect to F [7]. Let $r \ll n$ denote the rank of ∇_F. It follows that we can factor ∇_F as $\nabla_F = GJG^T$, where G is an $n \times r$ matrix and J is a signature matrix of the form $J = (I_p \oplus -I_q)$, $p+q = r$. The factorization $\nabla_F = GJG^T$ is highly nonunique since G can be replaced by $G\Theta$ for any $J-$unitary matrix Θ ($\Theta J \Theta^T = J$). A matrix R is therefore structured with respect to the displacement operation defined by (11.1) if it satisfies a displacement equation of the form

$$R - FRF^T = GJG^T \ , \tag{11.2}$$

with a low rank matrix G. Equation (11.2) uniquely defines R iff the diagonal entries of F satisfy $1 - f_i f_j \neq 0$ for all i, j. This condition is assumed throughout the paper, although it can be relaxed in some instances [7].

The pair (G, J) is said to be a generator pair for R since, along with F, it completely identifies R. Note however that, while R has n^2 entries, the matrix G has nr entries and r is usually much smaller than n. Therefore, algorithms that operate on the entries of G, with the purpose of obtaining a triangular factorization for R, will generally be an order of magnitude faster than algorithms that operate on the entries of R itself. The generalized Schur algorithm is one such fast $O(rn^2)$ procedure, which receives as input data the matrices (F, G, J) and provides as output data the Cholesky factor of R [7].

3 GENERALIZED SCHUR ALGORITHM

We focus here, for simplicity, on matrices R with displacement rank 2, say

$$R - FRF^T = \begin{bmatrix} u_1 & v_1 \end{bmatrix} \begin{bmatrix} 1 & 0 \\ 0 & -1 \end{bmatrix} \begin{bmatrix} u_1 & v_1 \end{bmatrix}^T, \tag{11.3}$$

where u_1 and v_1 denote the $n \times 1$ column vectors of G. Extensions of the discussion to the $r > 2$ case are possible and will be discussed elsewhere. The diagonal entries of F are also assumed to be strictly inside the open unit disc ($|f_i| < 1$). In this case, the matrix F is said to be stable. Also, since G is highly nonunique, it can always be chosen in proper form,

$$G = \begin{bmatrix} x & 0 \\ x & x \\ \vdots & \vdots \\ x & x \end{bmatrix}. \tag{11.4}$$

That is, the top entry of v_1, v_{11}, can always be chosen to be zero.

GENERALIZED SCHUR ALGORITHM. *Start with* (u_1, v_1) *and repeat for* $i \geq 1$:

1. *Compute* $\Phi_i = (F - f_i I)(I - f_i F)^{-1}$ *and form* $\begin{bmatrix} \Phi_i u_i & v_i \end{bmatrix}$. *The top* i *entries of* $\Phi_i u_i$ *and* v_i *are zero.*

2. *Apply a hyperbolic rotation* Θ_i *in order to annihilate the* $(i+1)$ *entry of* v_i. *Denote the resulting column vectors by* (u_{i+1}, v_{i+1}):

$$G_{i+1} = \begin{bmatrix} u_{i+1} & v_{i+1} \end{bmatrix} = \begin{bmatrix} \Phi_i u_i & v_i \end{bmatrix} \Theta_i. \tag{11.5}$$

   G_{i+1} *will also be in proper form, with its top* i *rows equal to zero.*

3. *The* $i-th$ *column of the Cholesky factor* L *is given by*

$$l_i = \sqrt{1 - |f_i|^2}(I - f_i F)^{-1} u_i. \tag{11.6}$$

   The top $(i-1)$ *entries of* l_i *are zero.*

After n steps, the algorithm provides the Cholesky decomposition $R = \sum_{i=1}^{n} l_i l_i^T$. Moreover, let R_i denote the Schur complement of R with respect to its leading $i \times i$ submatrix (R_i is $(n-i) \times (n-i)$) and define the $n \times n$ embedding

$$\tilde{R}_i = \begin{bmatrix} 0 & 0 \\ 0 & R_i \end{bmatrix}.$$

It can be shown that $\tilde{R}_i - F\tilde{R}_i F^T = G_i J G_i^T$ [7]. In other words, G_i is a generator matrix for the ith Schur complement, which is therefore structured.

4 LIMITS TO NUMERICAL ACCURACY

Given a symmetric positive-definite matrix R (not necessarily structured), if its Cholesky factor is evaluated by any standard backward stable method that operates on the entries of R, e.g., Gaussian elimination, the corresponding error bound is given by $\|R - \hat{L}\hat{L}^T\| \leq c_1 \epsilon \|R\|$ [2]. A fundamental question that needs to be answered then is the following: given (F, G, J), but not R, how accurately can we expect to be able to compute the Cholesky factorization of R *irrespective* of the algorithm used (*slow or fast*)? To address this issue we note that just representing (F, G) in finite precision already induces round-off errors. This in turn imposes limits on how accurate an algorithm that employs (F, G) can be. We demonstrate this point by the following example.

Let F be a stable diagonal matrix with distinct entries $\{f_i\}$ and assume f_1 is the largest in magnitude. Let the entries of the column vectors u_1 and v_1 be given by $u_{i1} = (0.5)^{i-1}$, $v_{i1} = \gamma f_i u_{i1}$, where $0 < \gamma < 1$. The matrix R that solves (11.3) for the given (F, u_1, v_1) is symmetric positive-definite. Now define the perturbed vectors $\hat{u}_1$ and $\hat{v}_1$ with $\hat{u}_{11} = u_{11}(1+\delta)$, $\hat{u}_{i1} = u_{i1}$, $i \geq 2$, and $\hat{v}_1 = v_1$. That is, we only make a relative perturbation in the first entry of u_1 and keep all other entries of u_1 and v_1 unchanged. Here, δ is a small number (e.g., for round-off errors, $|\delta|$ is smaller than machine precision). Let $\hat{R}$ be the solution of the displacement equation with the perturbed generator, $\hat{R} - F\hat{R}F^T = \hat{G}J\hat{G}^T$, $\hat{G} = \begin{bmatrix} \hat{u}_1 & \hat{v}_1 \end{bmatrix}$, and introduce the error matrix $E = R - \hat{R}$. Then it can be shown that a *lower bound* on the norm of E is

$$\|E\| \geq |E_{11}| \geq \frac{3}{2}|\delta| \, \|(I - F \otimes F)^{-1}\| \, \|u_1\|^2.$$

By suitably choosing γ, the norm of R can be much smaller than the above bound. Indeed,

$$\|R\| \leq n \max_i \left[\frac{1-\gamma^2 f_i^2}{1-f_i^2} \, u_{ii}^2 \right],$$

which shows that as $\gamma \to 1$, the norm of R can be bounded by $n\|u_1\|^2$. In summary, we have shown that, at the same time, $\|R - \hat{R}\|$ can be larger than $|\delta| \, \|(I - F \otimes F)^{-1}\| \, \|u_1\|^2$ and $\|R\|$ can be much smaller than $\|(I - F \otimes F)^{-1}\| \, \|u_1\|^2$. Hence, in general, we can not expect the error norm, $\|R - \hat{L}\hat{L}^T\|$, for any algorithm (slow or fast) that uses (F, G, J) as input data (but not R), to be as small as $c_1|\delta| \, \|R\|$, for some constant c_1.

Therefore, we conclude that *irrespective* of the algorithm we use (*slow or fast*), if the input data is (F, G, J), then we can not expect a better bound than

$$\|R - \hat{L}\hat{L}^T\| \leq c_2\epsilon \, \|(I - F \otimes F)^{-1}\| \, \|u_1\|^2. \tag{11.7}$$

5 HYPERBOLIC ROTATIONS

We now discuss how to improve the numerical accuracy of the generalized Schur algorithm. We start by discussing reliable implementations for the hyperbolic rotations.

For this purpose, we denote the prearray $\begin{bmatrix} \Phi_i u_i & v_i \end{bmatrix}$ by $\bar{G}_{i+1}$ and rewrite (11.5) in the compact form $G_{i+1} = \bar{G}_{i+1}\Theta_i$. This expression shows that, in infinite precision, the generator matrices G_{i+1} and $\bar{G}_{i+1}$ must satisfy the fundamental requirement $G_{i+1}JG_{i+1}^T = \bar{G}_{i+1}J\bar{G}_{i+1}^T$. Obviously, this condition cannot be guaranteed in finite precision. Indeed, a naive implementation of the hyperbolic transformation Θ_i can lead to large errors. In finite precision, if we apply Θ_i directly to $\bar{G}_{i+1}$ we obtain a computed $\hat{G}_{i+1}$ such that $\|\hat{G}_{i+1}J\hat{G}_{i+1}^T - \bar{G}_{i+1}J\bar{G}_{i+1}^T\| \leq c_3\epsilon \, \|\bar{G}_{i+1}\|^2 \, \|\Theta_i\|^2$. This is not desirable since $\|\Theta_i\|$ can be large. One possible way to ameliorate the problem is to employ a mixed-downdating procedure [5, 6].

An alternative so-called OD procedure that we propose is based on using the SVD of a hyperbolic rotation Θ:

$$\Theta = \frac{1}{\sqrt{1-\rho^2}} \begin{bmatrix} 1 & -\rho \\ -\rho & 1 \end{bmatrix}, \quad \rho = \frac{\beta}{\alpha},$$

$$= \frac{1}{\sqrt{2}} \begin{bmatrix} 1 & 1 \\ -1 & 1 \end{bmatrix} \begin{bmatrix} \sqrt{\frac{\alpha+\beta}{\alpha-\beta}} & 0 \\ 0 & \sqrt{\frac{\alpha-\beta}{\alpha+\beta}} \end{bmatrix} \begin{bmatrix} 1 & -1 \\ 1 & 1 \end{bmatrix} \frac{1}{\sqrt{2}} = QDQ^T.$$

This guarantees $\|\hat{G}_{i+1} J \hat{G}_{i+1}^T - \bar{G}_{i+1} J \bar{G}_{i+1}^T\| \leq c_4 \epsilon \left(\|\bar{G}_{i+1}\|^2 + \|\hat{G}_{i+1}\|^2 \right)$.

Another proposal (with better numerical properties) is to employ what we refer to as the H procedure. Let $\begin{bmatrix} x_1 & y_1 \end{bmatrix} = \begin{bmatrix} x & y \end{bmatrix} \Theta$ and assume w.l.o.g $|x| > |y|$. The advantage of the H procedure is that the computed quantities $\hat{x}_1$ and $\hat{y}_1$ satisfy $\begin{bmatrix} \hat{x}_1 + e_1 & \hat{y}_1 + e_2 \end{bmatrix} = \begin{bmatrix} x & y \end{bmatrix} \Theta$, with $|e_1| \leq c_5 \epsilon |\hat{x}_1|$, $|e_2| \leq c_6 \epsilon (|\hat{x}_1| + |\hat{y}_1|)$. The term x_1 is computed as

$$x_1 = \frac{|\alpha| x}{\sqrt{(\alpha-\beta)(\alpha+\beta)}} \left[1 - \frac{\beta}{\alpha} \frac{y}{x} \right],$$

where $\xi = 1 - \frac{\beta}{\alpha}\frac{y}{x}$ is evaluated to full relative accuracy as follows: If $\frac{\beta}{\alpha}\frac{y}{x} < 1/2$ then $\xi \leftarrow 1 - \frac{\beta}{\alpha}\frac{y}{x}$, else compute $d_1 = \frac{|\alpha|-|\beta|}{|\alpha|}$, $d_2 = \frac{|x|-|y|}{|x|}$, and $\xi \leftarrow d_1 + d_2 - d_1 d_2$. To compute y_1 we use the expression

$$y_1 = x_1 - \sqrt{\frac{\alpha+\beta}{\alpha-\beta}}\,(x-y).$$

From now on we shall denote by $\hat{u}_{i+1}$ and $\hat{v}_{i+1}$ the computed generator columns at step i, i.e., $\hat{G}_{i+1} = \begin{bmatrix} \hat{u}_{i+1} & \hat{v}_{i+1} \end{bmatrix}$, starting with $\bar{G}_i = \begin{bmatrix} \Phi_i \hat{u}_i & \hat{v}_i \end{bmatrix}$. Each step of the algorithm also requires multiplying the Blaschke matrix Φ_i by u_i.

6 THE BLASCHKE MATRIX

When F is stable and diagonal, Φ_i itself will be diagonal with $(\Phi_i)_{jj} = \frac{f_j - f_i}{1 - f_i f_j}$. In this case, the numerator of $(\Phi_i)_{jj}$ can be computed to full relative accuracy since $fl(f_j - f_i) = (f_j - f_i)(1+\delta_1)$, while the denominator $x_{ij} = (1 - f_i f_j)$ can be evaluated as follows. If $f_i f_j < 1/2$ then $x_{ij} \leftarrow 1 - f_i f_j$, else compute $d_j = 1 - |f_j|$, $d_i = 1 - |f_i|$, and $x_{ij} \leftarrow d_i + d_j - d_i d_j$. It can be shown that this guarantees $\hat{x}_{ij} = x_{ij}(1 + 33\delta_{11})$. Therefore, $\Phi_i \hat{u}_i$ can be computed to component-wise relative accuracy. As for the l_i's, define $\bar{l}_i = \sqrt{1-f_i^2}\,(I - f_i F)^{-1} \hat{u}_i$. If we use the expression

$$(\bar{l}_i)_j = \frac{\sqrt{(1-f_i)(1+f_i)}}{1 - f_i f_j} (\hat{u}_i)_j,$$

to compute $\bar{l}_i$, with the technique explained above for the denominator $(1 - f_i f_j)$, then we can show that $(\hat{l}_i)_j = (\bar{l}_i)_j(1 + c_7\delta_{13})$.

When F is strictly lower-triangular, the expression for Φ_i collapses to $\Phi_i = F$. In this case we use the standard matrix-vector multiplication and obtain $\|fl(F\hat{u}_i) - F\hat{u}_i\| \leq c_8\epsilon\|F\| \, \|\hat{u}_i\|$. Also, since $f_i = 0$, $\bar{l}_i = \hat{u}_i = \hat{l}_i$.

In summary, it follows that for diagonal and for strictly lower-triangular F we can write $\|fl(\Phi_i\hat{u}_i) - \Phi_i\hat{u}_i\| \leq c_9\epsilon\|\Phi_i\| \, \|\hat{u}_i\|$.

Finally, we remark that condition (11.11) on the matrix R guarantees the positive-definiteness of the successive Schur complements and, therefore, that $|\Phi_i\hat{u}_i|_{i+1} > |\hat{v}_i|_{i+1}$. This assures that the reflection coefficients will be smaller than one in magnitude, a condition that can be enforced in finite-precision as follows: If $|fl(\Phi_i\hat{u}_i)|_{i+1} < |\hat{v}_{i+1,i}|$ then $fl(\Phi_i\hat{u}_i)_{i+1} \leftarrow |\hat{v}_{i+1,i}|(1 + 3\epsilon)\text{sign}(fl(\Phi_i\hat{u}_i)_{i+1})$.

7 ERROR ANALYSIS

It follows from the procedures suggested above, for implementing the hyperbolic transformations and for evaluating $\Phi_i u_i$, that

$$\hat{u}_{i+1}\hat{u}_{i+1}^T - \hat{v}_{i+1}\hat{v}_{i+1}^T = \Phi_i\hat{u}_i\hat{u}_i^T\Phi_i^T - \hat{v}_i\hat{v}_i^T - M_{i+1}, \tag{11.8}$$

where the error matrices M_{i+1} satisfy

$$\|M_{i+1}\| \leq c_{10}\epsilon(\|\hat{u}_{i+1}\|^2 + \|\Phi_i\|^2\|\hat{u}_i\|^2 + \|\hat{v}_{i+1}\|^2 + \|\hat{v}_i\|^2). \tag{11.9}$$

If we define $\bar{R} = \sum_{i=1}^n \bar{l}_i\bar{l}_i^T$, $\hat{R} = \sum_{i=1}^n \hat{l}_i\hat{l}_i^T$, and $E = R - \hat{R}$, then the following error bound can be established:

$$\|E\| \leq c_{11}\epsilon\|(I - F \otimes F)^{-1}\| \, (2 + \|F\|^2)\left[\|\bar{R}\| + \sum_{j=1}^n \|\hat{u}_j\|^2\right]. \tag{11.10}$$

The derivation of this bound assumes that the algorithm does not break down. That is, at every iteration the reflection coefficients of the hyperbolic rotations are assumed to be strictly less than one in magnitude. This can be guaranteed by requiring the minimum eigenvalue of R to meet the lower bound

$$\lambda_{\min}(R) > c_{11}\epsilon\|(I - F \otimes F)^{-1}\| \, (2 + \|F\|^2)\left[\|\bar{R}\| + \sum_{j=1}^n \|\hat{u}_j\|^2\right]. \tag{11.11}$$

This ensures the positive-definiteness of the computed Schur complements.

The term $\|(I - F \otimes F)^{-1}\|$ in the error bound is expected from the perturbation analysis of Sec. 4. However, the presence of the norms of the successive generators makes the error bound larger than the bound suggested by the perturbation analysis, which only depends on the norm of the first generator matrix. The natural question then is how big can the norm of the generators be?

When F is strictly lower triangular, the error bound can be written in the alternate form

$$\|E\| \leq c_{12}\epsilon(2 + \|F\|^2)\left(\sum_{i=1}^{n} \|F^i\|^2\right)\left[\|\bar{R}\| + \left(\sum_{i=1}^{n} \|\hat{u}_i\|^2\right)\right],$$

and the u_i can be bounded by $\|u_i\|^2 \leq \|R\|$. This shows that, when F is contractive ($\|F\| \leq 1$), the error bound is as good as can be expected from the perturbation analysis of Sec. 4. If F is strictly lower triangular but non-contractive then the error norm can possibly depend on $\|(I - F \otimes F)^{-1}\|$.

When F is a stable diagonal matrix, the growth of the generators $\{\hat{u}_i\}$ can be further controlled [9]. Also, the bound in (11.10) may suggest that the norm of the error can become very large when the magnitude of the diagonal entries of F become close to one. But this is not necessarily the case since we can show that

$$\|E\| \leq c_{13}\epsilon \, \frac{\sum_{i=1}^{n} \|\bar{R}_i\|}{1 - \max_{j,i}\left(\rho_{j,i}^2\right)},$$

which is independent of the $\{f_i\}$. Here, $\rho_{j,i} = \hat{v}_{j,i}/\hat{u}_{j,i}$. In other words, if the coefficients $\rho_{j,i}$ are sufficiently smaller than one, then the algorithm will be backward stable irrespective of how close the $\{|f_i|\}$ are to one.

8 CONCLUDING REMARKS

For brevity, we have omitted several relevant points from our discussion. We briefly comment on these issues here.

To begin with, our analysis suggests that for $\|F\|$ sufficiently close to one, the error norm can become large. However, if our original motivation is to solve $Rx = b$, then the error can be improved by using iterative refinement [12].

Also, the analysis and results of this paper can be extended to positive-definite structured matrices with displacement ranks larger than 2, as well as to other forms of displacement structure, say the Hankel-like case $FR + RF^T = GJG^T$. We are also pursuing the extension of our results to general nonsymmetric structured matrices. Moreover, the results of this work suggest improvements to certain fast algorithms in adaptive filtering and state-space estimation in view of the connections of these algorithms to the Schur algorithm [7, 13, 14].

Finally, we remark that for indefinite or nonsymmetric matrices the LDU factorization is numerically difficult to compute without pivoting. Since pivoting can destroy the structure of a matrix, it is not always possible to incorporate it into a fast algorithm directly. Sometimes one can transform the matrix so that it is structured with respect to a diagonal F and then use partial pivoting during the Schur algorithm [10, 11]. Another way to approach this problem is to compute the QR factorization of the structured matrix rapidly. Several methods have been proposed earlier in the literature but none of them are numerically stable. In recent work, and motivated by the results of this paper, we have proposed a fast and backward stable Schur-type algorithm for solving linear systems of equations involving nonsymmetric Toeplitz (also quasi-Toeplitz and Toeplitz-like) matrices [15]. It is based on a modified fast QR factorization of the coefficient matrix T in $Tx = b$. The algorithm actually computes the factorization $T = \Delta(\Delta^{-1}Q)R$, where Δ is lower-triangular, $(\Delta^{-1}Q)$ is orthogonal and R is upper triangular. The solution x is then computed efficiently by using $x = R^{-1}(Q^T\Delta^{-T})\Delta^{-1}b$, which can be done in $O(n^2)$ operations, where n is the matrix dimension.

Acknowledgements

This work was partially supported by NSF under award no. MIP-9409319.

REFERENCES

[1] Kailath, T., "A theorem of I. Schur and its impact on modern signal processing,", *Operator Theory: Advances and Applications*, vol. 18, pp. 9–30, Birkhäuser, Boston, 1986.

[2] Golub, G. H. and Van Loan, C. F., *Matrix Computations*, The Johns Hopkins University Press, Baltimore, second ed., 1989.

[3] Cybenko, G., "The numerical stability of the Levinson-Durbin algorithm for Toeplitz system of equations," *SIAM J. Sci. Statis. Comput.*, vol. 1, pp. 303–319, 1980.

[4] Sweet, D. R., *Numerical Methods for Toeplitz Matrices*, PhD thesis, University of Adelaide, Adelaide, Australia, 1982.

[5] Bojanczyk, A. W., Brent, R. P., de Hoog, F. R., and Sweet, D. R., "On the stability of the Bareiss and related Toeplitz factorization algorithms," *SIAM J. Matrix Anal. Appl.*, vol. 16, pp. 40–57, 1995.

[6] Bojanczyk, A. W., Brent, R. P., van Dooren, P., and de Hoog, F. R., "A note on downdating the Cholesky factorization," *SIAM J. Sci. Stat. Comput.*, vol. 8, pp. 210–221, 1987.

[7] Kailath, T. and Sayed, A. H., "Displacement structure: Theory and applications," *SIAM Review*, vol. 37, no. 3, Sep. 1995.

[8] Kailath, T., Kung, S. Y., and Morf, M., "Displacement ranks of a matrix," *Bulletin of the American Mathematical Society*, vol. 1, no. 5, pp. 769-773, Sep. 1979.

[9] Gu, M., "Stable and efficient algorithms for structured systems of linear equations," *in preparation.*

[10] Heinig, G., "Inversion of generalized Cauchy matrices and other classes of structured matrices," in *Linear Algebra for Signal Processing*, IMA volumes in Mathematics and its Applications, vol. 69, pp. 63–81, 1995.

[11] Gohberg, I., Kailath, T., and Olshevsky, V., "Fast Gaussian elimination with partial pivoting for matrices with displacement structure," to appear in *Math. of Computation.*

[12] Chandrasekaran, S. and Sayed, A. H., "Stabilizing the fast generalized Schur algorithm," *submitted for publication.*

[13] Sayed, A. H., Kailath, T., and Lev-Ari, H., "Generalized Chandrasekhar recursions from the generalized Schur algorithm," *IEEE Transactions on Automatic Control*, vol. 39, no. 11, pp. 2265–2269, Nov. 1994.

[14] Sayed, A. H. and Kailath, T., "A state-space approach to adaptive RLS filtering," *IEEE Signal Processing Magazine*, vol. 11, no. 3, pp. 18-60, Jul. 1994.

[15] Chandrasekaran, S. and Sayed, A. H., "A fast and stable solver for non-symmetric structured systems," *in preparation.*

12

DISPLACEMENT STRUCTURE: TWO RELATED PERSPECTIVES

Hanoch Lev-Ari

Department of Electrical and Computer Engineering
Northeastern University
Boston, Massachusetts 02115

Dedicated to Tom Kailath on his sixtieth birthday,
in appreciation of our long-standing friendship and collaboration.

ABSTRACT

The matrix-theoretic and function-theoretic characterizations of matrices with a displacement structure are presented, and the relation between them is explored. The divided-difference expansion of a bivariate function is used to demonstrate the equivalence between the two characterizations.

1 INTRODUCTION

The theory of structured matrices has evolved from two distinct, yet mutually complementary perspectives, namely the matrix-theoretic perspective [1, 2, 3, 4, 5, 6, 7], and the function-theoretic (or transform-domain) perspective [8, 3, 9, 10, 11, 12]. The matrix-theoretic approach made it possible to extend the earlier work to non-Hermitian matrices, and to non-symmetric and even time-variant displacement structures [13, 14], with applications to factorization of composite matrices [4, 5] and to adaptive RLS filtering [15]. The function-theoretic approach made it possible to establish connections with lossless inverse scattering [16, 17], Bezoutians and root location [18], reproducing kernels [19], and interpolation/extension problems [20, 21].

The notion of displacement-structure can be characterized via either one of the two perspectives. The most general matrix-theoretic characterization is

$$\mathbf{A}\,\mathbf{R}\,\mathbf{C}^* - \mathbf{B}\,\mathbf{R}\,\mathbf{D}^* = \mathbf{G}\,J\,\mathbf{H}^* \tag{12.1}$$

where $\mathbf{A}, \mathbf{B}, \mathbf{C}, \mathbf{D}$ are arbitrary (given) lower-triangular matrices. Here the matrix $\mathbf{R}$ is said to have *generalized displacement structure,* and it can be efficiently characterized in terms of the matrices $\mathbf{G}$, $\mathbf{H}$, and J, provided that $\mathbf{G}, \mathbf{H}$ have significantly fewer columns than rows. The most general function-theoretic characterization of displacement structure is [10]

$$d(z,w)\, R(z,w) = G(z)\, J\, H^*(w) \tag{12.2a}$$

where $d(z,w)$ is an arbitrary (given) bivariate function, and $G(\cdot), H(\cdot)$ are row-vectors of functions. Here $R(z,w)$ is said to have *generalized displacement structure*, and it can be efficiently characterized in terms of $G(\cdot), H(\cdot)$ and J, since

$$R(z,w) = \frac{G(z)\, J\, H^*(w)}{d(z,w)} \tag{12.2b}$$

Furthermore, it can be shown that in order to admit an efficient factorization procedure of the Schur-Cholesky type, the bivariate function $d(z,w)$ must be representable as

$$d(z,w) = \alpha(z)\gamma^*(w) - \beta(z)\delta^*(w) \tag{12.3}$$

The symmetric version of this result (with $\gamma(z) = \alpha(z)$, and $\delta(z) = \beta(z)$) has been established in [10]. The nonsymmetric version (12.3) can be established by an extension of the argument used in [10].

Taking the constraint (12.3) into account, the function-theoretic characterization of displacement structure becomes

$$\alpha(z)R(z,w)\gamma^*(w) - \beta(z)R(z,w)\delta^*(w) = G(z)JH^*(w) \tag{12.4}$$

The objective of this paper is to establish the equivalence of the two characterizations (12.1) and (12.4). More precisely, we shall show that every expression of the form (12.1) can be obtained by a suitable manipulation, which we call a *bivariate divided-difference expansion*, of the expression (12.4).

One matrix-theoretic interpretation of (12.4) has been introduced in [3, 10]: by expanding (12.4) into a bivariate power series (i.e., an inverse bivariate $\mathcal{Z}$-transform) we obtain the expression

$$\alpha(\mathbf{Z})\, \mathbf{R}\, \gamma^*(\mathbf{Z}) - \beta(\mathbf{Z})\, \mathbf{R}\, \delta^*(\mathbf{Z}) = \mathbf{G}\, J\, \mathbf{H}^* \tag{12.5a}$$

where $\mathbf{G}, \mathbf{H}$ are obtained by vertically stacking the coefficients of the power series expansion of $G(\cdot), H(\cdot)$ (see, e.g., [10]), and where $\mathbf{Z}$ is the lower-shift matrix, with ones on the first subdiagonal and zeros elsewhere, viz.,

$$\mathbf{Z} = \{\, \delta_{i,j+1}\,;\ 0 \leq i,j < \infty \,\} \tag{12.5b}$$

The elements of $\mathbf{R}$ are the coefficients of the bivariate power series expansion of $R(z,w)$. Thus, we have obtained an expression of the form (12.1), in which the lower-triangular matrices

$$\mathbf{A} = \alpha(\mathbf{Z}), \quad \mathbf{B} = \beta(\mathbf{Z}), \quad \mathbf{C} = \gamma(\mathbf{Z}), \quad \mathbf{D} = \delta(\mathbf{Z}) \tag{12.5c}$$

all have a *Toeplitz structure,* as in [10]. Another choice is to *evaluate* (12.2b) at the point-pairs $(z,w) = (\zeta_i, \omega_j)$ for all $0 \le i, j < \infty$. This time, the matrix $\mathbf{R}$ consists of the entries $R(\zeta_i, \omega_j)$, and the matrix $\mathbf{G}$ (resp. $\mathbf{H}$) consists of the entries $G(\zeta_i)$ (resp. $H(\omega_j)$). The resulting matrix interpretation of (12.4) again has the form (12.1), but now with

$$\mathbf{A} = \operatorname{diag}\{\alpha(\zeta_i)\}, \quad \mathbf{B} = \operatorname{diag}\{\beta(\zeta_i)\}, \quad \mathbf{C} = \operatorname{diag}\{\gamma(\omega_i)\}, \quad \mathbf{D} = \operatorname{diag}\{\delta(\omega_i)\} \tag{12.6}$$

which are all diagonal matrices. We show in the sequel that these rather different matrix interpretations of (12.4) are two particular cases in a broad family of matrices which are all obtained from (12.4) by combining divided-difference expansions with similarity transformations.

2 SUMMARY OF MAIN RESULTS

The starting point for our derivation is the (univariate) *divided-difference expansion* of a function with respect to a sequence of complex points $\{\zeta_i\}$, viz.,

$$f(z) = \sum_{i=0}^{\infty} f_i \, p_i(z) \tag{12.7a}$$

where

$$p_0(z) \equiv 1\,, \quad \text{and} \quad p_i(z) = \prod_{j=0}^{i-1} (z - \zeta_j) \quad \text{for } i \ge 1 \tag{12.7b}$$

When $\zeta_i = 0$ for all i, this reduces to the well-known power series expansion of a function (of a complex variable), and when $\zeta_i = \zeta_0 \neq 0$ for all i, this is the well-known Taylor-series expansion at the point $z = \zeta_0$. The general case (i.e., arbitrary ζ_i) is sometimes known as a Newton-series expansion [22]. Several properties of this expansion are discussed in Sec. 3. Here we shall only observe that the sequence of coefficients $\{f_i\}$ can be interpreted as a representation of the function $f(z)$ in a basis spanned by the polynomials $\{p_i(z)\}$. We shall use the notation $\boldsymbol{V}\{f\}$ for the (infinite) column vector consisting of these coefficients. We show in Sec. 3 that

$$\boldsymbol{V}\{fg\} = \mathcal{L}\{f\}\, \boldsymbol{V}\{g\} \tag{12.8a}$$

where $\mathcal{L}\{f\}$ is a lower-triangular matrix, whose columns are the divided-difference expansions of $f(z)p_i(z)$ for $i = 0, 1, 2, \ldots$, viz.,

$$\mathcal{L}\{f\} = \Big[\boldsymbol{V}\{fp_0\} \;\; \boldsymbol{V}\{fp_1\} \;\; \boldsymbol{V}\{fp_2\} \;\; \cdots \Big] \tag{12.8b}$$

We shall call $\mathcal{L}\{f\}$ the *divided-difference matrix* associated with the function $f(z)$. This lower-triangular matrix can be interpreted as the representation, in the basis spanned by $\{p_i(z)\}$, of the linear operator associated with the multiplication by $f(z)$: each column of $\mathcal{L}\{f\}$ describes the representation in this basis of the function obtained by applying the "multiply-by-$f(z)$" operator to one of the basis polynomials.

The divided-difference expansion of a bivariate function such as $R(z,w)$ is defined in terms of basis functions of both variables, viz.,

$$R(z,w) = \sum_{i,j=0}^{\infty} r_{ij}\, p_i(z)\, q_j^*(w)$$

where $p_i(z)$ are the polynomials defined in (12.7), and $q_j(w)$ are defined in a similar fashion, but with a *different* sequence of complex points, say $\{\omega_j\}$. It will be convenient to arrange the expansion coefficients in a matrix $\mathbf{R}$, so that the (i,j)-th element of $\mathbf{R}$ is r_{ij}. Applying the bivariate divided-difference expansion to the function-theoretic expression (12.4), we obtain in Sec. 4 a derived matrix expression, viz.,

$$\mathcal{L}\{\alpha\}\,\mathbf{R}\,\mathcal{L}^*\{\gamma\} - \mathcal{L}\{\beta\}\,\mathbf{R}\,\mathcal{L}^*\{\delta\} = \mathbf{G}\,J\,\mathbf{H}^* \tag{12.9a}$$

where

$$\mathbf{G} = \boldsymbol{V}\{G\}\ , \quad \mathbf{H} = \boldsymbol{V}\{H\} \tag{12.9b}$$

Notice that $\mathbf{G}$, $\mathcal{L}\{\alpha\}$, $\mathcal{L}\{\beta\}$ are associated with $\{\zeta_i\}$, while $\mathbf{H}$, $\mathcal{L}\{\gamma\}$, $\mathcal{L}\{\delta\}$ are associated with $\{\omega_j\}$, as in (12.6). Furthermore, we can use arbitrary nonsingular lower-triangular matrices $\mathbf{T}_1, \mathbf{T}_2$ to transform this expression into the form

$$\mathbf{A}\,\left(\mathbf{T}_1\,\mathbf{R}\,\mathbf{T}_2^*\right)\,\mathbf{C}^* - \mathbf{B}\,\left(\mathbf{T}_1\,\mathbf{R}\,\mathbf{T}_2^*\right)\,\mathbf{D}^* = \left(\mathbf{T}_1\mathbf{G}\right)\,J\,\left(\mathbf{T}_2\mathbf{H}\right)^*$$

where

$$\mathbf{A} = \mathbf{T}_1\,\mathcal{L}\{\alpha\}\,\mathbf{T}_1^{-1},\ \ \mathbf{B} = \mathbf{T}_1\,\mathcal{L}\{\beta\}\,\mathbf{T}_1^{-1},\ \ \mathbf{C} = \mathbf{T}_2\,\mathcal{L}\{\gamma\}\,\mathbf{T}_2^{-1},\ \ \mathbf{D} = \mathbf{T}_2\,\mathcal{L}\{\delta\}\,\mathbf{T}_2^{-1} \tag{12.10a}$$

This results in the matrix-theoretic expression

$$\mathbf{A}\,\mathbf{R}\,\mathbf{C}^* - \mathbf{B}\,\mathbf{R}\,\mathbf{D}^* = \mathbf{G}\,J\,\mathbf{H}^* \tag{12.10b}$$

with $\mathbf{A}, \mathbf{B}, \mathbf{C}, \mathbf{D}$ as in (12.10a), and where we have redefined the matrices $\mathbf{R}, \mathbf{G}, \mathbf{H}$ to absorb the effects of $\mathbf{T}_1, \mathbf{T}_2$.

It may appear that the four lower-triangular matrices $\{\mathbf{A}, \mathbf{B}, \mathbf{C}, \mathbf{D}\}$ in (12.10) are not entirely arbitrary because they are all similar to divided-difference matrices. Nevertheless, we show in Sec. 4 that:

> Every lower-triangular (non-derogatory) matrix is similar to a divided-difference matrix. Furthermore, the similarity matrix is also lower-triangular.

Thus, each one of the following pairs of lower-triangular matrices — $(\mathbf{A}, \mathbf{C})$, $(\mathbf{B}, \mathbf{D})$, $(\mathbf{A}, \mathbf{D})$, and $(\mathbf{B}, \mathbf{C})$ — can be chosen entirely at will, provided the derogatory case is avoided. Also, we observe that the very general matrix-theoretic characterization (12.1) is in fact redundant: generically, only a pair of arbitrary lower-triangular matrices is required for a complete characterization, rather than four arbitrary matrices. For instance, if both $\mathbf{A}$ and $\mathbf{C}$ are nonsingular, then (12.1) is equivalent to

$$\mathbf{R} - (\mathbf{A}^{-1}\mathbf{B})\,\mathbf{R}\,(\mathbf{C}^{-1}\mathbf{D})^* = (\mathbf{A}^{-1}\mathbf{G})\,J\,(\mathbf{C}^{-1}\mathbf{H})^*$$

which requires only two arbitrary lower-triangular matrices, namely $\mathbf{A}^{-1}\mathbf{B}$ and $\mathbf{C}^{-1}\mathbf{D}$, to characterize a particular type of displacement structure. A similar argument holds if: (i) $\mathbf{B}$ and $\mathbf{D}$ are both non-singular, or (ii) $\mathbf{A}$ and $\mathbf{D}$ are both non-singular, or (iii) $\mathbf{B}$ and $\mathbf{C}$ are both non-singular. Since a generic lower-triangular matrix is both non-singular and non-derogatory, our preceding discussion establishes the claim that the matrix-theoretic characterization (12.1) with *generic* $\mathbf{A}, \mathbf{B}, \mathbf{C}, \mathbf{D}$ can always be obtained by a suitable manipulation of the function-theoretic characterization (12.4).

3 DIVIDED-DIFFERENCE EXPANSIONS

The definition (12.7) of a divided-difference expansion of a function $f(z)$ can be rewritten as an "inner-product" expression, viz.,

$$f(z) = \boldsymbol{P}(z)\boldsymbol{V}\{f\} \tag{12.11a}$$

where

$$\boldsymbol{P}(z) := \begin{bmatrix} p_0(z) & p_1(z) & p_2(z) & \ldots \end{bmatrix} \tag{12.11b}$$

and where the column vector $\boldsymbol{V}\{f\}$ consists of the expansion coefficients $\{f_i\}$ as already introduced in Sec. 2. This expansion introduces a mapping from a space of functions to a space of (infinite) column vectors, viz., $f(z) \longrightarrow \boldsymbol{V}\{f\}$. In this section we present several properties of this mapping.

First, we need to establish a computational procedure for carrying out the mapping $f(z) \longrightarrow \boldsymbol{V}\{f\}$. The most direct method is by means of the so-called *divided-difference recursion* [22]

$$f_{i+1}(z) = \frac{f_i(z) - f_i(\zeta_i)}{z - \zeta_i}\,, \quad f_0(z) := f(z) \tag{12.12}$$

from which we determine the expansion coefficients $f_i \equiv f_i(\zeta_i)$.

In order to derive the multiplication formula (12.8), we first consider the multiplication of a single "basis function" $p_j(z)$ by an arbitrary $f(z)$ [23]. According to (12.11), the divided-difference expansion of this product satisfies

$$f(z)\,p_j(z) = \boldsymbol{P}(z)\,\boldsymbol{V}\{f p_j\}$$

Collecting together these expressions for $j = 0, 1, 2, \ldots,$ we obtain

$$f(z)\left[p_0(z)\;\; p_1(z)\;\; p_2(z)\;\; \ldots\right] = \boldsymbol{P}(z)\left[\boldsymbol{V}\{fp_0\}\;\; \boldsymbol{V}\{fp_1\}\;\; \boldsymbol{V}\{fp_2\}\;\; \ldots\right]$$

which, in view of (12.8a) and (12.11b), we can write in the compact form

$$f(z)\,\boldsymbol{P}(z) = \boldsymbol{P}(z)\,\mathcal{L}\{f\} \tag{12.13}$$

We will use this fundamental relation in the sequel to obtain several known, as well as new, results.

First, it follows from (12.13) that

$$\boldsymbol{P}(z)\,\mathcal{L}\{fg\} = f(z)\,g(z)\,\boldsymbol{P}(z) = f(z)\,\boldsymbol{P}(z)\,\mathcal{L}\{g\} = \boldsymbol{P}(z)\,\mathcal{L}\{f\}\,\mathcal{L}\{g\}$$

and, consequently, that

$$\mathcal{L}\{fg\} = \mathcal{L}\{f\}\,\mathcal{L}\{g\} \tag{12.14}$$

In particular, comparing the first columns on each side of (12.14) we obtain (12.8).

The relation (12.13) implies that $\boldsymbol{P}(z)$ is a *left-eigenvector* of the infinite lower-triangular matrix $\mathcal{L}\{f\}$ and that $f(z)$ is the corresponding *eigenvalue*. This relation holds for all z within the region of convergence of the divided-difference expansion (12.7). In particular, it holds for $z = \zeta_0, \zeta_1, \zeta_2, \ldots,$ viz.,

$$f(\zeta_i)\left[p_0(\zeta_i)\;\; \ldots\;\; p_i(\zeta_i)\;\; 0\;\; \ldots\right] = \left[p_0(\zeta_i)\;\; \ldots\;\; p_i(\zeta_i)\;\; 0\;\; \ldots\right]\mathcal{L}\{f\} \tag{12.15}$$

so that $\left[p_0(\zeta_i)\;\; \ldots\;\; p_i(\zeta_i)\;\; 0\;\; \ldots\right]$ is a left-eigenvector with $f(\zeta_i)$ as the corresponding eigenvalue. This also means that the *finite-length* $(1 \times (N+1))$ vector $\left[p_0(\zeta_i)\;\; \ldots\;\; p_i(\zeta_i)\;\; 0\;\; \ldots\;\; 0\right]$ is the left-eigenvector of the leading $(N+1) \times (N+1)$ principal submatrix of $\mathcal{L}\{f\}$, for all $N \geq i$.

When all the points $\zeta_0, \zeta_1, \ldots, \zeta_N$ are *distinct*, the $(N+1) \times (N+1)$ leading principal submatrix of $\mathcal{L}\{f\}$ can be diagonalized via a similarity transformation involving lower-triangular matrices only. Indeed, collecting (12.15) for $i = 0, 1, 2, \ldots,$ into a matrix expression, we find that

$$\text{diag}\,\{f(\zeta_i)\}\,\mathbf{L} = \mathbf{L}\,\mathcal{L}\{f\} \tag{12.16a}$$

where

$$\mathbf{L} = \begin{pmatrix} p_0(\zeta_0) & & & O \\ p_0(\zeta_1) & p_1(\zeta_1) & & \\ p_0(\zeta_2) & p_1(\zeta_2) & p_2(\zeta_2) & \\ \vdots & & & \ddots \end{pmatrix} \tag{12.16b}$$

As this relation involves only lower-triangular matrices it can be truncated to any desirable finite size by restricting it to leading principal submatrices.

4 MATRIX CHARACTERIZATION OF DISPLACEMENT STRUCTURE

Having introduced and analyzed the (univariate) divided-difference expansion in the preceding section, we are now in position to extend this concept to bivariate functions, and subsequently apply it to the function-theoretic characterization (12.2).

In analogy to the expression (12.11) for univariate expansions, the bivariate expansion $R(z,w) = \sum r_{ij}\, p_i(z)\, q_j^*(w)$ can also be expressed in compact form, viz., $R(z,w) = \boldsymbol{P}(z)\,\mathbf{R}\,\boldsymbol{Q}^*(w)$, where $\boldsymbol{Q}(w) = \begin{bmatrix} q_0(w) & q_1(w) & q_2(w) & \ldots \end{bmatrix}$. It follows that

$$f(z)\,R(z,w) = \begin{bmatrix} f(z)\,\boldsymbol{P}(z) \end{bmatrix} \mathbf{R}\,\boldsymbol{Q}^*(w) = \begin{bmatrix} \boldsymbol{P}(z)\,\mathcal{L}\{f\} \end{bmatrix} \mathbf{R}\,\boldsymbol{Q}^*(w)$$

where we used the property (12.13). In summary,

$$f(z)\;R(z,w)\;g^*(w) = \boldsymbol{P}(z)\;\begin{bmatrix} \mathcal{L}\{f\}\,\mathbf{R}\,\mathcal{L}^*\{g\} \end{bmatrix}\;\boldsymbol{Q}^*(w) \tag{12.17}$$

and applying this property to the function-theoretic expression (12.4) we obtain

$$\boldsymbol{P}(z)\;\begin{bmatrix} \mathcal{L}\{\alpha\}\,\mathbf{R}\,\mathcal{L}^*\{\gamma\} - \mathcal{L}\{\beta\}\,\mathbf{R}\,\mathcal{L}^*\{\delta\} \end{bmatrix}\;\boldsymbol{Q}^*(w) = \boldsymbol{P}(z)\;\begin{bmatrix} \mathbf{G}\,J\,\mathbf{H}^* \end{bmatrix}\;\boldsymbol{Q}^*(w)$$

which implies the matrix-theoretic expression (12.9). By setting $\zeta_i = 0 = \omega_j$ in (12.9), we obtain the expression (12.5), because in this case $\boldsymbol{V}\{\alpha\}$ corresponds to the power-series expansion of $\alpha(z)$ and, consequently, $\mathcal{L}\{\alpha\}$ is a lower-triangular *Toeplitz* matrix, i.e., $\mathcal{L}\{\alpha\} \equiv \alpha(\mathbf{Z})$, as in (12.5). On the other hand, if all $\{\zeta_i\}$ and all $\{\omega_j\}$ are distinct then, in view of the similarity relation (12.16), we can rewrite (12.9) in the form

$$\mathbf{A}\;\;(\mathbf{L}_1\;\mathbf{R}\,\mathbf{L}_2^*)\,\mathbf{C}^* - \mathbf{B}\;\;(\mathbf{L}_1\;\mathbf{R}\,\mathbf{L}_2^*)\,\mathbf{D}^* = (\mathbf{L}_1\,\mathbf{G})\;\;J\;\;(\mathbf{L}_2\,\mathbf{H})^*$$

with $\mathbf{A}$, $\mathbf{B}$, $\mathbf{C}$, $\mathbf{D}$ as in (12.6). Thus both the Toeplitz case (12.5) and the diagonal case (12.6) are special cases of (12.9).

Our last step is to establish (12.10) in the generic case: to this end it will suffice to show that given an arbitrary $(N+1) \times (N+1)$ lower-triangular matrix $\mathbf{A}$ and a set of appropriately selected complex points $\{\zeta_0, \zeta_1, \ldots, \zeta_N\}$, there exist a function $\alpha(z)$ and a (nonsingular) lower-triangular matrix $\mathbf{T}$, such that $\mathbf{A} = \mathbf{T}\;\mathcal{L}\{\alpha\}\;\mathbf{T}^{-1}$. Because of space limitations, we can only provide here a sketch of the proof. In general, one must select the points $\{\zeta_i\}$ so that $\mathbf{A}$ and $\mathcal{L}\{\alpha\}$ have the same Jordan structure. In particular, if $\mathbf{A}$ has distinct diagonal elements (and is non-derogatory), then it is diagonalizable, i.e., $\mathbf{A} \sim \text{diag}\,\{a_{ii}\}$. Also, since in this case we may select distinct $\{\zeta_i\}$, a polynomial $\alpha(z)$ can be (uniquely) determined by the requirement $a_{ii} = \alpha(\zeta_i)$,

$0 \leq i \leq N$, and, since $\mathcal{L}\{\alpha\}$ is similar to $\text{diag}\{\alpha(\zeta_i)\} \equiv$ diag $\{a_{ii}\}$ via a lower-triangular similarity, then it is also similar to $\mathbf{A}$. If $\mathbf{A}$ is not (triangularly) similar to a diagonal matrix, then the preceding argument (as well as property (12.16)) need to be extended by using Theorem 8.5 of [3], which states that every non-derogatory lower-triangular matrix can be transformed by lower-triangular similarity to a particular *bidiagonal* matrix.

REFERENCES

[1] T. Kailath, S.-Y. Kung and M. Morf, "Displacement Ranks of Matrices and Linear Equations," *J. Math. Anal. and Appl.*, Vol. 68, No. 2, pp. 395-407, 1979. See also *Bull. Amer. Math. Soc.*, Vol. 1, pp. 769-773, Sept. 1979.

[2] Y. Genin, P. Van Dooren, T. Kailath, J. Delosme and M. Morf, "On Σ-Lossless Transfer Functions and Related Questions", *Lin. Alg. Appl.*, Vol. 50, pp. 251-275, April 1983.

[3] H. Lev-Ari, "Nonstationary Lattice-Filter Modeling," Ph.D. Dissertation, Dept. of Elect. Eng., Stanford University, Dec. 1983.

[4] J. Chun, T. Kailath and H. Lev-Ari, "Fast Parallel Algorithm for QR and Triangular Factorization," *SIAM Journal on Scientific and Statistical Computing*, Vol. 8, No. 6, pp. 891-913, Nov. 1987.

[5] J. Chun and T. Kailath, "Displacement Structure for Hankel, Vandermonde, and Related (derived) Matrices," *Lin. Alg. Appl.*, Vol. 151, pp. 199-227, 1991.

[6] H. Lev-Ari and T. Kailath, "State-Space Approach to Factorization of Lossless Transfer Functions and Structured Matrices," *Linear Algebra and Its Applications*, Vol. 162-164, pp. 273-295, Feb. 1992.

[7] A.H. Sayed, "Displacement Structure in Signal Processing and Mathematics," Ph.D. Dissertation, Dept. of Elect. Eng., Stanford University, August 1992.

[8] P. Dewilde, A. Vieira and T. Kailath, "On A Generalized Szegö-Levinson Realization Algorithm for Optimal Linear Predictors Based on a Network Synthesis Approach," *IEEE Trans. Circ. Syst.*, Vol. CAS-25, pp. 663-675, Sept. 1978.

[9] H. Lev-Ari and T. Kailath "Lattice-Filter Parametrization and Modeling of Nonstationary Processes," *IEEE Trans. Inform. Thy.*, Vol. IT-30, pp. 2-16, January 1984.

[10] H. Lev-Ari and T. Kailath, "Triangular Factorization of Structured Hermitian Matrices," in *I. Schur Methods in Operator Theory and Signal Processing, Operator Theory: Advances and Applications*, I. Gohberg (ed.), Vol. 18, pp. 301-324, Birkhäuser Verlag, Basel, 1986.

[11] D. Alpay, P. Dewilde and H. Dym, "On the Existence and Construction of Solutions to the Partial Lossless Inverse Scattering Problem with Applications to

Estimation Theory," *IEEE Trans. Inform. Thy.*, Vol. 35, pp. 1184-1225, Nov. 1989.

[12] D. Pal and T. Kailath, "Fast Triangular Factorization of Hermitian Toeplitz and related Matrices with Arbitrary Rank Profile," *SIAM Journal on Matrix Analysis and Applications*, Vol. 14, pp. 1016-1042, Oct. 1993.

[13] A.H. Sayed and T. Kailath, "Displacement Structure: Theory and Applications," *SIAM Review*, Vol. 37, pp. 297-386, Sept. 1995.

[14] A.H. Sayed, H. Lev-Ari and T. Kailath, "Time-Varying Displacement Structure and Triangular Arrays," *IEEE Trans. Sig. Proc.*, Vol. 42, pp. 1052-1062, May 1994.

[15] A.H. Sayed and T. Kailath, "A State-Space Approach in Adaptive RLS Filtering," *IEEE Sig. Proc. Mag.*, Vol. 11, pp. 18-60, July 1994.

[16] P. Dewilde and H. Dym, "Lossless Inverse Scattering, Digital Filters, and Estimation Theory," *IEEE Trans. Inform. Theory*, Vol. IT-30, No. 4, pp. 644-662, July 1984.

[17] A.M. Bruckstein and T. Kailath, "An Inverse Scattering Framework for Several Problems in Signal Processing," *IEEE ASSP Magazine*, Vol. 4, No. 1, pp. 6-20, Jan. 1987.

[18] H. Lev-Ari, Y. Bistritz and T. Kailath, "Generalized Bezoutians and Families of Efficient Root-Location Procedures," *IEEE Trans. Circ. Syst.*, Vol. CAS-38, pp. 170-186, Feb. 1991.

[19] H. Dym, "On Reproducing Kernel Spaces, J-Unitary Matrix Functions, Interpolation and Displacement Rank," *Oper. Theory Adv. Appl.*, Vol. 41, pp. 173-239, Birkhäuser Verlag, Basel, 1989.

[20] H. Lev-Ari, "Extension Problems under the Displacement Structure Regime," in *Signal Processing, Part I: Signal Processing Theory*, L. Auslander, T. Kailath and S. Mitter (eds.), The IMA Volumes in Mathematics and Its Applications, Vol. 22, pp. 211-227, Springer-Verlag, 1990.

[21] A.H. Sayed, T. Kailath, H. Lev-Ari and T. Constantinescu, "Efficient Recursive Solutions of Rational Interpolation Problems," *Integral Equations and Operator Theory*, Vol. 20, pp. 84-118, Sept. 1994.

[22] L.M. Milne-Thomson, *The Calculus of Finite Differences*, MacMillan, New York, 1960.

[23] R. Ackner, "Fast Algorithms for Indefinite Matrices and Meromorphic Functions," Ph.D. Dissertation, Dept. of Electrical Engineering, Stanford University, Stanford, CA, Dec. 1991.

13

STRUCTURED TOTAL LEAST SQUARES FOR HANKEL MATRICES

Bart L.R. De Moor

Department of Electrical Engineering (ESAT/SISTA)
Katholieke Universiteit Leuven
Kardinaal Mercierlaan 94, B-3001 Leuven, Belgium
tel: 32/(0)16/321715, fax: 32/(0)16/321986
email: bart.demoor@esat.kuleuven.ac.be

To Thomas Kailath, on the occasion of his sixtieth birthday.
With great admiration and gratitude.

ABSTRACT

In this paper[1], we present some recent results on structured total least squares problems, the solution of which can be derived from what we have called a Riemannian singular value decomposition. This is a nonlinear generalization of the singular value decomposition. There are many applications in statistics, signal processing and system identification, but we will concentrate on Hankel matrices, showing how to approximate, in a least squares sense, a given Hankel matrix by a rank deficient one. The importance of this problem originates in the well known equivalence between rank deficient Hankel matrices, realization theory and linear systems. Another application is the least squares fit of a given matrix by an observability matrix. We discuss several properties and characterizations of the optimal solution and illustrate our results on a certain time series related to Thomas Kailath's scientific activities.

[1]Research supported by the Flemish Government (Concerted Research Action MIPS: Model-based Information Processing Systems), the National Fund for Scientific Research (G.0292.95 Matrix algorithms and differential geometry for adaptive signal processing, system identification and control), the Belgian Government (Interuniversity Attraction Poles IUAP-17: Modelling and Control of Dynamical Systems, IUAP-50: Automation in Design and Production) and the European Commission (Human Capital and Mobility Network SIMONET: System Identification and Modelling Network). Bart De Moor is an associate professor at the Katholieke Universiteit Leuven, a Senior Research Associate with the National Fund for Scientific Research and the main advisor on Science and Technology of the Flemish minister-president.

1 INTRODUCTION

The *total linear least squares problem* for a given matrix $A \in \mathcal{R}^{p\times q}$, which is of full column rank (without loss of generality, we assume throughout that $p \geq q$), is to find a rank deficient least squares approximation $B \in \mathcal{R}^{\surd\times\mathrm{II}}$:

$$\min_{B\in\mathcal{R}^{\surd\times\mathrm{II}},\ \mathrm{rank}(\mathcal{B})=\mathrm{II}-\infty} \|A-B\|_F \ , \tag{13.1}$$

where $\|.\|_F$ is the matrix Frobenius norm. It is well known that the solution can be found from the singular value decomposition (SVD) of the matrix A, by removing from its dyadic decomposition, the term corresponding to the smallest singular value [12] [14] [23]. While the problem is certainly not new[2], the technique has quite recently been popularized in the mathematical engineering community by the books [13] [22].

Often, for applications in statistics, signal processing and system identification, the objective function $\|A-B\|_F^2 = \sum_{i=1}^{p}\sum_{j=1}^{q}(a_{ij}-b_{ij})^2$ is extended with 'elementwise' nonnegative weights w_{ij} and, in addition, the approximating rank deficient matrix B is required to have some structure, such as for instance a Hankel or Toeplitz structure, in which case we have *the structured total least squares problem*:

$$\min_{B\in\mathcal{R}^{\surd\times\mathrm{II}},\ \mathrm{rank}(\mathcal{B})=\mathrm{II}-\infty\ ,\ \mathcal{B}\ \mathrm{structured}} \sum_{i=1}^{p}\sum_{j=1}^{q}(a_{ij}-b_{ij})^2 w_{ij} \ . \tag{13.2}$$

While in this paper, we will concentrate on Hankel matrices, most of our results carry through (mutatis mutandis) to so-called *affine matrix structures*, i.e. structured matrices B that can be written as an affine (linear) combination of a given set $\{B_k; k = 0, \dots, N\}$ of basis matrices as $B = B_0 + B_1\beta_1 + \dots + B_N\beta_N$. Here, the real coefficients $\beta_k, k = 1, \dots, N$ 'parametrize' the structured matrix B. Examples of such structured matrices are (centro- and per-) symmetric matrices, (block-) Hankel, (block-) Toeplitz, (block-) circulant, Brownian, Hankel + Toeplitz matrices, matrices with a certain zero structure (sparsity pattern), etc ....

This paper treats the optimal least squares approximation of a given Hankel matrix by a rank deficient one. We have omitted all proofs and have kept the discussion informal. For more details and proofs, we will refer to the literature. In Section 2, we elaborate on the relation between rank deficient Hankel matrices, linear systems and realization theory. We also give a small motivating example on rank deficient approximation of Hankel matrices from given data, a problem which is one specific instance of a structured total least

[2] In the statistics community, total least squares is known as 'orthogonal regression' and 'errors-in-variables', with first references going back to the previous century [2] [3] [19].

squares problem. In Section 3, we discuss the solution of structured total least squares problems in terms of a so-called Riemannian singular value decomposition, which is a nonlinear generalization of the SVD. We present its structure for the Hankel case and discuss several characterizations and properties of the optimal solution. In Section 4, a numerical example is given, where we model the cumulative number of publications of Thomas Kailath. In Section 5, we show how the least squares approximation of a given data matrix by an observability matrix, also leads to a structured total least squares problem, this time for a block Hankel matrix. Finally, some summarizing conclusions and perspectives for future research are presented in Section 6.

2 HANKEL MATRICES AND REALIZATION THEORY

Hankel matrices play an important role in linear system theory, modelling and identification, ever since the path breaking work by many authors in the seventies and eighties on realization theory for linear systems. One of the central problems is how to obtain a state space model of a linear system from observed impulse response data.

An intriguing result here states that the rank deficiency of a Hankel matrix with data is equivalent with the fact that these data can be modelled as the impulse response of a linear dynamic time-invariant system. This basic result links the properties of systems being linear and time-invariant to the field of (numerical) linear algebra, in which the rank of a matrix, and also its structure (such as Hankel structure) are central notions. Indeed, the minimal system order (i.e. the minimum number of states) for the 'realization' of the given data sequence as a linear system, is exactly equal to the rank of the Hankel matrix (see e.g. [15]) (It seems that this result was already known to Kronecker in the context of rational approximation).

In practical situations however, where measurements are obtained using finite precision sensors, data are always observed inaccurately, i.e. they are perturbed by measurement noise. Therefore, any $p \times q$ Hankel matrix with $p \geq q$, constructed from these data, either directly or after a deconvolution, will be of full column rank q. It is then meaningful to approximate the 'observed' Hankel matrix A, by a least squares rank deficient approximation B. However, just calculating the SVD of the given Hankel matrix A, and then putting to zero the q-th singular value (the smallest one), will give the best least squares rank deficient approximation, but the reconstructed matrix will not be a Hankel matrix and hence, 'Kronecker's' result cited above does not apply. Let us illustrate this point with a simple numerical example. Consider the 3×2 Hankel matrix A, which obviously is of rank 2 (its 2 singular values being $\sigma_1 = 6.5468$ and $\sigma_2 = 0.37415$), and its best least squares rank deficient approximation C, which is obtained from the SVD of A by putting its smallest singular value to zero:

$$A = \begin{pmatrix} 1 & 2 \\ 2 & 3 \\ 3 & 4 \end{pmatrix} , \quad C = \begin{pmatrix} 1.2608 & 1.8193 \\ 2.0534 & 2.9630 \\ 2.8460 & 4.1067 \end{pmatrix} .$$

Obviously, C, while being of rank 1, is **not** a Hankel matrix. Next consider the following two rank one Hankel matrices:

$$B_1 = \begin{pmatrix} 1.3299 & 1.9448 \\ 1.9448 & 2.8441 \\ 2.8441 & 4.1593 \end{pmatrix} , \quad B_2 = \begin{pmatrix} 1.2734 & 1.8827 \\ 1.8827 & 2.7835 \\ 2.7835 & 4.1155 \end{pmatrix} , \tag{13.3}$$

The first one is the solution to the optimization problem

$$\min_{B \text{ , rank}(B)=1 \text{ , } B \text{ Hankel}} \|A - B\|_F ,$$

and can be found using our power method for the Riemannian SVD, discussed below. Its largest singular value is equal to 6.5430 which is not too different from the largest singular value of the original A. Note that due to the Hankel structure, certain elements are counted twice in the objective function $\|A - B\|_F$. In the case that we only count them once, the least squares rank one approximation, which can also be obtained from the Riemannian SVD by introducing the appropriate weights, is given by B_2, the largest singular value of which is 6.4144.

So far for the small example, which illustrates the fact that the singular value decomposition does not preserve the Hankel structure (nor does it with any other structure, except for symmetry) when used as a rank deficient approximation tool. Existing realization methods for 'inaccurate' data, such as the SVD based methods of Zeiger-McEwen [24] and Kung [16] are therefore heuristic, although they often seem to work surprisingly well in practice. Other authors, such as Cadzow [4], have tried to retain the idea of a rank deficient approximation via the SVD while preserving the Hankel structure. The resulting iteration converges, but not to a fixed point which is optimal in any sense (as shown in [8] where we also show that 'classical' iterative methods, such as Steiglitz-McBride and Iterative Quadratic Maximum Likelihood, do not converge to an optimal solution).

In this paper, we formulate and solve the following least squares realization problem:

Given a sequence of scalar data points $a_i \in \mathcal{R}, \rangle = \infty, \ldots, \mathcal{N}$ (collected in a vector $a \in \mathcal{R}^{\mathcal{N}}$) and a user-defined integer $q \leq (N+1)/2$. Find the impulse response $b_i, i = 1, \ldots, N$ (components of the vector $b \in \mathcal{R}^{\mathcal{N}}$) of a linear time-invariant system of order at most $q-1$, which is nearest to the given data in a least squares sense.

Mathematically, we have the following problem:

Structured Total Least Squares problem:

$$\min_{b_i, i=1,\ldots,N, y \in \mathcal{R}^{\text{II}}} \sum_{i=1}^{N} (a_i - b_i)^2 \text{ subject to } \begin{cases} B\,y = 0\,, \\ y^T y = 1\,, \\ B \text{ is a } p \times q \\ \text{Hankel matrix} \\ \text{with } N = p + q - 1 \end{cases} \tag{13.4}$$

Observe that the rank deficiency of B is guaranteed by the existence of a non-trivial vector y in its null space, and from the Hankel structure of B it is guaranteed that the data b_i, which are the elements of B, can be realized exactly by the impulse response of a linear time-invariant system of order at most $q-1$. As a matter of fact, the $q-1$ poles of the approximating linear system could be found from the components of the vector y if it were known, because they are the coefficients of the characteristic polynomial of the underlying linear system. Observe also that the number q of columns of B, is a *user-defined integer* that determines the order of the 'approximating' linear system, whose minimal order will be at most $q-1$. This implies that the user is free to specify q (as long as $p \geq (N+1)/2$ or equivalently $(N+1)/2 \geq q$, a condition that guarantees that $p \geq q$).

3 THE RIEMANNIAN SVD

Problem (13.4) is a constrained optimization problem and necessary conditions for an optimal solution in the unknowns $b_i, i = 1, \ldots, N$, $y \in \mathcal{R}^{\text{II}}$ can be obtained by introducing a vector $l \in \mathcal{R}\surd$ of Lagrange multipliers for the p constraints, $By = 0$, and a scalar Lagrange multiplier λ for the constraint $y^T y = 1$. From the Lagrangian, by setting all the derivatives to zero, one obtains a set of nonlinear equations in the unknowns $b \in \mathcal{R}^{\mathcal{N}}$, $y \in \mathcal{R}^{\text{II}}$, $l \in \mathcal{R}\surd$ and $\lambda \in \mathcal{R}$ (see [7] [8] for details). It turns out that $\lambda = 0$ and that the vector b can be eliminated. After some rescalings and renormalizations, one obtains a set of nonlinear equations, which was called the Riemannian SVD in [11]:

Riemannian SVD for Hankel matrices

$$\begin{aligned} A\,v &= D_v\,u\,\tau\,, & u^T D_v u = 1\,,\\ A^T\,u &= D_u\,v\,\tau\,, & v^T D_u v = 1\,,\\ & & v^T v = 1\,. \end{aligned} \qquad (13.5)$$

Here $u \in \mathcal{R}^{\surd}$ and $v \in \mathcal{R}^{\Pi}$ are a left, resp. right singular vector and $\tau \in \mathcal{R}$ is the singular value. $D_u \in \mathcal{R}^{\Pi\times\Pi}$ and $D_v \in \mathcal{R}^{\surd\times\surd}$ are banded symmetric Toeplitz positive definite matrix functions, the elements of which are quadratic in the components of u, resp. v as

$$D_u = T_u T_u^T\,,\; D_v = T_v T_v^T\,,$$

where $T_u \in \mathcal{R}^{\surd\times(\surd+\Pi-\infty)}$ is a banded Toeplitz matrix with the components of u as

$$T_u = \begin{pmatrix} u_1 & u_2 & \dots\dots & u_{p-1} & u_p & 0 & \dots & 0 & 0\\ 0 & u_1 & u_2\ \dots & u_{p-2} & u_{p-1} & u_p & \dots & 0 & 0\\ \dots & \dots & \dots & \dots & \dots & \dots & \dots & \dots & \dots\\ 0 & 0 & 0\ \dots & 0 & 0 & u_1 & \dots & \dots & u_p \end{pmatrix} \qquad (13.6)$$

and $T_v \in \mathcal{R}^{\surd\times(\surd+\Pi-\infty)}$ is constructed similarly from the components of v.

The relation between this Riemannian SVD and the optimal least squares solution to the noisy realization problem, follows from the following observations.

Fact 1: The singular value squared is exactly equal to the value of the objective function in (13.4):

$$\tau^2 = \sum_{i=1}^{N} (a_i - b_i)^2\,,$$

which implies that we need to find the minimal τ satisfying (13.5).

Fact 2: The structured rank deficient least squares approximation B can be reconstructed as

$$B = A - \begin{pmatrix} u_1 & u_2 & \dots\dots\dots & u_p & 0 & \dots & 0 & 0 & 0\\ u_2 & u_3 & \dots\dots & u_p & 0 & 0 & \dots\ 0 & 0 & u_1\\ u_3 & u_4 & \dots\ u_p & 0 & 0 & 0 & \dots\ 0 & u_1 & u_2\\ \dots & \dots & \dots & \dots & \dots & \dots & \dots & \dots & \dots\\ u_p & 0 & \dots\dots & 0 & u_1 & \dots\dots\dots & & u_{p-2} & u_{p-1} \end{pmatrix}$$

$$\times \tau \times \begin{pmatrix} v_1 & v_2 & \dots & v_q \\ 0 & v_1 & \dots & v_{q-1} \\ 0 & 0 & \dots & v_{q-2} \\ \dots & \dots & \dots & \dots \\ 0 & 0 & \dots & v_1 \\ \dots & \dots & \dots & \dots \\ v_q & 0 & \dots & 0 \\ v_{q-1} & v_q & \dots & 0 \\ \dots & \dots & \dots & \dots \\ v_2 & v_3 & \dots & 0 \end{pmatrix} \tag{13.7}$$

which means that the difference $A - B$ is the product of a Hankel matrix with the elements of u, a Toeplitz matrix with the elements of v (note the 'circulant' structure in both matrices) and the scalar τ [3].

Fact 3: There is an interpretation to the singular vectors u and v. The right singular vector v satisfies

$$Bv = 0 \ , \ v^T v = 1 \ . \tag{13.8}$$

and therefore, its components are the coefficients of the characteristic polynomial of the approximating linear system (Actually, v can be taken equal to y in (13.4)). The left singular vector u is the vector of Lagrange multipliers l, normalized so as to satisfy the norm constraint in (13.5). This implies that u can be used in analysing the first order sensitivity properties of the structured total least squares optimization problem.

Fact 4: It can be shown that the optimal approximating impulse response, which is contained in the vector b, and the original data vector a, satisfies the following orthogonality property: $(a-b)^T b = 0$. As this is a (structured) least squares approximation problem, it is not too surprising to see that the vector of residuals $a - b$ is orthogonal to the vector of approximants b.

Fact 5: The minimal singular value $\tau_{\min}$ and the corresponding left and right singular vectors u and v, and hence the optimal Hankel matrix B in (13.7), can be calculated by an algorithm, described in [7] [8]. The source of inspiration for this algorithm is the good old power method [25] for calculating the dominant

[3] It is interesting to observe that every $p \times q$ Hankel matrix A of full column rank q can be factorized in such a Hankel $\times$ Toeplitz product, i.e. when one puts $B = 0$ in (13.7), there exist $u \in \mathcal{R}^{\surd}$, $v \in \mathcal{R}^{\text{II}}$ and $\tau \in \mathcal{R}$ such that A is factorized as prescribed. Such a triplet (u, τ, v) will satisfy the Riemannian SVD equations (13.5), with $\tau = \sqrt{\sum_{i=1}^{N} a_i^2}$ but is not optimal in any sense. For instance, for the 3×2 Hankel matrix A in Section 2, we have $u^T = (0.35235, 0.12310, 0.85386)$,$v^T = (0.51816, 0.85529)$, $\tau = 5.4772 = \sqrt{(1+2^2+3^2+4^2)}$, which satisfies the equations (13.5). This solution however does not correspond to the minimizing one.

eigenvalue and corresponding eigenvector of a symmetric matrix and seems to work well in practice [4]. Its asymptotic convergence rate is linear, as asymptotically it behaves exactly like the power method. A numerical example is given in Section 4.

Fact 6: When D_v is invertible, it is straightforward to show that

$$\tau^2 = v^T A^T D_v^{-1} A v \;, \tag{13.9}$$

which would be a quadratic form in v, if D_v were a matrix independent of v, which is not the case though. Yet, this observation is the basis of the *constrained total least squares* method, described in [1], which is basically applying Gauss-Newton to the nonlinear optimization problem. The equivalence with the problem (13.4) is discussed in [10].
Another possible algorithm is suggested in [21]. It starts from the observation that the Riemannian SVD can be rewritten as a generalized eigenvalue problem of the form

$$A^T D_v^{-1} A v = D_u v \tau^2 \;,\; v^T D_u v = 1 \;.$$

When fixing D_u and D_v, one can solve this problem for the minimal eigenvalue τ^2 and corresponding eigenvector v, determine a corresponding u, update D_u and D_v and then restart the whole process until convergence [5]. Yet, our experience shows that convergence here occurs less often than with our power method. An intuitive explanation for this is that the power method only does one little step towards the minimal τ and corresponding eigenvector v, for fixed D_u and D_v instead of going 'all the way'. In a certain sense, the power method updates the matrices D_u and D_v more frequently than in the eigenvalue method outlined above. Of course, this is only an intuitive explanation which deserves further investigation.
In [6] we have discussed continuous time methods for structured total least squares problems.
Finally, in [8] we have investigated other methods, such as Cadzow's method and Iterative Quadratic Maximum Likelihood, with a demonstration that they deliver suboptimal results. In [17] we have shown the equivalence between Abatzoglou's [1] constrained total least squares solution and the structured total least squares approach presented here.

Fact 7: Let $\hat{a} \in \mathcal{R}^{\mathcal{N}}$ be the vector of 'exact' data (i.e. containing the impulse response of a linear system of order $q-1$), $\tilde{a} \in \mathcal{R}^{\mathcal{N}}$ be a vector of i.i.d. normally distributed zero mean white noise with variance σ^2 and let the observed data sequence be $a = \hat{a} + \tilde{a}$. The structured total least squares problem

[4] Several Matlab .m files are available from the author.

[5] Remark that, when in (13.5) D_u and D_v would be 'constant' matrices (i.e. independent of u and v), then, the Riemannian SVD would coincide with the Restricted SVD discussed in [5].

(13.4) is then the maximum likelihood formulation of the realization problem. It's interesting to observe that one can also motivate the objective function (13.9) in a heuristic way as follows. Let $A = \hat{A} + \tilde{A}$ be the $p \times q$ Hankel matrix with observed data (with obvious definitions for $\hat{A}$ and $\tilde{A}$ as Hankel matrices). Let's assume that $\text{rank}(\hat{A}) = q - 1$, $\text{rank}(A) = \text{rank}(\tilde{A}) = q$. Let $v \in \mathcal{R}^{\text{II}}$ be a vector in the null space of $\hat{A}$. Then

$$Av = \hat{A}v + \tilde{A}v = \tilde{A}v = e \ ,$$

where the vector $e \in \mathcal{R}\surd$ can be considered as a vector with the equation errors. It seems natural to minimize the so-called equation error, which is the norm of e. However, we have to take into account the correlation structure of the elements of e, which is induced by the Hankel structure in the matrix $\tilde{A}$. The covariance matrix is obtained as

$$\mathbf{E}(e.e^T) = \mathbf{E}(\tilde{A}.v.v^T.\tilde{A}^T) = D_v \sigma^2 \ ,$$

which is precisely the banded symmetric positive definite Toeplitz matrix of the Riemannian SVD (13.5), up to within the scalar σ^2, which is the white noise variance. Since we now know the structure of the noise covariance matrix, we can minimize the weighted equation error as

$$e^T D_v^{-1} e = v^T A^T D_v^{-1} A v \ ,$$

which is exactly the expression (13.9). Actually, this heuristic insight leads to a 'quick and dirty' way of calculating the structure of D_u and D_v for any structured total least squares problem of the form (13.2) as long as the matrix structure involved is affine.

Fact 8: Another interesting property is the following

$$\sum_{i=1}^{N}(a_i - b_i) = (\sum_{i=1}^{p} u_i)\ \tau\ (\sum_{i=1}^{q} v_i) \ .$$

4 A NUMERICAL EXAMPLE: PREDICTING THE CUMULATIVE NUMBER OF TK'S PUBLICATIONS

Consider the data sequence of Figure 13.1 which is the cumulative number of journal publications of Thomas Kailath, starting from 1960 (actually the year when I was born) up to 1995. There are 36 data points. We decide to fit these data by impulse responses of linear systems of increasing order, starting with order 1 (in which case the data Hankel matrix A of the Riemannian SVD is $(p = 35) \times (q = 2)$ [6], up to order 17 (where $p = 19$ and $q = 18$). Therefore, we have calculated the minimal singular value of the Riemannian SVD (13.5), for $p \times q$ data Hankel matrices A, where q varies from 2 up to 17. As can be expected, we get a better and better approximation as the order of the 'approximating' system increases. As is obvious from Figure 13.1, all of these realizations are unstable. For instance, the poles for the fifth-order approximation, which corresponds to the case $p = 31, q = 6$ are $\{1.0972, 1.0062 \pm 0.20505j, 0.91204 \pm 0.63262j\}$. They can easily be obtained by rooting the characteristic polynomial, the coefficients of which are given by the components of the right singular vector v of the Riemannian SVD (13.5).

5 LEAST SQUARES APPROXIMATION BY OBSERVABILITY MATRICES

A structured total least squares problem that often occurs in signal processing (e.g. in Direction of Arrival problems such as ESPRIT) or in system identification (subspace methods such as N4SID), is the following:
Given a data matrix $A \in \mathcal{R}^{\Updownarrow\times\backslash}$, $m > n$. Find the least squares approximation $G \in \mathcal{R}^{\Updownarrow\times\backslash}$ that is an observability matrix, i.e. a matrix for which the i-th row $g_i^T \in \mathcal{R}^{\infty\times\backslash}$ is given by $g_i^T = h^T.H^{i-1}$ where $h \in \mathcal{R}^{\backslash}$ and $H \in \mathcal{R}^{\backslash\times\backslash}$.
Stated as an optimization problem, this becomes:

$$\min_{h\in\mathcal{R}^{\backslash},\ \mathcal{H}\in\mathcal{R}^{\backslash\times\backslash}} \|A - \begin{pmatrix} h^T \\ h^T H \\ h^T H^2 \\ \vdots \\ h^T H^{m-1} \end{pmatrix} \|_F^2 \,. \tag{13.10}$$

One could also introduce elementwise weights into the objective function but we will consider the unweighted case only. We now show that this problem is equivalent with a structured total least squares problem, this time for a block Hankel matrix, as follows. Let the components of the vector $y \in \mathcal{R}^{\backslash+\infty}$ be the

[6]Actually, an impulse response of a first order linear system can be parametrized by two real numbers β and λ so that the optimization problem becomes an unconstrained one: $\min_{\beta,\lambda} \sum_{i=1}^{N}(a_i - \beta\lambda^{i-1})^2$. Setting to zero the derivative with respect to β, results in $\sum_{i=1}^{N}(a_i - \beta\lambda^{i-1})\lambda^{i-1} = 0$, and the derivative w.r.t. λ gives $\sum_{i=2}^{N}(a_i - \beta\lambda^{i-1})(i-1)\lambda^{i-2} = 0$. Observe that β appears linearly in both equations. When eliminated from one and substituted in the other, one obtains a polynomial in λ, which can be rooted. One of its real roots will generate the minimizing first order impulse response (see [8] for details). For instance, for the example (13.3), one then obtains the polynomial of degree 7 in λ: $3\lambda^7 + 0\lambda^6 + 3\lambda^5 - 6\lambda^4 - \lambda^3 - 12\lambda^2 - 5\lambda - 2 = 0$, which has 3 pairs of complex conjugated roots and only 1 real one, which is 1.4785, corresponding to the 3×2 optimal Hankel matrix B in (13.3)

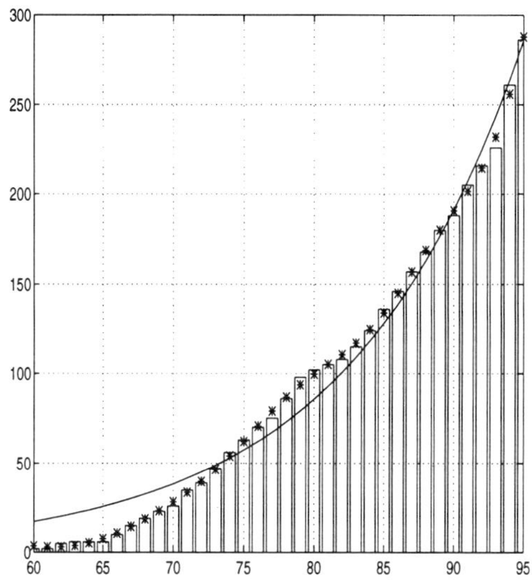

Figure 13.1 The x-axis has the years 1960 up to 1995. The y-axis is the cumulative number of TK's journal publications, which are shown in the bar diagram. The first order least squares impulse response approximation (full line) overmodels TK's publications in his younger years. The 5-th order approximation, shown in stars '*', gives an almost perfect fit. It is our hope and wish that this 5-th order linear time-invariant system is a robust and reliable prediction for TK's production well into the 21-st century.

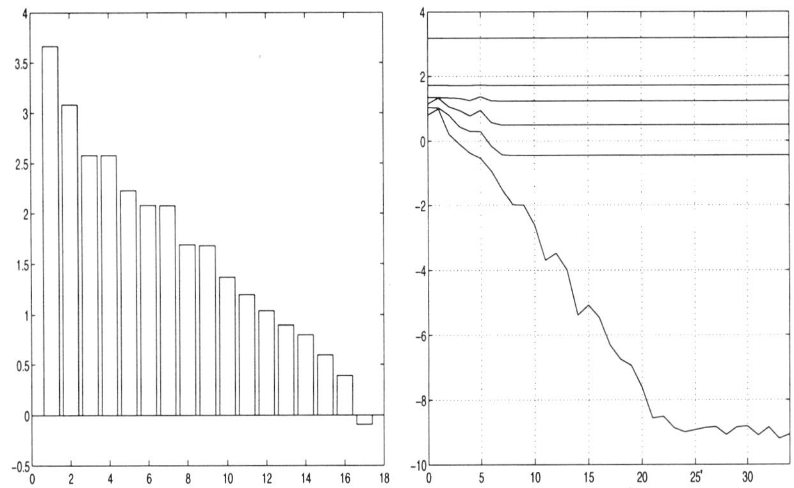

Figure 13.2 **Left plot**: The x-axis is the system order of the optimal impulse response, which is specified by the user by fixing q. The y-axis is the logarithm (base 10) of the minimizing singular value squared $\tau^2_{\min}$ of the Riemannian SVD (13.5), which is also the minimum value of the objective function of the Structured TLS problem (13.2). As can be expected, increasing the order of the approximating system, decreases the norm of the error. **Right plot**: Typical convergence pattern of the logarithm (base 10) of the singular values of the approximating Hankel matrix B, as a function of the iteration number, shown here for $q = 6$ (5-th order approximation). Observe that $q - 1$ singular values are approximately constant, while the smallest singular value goes to zero. The asymptotic convergence is linear, because asymptotically the algorithm behaves like the power method. For this example, numerical convergence is reached after $\pm$ 25 iterations.

coefficients of the characteristic polynomial of the matrix H. Then, it is easy to show that

$$T_y \begin{pmatrix} h^T \\ h^T H \\ \vdots \\ h^T H^{m-1} \end{pmatrix} = 0 \, ,$$

where $T_y \in \mathcal{R}^{(\Updownarrow-\backslash)\times\Updownarrow}$ is a banded Toeplitz matrix as in (13.6). If we put $g_i^T = h^T H^{i-1}$, then this relation is equivalent with

$$T_y G = 0 \Longleftrightarrow \begin{pmatrix} g_1 & g_2 & \cdots & g_n & g_{n+1} \\ g_2 & g_3 & \cdots & g_{n+1} & g_{n+2} \\ g_3 & g_4 & \cdots & g_{n+2} & g_{n+3} \\ \cdots & \cdots & \cdots & \cdots & \cdots \\ g_{m-n} & g_{m-n+1} & \cdots & g_{m-1} & g_m \end{pmatrix} . \, y = 0 \, , \qquad (13.11)$$

where G is a matrix for which the i-th row is g_i^T. The block matrix to the right is a block Hankel matrix with $m-n$ block rows, each of dimension n, and $n+1$ columns, which will be denoted by $B \in \mathcal{R}^{(\Updownarrow-\backslash)\backslash\times(\backslash+\infty)}$. Such a matrix is an affinely structured matrix, which is required to be rank deficient. Hence, we have shown that the least squares approximation of a given matrix A by an observability matrix as in (13.10) is a structured total least squares problem

$$\min_{\substack{g_i \in \mathcal{R}^{\backslash}, \rangle = \infty, \ldots, \Updownarrow \, , \\ y \in \mathcal{R}^{\backslash+\infty}}} \|A - G\|_F^2 \text{ subject to } \begin{cases} B\, y = 0 \, , \\ y^T y = 1 \, , \\ B \text{ block Hankel as in (13.11).} \end{cases}$$

The solution can be obtained from a Riemannian SVD as in (13.5), but with metric matrices D_u and D_v, whose structure is different from the ones in (13.5). It can easily be obtained, for instance with the heuristic 'noise' trick, outlined under 'Fact 7' above.

6 CONCLUSIONS

In this short paper, we have discussed the total least squares approximation of Hankel matrices, with as an additional constraint that the approximant be Hankel too. The insights here have been elaborated upon in detail in [7] [8] [9] to which we refer the reader for more details. Actually, the Riemannian SVD is quite a general result for rank deficient matrix approximation for weighted and structured matrix problems, as long as the matrix structure is affine. For instance, in [7], we also show how instead of minimizing the sum of absolute errors squared in the objective function, one can also minimize (a first order approximation to) the sum of *relative* errors squared. If the data a_i are already

samples from an impulse response of a (high-order, say r) linear system and if the number of rows $p \to \infty$ while $q < r$ is fixed, the Hankel approximation problem treated here is equivalent to the H_2 model reduction problem. Z-domain iterations have been described in [6] [20]. In [9] [10] we have discussed the application of this framework for the identification of linear dynamic systems from input-output data, when both the inputs and the outputs are corrupted by noise (errors-in-variables).
The structured total least squares problem could be treated as a nonlinear constrained optimization problem and 'classical' minimization algorithms could be applied to come up with numerical solutions. Yet we feel that our interpretation in terms of a Riemannian SVD reveals that there is quite some structure in the solution, which is reminiscent of the solution of the unstructured case. We hope to reveal in the future more useful properties which maybe ultimately lead to a globally convergent algorithm.

Acknowledgements

From our first meeting in the Roman terms of Bath, England, at an evening reception of the Conference on Mathematics and Signal Processing in September 1985, over our discussion in Valencia, Spain, at the International Conference on Linear Algebra in September 1987, 15 minutes before TK was going to address the plenary audience about displacement ranks, over my post-doc period in Stanford from August 1988 till September 1989, over several visits of me and my family to Stanford and of TK to Leuven, to the fabulous 5-day birthday party in Stanford in June 1995, is a period of only 10 years in which I have learned to appreciate Tom as a brilliant teacher, a devoted researcher and a friendly and caring person.
My students (Johan David, Jeroen Dehaene, Philippe Lemmerling, Christiaan Moons, Geert Schelfhout, Peter Van Overschee), who in one way or another contributed to the work presented here, and I, we all sincerely hope that the explicit instability of Figure 13.1 may be an appropriate model for TK's scientific production over a considerable time horizon in the future !

REFERENCES

[1] Abatzoglou T.J., Mendel J.M., Harada G.A., *The constrained total least squares technique and its application to harmonic superresolution*, IEEE Transactions on Signal Processing, SP-39, no.5, pp. 1070-1087, 1991.

[2] Adcock R.J. *Note on the method of least squares.* The Analyst, Vol IV, no.6, Nov. 1877, pp.183-184.

[3] Adcock R.J. *A problem in least squares.* The Analyst, March 1878, Vol V, no.2, pp.53-54.

[4] Cadzow J., *Signal enhancement: a composite property mapping algorithm*, IEEE Trans. on Acoustics, Speech, and Signal Processing, vol. ASSP-36, pp. 49-62, 1988.

[5] De Moor B., Golub G.H. *The restricted singular value decomposition: properties and applications.* Siam Journal on Matrix Analysis and Applications, Vol.12, no.3, July 1991, pp.401-425.

[6] De Moor B., Van Overschee P., Schelfhout G. *H_2-model reduction for SISO systems.* Proc. of the 12th World Congress International Federation of Automatic Control, Sydney, Australia, July 18-23 1993, Vol. II pp.227-230. [7]

[7] De Moor B. *Structured total least squares and L_2 approximation problems.* Special issue of Linear Algebra and its Applications, on Numerical Linear Algebra Methods in Control, Signals and Systems (eds: Van Dooren, Ammar, Nichols, Mehrmann), Volume 188-189, July 1993, pp.163-207.

[8] De Moor B. *Total least squares for affinely structured matrices and the noisy realization problem* . IEEE Transactions on Signal Processing, Vol.42, no.11, November 1994.

[9] De Moor B. *Dynamic Total Linear Least Squares.* SYSID '94, Proc. of the 10th IFAC Symposium of System Identification, 4-6 July 1994, Copenhagen, Denmark. Vol.3., pp.159-164.

[10] De Moor B., Roorda B. *L_2-optimal linear system identification: Dynamic total least squares for SISO systems.* Proc. of 33rd IEEE CDC, Florida, December 1994.

[11] De Moor B. *The Riemannian Singular Value Decomposition.* Proc. 3rd Int. Workshop on SVD and Signal Processing, 1995, pp. 61-78, Elsevier Science Publisher.

[12] Eckart C., Young G. *The approximation of one matrix by another of lower rank.* Psychometrika, 1, pp.211-218, 1936.

[13] Golub G.H., Van Loan C. *Matrix Computations.* Johns Hopkins University Press, Baltimore 1989 (2nd edition).

[14] Householder A.S., Young G. *Matrix approximation and latent roots.* Americ. Math. Monthly, 45, pp.165-171, 1938.

[15] Kailath T. *Linear Systems.* Prentice Hall, Englewood Cliffs, New Jersey, 1980.

[7]Reprinted in the Automatic Control World Congress 1993 5-Volume Set, Volume 1: Theory.

[16] Kung S.Y. *A new identification method and model reduction algorithm via the singular value decomposition.* 12th Asilomar Conference on Circuits, Systems and Computers, pp.705-714, Asilomar, California, 1978.

[17] Lemmerling P., De Moor B., Van Huffel S. *On the equivalence of constrained and structured total least squares.* ESAT-SISTA Report 1995-55, Department of Electrical Engineering, Katholieke Universiteit Leuven, Belgium, October 1995.

[18] Nagelkerke N.J.D., *Maximum Likelihood Estimation of Functional Relationships.* Lecture Notes in Statistics, Ed. J. Berger et al., Springer-Verlag, 1992.

[19] Pearson K. *On lines and planes of closest fit to systems of points in space.* Phil. Mag., 2, 6-th series, pp.559-572.

[20] Schelfhout G. *Model reduction for control design.* PhD Thesis, Department of Electrical Engineering, Katholieke Universiteit Leuven, Belgium, November 1995.

[21] Van hamme H. *Identification of linear systems from time- or frequency domain measurements.* PhD Thesis, April 1992, Department of Electrical Engineering, Vrije Universiteit Brussel, Belgium, pp. 294.

[22] Van Huffel S., Vandewalle J. *The Total Least Squares Problem: Computational Aspects and Analysis.* Frontiers in Applied Mathematics 9, SIAM, Philadelphia, 300 pp., 1991.

[23] Young G. *Matrix approximation and subspace fitting.* Psychometrika, vol.2, no.1, March 1937, pp.21-25.

[24] Zeiger H., McEwen A. *Approximate linear realizations of given dimension via Ho's algorithm.* IEEE Transactions on Automatic Control, 19, pp.153, 1974.

[25] Wilkinson J. *The algebraic eigenvalue problem.* New York, Oxford University Press, 1965.

14

J–LOSSLESS CONJUGATION FOR DISCRETE–TIME SYSTEMS AND ITS SEQUENTIAL STRUCTURE

Waree Kongprawechnon and Hidenori Kimura

Department of Mathematical Engineering and Information Physics
The University of Tokyo, Bunkyo-ku
Tokyo 113, Japan

Dedicated with appreciation to Tom Kailath on the occasion of his sixtieth birthday.

ABSTRACT

J–lossless conjugation is a powerful tool for carrying out the J–lossless factorization which is a crucial part of solving H^∞ control problem. In this paper, a necessary and sufficient condition for the existence of J–lossless conjugation is obtained which is represented in terms of a Riccati equation. Its sequential structure is exploited which is closely related to the celebrated factorization theory of J–lossless systems.

1 INTRODUCTION

It is well-known that each stable transfer function $G(z)$ is represented as the cascade of a lossless (inner) system $\Theta(z)$ and a unimodular (outer) system $\Pi(z)$, i.e.,

$$G(z) = \Theta(z)\Pi(z) \tag{14.1}$$

This is usually refered to as *inner–outer factorization.* If $G(z)$ is not stable, the factorization (14.1) does not exist, unless we allow $\Theta(z)$ to be a *J-lossless system*, which is a generalization of lossless systems. In that case, we call (14.1) a *J-lossless factorization.* The J–lossless factorization is not only a generalization of the inner-outer factorization, but it includes a spectral factorization of positive systems [6]. The H^∞ control problem was shown to be reduced to the J–lossless factorization [8]. The J–lossless conjugation was introduced in [4] as a powerful tool to compute the J–lossless factorization for continuous–time systems. It is closely related to the classical interpolation theory of Pick and Nevanlinna [5]. Some interesting properties of J–lossless conjugation related to the classical interpolation problem was shown in [5].

This paper is concerned with an extension of J–lossless conjugation to discrete–systems. The discrete–time case is much more complicated than the continuous–time case, because of the complicated representation of the discrete–time conjugate system. Nevertheless, it is shown that the complete analogue of the continuous–time case holds in a more complicated form.

In Section 2, we briefly review some preliminary results. We derive an existence condition for a J–lossless conjugation in terms of a Riccati equation in Section 3. Section 4 is devoted to the exploitation of the sequential J–lossless conjugation which is essentially equivalent to the Nevanlina algorithm.

Notations:

$$R^{\sim}(z) = R^T(\frac{1}{z})$$

$$R^*(z) = R^T(\overline{z})$$

$$G(z) = C(zI - A)^{-1}B + D := \left[\begin{array}{c|c} A & B \\ \hline C & D \end{array}\right]$$

$\sigma(A)$	:	the largest singular value of a matrix A.
$\boldsymbol{R}_{n\times n}$	:	the set of all $n \times n$ real constant matrices.
$\boldsymbol{RL}^{\infty}_{m\times r}$	:	the set of all $m \times r$ rational matrices without poles on the unit circle.
$\boldsymbol{RH}^{\infty}_{m\times r}$	:	the set of all $m \times r$ rational stable proper matrices.
$\boldsymbol{BH}^{\infty}_{m\times r}$	:	the subset of $\boldsymbol{RH}^{\infty}_{m\times r}$ whose norm is less than 1.

2 PRELIMINARIES

A matrix $S \in \boldsymbol{R}_{2n\times 2n}$ is called *symplectic* if

$$S^T \Sigma S = \Sigma \tag{14.2}$$

where

$$\Sigma = \begin{bmatrix} 0 & I_n \\ -I_n & 0 \end{bmatrix}.$$

It is easy to see that if $\lambda \neq 0$ is an eigenvalue of S, so is λ^{-1}. A symplectic matrix S generates a Riccati equation

$$XS_{11} - S_{22}X + XS_{12}X - S_{21} = 0. \tag{14.3}$$

A solution X of (14.3) which stabilizes

$$\hat{A} := S_{11} + S_{12}X \tag{14.4}$$

is denoted by

$$X = Ric(S). \tag{14.5}$$

If X is non-singular, then

$$X^{-1} = Ric(\begin{bmatrix} S_{22} & S_{21} \\ S_{12} & S_{11} \end{bmatrix}) = Ric(\begin{bmatrix} 0 & I_n \\ I_n & 0 \end{bmatrix} S \begin{bmatrix} 0 & I_n \\ I_n & 0 \end{bmatrix}). \tag{14.6}$$

Let J_{mr} be a sign matrix defined as

$$J_{mr} = \begin{bmatrix} I_m & 0 \\ 0 & -I_r \end{bmatrix}. \tag{14.7}$$

The notion of J_{mr}–lossless matrix below plays a fundamental role in various fields of linear system theory (see, e.g., [2]).

Definition 14.1 *A matrix $\Theta(z)$ with $m+r$ rows and $m+r$ columns is said to be J_{mr}–lossless if*

$$\Theta^{\sim}(z) J_{mr} \Theta(z) = J_{mr}, \qquad \forall z \tag{14.8}$$

$$\Theta^{*}(z) J_{mr} \Theta(z) \leq J_{mr}. \qquad \forall |z| \geq 1 \tag{14.9}$$

If the sizes of the blocks of $\Theta(z)$ is irrelevant or clear from the context, we suppress the subscripts m and r and denote simply as *J–lossless matrix.* The J-lossless matrix represents the transfer function of a J–lossless system which is in some sense a generalization of the lossless system where r is assumed to be zero in (14.7), or equivalently, $J_{mr} = I_m$. It is well–known that J–losslessness is a *chain–scattering representation* of losslessness [6].

A state–space characterization of J–lossless matrices in terms of the state–space representation

$$\Theta(z) = \left[\begin{array}{c|c} A & B \\ \hline C & D \end{array}\right] \tag{14.10}$$

is well–known.

Theorem 14.1 [2] *Let $J = J_{mr}$ and let $\Theta(z)$ in (14.10) be a minimal realization of a proper transfer function. Then $\Theta(z)$ is J–lossless, if and only if there exists a positive definite matrix P satisfying*

$$\begin{bmatrix} A^T & C^T \\ B^T & D^T \end{bmatrix} \begin{bmatrix} P & 0 \\ 0 & J \end{bmatrix} \begin{bmatrix} A & B \\ C & D \end{bmatrix} = \begin{bmatrix} P & 0 \\ 0 & J \end{bmatrix}. \tag{14.11}$$

Moreover, the matrix P is unique.

3 J–LOSSLESS CONJUGATION

The *J–lossless conjugation*, the *conjugation* by a J–lossless matrix, was introduced in [4] as a tool for computing J–lossless factorization. The idea of conjugation is very simple. It is just the cancellation of undesirable poles of a system by zeros of a system connected cascade way.

Definition 14.2 *A J–lossless matrix $\Theta(z)$ is said to be a stabilizing (anti-stabilizing) J–lossless conjugator of $G(z)$, if and only if*

(i) $H(z) := G(z)\Theta(z)$ is stable (anti-stable),

(ii) $deg(\Theta(z))$ is equal to the number of unstable (stable) poles of $G(z)$, multiplicity included.

The property (i) implies that all the unstable poles of $G(z)$ must be cancelled out by the zeroes of $\Theta(z)$ in the case of stabilizing conjugation. The property (ii) implies that the degree of $\Theta(z)$ is minimal for achieving (i). No pole of $\Theta(z)$ is cancelled out by a zero of $G(z)$.

Let $G(z)$ be a transfer function with $m+r$ inputs and $m+r$ outputs whose state space realization is given by

$$G(z) = \left[\begin{array}{c|c} A & B \\ \hline C & D \end{array}\right]. \tag{14.12}$$

We make the following assumptions:
(A_1) A has no eigenvalues on the unit circle.
(A_2) A is invertible.
The assumption (A_1) is crucial, while (A_2) is purely technical.

Due to (A_2), we have

$$G^{\sim}(z) = \left[\begin{array}{c|c} A^{-T} & -A^{-T}C^T \\ \hline B^TA^{-T} & D^T - B^TA^{-T}C^T \end{array}\right]. \tag{14.13}$$

If (A_2) fails to hold, we must use a descriptor form to represent a state–space form of the conjugate $G^{\sim}(z)$ of $G(z)$. The representation (14.13) is much more complicated than the corresponding one for a continuous–time systems. All the complications in discrete–time systems stem from (14.13).

The concatenation rule gives

$$\begin{aligned} G(z)JG^{\sim}(z) &= \left[\begin{array}{cc|c} A & -BJB^TA^{-T} & BJ(D^T - B^TA^{-T}C^T) \\ 0 & A^{-T} & A^{-T}C^T \\ \hline C & -DJB^TA^{-T} & DJ(D^T - B^TA^{-T}C^T) \end{array}\right] \\ &= \left[\begin{array}{cc|c} A^{-T} & 0 & A^{-T}C^T \\ -BJB^TA^{-T} & A & BJ(D^T - B^TA^{-T}C^T) \\ \hline -DJB^TA^{-T} & C & DJ(D^T - B^TA^{-T}C^T) \end{array}\right], \end{aligned} \tag{14.14}$$

where $J = J_{mr}$. Let us denote the two A–matrices of the above realizations by

$$S := \begin{bmatrix} A & -BJB^TA^{-T} \\ 0 & A^{-T} \end{bmatrix}, \tag{14.15}$$

$$\bar{S} := \begin{bmatrix} A^{-T} & 0 \\ -BJB^T A^{-T} & A \end{bmatrix}. \tag{14.16}$$

It is not difficult to see that both S and $\bar{S}$ satisfy (14.2). Hence, they are symplectic.

The symplectic matrix (14.15) generates a Riccati equation

$$XA - A^{-T}X - XBJB^T A^{-T}X = 0, \tag{14.17}$$

according to (14.3). If $X = Ric(S)$ exists, it stabilizes

$$\hat{A} := A - BJB^T A^{-T}X, \tag{14.18}$$

due to (14.4).

The Ricccati equation (14.17) can be written in a form without using the inverse of A. From (14.17), it follows that

$$XA = (I + XBJB^T)A^{-T}X. \tag{14.19}$$

We can easily see that $I + XBJB^T$ is non–singular. Indeed, if $\xi^T(I + XBJB^T) = 0$, it follows, from (14.19), that, $\xi^T XA = 0$. Due to (A_2), it implies that $\xi X = 0$. Hence $\xi^T = 0$. Therefore, $I + XBJB^T$ is invertible. We have, from (14.19),

$$X = A^T(I + XBJB^T)^{-1}XA, \tag{14.20}$$

which is further rewritten as

$$X = A^T XA - A^T XBJB^T(I + XBJB^T)^{-1}XA. \tag{14.21}$$

This is a familiar form in control literature appearing in the discrete–time LQG control theory[3][7].

The matrix $\hat{A}$ in (14.18) can be written as

$$\begin{aligned} \hat{A} &= A - BJB^T(I + XBJB^T)^{-1}XA \\ &= A + BF \end{aligned} \tag{14.22}$$

where

$$F := -JB^T(I + XBJB^T)^{-1}XA. \tag{14.23}$$

Now, we derive an existence condition of the stabilizing J–lossless conjugation in terms of a symplectic matrix S.

Theorem 14.2 *A state–space realization* (14.12) *is assumed to satisfy the assumption* (A_1) *and* (A_2). *It allows a stabilizing J–lossless conjugation, iff $X = Ric(S)$ exists and is non–negative definite, i.e., there exists a solution*

$X \geq 0$ of (14.21) which stabilizes (14.22). In that case, a stabilizing J–lossless conjugator is given by

$$\Theta(z) = \Theta_0(z) D_c \tag{14.24}$$

where

$$\Theta_0(z) = \left[\begin{array}{c|c} \hat{A} & B \\ \hline F & I \end{array}\right], \tag{14.25}$$

and D_c is any matrix satisfying

$$D_c^T (B^T X B + J) D_c = J. \tag{14.26}$$

The matrices $\hat{A}$ and F are given by (14.22) and (14.23), respectively. The conjugated matrix is given by

$$G(z)\Theta(z) = \left[\begin{array}{c|c} A + BF & B \\ \hline C + DF & D \end{array}\right] D_c. \tag{14.27}$$

Remark: Theorem 14.2 depends only on A and B in the realization (14.12) of $G(z)$. Therefore, we sometime say that $\Theta(z)$ is a J–lossless conjugator of (A, B).

Example Let

$$G(z) = \left[\begin{array}{cc|cc} 2 & 0 & 1 & \alpha \\ 0 & \frac{1}{4} & 0 & 1 \\ \hline 1 & 0 & 0 & 0 \\ 0 & 1 & 0 & 0 \end{array}\right],$$

where α is a real parameter. For $J = J_{1,1}$, the Riccati equation (14.17) is given in this case by

$$X \begin{bmatrix} 2 & 0 \\ 0 & \frac{1}{4} \end{bmatrix} - \begin{bmatrix} \frac{1}{2} & 0 \\ 0 & 4 \end{bmatrix} X - X \begin{bmatrix} (1-\alpha^2) & -\alpha \\ -\alpha & -1 \end{bmatrix} \begin{bmatrix} \frac{1}{2} & 0 \\ 0 & 4 \end{bmatrix} X = 0.$$

The stabilizing solution is given by

$$X = \begin{bmatrix} \delta & 0 \\ 0 & 0 \end{bmatrix}, \quad \delta := \frac{3}{1-\alpha^2},$$

$$\hat{A} = \begin{bmatrix} \frac{1}{2} & 0 \\ \frac{\delta\alpha}{2} & \frac{1}{4} \end{bmatrix}.$$

Therefore, $G(z)$ has a stabilizing $J_{1,1}$–lossless conjugaton iff $|\alpha| < 1$. In that case, the conjugator is given by

$$\Theta(z) = \left[\begin{array}{cc|cc} \frac{1}{2} & 0 & \delta & \delta\alpha \\ 0 & 4 & 0 & 0 \\ \hline -\frac{1}{2} & 0 & 1 & 0 \\ \frac{\alpha^2}{2} & 4 & 0 & 1 \end{array}\right] D_c = \left[\begin{array}{c|cc} \frac{1}{2} & \delta & \delta\alpha \\ \hline -\frac{1}{2} & 1 & 0 \\ \frac{\alpha^2}{2} & 0 & 1 \end{array}\right] D_c,$$

where, we chose

$$D_c = \frac{1}{3}\begin{bmatrix} 3+\delta & \alpha\delta \\ \alpha\delta & \alpha^2\delta + 3 \end{bmatrix}.$$

The conjugated system is given by

$$G(z)\Theta(z) = \left[\begin{array}{cc|cc} \frac{1}{2} & 0 & 1 & \alpha \\ \frac{\alpha\delta}{2} & \frac{1}{4} & 0 & 1 \\ \hline 1 & 0 & 0 & 0 \\ 0 & 1 & 0 & 0 \end{array}\right] D_c.$$

□

Though the proof of Theorem 14.2 is found in Appendix, we state some properties of the Riccati equation (14.21) which are interesting in their own light. From (14.21) and (14.22), it follows that $X\hat{A} = A^{-T}X$. Hence, we have

$$XA = \hat{A}^{-T}X. \tag{14.28}$$

Due to (14.20) and (14.23), $F = -JB^TA^{-T}X$. Therefore, (14.28) impiles that

$$F = -JB^TX\hat{A}. \tag{14.29}$$

The relation (14.28) impiles that $X = \hat{A}^TXA = \hat{A}^TX(\hat{A} - BF) = \hat{A}^TX\hat{A} - \hat{A}^TXBF$. From (14.29), it follows that

$$X = \hat{A}^TX\hat{A} + F^TJF. \tag{14.30}$$

This is an alternative representation of the Riccati equation (14.21).

Lemma 14.1 *The matrix $\Theta_0(z)$ given by (14.25) satisfies*

$$\Theta_0^{\sim}(z)J\Theta_0(z) = J + B^TXB. \tag{14.31}$$

Proof Straightforward computations yield

$$\begin{aligned}\Theta_0^{\sim}(z)J\Theta_0(z) &= (I + B^T(z^{-1}I - \hat{A}^T)^{-1}F^T)J(I + F(zI - \hat{A})^{-1}B), \\ &= J - B^T(z^{-1}I - \hat{A}^T)^{-1}\hat{A}^T XB - B^T X\hat{A}(zI - \hat{A})^{-1}B \\ &\quad + B^T(z^{-1}I - \hat{A}^T)^{-1}(X - \hat{A}^T X\hat{A})(zI - \hat{A})^{-1}B. \qquad (14.32)\end{aligned}$$

Using the identity $X - \hat{A}^T X\hat{A} = (z^{-1}I - \hat{A}^T)X(zI - \hat{A}) + \hat{A}^T X(zI - \hat{A}) + (z^{-1}I - \hat{A}^T)X\hat{A}$, we can prove the relation (14.31) immediately. □

Lemma 14.2 *The stable eigenspace of A is equal to Ker X.*

Proof Let λ be an eigenvalue of A and x the associated eigenvector. Due to (14.28),

$$0 = X(A - \lambda I)x = -(\lambda I - \hat{A}^{-T})Xx. \qquad (14.33)$$

If $|\lambda| < 1$, then $Xx = 0$. Indeed, if $Xx \neq 0$, λ is an eigenvalue of $\hat{A}^{-T}$, or equivalently, λ^{-1} is an eigenvalue of $\hat{A}$, which contradicts the stability of $\hat{A}$. In a similar way, we can show that $|\lambda| > 1$ implies $Xx = 0$. Thus, the assertion has been proven. □

Remark: If the realization (14.12) of $G(z)$ is decomposed into the stable and the anti–stable portions, i.e.,

$$G(z) = \left[\begin{array}{cc|c} A_+ & 0 & B_+ \\ 0 & A_- & B_- \\ \hline C_+ & C_- & D \end{array}\right], \qquad (14.34)$$

where A_+^{-1} and A_- are stable, Lemma 14.2 impiles that the stabilizing solution X of (14.17) is of the form

$$X = \begin{bmatrix} X_+ & 0 \\ 0 & 0 \end{bmatrix}, \qquad (14.35)$$

where X_+ is a stabilizing solution of

$$X_+A_+ + A_+^{-T}X_+ - X_+B_+JB_+^TA_+^{-T}X_+ = 0. \qquad (14.36)$$

Straightforward computation using (14.22) and (14.23) yields

$$F = \begin{bmatrix} F_+ & 0 \end{bmatrix} \qquad (14.37)$$

$$F_+ = -JB_+^T(I + X_+B_+JB_+^T)^{-1}X_+A_+ \qquad (14.38)$$

and

$$\hat{A} = \begin{bmatrix} A_+ + B_+F_+ & 0 \\ 0 & A_- \end{bmatrix}. \qquad (14.39)$$

Therefore,

$$\Theta_0(z) = \left[\begin{array}{cc|c} A_+ + B_+F_+ & 0 & B_+ \\ 0 & A_- & B_- \\ \hline F_+ & 0 & I \end{array}\right] = \left[\begin{array}{c|c} A_+ + B_+F_+ & B_+ \\ \hline F_+ & I \end{array}\right], \tag{14.40}$$

$$G(z)\Theta_0(z) = \left[\begin{array}{cc|c} A_+ + B_+F_+ & 0 & B_+ \\ 0 & A_- & B_- \\ \hline C_+ + DF_+ & C_- & D \end{array}\right] D_c. \tag{14.41}$$

The relations (14.40) and (14.41) imply that the stabilizing J–lossless conjugation conjugates only the anti–stabilizing portion of $G(z)$. The stable portion is left "untouched".

4 SEQUENTIAL J–LOSSLESS CONJUGATION

Due to the remark at the end of the preceding section, the stabilizing J–lossless conjugation is reduced essentially to the case where $G(z)$ is anti–stable, i.e., A^{-1} is stable in the realization (14.12). In this case, we can carry out the J–lossless conjugation sequentially according to the polar decomposition of the A–matrix. The sequential J–lossless conjugation which is the subject of this section is closely related to the factorization of J–lossless systems [2] which dates back to Potapov [9]. It was used in the cascade synthesis of digital filters by lossless inverse scattering method [1][10].

We first note that the J–lossless factorization becomes simpler when $G(z)$ is anti–stable, as is shown below.

Lemma 14.3 *Assume that the system $G(z)$ given by (14.12) is anti–stable, i.e., A^{-1} is stable in (14.12). Then, it admits a stabilizing J–lossless conjugation, iff there exists $Y = Ric(\bar{S})$ which is positive definite, where $\bar{S}$ is given by (14.16), i.e., a solution $Y > 0$ of the equation*

$$Y = AYA^T - BJB^T \tag{14.42}$$

exists which stabilizes

$$\hat{A} = A - BJB^TA^{-T}Y^{-1}. \tag{14.43}$$

In that case, a stabilizing J–lossless conjugator of $G(z)$ is given by (14.24), (14.25) where F is given by

$$F = -JB^T(Y + BJB^T)^{-1}A. \tag{14.44}$$

Proof: Due to Lemma 14.2, the solution X of (14.17), if exists, must be positive definite. Taking $Y = X^{-1}$ vertifies the assertion. □

Remark: The equation (14.42) is of Lyapunov type and is usually refered to as *Stein equation.*

Now, we assume that an anti–stable transfer function $G(z)$ has a realization

$$G(z) = \left[\begin{array}{cc|c} A_1 & 0 & B_0 \\ 0 & A_2 & B_1 \\ \hline C_1 & C_2 & D \end{array}\right]. \tag{14.45}$$

If $G(z)$ has a stabilizing J–lossless factorization, there exists a matrix $Y > 0$ such that

$$Y = \begin{bmatrix} A_1 & 0 \\ 0 & A_2 \end{bmatrix} Y \begin{bmatrix} A_1^T & 0 \\ 0 & A_2^T \end{bmatrix} - \begin{bmatrix} B_1 \\ B_2 \end{bmatrix} J \begin{bmatrix} B_1^T & B_2^T \end{bmatrix}. \tag{14.46}$$

Writing

$$Y = \begin{bmatrix} Y_{11} & Y_{12} \\ Y_{12}^T & Y_{22} \end{bmatrix}, \tag{14.47}$$

we see that

$$Y_{11} = A_1 Y_{11} A_1^T - B_1 J B_1^T. \tag{14.48}$$

Since $Y_{11} > 0$, Lemma 14.3 impiles that the pair (A_1, B_1) allows a stabilizing J–lossless conjugator $\Theta_1(z)$ given by

$$\Theta_1(z) = \left[\begin{array}{c|c} A_1^{-T} & Y_{11}^{-1} B_1 \\ \hline -J B_1^T A_1^{-T} & I \end{array}\right] D_{c_1}, \tag{14.49}$$

where D_{c_1} is any matrix satisfying

$$D_{c_1}^T (J + B_1^T Y_{11}^{-1} B_1) D_{c_1} = J. \tag{14.50}$$

Straightforward computation using the concatenation rule gives

$$G(z)\Theta_1(z) = \left[\begin{array}{ccc|c} A_1 & 0 & -B_1 J B_1^T A_1^{-T} & B_1 \\ 0 & A_2 & -B_2 J B_1^T A_1^{-T} & B_2 \\ 0 & 0 & A_1^{-T} & Y_{11}^{-1} B_1 \\ \hline C_1 & C_2 & -D J B_1^T A_1^{-T} & D \end{array}\right] D_{c_1}. \tag{14.51}$$

Taking a similarity transformation with the transformation matrix

$$T = \begin{bmatrix} I & 0 & -Y_{11} \\ -Y_{12}^T & I & Y_{12}^T Y_{11} \\ 0 & 0 & I \end{bmatrix}, \qquad T^{-1} = \begin{bmatrix} I & 0 & Y_{11} \\ Y_{12}^T & I & 0 \\ 0 & 0 & I \end{bmatrix}$$

and eliminating an uncontrollable mode, we have

$$G(z)\Theta_1(z) = \left[\begin{array}{cc|c} A_1^{-T} & 0 & Y_{11}^{-1}B_1 \\ 0 & A_2 & -Y_{12}^TY_{11}^{-1}B_1 + B_2 \\ \hline L & C_2 & D \end{array}\right] D_{c_1}, \tag{14.52}$$

where $L = C_2Y_{12}^T + C_1Y_{11} - DJB_1^TA_1^{-T}$. Comparison of the above identity with (14.45) shows that $\Theta_1(z)$ partially conjugates $G(z)$, namely, A_1–portion in (14.45) is conjugated.

The next step is to congugate the remaining portion $(A_2, (B_2 - Y_{12}^TY_{11}^{-1}B_1)D_{c_1})$. To see this pair has a stabilizing J–lossless conjugation, we rewrite the equation (14.46) by multiplying both side by

$$U = \begin{bmatrix} I & 0 \\ -Y_{12}^TY_{11}^{-1} & I \end{bmatrix}, \qquad U^T = \begin{bmatrix} I & -Y_{11}^{-1}Y_{12} \\ 0 & I \end{bmatrix}. \tag{14.53}$$

Writing

$$U\begin{bmatrix} A_1 & 0 \\ 0 & A_2 \end{bmatrix}U^{-1} = \begin{bmatrix} A_1 & 0 \\ A_{21} & A_2 \end{bmatrix} \tag{14.54}$$

$$UYU^T = \begin{bmatrix} Y_{11} & 0 \\ 0 & \Delta_2 \end{bmatrix}, \tag{14.55}$$

where $A_{21} = A_2Y_{12}^TY_{11}^{-1} - Y_{12}^TY_{11}^{-1}A_1$ and $\Delta_2 = Y_{22} - Y_{12}^TY_{11}^{-1}Y_{12}$, we have

$$A_{21}Y_{11}A_1^T = (B_2 - Y_{12}^TY_{11}^{-1}B_1)JB_1^T, \tag{14.56}$$

$$\Delta_2 = A_{21}Y_{11}A_{21}^T + A_2\Delta_2A_2^T - (B_2 - Y_{12}^TY_{11}^{-1}B_1)J(B_2 - Y_{12}^TY_{11}^{-1}B_1)^T. \tag{14.57}$$

Therefore, we have

$$A_{21}Y_{11}A_{21}^T = (B_2 - Y_{12}^TY_{11}^{-1}B_1)JB_1^TA_1^{-T}Y_{11}^{-1}A_1^{-1}B_1J(B_2 - Y_{12}^TY_{11}^{-1}B_1)^T. \tag{14.58}$$

Using (14.48) and the relation $JB_1^T(I + Y_{11}^{-1}B_1JB_1^T)^{-1} = (J + B_1^TY_{11}^{-1}B_1)^{-1}B_1^T$, we have, from (14.57),

$$\Delta_2 = A_2\Delta_2A_2^T - (B_2 - Y_{12}^TY_{11}^{-1}B_1)(J + B_1^TY_{11}^{-1}B_1)^{-1}(B_2 - Y_{12}^TY_{11}^{-1}B_1)^T.$$

Due to (14.50), we have

$$\Delta_2 = A_2\Delta_2A_2^T - (B_2 - Y_{12}^TY_{11}^{-1}B_1)D_{c_1}JD_{c_1}^T(B_2 - Y_{12}^TY_{11}^{-1}B_1)^T. \tag{14.59}$$

Since $Y > 0$, the Schur complement Δ_2 of Y is also positive. Hence, we conclude that the stabilizing J–lossless conjugator Θ_2 of $G\Theta_1$, or of the pair

$(A_2, (B_2 - Y_{12}^T Y_{11}^{-1} B_1) D_{c_1})$, exists and is given by

$$\Theta_2(z) = \left[\begin{array}{c|c} A_2^{-T} & \Delta_2^{-1}(B_2 - Y_{12}^T Y_{11}^{-1} B_1) D_{c_1} \\ \hline -J D_{c_1}^T (B_2 - Y_{12}^T Y_{11}^{-1} B_1)^T A_2^{-T} & I \end{array}\right] D_{c_2}. \tag{14.60}$$

where D_{c_2} is any non-singular matrix satisfying

$$D_{c_2}^T \left(J + D_{c_1}^T (B_2 - Y_{12}^T Y_{11}^{-1} B_1)^T \Delta_2^{-1} (B_2 - Y_{12}^T Y_{11}^{-1} B_1) D_{c_1} \right) D_{c_2} = J.$$

Thus, $\Theta(z) = \Theta_1(z)\Theta_2(z)$ is a stabilizing J–lossless conjugator of $G(z)$, in other words, the stabilizing J–lossless conjugator $\Theta(z)$ of $G(z)$ is represented in the factorized form

$$\Theta(z) = \Theta_1(z)\Theta_2(z) \tag{14.61}$$

according to the polar decomposition of the A–matrix of a realization of $G(z)$ in (14.45) in the sense that $\Theta_i(z)$ has A_i^{-T} as its A–matrix. Also, as is shown in (14.59), the sequential conjugation corresponds to computation of the Schur complement Δ_2 of Y. The above procedure can be applied to a more general case where $G(z)$ is represented as

$$G(z) = \left[\begin{array}{cccc|c} A_1 & 0 & \dots & 0 & B_1 \\ 0 & A_2 & \dots & 0 & B_2 \\ \cdot & \cdot & \cdot & \cdot & \cdot \\ \cdot & \cdot & \cdot & \cdot & \cdot \\ 0 & 0 & \dots & A_l & B_l \\ \hline C_1 & C_2 & \dots & C_l & D \end{array}\right]. \tag{14.62}$$

The J–lossless conjugation is carried out successively resulting in the factorized representation

$$\Theta(z) = \Theta_1(z)\Theta_2(z)\dots\Theta_l(z) \tag{14.63}$$

such that $\Theta_i(z)$ has A_i^{-T} as its A–matrix. This procedure is actually equivalent to successively finding Schur complements of the solution $Y > 0$ of the equation

$$Y = \begin{bmatrix} A_1 & 0 & \dots & 0 \\ 0 & A_2 & \dots & 0 \\ \cdot & \cdot & \cdot & \cdot \\ \cdot & \cdot & \cdot & \cdot \\ 0 & 0 & \dots & A_l \end{bmatrix} Y \begin{bmatrix} A_1 & 0 & \dots & 0 \\ 0 & A_2 & \dots & 0 \\ \cdot & \cdot & \cdot & \cdot \\ \cdot & \cdot & \cdot & \cdot \\ 0 & 0 & \dots & A_l \end{bmatrix}^T - \begin{bmatrix} B_1 \\ B_2 \\ \cdot \\ \cdot \\ B_l \end{bmatrix} J \begin{bmatrix} B_1 \\ B_2 \\ \cdot \\ \cdot \\ B_l \end{bmatrix}^T \tag{14.64}$$

in an nested way. Mathematically, it is equivalent to the computation of a Cholesky decomposition of Y, i.e., the computation of $U_{ij}, (i > j)$ and $\hat{Y}_i$ such that

$$\begin{bmatrix} I & 0 & \dots & 0 \\ U_{21} & I & \dots & 0 \\ \cdot & \cdot & \cdot & \cdot \\ \cdot & \cdot & \cdot & \cdot \\ U_{l1} & U_{l2} & \dots & I \end{bmatrix} Y \begin{bmatrix} I & 0 & \dots & 0 \\ U_{21} & I & \dots & 0 \\ \cdot & \cdot & \cdot & \cdot \\ \cdot & \cdot & \cdot & \cdot \\ U_{l1} & U_{l2} & \dots & I \end{bmatrix}^T = \begin{bmatrix} \hat{Y}_1 & 0 & \dots & 0 \\ 0 & \hat{Y}_2 & \dots & 0 \\ \cdot & \cdot & \cdot & \cdot \\ \cdot & \cdot & \cdot & \cdot \\ 0 & 0 & \dots & \hat{Y}_l \end{bmatrix}. \tag{14.65}$$

The precise description of the sequential J–lossless conjugation is described as follows:

Theorem 14.3 *Assume that $G(z)$ is anti–stable and its realization is given in the form (14.62). If $G(z)$ allows a stabilizing J–lossless conjugation, then a J–lossless conjugator $\Theta(z)$ of $G(z)$ is represented in the factorized form*

$$\Theta(z) = \Theta_1(z)\Theta_2(z)....\Theta_l(z) \tag{14.66}$$

such that $\Theta_i(z)$ is a stabilizing J–lossless conjugator of the pair $(A_i, \hat{B}_i\hat{D}_{i-1})$ where

$$\begin{bmatrix} \hat{B}_1 \\ \hat{B}_2 \\ \vdots \\ \hat{B}_l \end{bmatrix} = \begin{bmatrix} I & 0 & \dots & 0 \\ U_{21} & I & \dots & 0 \\ \cdot & \cdot & \cdot & \cdot \\ \cdot & \cdot & \cdot & \cdot \\ U_{l1} & U_{l2} & \dots & I \end{bmatrix} \begin{bmatrix} B_1 \\ B_2 \\ \vdots \\ B_l \end{bmatrix}, \tag{14.67}$$

and D_i is a matrix satisfying

$$D_i^T(J + \hat{B}_i^T\hat{Y}_i^{-1}\hat{B}_i)D_i = J, \qquad D_0 = I. \tag{14.68}$$

The matrices $\hat{Y}_i$ in (14.65) satisfy the recursive relations

$$\hat{Y}_i = A_i\hat{Y}_iA_i^T - \hat{B}_i(J + \hat{B}_{i-1}^T\hat{Y}_{i-1}^{-1}\hat{B}_{i-1})^{-1}\hat{B}_i^T, \tag{14.69}$$

$$\hat{B}_0 = 0, \qquad \hat{Y}_0 = 0. \tag{14.70}$$

The above result can be used to factorize a stable J–lossless matrix. If $\Theta(z)$ is a stable J–lossless systems, then $\Theta^{-1}(z)$ is anti–stable because $\Theta^{-1}(z) = J\Theta^{\sim}(z)J$. The sequential J–lossless conjugation of $\Theta^{-1}(z)$ yields

$$\Theta^{-1}(z)\Theta_1(z)\Theta_2(z)...\Theta_l(z) = M, \tag{14.71}$$

where M is a constant J–lossless matrix. Thus, we obtain a factorization

$$\Theta(z) = \Theta_1(z)\Theta_2(z)...\Theta_l(z)M. \tag{14.72}$$

Example Consider a matrix $\Theta(z)$ given by

$$\Theta(z) = \left[\begin{array}{cc|cc} \frac{1}{2} & \frac{10}{9} & \frac{5}{12} & \frac{3}{4} \\ 0 & -\frac{1}{3} & 1 & 0 \\ \hline \frac{3}{4} & -\frac{5}{9} & -\frac{5}{24} & -\frac{3}{8} \\ 0 & \frac{2}{3} & \frac{1}{4} & \frac{5}{4} \end{array}\right]. \tag{14.73}$$

We can easily show that the above realization satisfies the equation (14.11) with

$$P = \begin{bmatrix} \frac{3}{4} & 0 \\ 0 & \frac{8}{9} \end{bmatrix}. \tag{14.74}$$

Hence, $\Theta(z)$ is a $J_{1,1}$–lossless matrix. Straightforward computation yields

$$\Theta^{-1}(z) = \Psi(z) = \left[\begin{array}{cc|cc} 2 & 0 & \frac{3}{2} & 0 \\ 0 & -3 & \frac{15}{2} & \frac{9}{4} \\ \hline -\frac{15}{2} & -\frac{8}{3} & -\frac{15}{2} & -\frac{9}{4} \\ \frac{3}{2} & 0 & \frac{3}{2} & \frac{5}{4} \end{array}\right]. \tag{14.75}$$

According to the algorithm shown above, we first conjugate the pair $\left(2, \begin{bmatrix} \frac{3}{2} & 0 \end{bmatrix}\right)$. The resulting stabilizing J–lossless conjugator is given by

$$\Theta_1(z) = \left[\begin{array}{c|cc} \frac{1}{2} & 1 & 0 \\ \hline -\frac{3}{4} & \frac{1}{2} & 0 \\ 0 & 0 & 1 \end{array}\right]. \tag{14.76}$$

The conjugated system is calculated to be

$$\Psi(z)\Theta_1(z) = \left[\begin{array}{cc|cc} \frac{1}{2} & 0 & 1 & 0 \\ -\frac{45}{8} & -3 & \frac{15}{4} & \frac{9}{4} \\ \hline 0 & \frac{8}{3} & -\frac{15}{4} & -\frac{9}{4} \\ 0 & 0 & \frac{3}{4} & \frac{5}{4} \end{array}\right]. \tag{14.77}$$

We extract a constant J–lossless matrix

$$\Theta_2 = \frac{1}{4}\begin{bmatrix} 5 & -3 \\ -3 & 5 \end{bmatrix} \tag{14.78}$$

from the D–matrix of the conjugated system (14.77) to simplify the expression. This gives

$$\Psi(z)\Theta_1(z)\Theta_2 = \left[\begin{array}{cc|cc} \frac{1}{2} & 0 & \frac{5}{4} & -\frac{3}{4} \\ -\frac{45}{8} & -3 & 3 & 0 \\ \hline 0 & \frac{8}{3} & -3 & 0 \\ 0 & 0 & 0 & 1 \end{array}\right]. \tag{14.79}$$

Now, we conjugate the pair $(-3, [3\ 0])$. The stabilizing J–lossless conjugator $\Theta_3(z)$ is given by

$$\Theta_3(z) = \left[\begin{array}{c|cc} -\frac{1}{3} & \frac{8}{9} & 0 \\ \hline 1 & \frac{1}{3} & 0 \\ 0 & 0 & 1 \end{array}\right]. \tag{14.80}$$

The conjugated system becomes

$$\Psi(z)\Theta_1(z)\Theta_2\Theta_3(z) = \begin{bmatrix} -1 & 0 \\ 0 & 1 \end{bmatrix} \tag{14.81}$$

Thus, $\Psi^{-1}(z) = \Theta(z)$ is represented in represented in the factorized form as

$$\begin{aligned}\Theta(z) &= \Theta_1(z)\Theta_2\Theta_3(z)\begin{bmatrix}-1 & 0\\ 0 & 1\end{bmatrix}\\ &= \begin{bmatrix}l_1(z) & 0\\ 0 & 1\end{bmatrix}\frac{1}{4}\begin{bmatrix}5 & -3\\ -3 & 5\end{bmatrix}\begin{bmatrix}l_2(z) & 0\\ 0 & 1\end{bmatrix},\end{aligned} \tag{14.82}$$

where

$$l_1(z) = \frac{z-2}{2z-1}, \qquad l_2(z) = -\frac{z+3}{3z+1}. \tag{14.83}$$

The factorized form corresponds to the well–known lattice realization [1][10], as is illustrated in Fig 14.1.

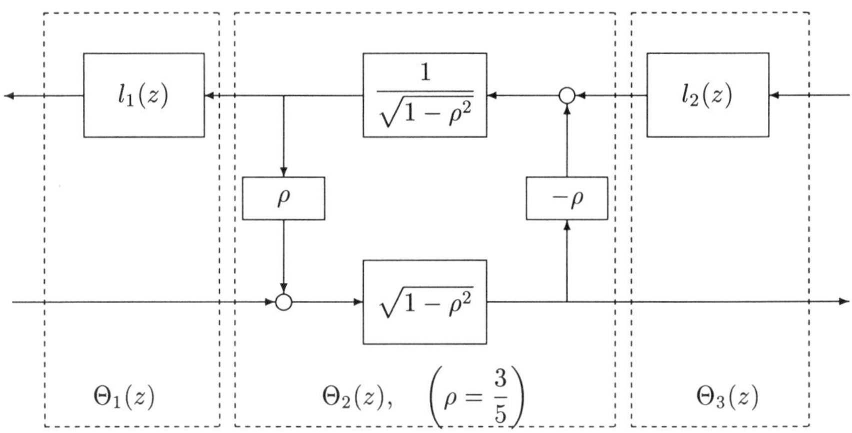

Figure 14.1 Lattice Realization of Factorization (14.82).

5 CONCLUSION

The existence condition for a stabilizing J–lossless conjugation has been derived for general discrete–time systems. The condition is represented in terms of a Riccati equation which is associated with the A–matrix of the realization of the concatenation of the system considered and its conjugate. It is shown that the existence condition is reduced to the solvability of a Lyapunov type equation, normally called Stein equation, in the case where the system to be conjugated is anti–stable. The sequential structure of J–lossless conjugation has been exploited and an algolithm of sequential J–lossless conjugation has been derived which turns out to be equivalent to the Cholesky decomposition of the solution of Stein equation.

APPENDIX A

PROOF OF THEOREM 14.2

First, we assume that (A, B) is controllable. Suppose that there exists a stabilizing J–lossless conjugator $\Theta(z)$ of $G(z)$ with a minimal realization

$$\Theta(z) = \left[\begin{array}{c|c} A_c & B_c \\ \hline C_c & D_c \end{array}\right]. \tag{A.1}$$

From the concatenation rule, it follows that

$$H(z) = G(z)\Theta(z) = \left[\begin{array}{cc|c} A & BC_c & BD_c \\ 0 & A_c & B_c \\ \hline C & DC_c & DD_c \end{array}\right]. \tag{A.2}$$

Since A has no eigenvalue on the unit circle, the A–matrix of the above realization can be decomposed into the the stable part and the anti-stable part by a similarity transformation

$$\begin{bmatrix} A & BC_c \\ 0 & A_c \end{bmatrix} \begin{bmatrix} M_1 & M_2 \\ N_1 & N_2 \end{bmatrix} = \begin{bmatrix} M_1 & M_2 \\ N_1 & N_2 \end{bmatrix} \begin{bmatrix} A_1 & 0 \\ 0 & A_2 \end{bmatrix}. \tag{A.3}$$

Here A_1 is stable and A_2 is anti-stable. Note that A_c is stable and $A_1 \in \boldsymbol{R}_{n\times n}$ with $n = deg(G)$. Since $H(z)$ is stable and no pole of $\Theta(z)$ is cancelled out by a zero of $G(z)$, the anti-stable portion of (A.2) must be uncontrollable. Therefore, the B–matrix of (A.2) must be included in the stable eigenspace of the A–matrix of (A.2). This implies that there exists a matrix B_1 such that

$$\begin{bmatrix} BD_c \\ B_c \end{bmatrix} = \begin{bmatrix} M_1 \\ N_1 \end{bmatrix} B_1. \tag{A.4}$$

The matrix M_1 is non-singular. Indeed, if $\xi^T M_1 = 0$ for some $\xi^T \neq 0$, (A.3) and (A.4) imply that $\xi^T B = 0$, $\xi^T A M_1 = 0$. This obviously contradicts the assumption that (A, B) is controllable. Hence M_1^{-1} exists.

Let $S := N_1 M_1^{-1}$. From (A.3) and (A.4), it follows that

$$B_c = SBD_c, \tag{A.5}$$
$$A_1 = M_1^{-1}(A + BC_c S)M_1. \tag{A.6}$$

Since $\Theta(z)$ is J–lossless matrix by the assumption, Theorem 14.1 implies that

$$A_c^T P A_c + C_c^T J C_c = P, \tag{A.7}$$
$$B_c^T P A_c + D_c^T J C_c = 0, \tag{A.8}$$
$$B_c^T P B_c + D_c^T J D_c = J \tag{A.9}$$

for some $P > 0$. Due to (A.8), we have

$$C_c = -JB^T S^T P A_c. \tag{A.10}$$

From (A.6) and (A.10), it follows that $A_c S = N_1 A_1 M_1^{-1} = S(A + BC_c S) = S(A - BJB^T S^T P A_c S) = SA - SBJB^T S^T P A_c S$. Then

$$A_c S = (I + SBJB^T S^T P)^{-1} SA \tag{A.11}$$

where we take $X = S^T P S$. From (A.7), (A.10) and (A.11), it follows that

$$A^T X A - A^T X BJB^T (I + XBJB^T)^{-1} XA = X. \tag{A.12}$$

Also,

$$\begin{aligned} A + BC_c S &= M_1 A_1 M_1^{-1} \\ &= A - BJB^T (I + XBJB^T)^{-1} XA. \end{aligned} \tag{A.13}$$

Hence, $\hat{A} := A - BJB^T(I + XBJB^T)^{-1} XA$ is stable. This establishes the first assertion.
From $(A.5)$ and $(A.10)$, it follows that

$$\Theta(z) = \left[\begin{array}{c|c} A_c & B_c \\ \hline C_c & C_c \end{array}\right] = \left[\begin{array}{c|c} A_c & SB \\ \hline -JB^T S^T P A_c & I \end{array}\right] D_c$$

Due to $(A.11)$, $(zI - A_c)^{-1} S = S(zI - \hat{A})^{-1}$. It follows that

$$\begin{aligned} \Theta(z) &= \left[\begin{array}{c|c} (I + BJB^T X)^{-1} A & B \\ \hline -JB^T A^{-T} X & I \end{array}\right] D_c, \\ &= \left[\begin{array}{c|c} A^{-T} & XB \\ \hline -JB^T A^{-T} & I \end{array}\right] D_c. \end{aligned} \tag{A.14}$$

which vertify (14.25). The second equality is due to $X(zI - \hat{A})^{-1} = (zI - A^{-T})^{-1} X$.

From the concatenation rule, it follow that

$$G(z)\Theta(z) = \left[\begin{array}{cc|c} A & -BJB^TA^{-T} & B \\ 0 & A^{-T} & XB \\ \hline C & -DJB^TA^{-T} & D \end{array}\right] D_c. \tag{A.15}$$

Taking the similarity tranformation with

$$T = \begin{bmatrix} I & 0 \\ -X & I \end{bmatrix}, \qquad T^{-1} = \begin{bmatrix} I & 0 \\ X & I \end{bmatrix}$$

yield (14.27). Finally, from (A.5) and (A.9), D_c is any non-singular matrix satisfying

$$D_c^T(B^TXB + J)D_c = J. \tag{A.16}$$

The existence of such D_c is guaranteed by (14.31) if we take $D_c = \Theta_0(\lambda)^{-1}$ for $|\lambda| > 1$.

To prove the converse, assume that there exists a solution of $X = Ric(S) \geq 0$. It is sufficient to prove that $\Theta(z)$ given in (A.14) is J–lossless matrix and $deg(\Theta(z))$ is equal to the number of unstable eigenvalues of A, in view of Definition 14.2. (See Lemma 14.2.) It is easy to see that $\Theta(z)$ given in (14.25) is J–lossless by using Theorem 14.1. □

REFERENCES

[1] Dewilde P and Dym H., "Lossless inverse scattering, estimation theory and digital filters," IEEE Transactions Information Theory, 1984, pp. 644–662

[2] Genin Y., Dooren P.V., Kailath, T., Delsome, J-M., and Morf, M.,"On Σ–lossless transfer function and related questions," Linear Algebra and Its Applications, 1983, pp. 251–275

[3] Gu, D. W., Tsai, M. C., O'Young, S. D., and Postlethwaite, I., "Stace–space formulae for discrete–time H^∞ optimization," International Journal of Control, 1989, pp. 1683–1723.

[4] Kimura H.,"Conjugation, interpolation and model–matching in H^∞," International Journal of Control, 1989, pp. 269–307

[5] Kimura H.,"J–lossless conjugation and classical interpolation problems," Archiv für Elektronik und Übertragungstechnik, vol 49, 1995, pp. 313-319

[6] Kimura H.,"Chain–scattering representation, J–lossless factorization and H^∞ control," Journal of Mathematical Systems, Estimation and Control, 1995, pp. 203–255

[7] Laub, A. J., "A Schur method for solving algebraic Riccati equations," IEEE Transactions on Automatic Control, 1979, pp. 913–921

[8] Liu, K. Z., and Mita, T., "Conjugation and H^∞ control of discrete–time systems," International Journal of Control, 1989, pp. 1435–1460

[9] Potapov, V., "The multiplicative structure of J–contractive matrix functions," Tran. AMS. Series2, Vol 15, 1960, pp. 131–243

[10] Rao, K., and Kailath, T., "Orthogonal digital filters for VLSI implementation," IEEE Transactions on Circuits & Systems, Vol 31, 1984, pp. 933–945

15

SEMIDEFINITE PROGRAMMING RELAXATIONS OF NON-CONVEX PROBLEMS IN CONTROL AND COMBINATORIAL OPTIMIZATION

Stephen Boyd and Lieven Vandenberghe

Information Systems Laboratory
Department of Electrical Engineering
Stanford University
Stanford, CA 94305-4055

In celebration of Tom Kailath's contributions, on his 60th birthday. Dedicated to Tom Kailath: mentor, model, colleague, teacher, friend.

ABSTRACT

We point out some connections between applications of semidefinite programming in control and in combinatorial optimization. In both fields semidefinite programs arise as convex relaxations of NP-hard quadratic optimization problems. We also show that these relaxations are readily extended to optimization problems over bilinear matrix inequalities.

1 SEMIDEFINITE PROGRAMMING

In a *semidefinite program* (SDP) we minimize a linear function of a variable $x \in \mathbf{R}^m$ subject to a matrix inequality:

$$\begin{array}{ll} \text{minimize} & c^T x \\ \text{subject to} & F(x) \geq 0 \end{array} \tag{15.1}$$

where

$$F(x) \stackrel{\Delta}{=} F_0 + \sum_{i=1}^{m} x_i F_i.$$

The problem data are the vector $c \in \mathbf{R}^m$ and $m+1$ symmetric matrices $F_0, \ldots, F_m \in \mathbf{R}^{n \times n}$. The inequality sign in $F(x) \geq 0$ means that $F(x)$ is positive semidefinite, *i.e.*, $z^T F(x) z \geq 0$ for all $z \in \mathbf{R}^n$. We call the inequality $F(x) \geq 0$ a *linear matrix inequality* (LMI).

Semidefinite programs can be regarded as an extension of linear programming where the componentwise inequalities between vectors are replaced by matrix inequalities, or, equivalently, the first orthant is replaced by the cone of positive semidefinite matrices. Semidefinite programming unifies several standard problems (*e.g.,* , linear and quadratic programming), and finds many applications in engineering and combinatorial optimization (see [Ali95], [BEFB94], [VB96]). Although semidefinite programs are much more general than linear programs, they are not much harder to solve. Most interior-point methods for linear programming have been generalized to semidefinite programs. As in linear programming, these methods have polynomial worst-case complexity, and perform very well in practice.

2 SEMIDEFINITE PROGRAMMING AND COMBINATORIAL OPTIMIZATION

Semidefinite programs play a very useful role in non-convex or combinatorial optimization. Consider, for example, the quadratic optimization problem

$$\begin{array}{ll} \text{minimize} & f_0(x) \\ \text{subject to} & f_i(x) \leq 0, \ i = 1, \ldots, L \end{array} \tag{15.2}$$

where $f_i(x) = x^T A_i x + 2b_i^T x + c_i$, $i = 0, 1, \ldots, L$. The matrices A_i can be indefinite, and therefore problem (15.2) is a very hard, non-convex optimization problem. For example, it includes all optimization problems with polynomial objective function and polynomial constraints (see [NN94, §6.4.4], [Sho87]).

For practical purposes, *e.g.,* , in branch-and-bound algorithms, it is important to have good and cheaply computable lower bounds on the optimal value of (15.2). Shor and others have proposed to compute such lower bounds by solving the semidefinite program (with variables t and τ_i)

$$\begin{array}{ll} \text{maximize} & t \\ \text{subject to} & \begin{bmatrix} A_0 & b_0 \\ b_0^T & c_0 - t \end{bmatrix} + \tau_1 \begin{bmatrix} A_1 & b_1 \\ b_1^T & c_1 \end{bmatrix} + \cdots + \tau_L \begin{bmatrix} A_L & b_L \\ b_L^T & c_L \end{bmatrix} \geq 0 \\ & \tau_i \geq 0, \quad i = 1, \ldots, L. \end{array} \tag{15.3}$$

One can easily verify that this semidefinite program yields lower bounds for (15.2). Suppose x satisfies the constraints in the nonconvex problem (15.2), *i.e.,*

$$f_i(x) = \begin{bmatrix} x \\ 1 \end{bmatrix}^T \begin{bmatrix} A_i & b_i \\ b_i^T & c_i \end{bmatrix} \begin{bmatrix} x \\ 1 \end{bmatrix} \leq 0$$

for $i = 1, \ldots, L$, and t, $\tau_1, \ldots, \tau_L$ satisfy the constraints in the semidefinite program (15.3). Then

$$\begin{aligned} 0 &\leq \begin{bmatrix} x \\ 1 \end{bmatrix}^T \left(\begin{bmatrix} A_0 & b_0 \\ b_0^T & c_0 - t \end{bmatrix} + \tau_1 \begin{bmatrix} A_1 & b_1 \\ b_1^T & c_1 \end{bmatrix} + \cdots + \tau_L \begin{bmatrix} A_L & b_L \\ b_L^T & c_L \end{bmatrix} \right) \begin{bmatrix} x \\ 1 \end{bmatrix} \\ &= f_0(x) - t + \tau_1 f_1(x) + \cdots + \tau_L f_L(x) \\ &\leq f_0(x) - t. \end{aligned}$$

Therefore $t \leq f_0(x)$ for every feasible x in (15.2), as desired. Problem (15.3) can also be derived via Lagrangian duality; for a deeper discussion, see Shor [Sho87], or Poljak, Rendl, and Wolkowicz [PRW94].

Most semidefinite relaxations of NP-hard combinatorial problems seem to be related to the semidefinite program (15.3), or the related one,

$$\begin{array}{ll} \text{minimize} & \mathbf{Tr} X A_0 + 2b_0^T x + c_0 \\ \text{subject to} & \mathbf{Tr} X A_i + 2b_i^T x + c_i \leq 0, \quad i = 1, \ldots, L \\ & \begin{bmatrix} X & x \\ x^T & 1 \end{bmatrix} \geq 0, \end{array} \tag{15.4}$$

where the variables are $X = X^T \in \mathbf{R}^{k \times k}$ and $x \in \mathbf{R}^k$. It can be shown that (15.4) is the semidefinite programming *dual* of Shor's relaxation (15.3); the two problems (15.3) and (15.4) yield the same bound.

Note that the constraint

$$\begin{bmatrix} X & x \\ x^T & 1 \end{bmatrix} \geq 0 \tag{15.5}$$

is equivalent to $X \geq xx^T$. The semidefinite program (15.4) can therefore be directly interpreted as a relaxation of the original problem (15.2), which can be written as

$$\begin{array}{ll} \text{minimize} & \mathbf{Tr} X A_0 + 2b_0^T x + c_0 \\ \text{subject to} & \mathbf{Tr} X A_i + 2b_i^T x + c_i \leq 0, \quad i = 1, \ldots, L \\ & X = xx^T. \end{array} \tag{15.6}$$

The only difference between (15.6) and (15.4) is the replacement of the (nonconvex) constraint $X = xx^T$ with the convex relaxation $X \geq xx^T$. It is also interesting to note that the relaxation (15.4) becomes the original problem (15.6) if we add the (nonconvex) constraint that the matrix on the left hand side of (15.5) is rank one.

As an example, consider the $(-1, 1)$-quadratic program

$$\begin{array}{ll} \text{minimize} & x^T A x + 2b^T x \\ \text{subject to} & x_i^2 = 1, \quad i = 1, \ldots, k, \end{array} \tag{15.7}$$

which is NP-hard. The constraint $x_i \in \{-1, 1\}$ can be written as the quadratic equality constraint $x_i^2 = 1$, or, equivalently, as two quadratic inequalities $x_i^2 \leq 1$ and $x_i^2 \geq 1$. Applying (15.4) we find that the semidefinite program in $X = X^T$ and x

$$\begin{array}{ll} \text{minimize} & \mathbf{Tr} XA + 2b^T x \\ \text{subject to} & X_{ii} = 1, \quad i = 1, \ldots, k \\ & \begin{bmatrix} X & x \\ x^T & 1 \end{bmatrix} \geq 0 \end{array} \tag{15.8}$$

yields a lower bound for (15.7). In a recent paper on the MAX-CUT problem, which is a specific case of (15.7) where $b = 0$ and the diagonal of A is zero, Goemans and Williamson have *proved* that the lower bound from (15.8) is at most 14% suboptimal (see [GW94] and [GW95]). This is much better than any previously known bound. Similar strong results on semidefinite programming relaxations of NP-hard problems have been obtained by Karger, Motwani, and Sudan [KMS94].

The usefulness of semidefinite programming in combinatorial optimization was recognized more than twenty years ago (see, *e.g.*, , Donath and Hoffman [DH73]). Many people seem to have developed similar ideas independently. We should however stress the importance of the work by Grötschel, Lovász, and Schrijver [GLS88, Chapter 9], [LS91] who have demonstrated the power of semidefinite relaxations on some very hard combinatorial problems. The recent development of efficient interior-point methods has turned these techniques into powerful practical tools; see Alizadeh [Ali92b, Ali91, Ali92a], Kamath and Karmarkar [KK92, KK93], Helmberg, Rendl, Vanderbei and Wolkowicz [HRVW94].

For a more detailed survey of semidefinite programming in combinatorial optimization, we refer the reader to the recent paper by Alizadeh [Ali95].

3 SEMIDEFINITE PROGRAMMING AND CONTROL THEORY

Semidefinite programming problems arise frequently in control and system theory; Boyd, El Ghaoui, Feron and Balakrishnan catalog many examples in [BEFB94]. We will describe one simple example here.

Consider the *differential inclusion*

$$\frac{dx}{dt} = Ax(t) + Bu(t), \qquad y(t) = Cx(t), \qquad |u_i(t)| \leq |y_i(t)|, \; i = 1, \ldots, p \tag{15.9}$$

where $x(t) \in \mathbf{R}^l$, $u(t) \in \mathbf{R}^p$, and $y(t) \in \mathbf{R}^p$. In the terminology of control theory, this is described as a linear system with uncertain, time-varying, unity-bounded, diagonal feedback.

We seek an invariant ellipsoid, *i.e.*, an ellipsoid $\mathcal{E}$ such that for any x and u that satisfy (15.9), $x(T) \in \mathcal{E}$ implies $x(t) \in \mathcal{E}$ for all $t \geq T$. The existence of such an ellipsoid implies, for example, that all solutions of the differential inclusion (15.9) are bounded.

The ellipsoid $\mathcal{E} = \{x \mid x^T P x \leq 1\}$, where $P = P^T > 0$, is invariant if and only if the function $V(t) = x(t)^T P x(t)$ is nonincreasing for any x and u that satisfy (15.9). In this case we say that V is a quadratic Lyapunov function that proves stability of the differential inclusion (15.9).

We can express the derivative of V as a quadratic form in $x(t)$ and $u(t)$:

$$\frac{d}{dt}V(x(t)) = \begin{bmatrix} x(t) \\ u(t) \end{bmatrix}^T \begin{bmatrix} A^T P + PA & PB \\ B^T P & 0 \end{bmatrix} \begin{bmatrix} x(t) \\ u(t) \end{bmatrix}. \tag{15.10}$$

We can express the conditions $|u_i(t)| \leq |y_i(t)|$ as the quadratic inequalities

$$u_i^2(t) - y_i^2(t) = \begin{bmatrix} x(t) \\ u(t) \end{bmatrix}^T \begin{bmatrix} -c_i^T c_i & 0 \\ 0 & E_{ii} \end{bmatrix} \begin{bmatrix} x(t) \\ u(t) \end{bmatrix} \leq 0, \qquad i = 1, \ldots, p, \tag{15.11}$$

where c_i is the ith row of C, and E_{ii} is the matrix with all entries zero except the ii entry, which is 1.

Putting it all together we find that $\mathcal{E}$ is invariant if and only if (15.10) holds whenever (15.11) holds. Thus the condition is that one quadratic inequality should hold whenever some other quadratic inequalities hold, *i.e.*, :

$$\text{for all } z \in \mathbf{R}^{l+p}, \quad z^T T_i z \leq 0, \ i = 1, \ldots, p \implies z^T T_0 z \leq 0 \tag{15.12}$$

where

$$T_0 = \begin{bmatrix} A^T P + PA & PB \\ B^T P & 0 \end{bmatrix}, \qquad T_i = \begin{bmatrix} -c_i^T c_i & 0 \\ 0 & E_{ii} \end{bmatrix}, \ i = 1, \ldots, p.$$

In the general case, simply verifying that (15.12) holds for a given P is very difficult. But an obvious sufficient condition is

$$\text{there exist } \tau_1 \geq 0, \ldots, \tau_p \geq 0 \text{ such that } T_0 \leq \tau_i T_1 + \cdots + \tau_p T_p. \tag{15.13}$$

Replacing the condition (15.12) with the stronger condition (15.13) is called the $\mathcal{S}$-procedure in the Soviet literature on control theory, and dates back at least to 1944 (see [BEFB94, p.33], [FY79], [LP44]). Note the similarity between Shor's bound (see (15.2) and (15.3)) and the $\mathcal{S}$-procedure ((15.12) and (15.13)). Indeed Shor's bound is readily derived from the $\mathcal{S}$-procedure, and vice versa.

Returning to our example, we apply the $\mathcal{S}$-procedure to obtain a sufficient condition for invariance of the ellipsoid $\mathcal{E}$: for some $D = \mathbf{diag}(\tau_1, \ldots, \tau_p)$,

$$\begin{bmatrix} A^T P + PA + C^T D C & PB \\ B^T P & -D \end{bmatrix} \leq 0. \tag{15.14}$$

This is a linear matrix inequality in the (matrix) variables $P = P^T$ and (diagonal) D. Hence, by solving a semidefinite feasibility problem we can find an invariant ellipsoid (if the problem is feasible). One can also optimize various quantities over the feasible set; see [BEFB94]. Note that (15.14) is really a convex relaxation of the condition that $\mathcal{E}$ be invariant, obtained via the $\mathcal{S}$-procedure.

The close connections between the $\mathcal{S}$-procedure, used in control theory to form semidefinite programming relaxations of hard control problems, and the various semidefinite relaxations used in combinatorial optimization, do not appear to be well known.

4 EXTENSION TO BILINEAR MATRIX INEQUALITIES

We now consider an extension of the SDP (15.1), obtained by replacing the linear matrix inequality constraints by *bilinear* (or bi-affine) matrix inequalities (BMIs),

$$\begin{array}{ll} \text{minimize} & c^T x \\ \text{subject to} & F_0 + \sum_{i=1}^{m} x_i F_i + \sum_{j,k=1}^{m} x_j x_k G_{jk} \geq 0. \end{array} \tag{15.15}$$

The problem data are the vector $c \in \mathbf{R}^m$ and the symmetric matrices F_i, $G_{jk} \in \mathbf{R}^{n\times n}$.

Bilinear matrix inequality problems are NP-hard, and include a wide variety of control problems (see, *e.g.*, , [GLTS94], [GSP94], [SGL94], [GTS+94], [GSL95]). They also include all quadratic problems (when the matrices in (15.15) are diagonal), all polynomial problems, all $\{0,1\}$ and integer programs, etc.

Several heuristic methods for BMI problems have been presented in the literature cited above and reported to be useful in practice. Our purpose here is to point out that the semidefinite relaxations for quadratic problems can be easily extended to bilinear matrix inequalities.

We first express the BMI problem as

$$\begin{array}{ll} \text{minimize} & c^T x \\ \text{subject to} & F_0 + \sum_{i=1}^{m} x_i F_i + \sum_{j,k=1}^{m} w_{jk} G_{jk} \geq 0 \\ & w_{jk} = x_j x_k, \quad j,k = 1,\ldots,m, \end{array}$$

and then relax the second constraint as an LMI

$$\begin{array}{ll} \text{minimize} & c^T x \\ \text{subject to} & F_0 + \sum_{i=1}^{m} x_i F_i + \sum_{j,k=1}^{m} w_{jk} G_{jk} \geq 0 \\ & \begin{bmatrix} W & x \\ x^T & 1 \end{bmatrix} \geq 0. \end{array}$$

This is an SDP in the variables W, x. Its optimal value is a lower bound for the optimal value of problem (15.15). As in Section 2, this SDP can also be interpreted as the dual of the Lagrangian relaxation of problem (15.15).

We should point out that the SDP relaxation may require some manipulation (*e.g.*, , when some of the matrices G_{ii} are zero), just as in the general indefinite quadratic programming case.

We do not yet have any numerical experience with this SDP relaxation of the BMI problem.

5 CONCLUSION

The simultaneous discovery of semidefinite programming applications in control and combinatorial optimization is remarkable and raises several interesting questions. For example,

- can we obtain Goemans and Williamson-type results in control theory, *i.e.*, solve a (polynomial-time) SDP and get a *guaranteed* bound on suboptimality?
- what is the practical performance of semidefinite relaxations in nonconvex quadratic or BMI problems?

REFERENCES

[Ali91] F. Alizadeh. *Combinatorial optimization with interior point methods and semi-definite matrices.* PhD thesis, Univ. of Minnesota, October 1991.

[Ali92a] F. Alizadeh. Combinatorial optimization with semidefinite matrices. In *Proceedings of second annual Integer Programming and Combinatorial Optimization conference.* Carnegie-Mellon University, 1992.

[Ali92b] F. Alizadeh. Optimization over the positive-definite cone: interior point methods and combinatorial applications. In Panos Pardalos, editor, *Advances in Optimization and Parallel Computing.* North-Holland, 1992.

[Ali95] F. Alizadeh. Interior point methods in semidefinite programming with applications to combinatorial optimization. *SIAM Journal on Optimization*, 5(1):13–51, February 1995.

[BEFB94] S. Boyd, L. El Ghaoui, E. Feron, and V. Balakrishnan. *Linear Matrix Inequalities in System and Control Theory*, volume 15 of *Studies in Applied Mathematics.* SIAM, Philadelphia, PA, June 1994.

[DH73] W. E. Donath and A. J. Hoffman. Lower bounds for the partitioning of graphs. *IBM Journal of Research and Development*, 17:420–425, 1973.

[FY79] A. L. Fradkov and V. A. Yakubovich. The $\mathcal{S}$-procedure and duality relations in nonconvex problems of quadratic programming. *Vestnik Leningrad Univ. Math.*, 6(1):101–109, 1979. In Russian, 1973.

[GLS88] M. Grötschel, L. Lovász, and A. Schrijver. *Geometric Algorithms and Combinatorial Optimization*, volume 2 of *Algorithms and Combinatorics.* Springer-Verlag, 1988.

[GLTS94] K. C. Goh, J. H. Ly, L. Turand, and M. G. Safonov. μ/k_m-synthesis via bilinear matrix inequalities. In *Proceedings of the 33rd IEEE Conference on Decision and Control*, December 1994.

[GSL95] K. C. Goh, M. G. Safonov, and J. H. Ly. Robust synthesis via bilinear matrix inequalities. 1995. Submitted to the *International Journal of Robust and Nonlinear Control.*

[GSP94] K. C. Goh, M. G. Safonov, and G. P. Papavassilopoulos. A global optimization approach for the BMI problem. In *Proceedings of the 33rd IEEE Conference on Decision and Control*, December 1994.

[GTS+94] K. C. Goh, L. Turan, M. G. Safonov, G. P. Papavassilopoulos, and J. H. Ly. Biaffine matrix inequality properties and computational methods. In *Proc. American Control Conf.*, pages 850–855, 1994.

[GW94] M. Goemans and D. Williamson. .878-approximation algorithm for max-cut and max-2sat. Technical report, MIT, 1994.

[GW95] M. Goemans and D. Williamson. Improved approximation algorithms for maximum cut and satisfiability problems using semidefinite programming. Technical report, Department of Mathematics, MIT, 1995.

[HRVW94] C. Helmberg, F. Rendl, R. J. Vanderbei, and H. Wolkowicz. An interior-point method for semidefinite programming. Manuscript. Program in Statistics and Operations Research, Princeton University, 1994.

[KK92] A. Kamath and N. Karmarkar. A continuous approach to compute upper bounds in quadratic maximization problems with integer constraints. In C. Floudas and P. Pardalos, editors, *Recent Advances in Global Optimization*, pages 125–140. Princeton University Press, Oxford, 1992.

[KK93] A. Kamath and N. Karmarkar. An $O(nL)$ iteration algorithm for computing bounds in quadratic optimization problems. In P. M. Pardalos, editor, *Complexity in Numerical Optimization*, pages 254–268. World Scientific Publishing Co., 1993.

[KMS94] D. Karger, R. Motwani, and M. Sudan. Approximate graph coloring by semidefinite programming. Technical report, Department of Computer Science, Stanford University, 1994.

[LP44] A. I. Lur'e and V. N. Postnikov. On the theory of stability of control systems. *Applied mathematics and mechanics*, 8(3), 1944. In Russian.

[LS91] L. Lovász and A. Schrijver. Cones of matrices and set-functions and 0-1 optimization. *SIAM J. on Optimization*, 1(2):166–190, 1991.

[NN94] Yu. Nesterov and A. Nemirovsky. *Interior-point polynomial methods in convex programming*, volume 13 of *Studies in Applied Mathematics*. SIAM, Philadelphia, PA, 1994.

[PRW94] S. Poljak, F. Rendl, and H. Wolkowicz. A recipe for best semidefinite relaxation for $(0, 1)$-quadratic programming. Technical Report CORR 94-7, University of Waterloo, 1994.

[SGL94] M. G. Safonov, K. C. Goh, and J. H. Ly. Control system synthesis via bilinear matrix inequalities. In *Proc. American Control Conf.*, pages 45–49, 1994.

[Sho87] N. Z. Shor. Quadratic optimization problems. *Soviet Journal of Circuits and Systems Sciences*, 25(6):1–11, 1987.

[VB96] L. Vandenberghe and S. Boyd. Semidefinite programming. *SIAM Review*, 38(1):49-95, 1996.

16

COOPERATIVE CLEANERS : A STUDY IN ANT ROBOTICS

Israel A. Wagner and Alfred M. Bruckstein

Computer Science Department
Technion - Israel Institute of Technology
Haifa, Israel 32000

To Tom Kailath, teacher and friend, with appreciation and best wishes for the next sixty great years!

ABSTRACT

In the world of living creatures, "simple minded" animals often cooperate to achieve common goals with amazing performance. One can consider this idea in the context of robotics, and suggest models for programming goal-oriented behavior into the members of a group of simple robots lacking global supervision. This can be done by controlling the local interactions between the robot agents, to have them jointly carry out a given mission. As a test case we analyze the problem of many simple robots cooperating to clean the dirty floor of a non-convex region in $\mathbf{Z}^2$, using the dirt on the floor as the main means of inter-robot communication.

1 INTRODUCTION

In the world of living creatures, "simple minded" animals like ants or birds cooperate to achieve common goals with surprising performance. It seems that these animals are "programmed" to interact locally in such a way that the desired global behavior is likely to emerge even if some individuals of the colony die or fail to carry out their task for some other reasons. It is suggested to consider a similar approach to coordinate a group of robots without a central supervisor, by using only local interactions between the robots. When this, decentralized approach is used, much of the communication overhead (characteristic to centralized systems) is saved, the hardware of the robots can be fairly simple, and better modularity is achieved. A properly designed system should achieve reliability through redundancy. Significant research effort has been invested during the last few years in design and simulation of multi-agent systems (e.g. [20], [15], [4], [19]). Unfortunately the geometrical theory of such multi-agent systems is far from being satisfactory, as has been pointed out in [2] and many other papers.

Our interest is focused on developing the mathematical tools necessary for the design and analysis of such systems. In a recent report [21] we have shown that a number of agents can arrange themselves equidistantly in a row via a sequence of linear adjustments, based on a simple "local" interaction. The convergence of the configuration to the desired one is exponentially fast. A different way of cooperation between agents is inspired by the behavior of ant-colonies, and is described in [6]. There it was proved that a sequence of ants engaged in deterministic chain pursuit will find the shortest (i.e. straight) path from the ant-hill to the food source, using only local interactions. In [7] we investigate the behavior of a group of agents on $\mathbf{Z}^2$, where each ant-like agent is pursuing his predecessor, according to a discrete biased-random-walk model of pursuit on the integer grid. We proved there that the average paths of such a sequence of a(ge)nts engaged in a chain of probabilistic pursuit converge to the straight line between the origin and destination, and this happens exponentially fast.

In this paper we investigate a more complicated question concerning the behavior of many agents cooperating to explore an unknown area (for purposes of cleaning, painting, etc.), where each robot/agent can only see his near neighborhood, and the only way of inter-robot communication is by leaving traces on the common ground and sensing the traces left by other robots. We present an algorithm for cleaning a dirty area that guarantees task completion (unless *all* robots die) and prove an upper bound on the time complexity of this algorithm. We also show simulation results of the algorithm on several types of regions. These simulations indicate that the precise time depends on the number of robots, their initial locations, and the shape of the region.

In the spirit of [3], we consider simple robots with only a bounded amount of memory (i.e. a *finite-state-machine*). Generalizing an idea from computer graphics, [10], we preserve the connectivity of the "dirty" region by allowing an agent to clean only so called *non-critical* points, points that do not disconnect the graph of dirty grid points. This ensures that the robots will stop only upon completing their mission. An important advantage of this approach, in addition to the simplicity of the agents, is fault-tolerance: even if almost all the agents cease to work before completion, the remaining ones will eventually complete the mission. We prove the correctness of the algorithm as well as an upper bound on its running time, and show how our algorithm can be extended for regions with obstacles.

The rest of the paper is organized as follows: in Section 2 the formal problem is defined as well as the aims and assumptions involved. The algorithm is presented in Section 3, where its time-complexity is analyzed, and some simulation examples are shown. Section 4 is devoted to the problem of exploring/cleaning regions with obstacles, and is followed by some examples from our simulation runs. We conclude the paper with a discussion and by pointing out several connections to related work.

2 COOPERATIVE CLEANING: PROBLEM STATEMENT

The problem we address is the following: we are given a region R_0 to be explored, which is assumed to be a connected subset of the integer grid, $\mathbf{Z}^2$, and a set of k robots initially located at $\mathbf{r}(0) = (r_1(0), r_2(0), \ldots, r_k(0))$. We want these robots to clean R_0, i.e. initially all points of R_0 are assumed to be dirty ("off",0) and must be turned clean ("on",1). The system should evolve according to a rule that can be described as

$$(\mathbf{R}(t+1), \mathbf{r}(t+1)) = f(\mathbf{R}(t), \mathbf{r}(t))$$

where $\mathbf{R}(t)$ is the *status map* of R_0 at time t - indicating the region still to be cleaned by 0's and designating those points that have already been cleaned by 1. Initially all bits are set to indicate the status of all points in R_0 as 0.

We seek for a rule f that will fulfill the following aims

1. **Full coverage:** each point in R_0 is eventually cleaned.
2. **Agreement on Completion:** After the whole region has been cleaned, there should eventually be a time when all the robots that are still active become *aware* of this fact (and hence can go and take a nap or proceed to the next mission).
3. **Efficiency:** in time and space (memory usage of each robot).
4. **Fault Tolerance:** if some of the robots die and hence stop working before completion, the others should be able to carry on and complete the mission.

We make the following assumptions

1. A robot can move one grid unit at a time, from a grid point to an adjacent one.
2. Each point in R_0 represents a room and several robots may occupy the same room at the same time.
3. There is no prior knowledge of the shape of R_0 except that it is assumed to be simply connected [1].
4. There is no central control; it is a 'de-centralized' system with no leader (i.e. all agents/robots are identical).

[1] We shall later show how to deal with non-simple regions too.

5. A robot can see and sense the status of his 8-neighborhood. (i.e. the 3×3 square around its current location).

3 A MULTI-AGENT CLEANING ALGORITHM

To solve the cooperative cleaning problem, we suggest the protocol **CLEAN** (see Figure 16.1), to be programmed into and executed by each robot at every (discrete) point of time.

Each robot has, by assumption, a bounded amount of memory (i.e. it is a finite automaton). The connectivity of the region still to be cleaned, R, is preserved by allowing a robot to clean only *non-critical* points, i.e. points that do not disconnect the current graph of dirty grid points. This way, it is also guaranteed that the robots will stop only upon completing their mission. An important advantage of this approach, in addition to the simplicity of the agents, is fault-tolerance: even if almost all the agents cease to work before completion, the remaining agents will take over their responsibilities and complete the mission. The "pivot" is a special point on the boundary of R, ∂R (∂R is defined as the set $\{(u,v) \,|(u,v) \in R \text{ and } (u,v) \text{ has an 8-neighbor in } \mathbf{Z}^2 \setminus R\}$), and all robots are assumed to be initially located there. This point (denoted p_0) is artificially set to be critical during the execution, hence it is also guaranteed to be the last point cleaned. Completion of the mission can therefore be concluded from the fact that all (working) robots are back at p_0 with no dirty neighbors to go to, thereby reporting on completion of their individual missions. Note that the pivot is not necessary for the algorithm to work well - it just makes life easier for the user since one then knows where to find the robots after cleaning has been completed. If we do not start with all robots at the pivot and force p_0 to be critical, the location of the robots upon completion will generally not be known in advance. We denote the initial dirty region by R_0 and the remaining dirty region by R; our aim is to finish with $R = \phi$.

The introduction of the notion of critical points makes time-complexity analysis significantly harder since a critical point may be visited several times before its cleaning. We conjecture that the algorithm proposed is efficient, and additional robots will speed up the cleaning process only up to a limit. In order to give this argument a rigorous form, we need some definitions. We denote the area of R by $a(R)$. In the "grid" context, this area was defined as the number of grid points in R. Actually, R defines a dichotomy of $\mathbf{Z}^2$ into R and $\overline{R} = \mathbf{Z}^2 \setminus R$. The *border* of R, denoted ∂R, was defined as the set of points in R which have 8-neighbors in $\overline{R}$. We define the circumference of R, $c(R)$, to be the area of ∂R. A *path* in R will be defined as a sequence $(p_1, p_2, \ldots, p_m)$ of points in R such that any two consecutive points are 4-connected in R. (Two points are 4-connected if the Manhattan distance between them is one). We now formalize a notion of R's "fatness" - i.e. its maximal width - as following. Consider $u \in R$, a point in R. A *string* of u is a simple path in R that starts at u and ends at a non-critical border point. The *depth* of u, denoted by $w(R, u)$, will be defined as the shortest string of u (unless u is itself critical - then its depth will be defined to be zero). The *length* of a string is defined as the number of points in it. The *total width* of R, denoted $w(R)$, will be defined as the maximal depth among all points of R: $w(R) = \max_{u \in R}\{w(R, u)\}$. In the **CLEAN** algorithm, the connectivity of the dirty shape is always preserved, since a point is never erased as long as it is a critical one. We shall assume that some (arbitrary) point p_0 on the boundary of R is a starting point for our agents, and we shall call it a *pivot*. The longest in-region distance between p_0 and any other point in R will be referred to as $l(R)$, the *length* of R:

$$l(R) = \max_{v \in R}\{d_R(p_0, v)\}$$

where $d_R(x, y)$ is the distance (shortest path within R) between x and y. See Figure 16.3 for an illustration of the above definitions.

With these concepts defined, we have the following results.

Lemma 16.1 *If $a(R) > 0$ then $w(R) \geq 1$, that is: any finite simple region has at least two non-critical points on its boundary.*

Proof: The boundary ∂R is a connected graph, hence it has a spanning tree. In such a tree there are at least two vertices with degree 1; these vertices are necessarily not critical in R, since any boundary point which is critical in R is also critical in ∂R. □

Theorem 16.1 *A group of robots executing the* **CLEAN** *protocol will eventually clean a simply connected region and stop together at p_0, the pivot set by the "INITIALIZE" phase of the algorithm.*

Proof: While R has not yet been cleaned, $a(R) > 0$ and hence, by Lemma 16.1 there is a non-critical point on ∂R. Since the robots obey the **CLEAN** rule, at least one of them will arrive to a non-critical point once in a period of $a(R)$ (since $c(R) \leq a(R)$), and erase this point from R, thus reducing $a(R)$ by 1. We conclude that after no more than $a^2(R)$ units of time we shall have $a(R) = 0$. The fact that all robots will meet at the same point is implied by the following two rules that are implemented in the **CLEAN** algorithm:

- **rule 1:** R is always being kept connected. (We never clean a room that has no clean neighbor and never clean a critical room either).
- **rule 2:** The pivot p_0 is cleaned only when no other dirty points are left in R.

This completes the proof of Theorem 16.1. □

Eventually, as the agents progress in their job, each point becomes non-critical and is cleaned. The point last cleaned is the pivot p_0, since we (artificially) keep it critical until all other points have been cleaned. Let us denote the j'th robot time of its i'th visit to p_0 by τ_j^i. The i'th *tour* of the boundary by robot j, denoted T_j^i, is the path traversed by robot j between two consecutive visits at p_0, namely:

$$T_j^i = [r_j(\tau_j^i), r_j(\tau_j^i + 1), r_j(\tau_j^i + 2), \ldots, r_j(\tau_j^{i+1})].$$

The path of robot j can then be decomposed into a series of tours:

$$\text{path}(j) = T_j^1, T_j^2, \ldots, T_j^M$$

where $r_j(t)$ is the location of j at time t, τ_j^i is the start-time of T_j^i, the i'th tour of the boundary by r_j, and M is the total number of tours. In the **CLEAN** algorithm, the inter-robot order never changes; that is - there is exactly one visit of j at p_0 between two consecutive visits of i at this point, for any i, j such that $1 \leq i, j \leq k$. Note that due to Lemma 16.1, and the order-preservation property of the **CLEAN** algorithm, the total number of tours is the same for all robots.

Hence we can order the tours by the order they visit the pivot p_0:

$$T_1^1, T_2^1, \ldots, T_k^1, T_1^2, T_2^2, \ldots, T_k^M$$

The next Lemma states that tour-cardinalities are non-increasing.

Lemma 16.2 *The cardinality of the set of points visited during a tour cannot be larger than the preceding one, namely*

$$v(T_{j+1}^m) \leq v(T_j^m) - 8$$

where $v(T)$ denotes the number of vertices in a tour T.

Proof: First we note that each tour is a simple, closed, rectilinear polygon. T_{j+1}^m, the m'th tour of robot r_j is created by tracking the previous tour T_j^m, while deleting any non-critical points along the way. Going along such a polygon, we either go straight, turn right or turn left. For the sake of simplicity, let us denote T_{j+1}^m by T' and T_j^m by T. It is easy to see (see Figure 16.4(a),(b)) that a right turn increases the tour-length by two and a left turn decreases it by two. (Going straight has no effect). But the tour is a simple rectilinear polygon, hence it always has four "left" turns more than "right" turns (This is a simple consequence of the "rotation index" Theorem (see, e.g. [8] pp. 396): If $\alpha : [0, 1] \to R^2$ is a plane, regular, simple, closed curve then $\int_0^1 k(s)ds = 2\pi$, where $k(s)$ is the curvature of $\alpha(s)$ and the curve is traversed in the positive direction (i.e. with the inside to the left of the walker)). The only exception occurs when one or more points along the tour are critical (see Figure 16.4(c)). In such a case, the critical points are just repeated in $V(T')$ but do not increase the overall size of the set. □

Note that the length of the tour may increase but its cardinality will never exceed that of the first one, namely $|\partial R|$.

Next we establish a link between the cardinality of a tour and its length:

Lemma 16.3 *The length of a tour never exceeds four times its cardinality, that is:*

$$c(T) \leq 4v(T) \tag{16.1}$$

where $c(T)$ is the length of tour T and $v(T)$ is the number of points in it.

Proof: The tour $T = (V, E)$ is a directed cyclic path traversing the vertices of $V = V(\partial R)$, the border of R, along the edges $E = E(\partial R)$ on the boundary. Observing that the maximal degree in R (and hence in ∂R) is 4, and that no edge in T is traversed more than twice, one can easily see that

$$c(T) \leq 2\,|E(T)| \leq \sum_{u \in V} d(u) \leq 4v(T)$$

□

The conclusion from Lemma 16.2 and Lemma 16.3 is that a tour length never exceeds twice the original circumference of the region. We next proceed to show that the total number of tours is bounded above. Let us define $w(T_j^m)$, the width of the m'th tour of robot j, as the width of the region surrounded by this tour. Then we have

Lemma 16.4 *The width of robots' tours is monotonically decreasing, that is:*

$$\forall j, m \quad 1 \leq j \leq k, \;\; 1 \leq m \leq M: \quad w(T_j^m) \leq \max\left\{0, w(T_{j-1}^m) - 1\right\}$$

Proof: By definition, $w(T_i^j)$ is the longest distance from an internal point of T_j^m to a non-critical point on its boundary. But, according to the **CLEAN**ing rule, robot j has cleaned all non-critical points from the boundary of T_{j-1}^m - hence its width has decreased by one, the only exception being when T_{j-1}^m is already equal to one. □

From Lemma 16.4 it clearly results that the number of tours, M, is bounded:

Corollary 16.1

$$M \leq \frac{w(R_0)}{k} + 1$$

Proof: According to Lemma 16.4 there are at most $\frac{w}{k}$ tours before the width reduces to 1. Once $w = 1$ all that remains from R_0 is a skeleton that can clearly be cleaned in a single tour. □

A relation between the length and duration of each tour is stated next.

Lemma 16.5 *The time it takes robot j to make its i'th tour is bounded above by 1.5 times the geometric length of the tour, namely:*

$$\tau_j^{i+1} - \tau_j^i = 1.5 \cdot c(T_j^i)$$

Proof: With the **CLEAN** procedure the robot follows its tour step-by-step, the only exception being when more than one robot enter the same point at the same time, going in the same direction. But, this cannot happen with more than 4 robots (since any point has no more than 4 neighbors), and, such collisions are resolved by releasing the 4 robots at 4 consecutive time points. The times that the four robots have to wait in this location are $0, 1, 2$ and 3 units of time, respectively. Hence the ratio of the "efficient" time to the total time is $\frac{1+2+3}{9} = \frac{2}{3}$, so the total touring-time of a robot cannot exceed 1.5 times the length of the tour. □

We are now ready to prove the main timing theorem:

Theorem 16.2 *Assume that k robots start at some boundary point p_0 of a simple connected region R_0 and work according to the* **CLEAN** *algorithm, and denote by t_k the time needed for this group to clean R_0. Then it holds that:*

$$\max\left\{2k, \left\lceil\frac{a}{k}\right\rceil, 2l\right\} \leq t_k \leq 2k + 6c\left(\frac{w}{k} + 1\right) \tag{16.2}$$

where a,c,l and w denote the area, circumference, length and width of of R_0, respectively.

Proof :

1. The lower bound is quite obvious - the left term $2k$ is the time needed to release the k robots from the pivot, $\frac{a}{k}$ is a lower bound on the time necessary to cover the region if the robots were optimally located at the beginning. The right term, $2l$, comes from the observation that at least one robot should visit (and return from) any point of R_0, including the one farthest from p_0, to which the distance is l.

2. According to Lemma 16.2 all robots stop simultaneously at p_0 after completion of M tours. Hence, we can estimate the total stopping time as the stopping time for any of the robots, say robot 1:

$$t_k = \tau_1^M = \tau_1^1 + \sum_{m=1}^{M} (\tau_1^{m+1} - \tau_1^m)$$

by Lemma 16.4, Corollary 16.1 and Lemma 16.5, we get:

$$t_k \leq M \cdot \max_{m=1,\ldots,M} \{c(T_1^m)\} \leq 6 \cdot \left(\frac{w(R)}{k} + 1\right) \cdot c(R) = 6\left(\frac{wc}{k} + c\right)$$

3. The additional term of $2k$ stands for the time of release - since all robots are initially concentrated in one point and they keep a separation of two time units between them, we need $2k$ units of time before all k robots become operational.

□

Using the above theorem we can bound the *speedup ratio*, defined as

$$t_k(R_0)/a(R_0),$$

which expresses the benefit of using k robots for a cleaning mission:

Corollary 16.2

$$\max\left\{\frac{2k}{a}, \frac{1}{k}, \frac{2l}{a}\right\} \leq \frac{t_k}{a} \leq \frac{2k}{a} + 6\left(\frac{wc}{ka} + \frac{c}{a}\right) \tag{16.3}$$

where a,w,c,l and k are as in Theorem 16.1.

An interesting result of Corollary 16.2 is that when $a >> k >> w$, i.e. the number of robots is large relative to the width but small compared to the area, then the speedup is bounded below by twice the ratio of the length and area, and bounded above by $6\frac{c}{a}$, three times the ratio of the area and circumference of R_0. Note here the similarity to the ratio $\frac{c}{\sqrt{a}}$, known as the *shape-factor.*

Another conclusion is that when we scale up the region by a factor of n, the area increases as n^2 but the width, length and circumference all increase as n so we get

Corollary 16.3

$$\text{as } n \to \infty, \qquad L_\infty \leq \frac{t_k}{a} \leq U_\infty$$

where

$$L_\infty = \frac{1}{k}, \qquad U_\infty = 3\frac{c_0 w_0}{k a_0}$$

and $a = a_0 n^2, c = c_0 n, w = w_0 n$ *are the scaled area, circumference and width, respectively.*

4 REGIONS WITH OBSTACLES

So far we have only dealt with simply connected regions, i.e. - regions with no "holes". In the case of a (connected) region with obstacles (i.e. holes) the simple **CLEAN**ing algorithm will not work, due to the following "cycle" problem: eventually, each obstacle will be surrounded by critical points, and there will be a time when *all* boundary points of R will be critical, contrary to the statement of Lemma 16.1. (We shall call such a situation *useless*, as opposed to the *useful* state when some points are cleaned during a tour). As a cure to this problem, we suggest to add an "odoring" feature to our cleaners; that is, a robot will be able to label a point on the floor by a small amount of "perfume". (This action may remind one of the pheromones left by ants during their walks). These labeled points will designate the external boundary of the dirty region. Upon getting to the useless state (detected by each robot due to no cleaning between two consecutive visits to the pivot) a robot will continue to traverse the boundary, but will now look for a point which has a "mixed" boundary - that is, one that has odor on one side but no odor on the other. The robot will clean this point (despite its "criticality" - it is not really critical since it is necessarily part of a cycle around an obstacle) and then continue as in a useful state. This will open one cycle, hence, if there are s obstacles, we will need s/k such tours before the region is completely clean. (See Figure 16.2 for the modified algorithm). On the other hand, Lemma 16.2 no longer holds (see Figure 16.6) since the boundary area can increase with time. However the boundary is always bounded above by the area. We now make the following

Conjecture 16.1 *Assume that k robots start at some boundary point of a non-simple connected region R with s obstacles in it, and work according to the* **CLEAN-WITH-OBSTACLES** *algorithm, and denote by t_k the time needed for this group to clean R. Then it holds that:*

$$\max\left\{2k, \left\lceil \frac{a}{k} \right\rceil, 2l\right\} \le t_k \le 2k + 1.5 \cdot a \left(\frac{w}{k} + \left\lceil \frac{s}{k} \right\rceil\right) \qquad (16.4)$$

where a,c,l and w denote the free area, circumference, length and width of of R, respectively, and s is the number of obstacles in R.

5 SIMULATION EXAMPLES

A simple motion rule is the core of the cleaning algorithm, as shown in Figures 16.1 and 16.2. Each robot checks if his location is critical. If not - he cleans this location. Then, if the robot is the only one at the current point, it looks around and goes to the rightmost free grid point that is closer than others to the border of the region. If a robot with higher priority occupies the same location, the robot waits.

We ran the algorithm on several shapes of regions and for numbers of robots varying from 1 to 20. See Figure 16.5 - 16.6 for some examples of the evolution of the layout with time. The gray level of each pixel designates the index of the robot that actually cleaned this point. The right side of Figure 16.5 shows the same region and number of robots as in the left side of this Figure, but with randomly chosen initial locations at the corners. It can be seen that the dirty region is cleaned in a similar way. It should be said here that all the theory we developed in the previous sections (up to a small additive constant) applies to the case where the robots are initially located in randomly selected points on the boundary of R_0 (rather than starting from p_0). Figure 16.6 shows the evolution of the **CLEAN-WITH-OBSTACLES** algorithm for the same shapes with four additional obstacles in each, with 2 robots (left) and 10 robots (right).

Figure 16.7 summarizes the timing results of many simulations, plotting the time (normalized by area) vs. number of robots, compared to the theoretical bounds. In Figure 16.8 we show the results for the same figures with additional obstacles, together with the conjectured theoretical bounds.

6 RELATED WORK AND DISCUSSION

Our cooperative **CLEAN**ing algorithm can be considered as a case of social behavior in the sense of [17], where one induces multi-agent cooperation by forcing the agents to obey some simple "social" guidelines. This raises the question what happens if one robot malfunctions. We have shown that if less than k robots stop, the others will take over their responsibilities. But what if some robots start to cheat ? Such adversaries will have catastrophic consequences, since a crazy robot may clean a critical point and disconnect the dirty region.

Another question of interest is the resolution of collisions between robots. In the **CLEAN** algorithm we resolve such a problem by giving each robot a priority measure depending on his previous location. But it is an interesting open question whether a coin flipping is better here.

The cleaning problem discussed is related to the geometric problem of pocket-machining, see [9] for details. An interesting problem of cleaning and maintenance of a system of pipes by an autonomous robot is discussed in [14]. The importance of cleaning hazardous waste by robots is described in [11].

Our approach is that cleaning is always done at the boundary. It is possible that a better efficiency will be achieved using other approaches:

1. Given that several neighbors are dirty, visit the non-critical ones first (even if not on the boundary). This approach is quite efficient for one robot, but can be a mess for several robots.

2. Once entering a large "room" (that is - upon passing from a critical area to a non-critical one) - designate the entrance by a special type of token,

so that other robots will enter only in case there is no other work to do. This approach guarantees that the robots will be distributed between the large rooms of the R-configuration. This is attractive if the region has such rooms of quite similar areas.

3. Quite a different idea is to divide the work into two phases - the first one of "distribution" - the robots locate themselves uniformly around the area. Then, in the second phase, each robot cleans around his "center". If the distribution is appropriate, there will be minimum of interactions between robots in the second phase.

We use the dirt on the floor as a means of inter-robot communication, but other ways for communication between agents have been suggested. One is to use heat trails for this end, as was reported in [16]. In [20], self-organization is achieved among a group of lunar robots that have to explore an unknown region, and bring special rock-samples back to the mother-spaceship, by programming each robot to drop a crumb at each point he visits and walk around at random with a bias toward the negative gradient of crumb concentration.

Another question of interest is how to guarantee covering, at least with high probability, without using any external signs, using only the inter-robot collisions as an indicator for a good direction to proceed.

It is of interest to notice here that an off-line version of the problem, that is: finding the shortest path that visits all grid points in R, where R is completely known in advance, is NP-hard even for a single robot. It is a corollary of the fact that Hamilton path in a non-simple grid-graph is NP-complete [12].

In summary, we would like to cite a statement made by a scientist after watching an ant making his laborious way across a wind-and-wave-molded beach[18]:

> An ant, viewed as a behaving system, is quite simple. The apparent complexity of its behavior over time is largely a reflection of the environment in which it finds itself.

Such a point of view, as well as the results of our simulations and analysis, make us believe that even simple, ant-like creatures, yield very interesting, adaptive and quite efficient goal oriented behavior.

*

```
Procedure INITIALIZE (r_1, r_2,...,r_k,p_0):
I.1) Choose a pivot p_0 ∈ ∂R_0;
I.2) Locate the robots at p_0;
I.3) For(t = 1;  t ≤ k;  t++)
    I.3.1) Let robot r_t start working at p_0
           according to the CLEAN protocol;
    I.3.2) Wait 2 units of time;
end INITIALIZE.
```

```
Protocol CLEAN(r,x,y):
A) if not(is-critical(x,y)) then   /* can I clean my current location ? */
           Set R(x,y) to 1;
B) if (x,y) has no dirty neighbors then STOP.
C) if [there are no other robots at (x,y)] or      /* plan the next move */
    [priority(r) is higher than the priorities
           of the other robots at (x,y)]
                then go to the rightmost neighbor of (x,y) on ∂R.
end CLEAN.
```

```
Function is-critical(x,y):                            /* criticality test */
if [(x,y) has two dirty 4-neighbors which are not connected
    via the 8-neighborhood of (x,y) (excluding (x,y) itself)]
    or [(x,y) = p_0]
           then return(TRUE);
                else return(FALSE);
end is-critical.
```

```
Function priority(r):                               /* priority measure */
(x_0, y_0) := r's previous location;
(x_1, y_1) := r's current location;
    return(2 · (x_0 − x_1) + (y_0 − y_1));
end priority.
```

Figure 16.1 The **CLEAN** protocol for robot r currently at (x, y), and its sub-functions. In step C, *rightmost* means: starting from the previous boundary point sweep the neighbors of (x, y) in a clockwise order until you find another boundary point

---

Procedure **INITIALIZE** ($\mathbf{r}_1$, $\mathbf{r}_2$,. . .,$\mathbf{r}_k$,$\mathbf{p}_0$):
I.1) Choose a pivot $p_0 \in \partial R_0$;
I.2) Locate the robots at p_0;
I.3) For each r, status(r):=**useful**;
I.4) For($t = 1$; $t \leq k$; $t++$)
    I.4.1) Let robot r_t start working at p_0
        according to the **CLEAN-WITH-OBSTACLES** protocol;
    I.4.2) Wait 2 units of time;
end **INITIALIZE**.

Protocol **CLEAN-WITH-OBSTACLES**(**r**,**x**,**y**):
A) if not(**is-critical**(x, y)) /* can I clean my current location ? */
    or [status(r)=**useless** and
        there is odor on only one side of (x, y)]
            then
                Set $R(x, y)$ to 1;
                Set odor(x, y) to 1;
B) if (x, y) has no dirty neighbors then STOP.
C) if [there are no other robots at (x, y)] or /* plan the next move */
    [**priority**(r) is higher than the priorities
        of the other robots at (x, y)]
            then go to the rightmost neighbor of (x, y) on ∂R.
D) if [(x, y) $= p_0$ and no point has been cleaned by r
    since previous visit to p_0]
        then status(r):=**useless**;
end **CLEAN-WITH-OBSTACLES**.

Figure 16.2 The **CLEAN-WITH-OBSTACLES** protocol. Note the use of *odor* to identify cycles. The **priority** and **is-critical** functions are the same as in **CLEAN** algorithm.

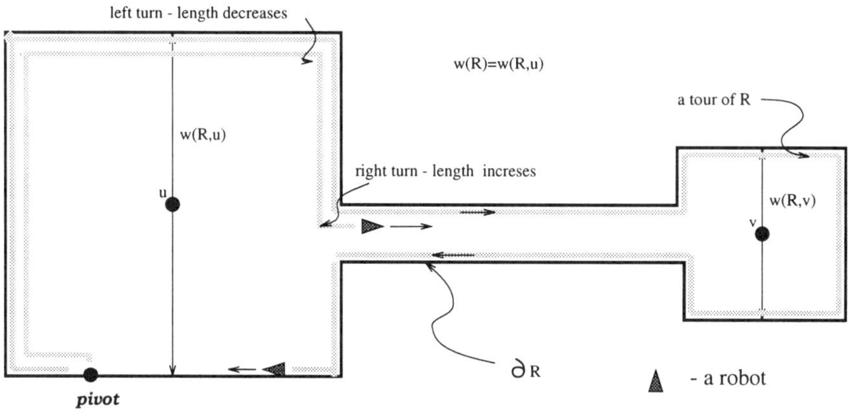

Figure 16.3 An illustration of the definitions involved in Theorem 16.1.

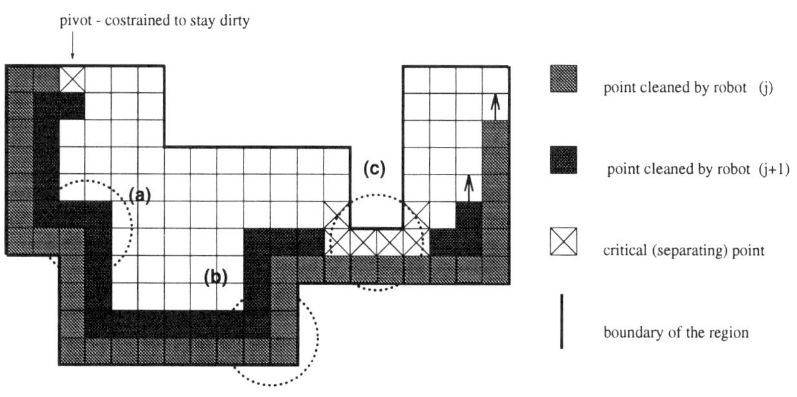

Figure 16.4 The effect of corners on tour-size may be either an increase (a,c) or decrease (b)

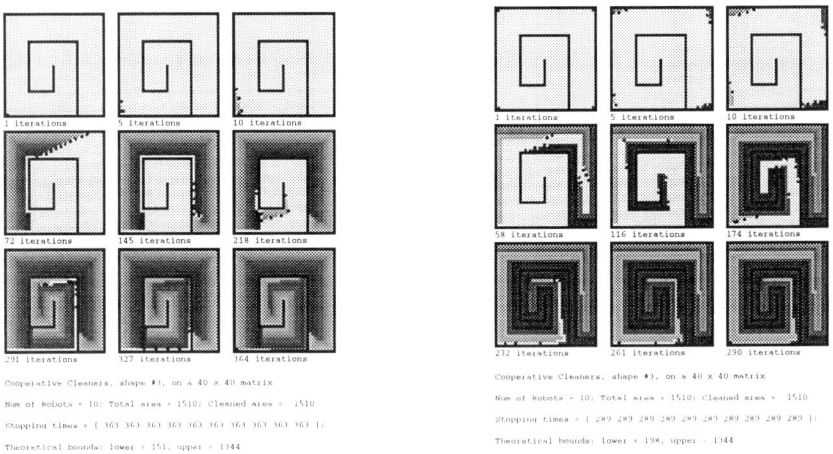

Figure 16.5 Cooperative cleaners: maze, 10 robots (left), and 10 robots with random initial locations on the boundary (right).

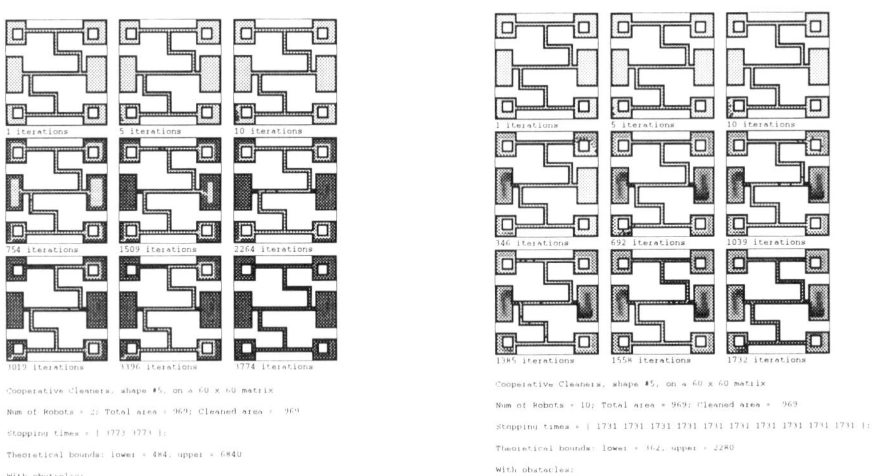

Figure 16.6 Cooperative cleaners: 2 robots, 6 rooms with obstacles (left), 10 robots, 6 rooms with obstacles (right).

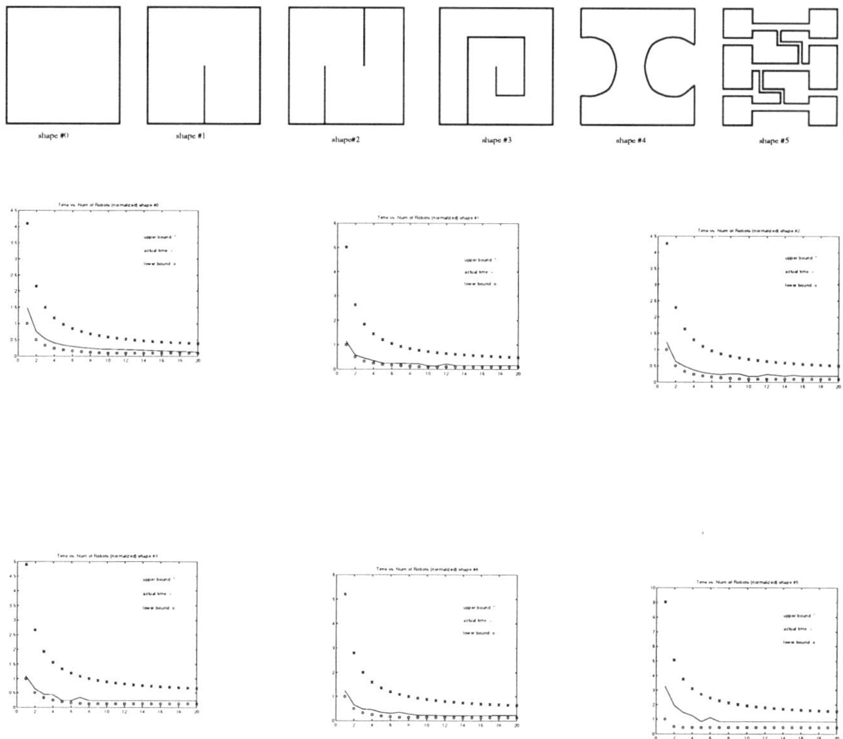

Figure 16.7 Various shapes tested in **CLEAN** simulations and the normalized time $\left(\frac{t_k}{a}\right)$ vs. number of robots (k) for each shape, together with the lower and upper bounds according to Theorem 16.2

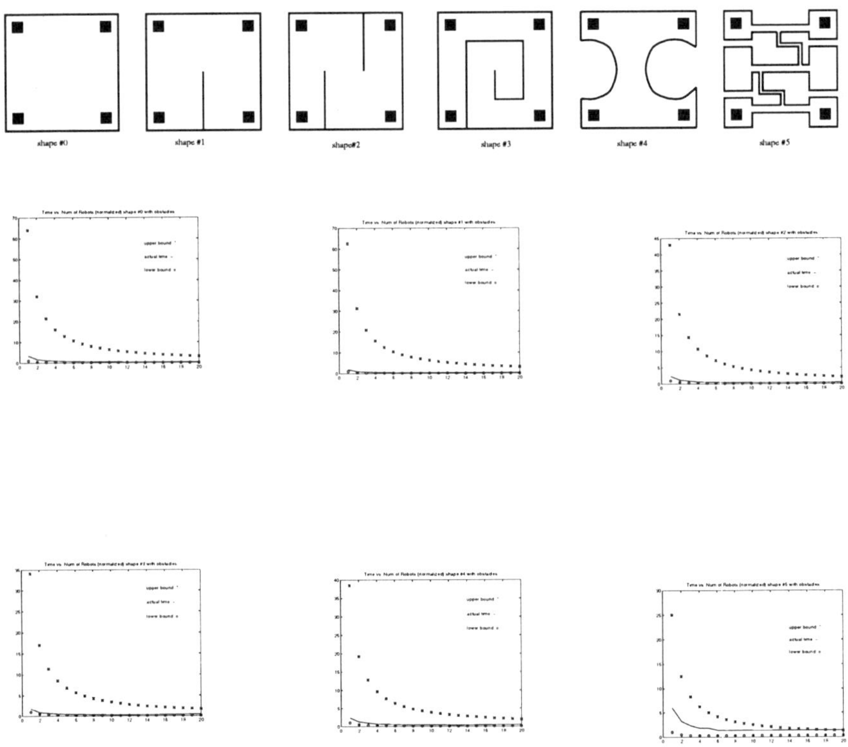

Figure 16.8 Various shapes with obstacles, tested in **CLEAN-WITH-OBSTACLES** simulations and the normalized time $\left(\frac{t_k}{a}\right)$ vs. number of robots (k) for each shape, together with the lower and upper bounds according to Theorem 2

REFERENCES

[1] B. Bollobas, *Graph Theory - an Introductory Course,* Springer-Verlag, 1990.

[2] G. Beni, J. Wang, "Theoretical problems for the realization of distributed robotic systems," *Proc. of the 1991 IEEE Internl. Conference on Robotics and Automation*, pp. 1914-1920, Sacramento, California, April 1991.

[3] V. Braitenberg, *Vehicles,* MIT Press 1984.

[4] R.A. Brooks, "Elephants Don't Play Chess," in *Designing Autonomous Agents,* P. Maes (Ed.), pp. 3-15, MIT Press/Elsevier, 1990.

[5] R.A. Brooks, "Intelligence without representation," *Artificial Intelligence,* 47 (1991) 139-159, Elsevier.

[6] A.M. Bruckstein, "Why the Ant Trails Look So Straight and Nice ," *The Mathematical Intelligencer,* vol. 15, No. 2, pp. 59-62, 1993.

[7] A.M. Bruckstein, C.L. Mallows and I.A. Wagner, "Probabilistic Pursuits on the Integer Grid," *Technical report CIS-9411, Center for Intelligent Systems,* Technion, Haifa, September 1994. Submitted to *SIAM Review.*

[8] M.P. Do-Carmo, *Differential Geometry of Curves and Surfaces,* Prentice-Hall, New-Jersey, 1976.

[9] M. Held, *On the Computational Geometry of Pocket Machining,* Lecture Notes in Computer Science,Springer-Verlag 1991.

[10] D. Henrich, "Space-efficient region filling in raster graphics," *The Visual Computer (1994),* 10:205-215, Springer-Verlag 1994.

[11] S. Hedberg, "Robots Cleaning Up Hazardous Waste," *AI Expert* , May 1995, pp. 20-24. Springer-Verlag 1994.

[12] A. Itai, C.H. Papadimitriou, J.L. Szwarefiter, "Hamilton Paths in Grid Graphs," *SIAM J. on Computing (1982),* 11:676-686

[13] *Instantiating Real-World Agents,* Papers from the AAAI 1993 Fall Symposium, Tech.Rep. FS-93-03, AAAI Press, Menlo Park, California.

[14] W. Neubauer, "Locomotion with Articulated Legs in Pipes or Ducts," *Robotics and Autonomous Systems,* 11:163-169. Elseveier 1993.

[15] S. Levy, *Artificial Life - The Quest for a New Creation,* Penguin Books, 1992.

[16] R.A. Russell, "Mobile robot guidance using a short-lived heat trail," *Robotica* , Vol. 11, 1993, pp. 427-431.

[17] Y. Shoham, M. Tennenholtz, "On Social Laws for Artificial Agent Societies: Off Line Design," to appear at *AI-Journal*, 1995.

[18] H.A. Simon, *The Sciences of the Artificial,* 2'nd ed., MIT Press, 1981.

[19] S. Sen, M. Sekaran, J. Hale, "Learning to Coordinate Without Sharing Information," *Proceedings of AAAI-94*, pp. 426-431.

[20] L. Steels, "Cooperation Between Distributed Agents Through Self-Organization," *Decentralized A.I. - Proc. of the 1'st European Workshop on Modeling Autonomous Agents in Multi-Agent World,* Y. DeMazeau, J.P. Muller (Eds.), pp. 175-196, Elsevier, 1990.

[21] I. A. Wagner and A. M. Bruckstein, "Row Straightening via Local Interactions," *Technical report CIS-9406, Center for Intelligent Systems,* Technion, Haifa, May 1994. Submitted to *SIAM J. on Matrix Analysis.*

17

FUNDAMENTAL ISSUES IN ATOMIC/NANOELECTRONIC COMPUTATION

M. P. Anantram and Vwani Roychowdhury

Dept. of Electrical Engineering
University of California
Los Angeles, CA 90095

Dedicated to Professor Thomas Kailath with great admiration and gratitude.

ABSTRACT

The past decade has seen tremendous experimental and theoretical progress in the field of mesoscopic devices and molecular self assembly techniques, leading to laboratory demonstration of many new device concepts. While these studies have been important from a fundamental physics perspective, it has been recognized by many that they may offer new insights into building a future generation of computing machines. This has recently led to a number of proposals for computing machines which use these new and novel device concepts. In this chapter, we explain the physical principles behind the operation of some of these proposals, primarily those based on conventional architectures. We then show that these proposals do not satisfy some fundamental requirements of a good computing model. The specific schemes discussed fall into the categories of **ground state** and **atomic** computing models. Finally, we propose a new scheme for atomic computing.

1 INTRODUCTION

The past decade has seen much experimental and theoretical study related to mesoscopic and atomic sized devices [1, 2, 3, 4]. Several groups in different countries have successfully built nanoscale devices, such as single-electron transistors and quantum dots, using both advanced lithographic techniques, as well as novel self-assembly methodologies, and have investigated the physics of these devices. The feature sizes of these devices are in the $1 - 100nm$ range.

An important question is : Can these new device concepts be used to build a future generation of computing machines? There have recently been many papers which address this question and they deal with proposals for both conventional and non conventional computing models. It is worth noting that architectures based on even the conventional models have remained rather elusive because it is widely believed in the solid state electronics community that the present strategies for integrating devices on a chip will be impractical for nanoelectronic devices due to their small size, low power handling capacity and low gain. The novel proposals which have recently been advanced, articulating primarily, visions of future atomic/nanoelectronic systems utilize unconventional architectures [5, 6, 7, 8, 9] and possess several characteristics suited to nanoelectronic/atomic implementation such as the absence of explicit interconnects and the convenience of loading (reading) the input (output) from the edges of the computing system. Some examples of the recently advanced non conventional computing models include neuromorphic [10] and quantum computing [8, 11, 12].

In this chapter, we mainly focus on the realization of conventional computing models using novel architectures involving atomic/nanoelectronic devices. We will begin with an explanation of the basic physics underlying two of the recently proposed paradigms, **ground state** and **atomic computing**. We will then show that these models do not fulfill some fundamental requirements of a good computing model. We will make concrete proposals for improvements, where possible.

Amongst the many features a good computing model should possess, the following three can be regarded as fundamental :

(i) the ability to interchange bit values between the basic units (bistable devices) in a manner independent of either the bit values being interchanged or the remaining bit values.

(ii) time efficient, unidirectional flow of information along the computational trajectory.

(iii) the ability to correct errors (we do not address issues related to error correction in this chapter).

In the novel paradigm of ground state computing [5, 6, 9], the computing architecture comprises locally-coupled arrays of basic units (Fig. 17.1). The basic units are bistable devices which represent binary values of 0 and 1 (section 2), and they interact with one another by Coulomb (or some such) interaction to form larger devices such as wires, logic gates and cellular automata (Fig. 17.1). Two features which distinguish their operation from conventional architectures are :

(a) They are edge driven. That is, the basic units on the edges form the input and output (Fig. 17.1). The interior units do not interact with the outside world.

(b) A computation is complete only when the computing machine relaxes to its ground state. The ground state depends only on the values of the input cells and not on the initial configuration of the remaining cells. The input cells are held fixed throughout the computation. The output to the computation is read from the cells marked 'output' only after the system has settled in its ground state (Fig. 17.1).

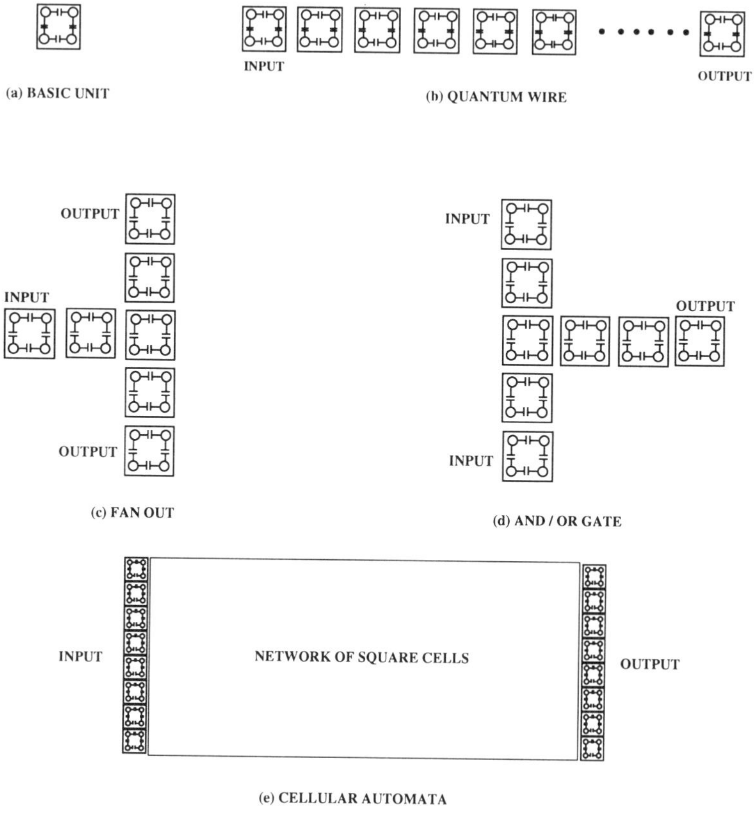

Figure 17.1 (a) Pictorial representation of the basic bistable unit. Construction of a (b) wire (c) fan out (d) AND / OR gates (d) using the basic unit (Ref. 4). The basic units interact with each other by Coulomb or some such interaction.

The requirement of evolution of the computing system to its ground state raises many important questions :

- Is a time efficient evolution of the computational trajectory to the ground state of the physical system involved feasible (requirement (ii) of page 310)? Or will the system be trapped by metastable states, thus delaying time efficient evolution to the ground state?

- Will the system remain in the ground state for a long enough time to be of computational use, before being excited to higher lying states?

There are no clear answers to these questions in the papers proposing ground state computing [5, 6, 9]. These papers only verified that the ground states of various structures correspond to logical operations. While there have been some hypotheses in literature that relaxation to the ground state will take place by a domino-type effect [6, 13], there have also been suggestions that a computing machine which relies on evolution to the ground state might not be feasible if metastable states are encountered in the computational trajectory (Fig. 17.2) [13, 14]. It is also of relevance to note that in other energy-based models of computation, such as the Ising spin-glass systems and the Hopfield Networks, meta stable states exist and play a definite role. The ground state in these schemes corresponds to an optimal solution for the problem under consideration and metastable states correspond to approximate solutions. In contrast, in the proposed models of ground state computing, metastable states (if they exist) do not have any useful computational interpretation.

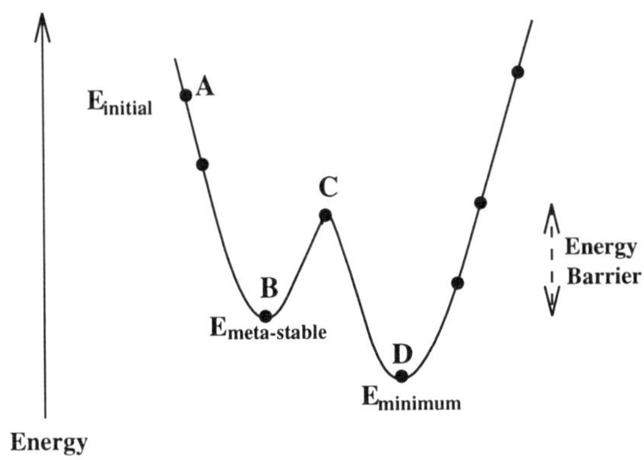

Figure 17.2 A system in the initial state (A) can reach the ground state (D) only by the series of single electron transitions, A $\rightarrow$ B $\rightarrow$ C $\rightarrow$ D. State B is a meta stable state as an energy barrier has to be surmounted to reach state C.

We have *resolved* the controversy involving the role of metastable states in ground state computing and have answered the above questions for the first time, by considering a specific model for ground state computing [15]. The salient features are discussed in section 2. We show that metastable states play an important role in determining the evolution of the computational trajectory, resulting in the time required for evolution to the ground state to be too large to be of computational use, even for the most advantageous choice of parameters.

Atomic Computing (Computing on an atomic chain) :

There has been excitement over the recent technological prospect of using an array of two level atoms as a computing machine, with the atoms serving as bistable devices [7, 8]. These proposals exploit the fact that the ground ($|0>$)

and excited ($|1>$) states of a two level atom can be used to represent binary values 0 and 1 as shown in Fig. 17.3 (a). ϵ^0 and ϵ^1 denote the ground and excited state energies. Switching between the two binary states ($|0>$ and $|1>$) can be accomplished by shining a π pulse. A π pulse has a resonant frequency equal to $\epsilon^1 - \epsilon^0$ and has an electric field profile shaped so as to cause an exact transition between states $|0>$ and $|1>$. Fig 17.3 (b) shows how to flip the state

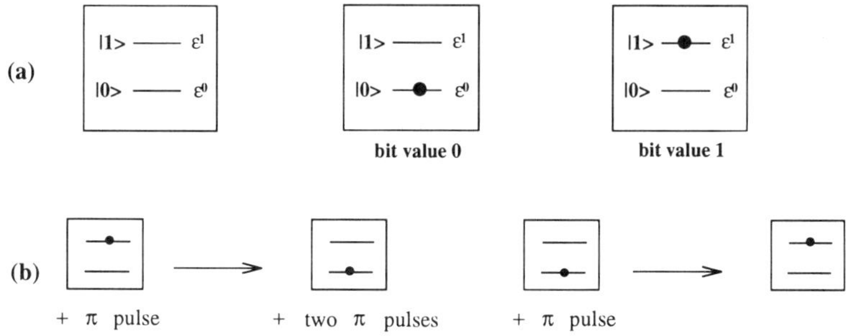

Figure 17.3 (a) A two level system. The ground and excited states states of the system can be used to represent binary bit values of 0 and 1. (b) By applying a π pulse with a resonant frequency equal to the resonant energy of the system, precise switching between binary numbers 0 and 1 can be accomplished.

of an isolated bistable device. To realize a computing machine, it is however necessary to have the ability to flip the state of a bistable device in a manner which depends on the bit values of a few other bistable devices. The interaction between atoms provides an adequate environment to achieve this. For example, consider two atoms A and B with distinct resonant frequencies interacting with each other. Now π pulses with distinct frequencies, $\omega^A_{B=0}$ and $\omega^A_{B=1}$ are required to flip the state of atom A, depending on the bit value stored in atom B. Infact, it can be shown that the basic logic operations such as AND, EX-OR and FAN OUT can be achieved using an array of three atoms with distinct resonant frequencies (section 3).

This line of thought was used in Ref. [8], which proposed that a polymer chain $A\ B\ C\ A\ B\ C\ ...\ A\ B\ C$ and a repertoire of π pulses comprises a computing machine (Fig. 17.4 (a)). Atoms A, B and C have distinct resonant frequencies. Such a system exhibits the following parallel feature : a π pulse would flip all atoms of a given kind with identical neighbors, throughout the chain. For example, in Fig. 17.4, if the bit values of A_2 and C_2 are identical to those of A_4 and C_4, then a π pulse which flips the bit value of atom B_2 will also flip the bit value of B_3. While the parallel feature can be used as an advantage in some applications, it is not a desirable feature in others (section 3). Specifically, the parallel feature proves to be the nemesis when it comes to satisfying requirement **(i)** discussed in page 310.

We propose a new computing model composed of two atomic chains which interact with each another (Fig. 17.4). This system does not suffer from the drawback discussed in the previous paragraph (section 3). Also, the parallel feature can be exploited when necessary in this model.

(a) $A_1\ B_1\ C_1\ A_2\ B_2\ C_2\ A_3\ B_3\ C_3 \cdots$

(b) $A_1\ B_1\ C_1\ A_2\ B_2\ C_2\ A_3\ B_3\ C_3 \cdots$
$\alpha\ \beta\ \gamma\ \alpha\ \beta\ \gamma\ \alpha\ \beta\ \gamma \cdots$

Figure 17.4 (a) A single polymer chain (b) Two interacting polymer chains : The various atoms of the chain on the top represent bistable devices. Only one atom of the chain on the bottom has a bit value of 1, the rest have a bit value of 0. This feature is used to uniquely mark one atom in the top chain. Note that the single polymer chain lacks this feature as all atoms of a given type (say B) are marked only by its neighbors (A and C) throughout the chain.

2 GROUND STATE COMPUTING

The previous studies of ground state computing only verified that the ground state of suitably designed coupled arrays of quantum dots corresponds to the basic logic operations. However, to resolve the controversy surrounding the role of metastable state in ground state computing models, it is necessary to study the dynamical evolution of the computational trajectory. In this section, we study this in the simplest network of quantum dots, a wire (Fig. 17.1). The quantum dots in our study are metallic type (continuum of energy levels) and have been successfully modeled by the orthodox theory of single electron tunneling [16]. The previous proposals of ground state computing considered dots with a single energy level using a Hubbard-type Hamiltonian and calculated only the energies of the ground state configurations [5, 6, 9]. These studies did not address the dynamics probably because the electron relaxation process in quantum dots with discrete energy levels is not well understood. Some recent studies however seem to indicate that the electron relaxation rate in quantum dots with discrete levels is orders of magnitude larger than in metallic type dots [17]. In which case, we expect our results to be indicative of dots with discrete levels also. The following specific issues are studied in this section :

- We first discuss the model used to study the dynamics of the computing system.
- Then, the existence of metastable states in the computational trajectory is established at zero temperature.
- By increasing the temperature, the ground state of the computing machine can always be accessed. However, for ground state computing to be feasible, it is necessary for the system to remain in the ground state for a long enough time period. Hence, we determine if there exists a region in the parameter space of the device dimensions and the temperature, where the system both relaxes to the ground state and remains there for a sufficiently long time period.
- Finally, it is verified if the time required to relax to the ground state is acceptable from a computational viewpoint, at least for the most advantageous choice of parameters.

2.1 The Model

Basic unit : The basic unit representing a bistable device, a *cell*, is shown in Fig. 17.5. It consists of four quantum dots along the edges of a square, which share two excess electrons. The four dots are separated by tunnel barriers whose resistances are much smaller than $\frac{h}{e^2}$ and in this limit, the dots have an integer number of electrons at any time. Two or more bistable devices interact via Coulomb interaction and this is modeled by capacitances. The non zero capacitances between the dots are shown in Fig. 17.1 (b). All other capacitances are zero. The intra and inter cell capacitances C and C', and the capacitance between the dots and the ground C_0 are identical in every cell of the wire. The tunnel resistances between the two dots along the sides of a cell are represented by R, and the tunnel resistances between dots along the diagonals of a cell and between dots of different cells are infinite.

bit value of 0 — bit value of 1

Figure 17.5 Representation of a single binary bit. The square cell consists of four quantum dots which share two electrons. Electrons occupying diagonally opposite dots represent binary values of 0 and 1.

The lowest energy (ground) state of a single cell is two fold degenerate and it consists of electrons aligned either along the left are right diagonals. The two degenerate ground states can be used to represent binary values 0 and 1 (Fig. 17.5). The ground state of a wire is also two fold degenerate and is shown in Fig. 17.6. Here the two excess electrons are aligned along the right (or left) diagonals in every cell.

System : The specific device studied here is the most fundamental one, a wire, which consists of an array of square cells (Fig. 17.1). The ground state is two fold degenerate and is shown in Fig. 17.6. Here the two excess electrons are aligned along the right (or left) diagonals in every cell.

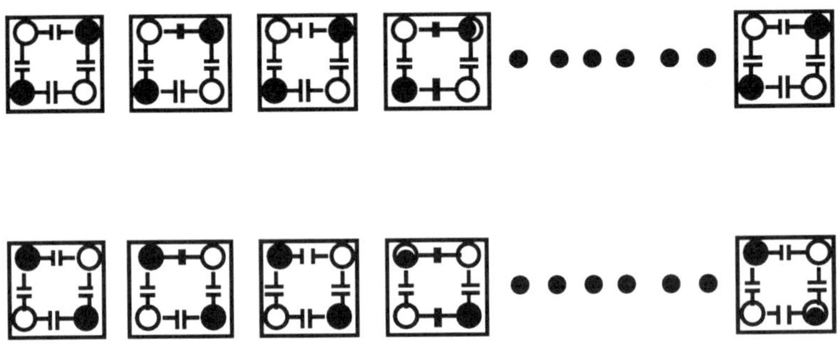

Figure 17.6 The ground state of a quantum wire is two fold degenerate. The two electrons are aligned either along the left or right diagonals as shown.

Initial state : Two excess electrons can be distributed in $4C_2$ ways in every cell. If there are N cells in the wire there are 6^N possible initial configurations. Our monte carlo simulator can handle any of these initial configurations. Here we will only consider the following initial state which is physical interesting. Consider the system to be in the ground state shown in Fig. 17.6 (a). In the initial state considered, the polarization of the left most cell alone is flipped (Fig. 17.7) and held fixed. The left most cell which represents the input is held fixed. If the wire relaxes to the ground state shown in Fig. 17.6 (b), then information is transmitted from the left most cell (the input) to the right most cell (the output).

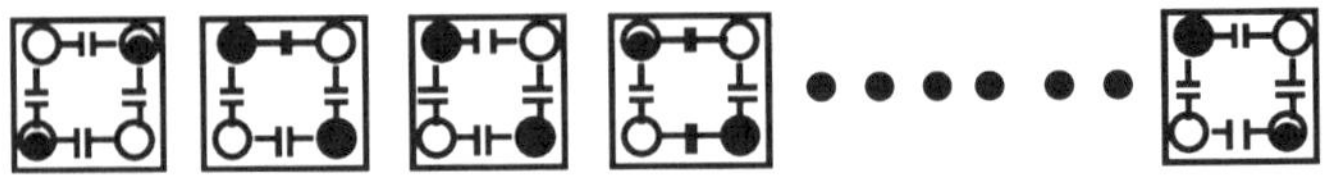

Figure 17.7 The initial state considered in this chapter.

Time evolution and the model

The system is set up in the initial state (Fig. 17.7). For the computation to be complete, the system should evolve into the ground state shown in Fig. 17.6.

The output is measured from the cell marked so after the system settles in the ground state.

The time evolution is modeled by the orthodox theory of single electron tunneling. [18] Here, a vector $Q(t)$ represents the charge in the quantum dots at time t. The charge configuration at the next time step $t + \delta t$ is computed using the monte carlo method which is briefly explained now. The transition probability for a single electron tunneling event where an electron tunnels from dot i to dot j is given by, [18]

$$\Gamma_{ij} = \frac{\Delta E_{ij} G}{C'} \frac{1}{exp(\frac{\Delta E_{ij}}{kT}) - 1}, \tag{17.1}$$

where, G and C' are the tunnel conductance and capacitance of the barrier between dots i and j respectively. $\Delta E_{ij} = E^a - E^b$, where E_a and E_b are total energies of the system after and before the tunneling event. Then a cumulative vector of all the tunneling probabilities,

$$S^p = \sum_{1}^{p} \Gamma_p, \tag{17.2}$$

is calculated. p labels the various tunneling events and can take values 1, 2, 3, .. ., M, where M represents the total number of tunneling events. A random number r is chosen between 0 and S^M. The charge vector Q in the next time step of the monte carlo simulation is determined by the single electron transition labeled n which satisfies $S^n < r < S^{n+1}$. The time taken for a single electron tunneling event is calculated using the expression $-\frac{\ln(r')}{S^M}$, where r' is a random number between 0 and 1 and S^M is the sum of all possible single electron transition probabilities from the initial state.

Metastable states : At zero temperature, a metastable state is reached if all the tunneling probabilities Γ_{ij} are zero (which implies that all $\Delta E_{ij} \geq 0$). At a finite temperature, we define the system to be trapped in a metastable state for a long time if S^M is smaller than $1e^{-8}$ $(RC)^{-1}$.

2.2 Zero Temperature Results

We first discuss the simple cases of wires with two and three cells. It is useful to remember that when (i) $C' >> C$, the interaction energy is large for electrons in neighboring cells to be adjacent to each other and (ii) $C' << C$, the interaction energy is large for electrons in the same cell to be on adjacent dots.

Two cell case : The initial state is shown in Fig. 17.8(a). The ground state energy is the two cell case of Fig. 17.6. When $C' >> C$, it is energetically favorable for the electron in dot 5 to tunnel to dot 7 (state B of Fig. 17.8 (b)) because the large interaction energy between the electrons in dots 2 and 5 can be avoided. The ground state is now easily reached by an electron tunneling from dots 8 to dot 6. When $C' << C$, the single electron tunneling events all lead to states where two electrons occupy adjacent dots of the right cell. All such states are energetically unfavorable in the $C' << C$ limit (the electrons like to align along the diagonals of a cell) and the system remains trapped in the initial state, which is a metastable state. So, there is a threshold value for $\frac{C'}{C}$ above which the system can reach the ground state and below which the system remains trapped in the initial state.

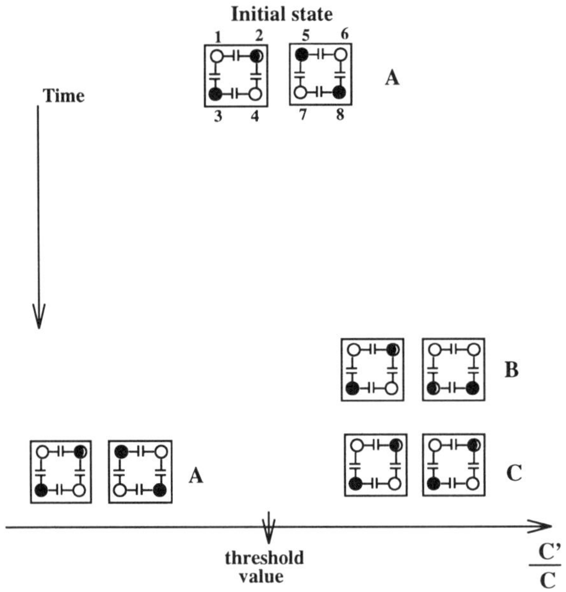

Figure 17.8 Evolution of the initial state as a function of $\frac{C'}{C}$.

Three cell case : Here, we have evaluated the total energies of various states to reach the following conclusions. For $C' >> C$, the interaction energy is very large for states where electrons are in adjacent dots of different cells. This interaction energy can be avoided only by the tunneling of an electron from dot 5 to dot 7 (Fig. 17.9). The remaining seven single electron tunneling events lead to total energies which are larger than that of the state A. Tunneling of an electron from dot 12 to dot 10 further reduces the total energy because it lowers the interaction between dots 8 and 12. The system is now trapped in state C as all the possible single electron tunneling events now only raises the total energy. So state C is a metastable state for $\frac{C'}{C}$ larger than threshold 2 (Fig. 17.9). In the other limit, where $\frac{C'}{C}$ is smaller than threshold value 1, as in the two cell case, the initial state is a metastable state. This is because all the eight possible single electron tunneling events lead to a state where electrons in either the second or the third cells are on adjacent dots. Such a configuration has a large energy in the $C' << C$ limit. For the parameter space in between thresholds 1 and 2, we find that the system is trapped in state B which is a metastable state (Fig. 17.9).

For cells of longer lengths (up to sixty cells), we have verified by monte carlo simulation that a linear wire is trapped in a metastable states at zero temperature.

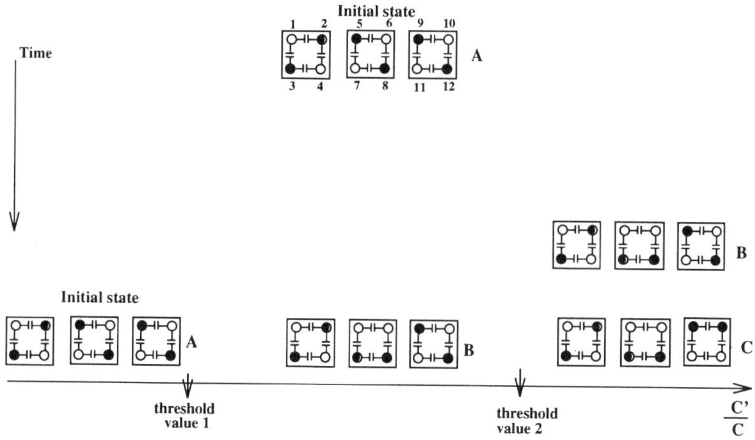

Figure 17.9 A plot of the various metastable states reached as a function of $\frac{C'}{C}$. The energies of various states have been calculated to verify the behavior.

2.3 Finite Temperature Results

We have so far shown that at zero temperature, ground state computing using a coupled array of quantum dots will suffer from problems associated with metastable states. We now ask, can the computing system escape from the metastable state, relax to the ground state and remain there for a long time period, at a finite temperature? At a high enough temperature, the system will fluctuate between various configurations including the ground state. For ground state computation to be feasible, it is however necessary for the system to remain in the ground state for a long enough time period. For this it is important that the temperature be large enough to *shake* the system out of a metastable state but not so large as to cause the system to leave the ground state in a short time period. We study the dynamics with the aim of determining if there is a regime of device parameters and temperature where this would happen. Then, we determine if the time required to relax to the ground state is acceptable from a computational viewpoint, at least for the most advantageous choice of parameters. This question can be addressed by a monte carlo simulation of the electron dynamics [18]. A detailed discussion of the dynamics will be given elsewhere [15] and we only summarize our main results.

We compute the time taken to reach the ground state as a function of the temperature for wires of various lengths. The time taken to reach the ground state decreases exponentially with increase in temperature (Fig. 17.10). For temperatures larger than the threshold kT_0, the transition probability to leave the ground state is larger than the threshold rate of $1e^{-8}$ $(RC)^{-1}$. This means that the system stays in the ground state only for a time smaller than $1e^8$ RC. This happens because the transition probabilities increase exponentially with temperature (Eq. (17.1)). We also find that the threshold temperature kT_0 does not depend significantly on the length of the wire [15]. The wires relax to the ground state and remain there for an average time which is larger than $1e^8$ RC only in the narrow parameter range, $\frac{C'}{C}$ from approximately 1.1 and 1.35, for wires having from three to sixty cells.

¿From Fig. 17.10 (a), we see that a wire having 45 cells reaches the ground state in a time of 10^8 RC. For an RC of about 0.1 pico seconds, this corresponds to a time of 10 micro seconds. This time is too long to be useful from a computational viewpoint.

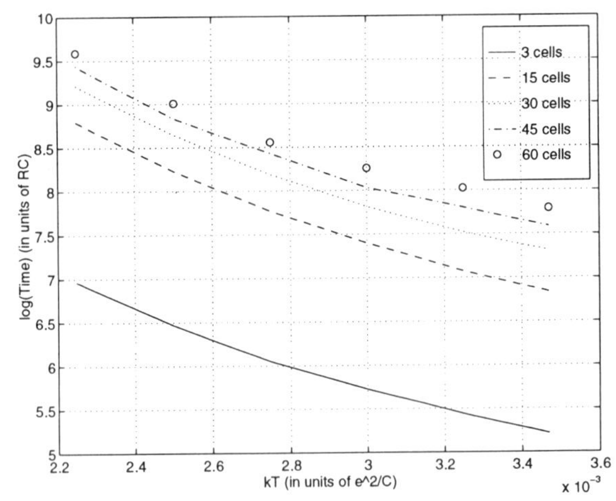

Figure 17.10 A plot of the time taken to reach the ground state versus temperature for wires of different lengths. The initial state is shown in Fig. 7. The value of $\frac{C'}{C}$ was chosen to be 1.3.

3 COMPUTING ON AN ATOMIC CHAIN

In the introduction, we explained the representation of binary values 0 and 1 using a two level atom. We will now address the following important issues related to building a computing machine using an array of atoms :

- **Construction of a logic gate from a set of interacting atoms.**
- **Where are the wires to connect to connect the 'atomic' gates?**
- **Is a single polymer chain** $A\ B\ C\ A\ B\ C\ ...\ A\ B\ C$ **a general purpose computing machine? What are its shortcomings?**
- **A new proposal to build an 'atomic' computer is suggested.**

3.1 A Three Atom Logic Gate

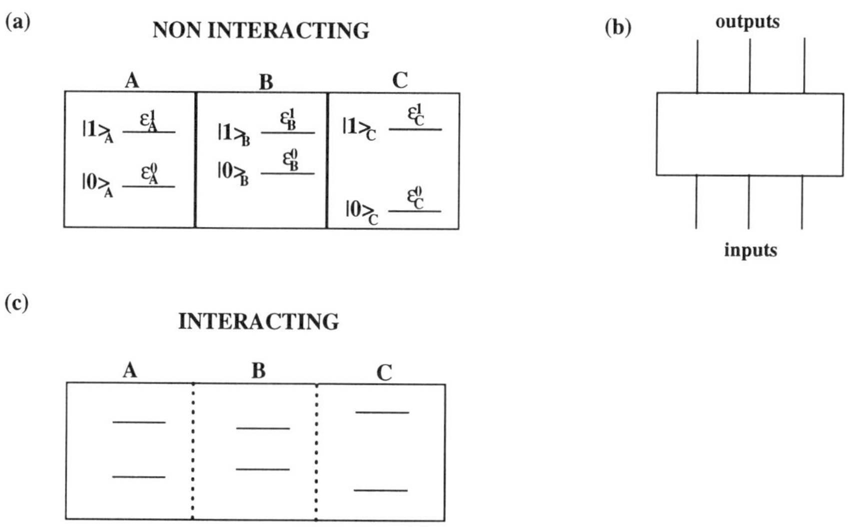

Figure 17.11 (a) A system of isolated atoms A, B and C. The resonant frequency of a atoms does not depend on the state of its neighbors. (b) A pictorial representation of a logic gate where the three output and input terminals are spatially separated from each other. (c) If atoms A, B and C interact via say the nearest neighbor interaction, the resonant frequency of an atom depends on the state of its nearest neighbors.

First, consider three two level atoms A, B and C isolated from one another (Fig. 17.11(a)). The (wave function, energy) of the ground and excited state of the individual atoms are represented by $(|0>_i, \epsilon_i^0)$ and $(|1>_i, \epsilon_i^1)$ respectively. The bit value of these atoms can be flipped using π pulses with resonant frequency, $\omega^i = \epsilon_i^1 - \epsilon_i^0$, where $i \in A,\ B,\ C$. Any 3-bit binary number can be represented by this system in much the same way as they are represented by three devices in a conventional logic gate. Three wires in a conventional logic gate can however be fed as the 'input' to a black box which performs a logic operation, giving an output (Fig. 17.11(b)). For the output to be different from the input, it is important that the input terminals interact inside the black box. Also, the different logic gates are realized by using different interactions (different circuits) in the black box. All this can be emulated by a series of two level atoms if they are allowed to interact (Fig. 17.11(c)). A good starting point to include interactions between the different atoms is the much studied *nearest neighbor interaction* of solid state physics. Here, it is assumed that every atom interacts with only its immediate left and right neighbors except for the end atoms which have only a single nearest neighbor. As a result of the nearest neighbor interaction, the resonant frequency required to flip the bit value of atom i depends on the bit values of its nearest neighbors, atoms $i-1$ and $i+1$,

$$\omega_A \; ----> \omega_0^A,\ \omega_1^A$$
$$\omega_B \; ----> \omega_{00}^B,\ \omega_{10}^B,\ \omega_{01}^B,\ \omega_{11}^B$$
$$\omega_C \; ----> \omega_0^C,\ \omega_1^C$$

Here, ω_A, ω_B, and ω_C are the resonant frequencies of the isolated atoms. The subscripts on the right hand side refer to the state (binary bit value) of the nearest neighbors. For example, ω_0^A is the resonant frequency required to flip the bit value of atom B when atom C has a bit value of 0. Similarly, ω_{01}^B is the resonant frequency required to flip the bit value of atom B when atoms A and C have bit values of 0 and 1 respectively.

Now a Toffoli gate[1], with atoms A and C as controls can be realized by applying a π pulse with frequency ω_{11}^B. The state of atom B is then inverted only if both atoms A and C have a bit value of 1.

An important difference between conventional and 'atomic' logic gates : There is an important difference between the spatial location of the input and output terminals of conventional logic gates and the system under consideration. In a conventional representation of a logic gate (Fig. 17.11(b)), the input and output terminals are spatially separated. On the other hand for the system in Fig. 17.11 (c), both the input and output terminals are represented by the same set of atoms. The bit values of the atoms at two different times, prior to and after shining the desired sequence of pulses represent the input and output.

[1] A Toffoli gate is a universal gate with equal number of input and output terminals. Other logical gates such as the AND, NAND etc. can be constructed from copies of it as discussed in [19].

3.2 The Wiring Paradigm in Atomic Computers

A general purpose computing machine can be viewed as a circuit of a large number of Toffoli gates. A 'wiring' scheme for a computing machine should possess the following rather obvious features :

(a) It must be possible to perform local operations on a few terminals at a time without changing the state of the remaining terminals

(b) It must be possible to 'wire' together terminals far away from each other so that a controlled operation such as the Toffoli gate can be realized using these terminals (Fig. 17.12) (requirement (i) of page 310).

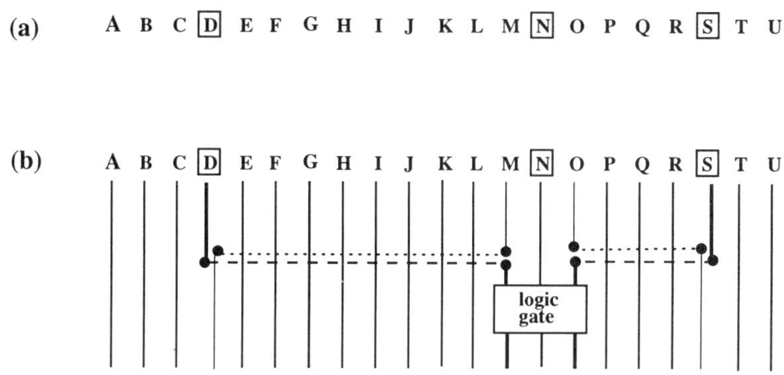

Figure 17.12 The terminals in this figures are labeled A through U. (a) The three 'boxed' terminals form the input to a logic gate in the next stage of the computation. How can the terminals be 'wired' in an array of interacting atoms. (b) In a conventional VLSI circuit this is done by routing the wires appropriately. The solid, dashed and dotted wires are patterned on three different layers of the circuit. So they are non intersecting except for vertical connection between the layers where necessary.

We have shown that the Toffoli operation can be performed on terminal i with terminals $i-1$ and $i+1$ as controls. So, point (a) above is satisfied and we will now discuss point (b). In conventional VLSI circuits, physical wires are 'routed' in and between different layers of the circuit to realize an arbitrary 'wiring' scheme. For example, in Fig. 17.12(b) wires **D**, **N** and **S** can be routed as shown to form the inputs to a logic gate. The solid, dotted and dashed lines are wires in three different layers of the circuit and they only intersect at the vertical connections between the layers, which are located at the solid dots. In an array of atoms, physical wires connecting the atoms does not make sense. However, the role of physical wires can be emulated, if the bit values of *any* two terminals can be interchanged in a systematic manner, which does not either change or depend on the detailed information of the bit values of the

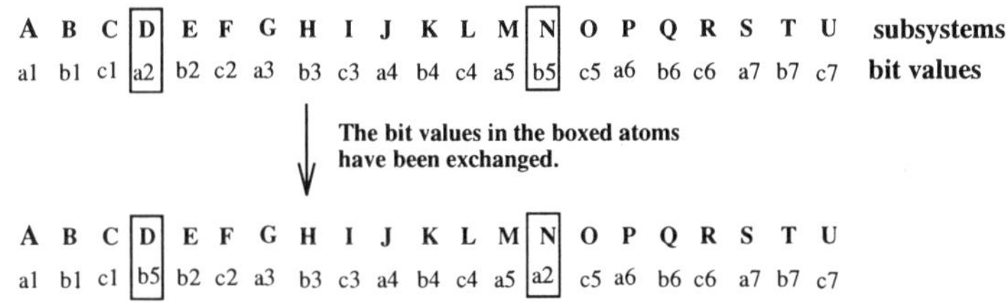

Figure 17.13 To accomplish a wiring scheme such as the one in the previous figure, it is necessary to be able to interchange the bit values of any two terminals far apart from each other (terminals D and N in this figure) without using and changing the bit values of the remaining terminals.

other atoms in the array (Fig. 17.13). Going a step further, we assert that this would be possible if the bit values of *adjacent* terminals can be interchanged in a systematic manner which does not either change or depend on the bit values of the remaining terminals. The interchange of bit values of a pair of terminals far away from each other can then be realized by a series interchanges of the bit values of adjacent atoms.

4 IS A POLYMER CHAIN $A\,B\,C\,A\,B\,C\,..\,A\,B\,C$ A GENERAL PURPOSE COMPUTING MACHINE?

We now consider an array of atoms $A\ B\ C\ A\ B\ C\ ..\ A\ B\ C$, where the constituent atoms A, B and C have different resonant frequencies (the subscripts in Fig. 17.4 (a) are only used to label the positions of the atoms). Ref. [8] proposed that such an atomic array and a repertoire of π pulses with appropriate resonant frequencies comprises a computing machine. We will show here that this cannot be the case because point (b) discussed above is not satisfied. If light pulses can be localized to the size of a single atom, the bit values of any two adjacent terminals can be interchanged as required by point (b), by shining the appropriate sequence of pulses. This follows because atoms A, B and C have different resonant frequencies and so distinct pulses are required to interchange the bit values of atoms i and $i+1$, and atoms i and $i-1$. Unfortunately, the typical sizes of individual atoms is in the range of tens of Angstroms and their excitation frequencies are usually in the meV range. At these frequencies, the spatial extent of a light pulse is of the size of many atoms put together. Then, it is relevant to consider schemes where the light pulses shine over the entire atomic array. Due to this unavoidable restriction, point (b) cannot be satisfied. This is because a series of pulses which purports to interchange the bit values of B_i and C_i will also interchange the bit values of all other B_j and C_j atoms, where j is a number labeling the various triplets of the array. The main drawback of the single chain model is that it lacks the ability to target (or uniquely mark) a single atom in the chain.

4.1 A New Proposal For an Atomic Computing Machine

Are there any schemes which would not suffer from the drawbacks of the $A\ B\ C\ A\ B\ C\ ...\ A\ B\ C$ chain? A system which does not suffer from the drawback of the previous section is a chain where all the constituent atoms are different from each other. The atoms all have distinct resonant frequencies and a pulse or sequence of pulses with a large spatial extent can be designed to satisfy points (a) and (b) of the previous subsection. This scheme does not suffer from any logical problems but there are two drawbacks. Firstly it is difficult to build a system where the constituent atoms are all different. And secondly, a repertoire of light pulses with too many resonant frequencies will be required.

The desirable features (a) and (b) (of the previous sub section) can be realized in a model consisting of two chains $A\ B\ C\ A\ B\ C\ ...\ A\ B\ C$ and $\alpha\ \beta\ \gamma\ \alpha\ \beta\ \gamma\ ...\ \alpha\ \beta\ \gamma$, which interact with each other (Fig. 17.4). Each atom in the top (bottom) chain interacts with only the atoms to its immediate left, right and bottom (top). The atoms of the first chain ($A\ B\ C\ A\ B\ C\ ...\ A\ B\ C$) comprise the terminals of the computer. The second chain ($\alpha\ \beta\ \gamma\ \alpha\ \beta\ \gamma\ ...\ \alpha\ \beta\ \gamma$) is used to mark a single atom in the first chain, a feature which the single chain model lacked. This is done by allowing only one of the atoms in the second chain to have the bit value of 1, the remaining atoms have a bit value of 0. This system has the following resonant frequencies :

$$\begin{aligned}
\omega^A & \ ---->\ \omega^{A:end}_{B;\alpha} \text{for the left most A of the array} \\
& \qquad\quad\ \omega^{A}_{CB;\alpha} \text{ for all other A} \\
\omega^B & \ ---->\ \omega^{B}_{AC;\beta} \\
\omega^C & \ ---->\ \omega^{C:end}_{B;\gamma} \text{ for the right most C of the array} \\
& \ ---->\ \omega^{C}_{BA;\gamma} \text{ for all other C} \\
\Omega^\alpha & \ ---->\ \Omega^{\alpha:end}_{\beta;A} \text{ for the left most } \alpha \text{ of the array} \\
& \qquad\quad\ \Omega^{\alpha}_{\gamma\beta;A} \text{ for all other } \alpha \\
\Omega^\beta & \ ---->\ \Omega^{\beta}_{\alpha\gamma;B} \\
\Omega^\gamma & \ ---->\ \Omega^{\gamma:end}_{\beta;C} \text{ for the right most } \gamma \text{ of the array} \\
& \ ---->\ \Omega^{\gamma}_{\beta\alpha;C} \text{ for all other } \gamma
\end{aligned}$$

As the left most A and α atoms and the right most C and γ atoms have only one nearest neighbor on the same chain, their resonant frequencies are different from the A, α, C and γ atoms in the interior of the chain.

Any computation can be done on this array by : (i) initializing the array to correspond to the input and (ii) applying a series of π pulses to perform the computation. The first step corresponds to initializing the input terminals of a computing machine and the second step is equivalent to passing these input values through a sequence of logical gates.

We end this discussion by clearly giving the rules for exchanging the bit values of any two atoms (requirement (i) of page 310) without using or changing the bit values of the remaining atoms.

Rules for exchanging the bit values of any two atoms :

Rule 1 (Marker Manipulation) Only one of the atoms in the second row is in the 1 state. This atom constitutes the marker. The marker can be moved to the right (left) by applying a sequence of pulses which will interchange the 1 state of the atom currently serving as the marker with the 0 state of the atom to its right (left). Further, what is important is that the marker can be moved independent of the bit values in Row 1.

The *pulse coding* for exchange of bit information between atoms i and $i+1$ is

$$[\,\omega^{i+1}_{11;0}\;\omega^{i+1}_{10;0}\;\omega^{i}_{01;0}\;\omega^{i}_{11;0}\,][\,\omega^{i+1}_{11;0}\;\omega^{i+1}_{10;0}\;\omega^{i}_{01;1}\;\omega^{i}_{11;1}\,]$$
$$[\,\omega^{i+1}_{11;1}\;\omega^{i+1}_{10;1}\;\omega^{i}_{01;0}\;\omega^{i}_{11;0}\,][\,\omega^{i+1}_{11;1}\;\omega^{i+1}_{10;1}\;\omega^{i}_{01;1}\;\omega^{i}_{11;1}\,]$$

$\omega^i_{xy;z}$ is used to represent the resonant frequency of atom i in Row 2 if its left, right and top (in Row 1) neighbors have bit values of x, y and z respectively. The pulse sequences in the first, second, third, and fourth bracket move the marker to the right if the atoms in Row 1 located right on top of atoms i and $i+1$ in Row 2 have a bit values of (0 and 0), (1 and 0), (0 and 1), and (1 and 1) respectively. By applying all four pulse sequences, the position of the marker will move one unit to the right, independent of the bit values of the atoms in Row 1. ω is used to represent the the excitation frequencies of atoms in Row 2. A similar sequence of pulses can move the marker to the left by one position.

Rule 2 (Exchange of any two bits - 'wiring') :

Bit values in Row 1 ... v_1 v_2 v_3 X_1 y_1 y_2 ... y_{N-1} y_N Z_1 w_1 w_2 w_3 ...

Bit values in Row 2 ... 0 0 0 1 0 0 ... 0 0 0 0 0 0 ...

can be changed to

Bit values in Row 1 ... v_1 v_2 v_3 y_N y_1 y_2 ... y_{N-1} X_1 Z_1 w_1 w_2 w_3 ...

Bit values in Row 2 ... 0 0 0 0 0 0 ... 0 1 0 0 0 0 ...

by the following sequence of pulses : (a) Place the marker at the location of X_1 using Rule 1. Interchange x_i with x_{i+1} in Row 1 if the marker is at location i and then move the marker to $i+1$ (using Rule 1). Do this N times and then move the marker one unit to the left. This leaves the marker at y_N. (b) Interchange x_i and x_{i-1} in Row 1 if the marker is at location i and then move the marker to $i-1$. Do this N times and then move the marker N steps to the right.

The *pulse coding* for exchange of bit information between atoms i and $i+1$ in Row 1 conditional to the fact that the marker is at position i in Row 2 is:

$$[\ \Omega^{i}_{01;1}\ \Omega^{i}_{11;1}\ (\text{Rule 1 (1R)})\ \Omega^{i+1}_{11;1}\ \Omega^{i+1}_{10;1}\ (\text{Rule 1 (1L)})\ \Omega^{i}_{01;1}\ \Omega^{i}_{11;1}\ ].$$

$\Omega^{i}_{xy;z}$ represents the resonant frequency of atom i in Row 1 if its left, right and bottom (in Row 2) neighbors have bit values of x, y and z respectively. (Rule 1 (1R [L])) means apply Rule 1 to move the marker one unit to the right [left].

5 CONCLUDING REMARKS

A general purpose atomic / nanoelectronic computing machine must have the following characteristics : **(i)** the ability to interchange bit values between the basic units in a manner independent of either the bit values being interchanged or the remaining bit values. **(ii)** time efficient, unidirectional flow of information along the computational trajectory. **(iii)** the ability to correct errors.

In this chapter, we resolved some controversial issues surrounding points (i) and (ii) above in the recently proposed paradigms of **ground state** and **atomic computing**. We have also proposed a new scheme for atomic computing which does not suffer from the drawbacks of a single chain model.

Conclusion 1 : The controversy surrounding the role of metastable states in ground state computation has been resolved for a particular model involving an array of metallic type dots. We find that ground state computing models can be plagued by metastable states. We show that, at zero temperature, the computing machine can be trapped by metastable states for very long time periods, thus hindering relaxation to the ground state of the machine. In terms of computation, this implies inefficient flow of information from the input to the output. By raising the temperature, the ground state can be reached. However, for ground state computing to be feasible, it is necessary for both the computing machine to remain in the ground state for a long time period and for the computational trajectory to lead to the ground state in a short time period. For the model considered, we have shown that there is a narrow parameter regime of the device dimensions and the temperature, where relaxation to the ground state is possible for long time periods. Unfortunately, the physical time required to relax to the ground state is too large to be of any computational use.

There have been some studies indicating that the relaxation time for electrons in quantum dots with discrete energy levels is orders of magnitude larger than for metallic type dots with a continuum of energy levels [17]. So, even though we study a model with metallic type dots, we expect our results to be valid in the limit of dots with discrete energy levels also.

Conclusion 2 : The recently proposed computing paradigms, which use two level atoms as bistable devices, presents us with a future direction in building 'atomic' computers [8]. We showed that a single polymer chain of the form $A\ B\ C\ A\ B\ C\ ...\ A\ B\ C$, does not comprise a general purpose computing machine because an arbitrary 'wiring' scheme cannot be realized in this system. The reason for this is related to the inability to interchange the bit values of two adjacent bistable devices (atoms) in a manner consistent with requirement (i). We propose a new model for 'atomic' computers consisting of two arrays of atoms interacting with one another, and show that requirement (i) is satisfied by this model. We present explicit rules using π pulses, to interchange bit values of two atoms in a manner consistent with requirement (i).

Acknowledgements

This work was supported in part by NSF Grants No. ECS-9308814 and No. ECS-9523423.

REFERENCES

[1] C. W. J. Beenakker and H. van Houten. *Quantum Transport in Semiconductor Nanostructures, Solid State Physics, vol. 44, Eds. H. Ehrenreich and D. Turnbull.* Academic Press, Inc., U S A, 1991.

[2] Claude Weisbuch and Borge Vinter. *Quantum Semiconductor Structures, Fundamentals and Applications.* Academic Press, Inc., San Diego, U.S.A, 1991.

[3] Hermann Grabert and Michel H. Devoret. *Single Charge Tunneling, Coloumb Blockade Phenomena in Nanostructures.* Plenum Press, New York, U.S.A, 1992.

[4] In Ari Aviram, editor, *Molecular electronics – science and technology : St. Thomas, Virgin Islands 1991*, New York, U.S.A, 1992. American Institute of Physics.

[5] P. Bakshi, D. A. Broido, and K. Kempa. *J. Appl. Phys.*, 70:5150, 1991.

[6] Craig Lent, P. Douglas Tougaw, Wolfgang Porod, and Gary H. Bernstein. Quantum cellular automata. *Nanotech.*, 4:49, 1993.

[7] H. Korner and G. Mahler. Optically driven quantum networks : Applications in molecular electronics. *Phys. Rev. B*, 48(4):2335, 1993.

[8] S. Llyod. A potentially realizable quantum computer. *Science*, 261:1569, 1993.

[9] S. Bandyopadhyay, B. Das, and A. E. Miller. Supercomputing with spin polarized single electrons in a quantum coupled architecture. *Nanotechnology*, 5:113, 1994.

[10] Vwani Roychowdhury, D. B. Janes, S. Bandyopadhyay, and Xiaodong Wang. Collective computational acitivity in in self-assembled arrays of quantum dots: A novel neuromorphic architecture for nanoelectronics. *To Appear in IEEE transactions on electron devices.*

[11] Proceedings of the workshop on physics and computation physcomp '94. In *IEEE International Electron Devices Meeting*, Los Alamitos, CA, 1994. IEEE Computer Society Press.

[12] J. I. Cirac and P. Zoller. Quantum computation with cold trapped ions. *Phys. Rev. Lett.*, 74(20):4091, 1995.

[13] S. Bandyopadhyay. *Preprint.*

[14] Rolf Landauer. *Phil. Trans. Royal Soc. London, Ser. A.* To Appear in.

[15] M. P. Anantram and V. Roychowdhury. Preprint.

[16] D. V. Averin and K. K. Likharev. Single electronics : A correlated transfer of single electrons and cooper pairs in systems of small tunnel junctions. In B. L. Altshuler, P. A. Lee, and R. A. Webb, editors, *Mesoscopic Phenomena in Solids*, U. S. A, 1991. Elsevier Science Publishers.

In B. L. Altshuler, P. A. Lee, and R. A. Webb, editors, *Mesoscopic Phenomena in Solids*, U. S. A, 1991. Elsevier Science Publishers.

[17] H. Benisty. Reduced electron-phonon relaxation rates in quantum-box systems: Theoretical analysis. *Phys. Rev. B*, 51(19):13281, 1995.

[18] N. S. Bakhalov, G. S. Kazacha, K. K. Likharev, and S. I. Serdyokova. *Sov. Phys. JETP*, 68(3):581, 1989.

[19] Richard P. Feynman. Quantum mechanical computers. *Foundations of Physics*, 16(6):507, 1986.

PART III

LINEAR SYSTEMS AND CONTROL

18

BUILDING SPECIAL LINEAR SYSTEM REALIZATIONS OF SPECIAL TRANSFER FUNCTIONS

Brian D.O. Anderson

Department of Systems Engineering
Research School of Information Sciences and Engineering
Australian National University, Canberra ACT 0200, Australia

This paper is dedicated to Professor Thomas Kailath, in thanks for the inspiration provided from his papers and books over three decades.

ABSTRACT

Linear system realization problems require the construction of a linear system realization, or a quadruple of real matrices F, G, H, J, of a given real rational $W(s)$, so that $W(s) = J + H'(sI - F)^{-1}G$. When $W(s)$ has certain properties, corresponding to symmetry, passivity, or losslessness, it is possible to select a realization that closely (but not necessarily exactly) reflects these properties. We review such possibilities, then consider a new one, initially for discrete time systems: if the impulse response associated with W is nonnegative, can the entries of the realizing matrices be chosen to be nonnegative also? Necessary and barely differing sufficient conditions are presented for this to be the case.

1 INTRODUCTION

Linear system theory results of two or three decades ago are often focussed on the question of relating rational transfer function matrices, and state-variable realizations of these transfer function matrices. Thus with $W(s)$ a real rational matrix in s, normally with $W(\infty)$ finite, a quadruple of real matrices $\{F, G, H, J\}$ is a realization of $W(s)$ when

$$W(s) = J + H'(sI - F)^{-1}G \tag{18.1}$$

Of course, the underlying state-variable equations are

$$\dot{x} = Fx + Gu \qquad y = H'x + Ju \tag{18.2}$$

and the Laplace transforms $U(s)$ and $Y(s)$ of $u(.)$ and $y(.)$ [with zero initial conditions] are related by $Y(s) = W(s)U(s)$.

Given F, G, H, and J, it is obvious that $W(s)$ is uniquely defined. Given $W(s)$, there are an infinity of possible quadruples $\{F, G, H, J\}$. The investigation of the one-to-many mapping $W(s) \to \{F, G, H, J\}$ has generated a wealth of results. One group of results focuses on minimal dimension realizations, viz those for which F, which is square, has least dimension. Here, there are elegant answers to questions like: how can the minimal dimension be characterized in terms of $W(s)$? How are minimal dimension realizations related? How can minimal dimension realizations be found? What minimal parametrizations for minimal dimension realizations can be found? These sorts of questions are extensively treated in [1].

Other questions stem from trying to carry over certain properties which $W(s)$ may have to similar properties of the associated realization. For example, we can ask what can be said about the realization, or existence of special realizations, in case W has symmetry, passivity or losslessness properties, eg

$$W(s) = W'(s) \tag{18.3}$$

$$W(s) + W'(-s) \geq 0 \qquad \text{or} \qquad I - W'(-s)W(s) \geq 0 \qquad \text{for } Re[s] \geq 0 \tag{18.4}$$

$$W(s) + W'(-s) = 0 \qquad \text{or} \qquad W'(-s)W(s) = I \tag{18.5}$$

The properties of the transfer function matrix exhibited in (18.3) through (18.5) often arise in physical systems, for reasons explained by physics. Thus the symmetry condition (18.3) has its roots in observations of Maxwell on the behaviour of electromagnetic systems, and of structural systems [2,3]. The condition (18.4) can be linked to results of Boltzmann, Johnson and Nyquist on noise in electrical networks [4] (more generally encapsulated by physicists in the fluctuation-dissipation theorem [5]). Physical systems typically have internal structure, and realizations have to deal with internal structure. Properties of special realizations of $W(s)$ satisfying one of (18.3) through (18.5) will often reflect what physics has to say about the internal structure of physical systems. In the next section, we will reveal the properties of realizations $\{F, G, H, J\}$ that are securable in the light of one of (18.3) through (18.5). The third section considers a further input-output property and the question of how it can be reflected in realizations. In particular, we shall suppose that $W(s)$ has a nonnegative impulse response. We shall report on recent results for this problem. They should not be viewed in isolation, but as further results along the lines of those summarized in section 2.

2 SYMMETRY AND LIKE PROPERTIES

2.1 Symmetric $W(s)$

Let $W(s) = W'(s)$. Suppose that $\{F, G, H, J\}$ is a minimal realization. From (18.1) it follows that $\{F', G, H, J'\}$ is a minimal realization. Then from a standard theorem connecting minimal realizations [1] there exists a unique nonsingular T such that

$$TFT^{-1} = F' \qquad TG = H \qquad H'T^{-1} = G \qquad J = J' \tag{18.6}$$

It is easily checked that these equations imply the same equations with T replaced by T' and thus $T = T'$. Writing $T = V'\sum V$ where V is nonsingular and $\sum = diag[I_m, -I_p]$ and setting $F_o = VFV^{-1}$, $G_o = VG$, $H_o' = H'V^{-1}$ gives a minimal realization $\{F_o, G_o, H_o, J\}$ where

$$\begin{bmatrix} \sum & O \\ O & I \end{bmatrix} \begin{bmatrix} F_o & G_o \\ H_o' & J \end{bmatrix} = \begin{bmatrix} F_o & G_o \\ H_o' & J \end{bmatrix}' \begin{bmatrix} \sum & O \\ O & I \end{bmatrix} \tag{18.7}$$

This is the nearest one can get to a symmetric realization. The integer $m - p$ is the Cauchy index of $W(s)$, [6]. When a network of inductors, capacitors, transformers and resistors is modelled with state-vector corresponding to inductor currents and capacitor voltages, the associated realization obeys (18.7), with m, p corresponding to the numbers of the two types of energy storage elements.

2.2 Passivity

Equations (18.4) are two ways of describing passivity: consider a network with m ports, and with port voltage and current vectors v and i. Let $W(s)$ be the transfer function taking currents to voltages. If for all $i(.)$,

$$\int_{t_o}^{\infty} v'idt \geq 0 \tag{18.8}$$

whenever $i(t) = 0$ for $t \leq t_o$ and the network is unexcited at the time $t = t_o$, the network is passive and the first of (18.4) holds, [7]. The other inequality follows if excitation and responses are defined a little differently, using incident and reflected waves, an idea which is very common in microwave circuits [8] and in digital filtering where the transfer function is a discrete time entity (z-transform) [9]. The main result [10] is that a minimal realization $\{F_o, G_o, H_o, J\}$ of a transfer function matrix satisfying the first (18.4) can always be found so that

$$\begin{bmatrix} -F_o & G_o \\ -H_o' & J \end{bmatrix} + \begin{bmatrix} -F_o & G_o \\ -H_o' & J \end{bmatrix}' \geq 0 \tag{18.9}$$

The reflecting of the second (18.4) in a realization is more transparent when the corresponding discrete-time problem is considered. In discrete-time, one is concerned with underlying state-variable difference equations

$$x_{k+1} = Fx_k + Gu_k \qquad y_k = H'x_k + Ju_k \tag{18.10}$$

and when z-transforms $U(z)$ and $Y(z)$ are given by $U(z) = \sum u_k z^{-k}$, $Y(z) = \sum y_k z^{-k}$,there holds $Y(z) = W(z)U(z)$ where

$$W(z) = J + H'(zI - F)^{-1}G \tag{18.11}$$

Replacing the second of (18.4) there is

$$I - W'(z^{-1})W(z) \geq 0 \qquad \text{for } |z| \geq 1$$

and the main result [10] is that there exists a minimal realization

$$\{F_o, G_o, H_o, J_o\}$$

such that

$$I - \begin{bmatrix} F_o & G_o \\ H_o' & J \end{bmatrix}' \begin{bmatrix} F_o & G_o \\ H_o' & J \end{bmatrix} \geq 0 \tag{18.12}$$

2.3 Losslessness

In lossless systems inequality (18.8), under the constraint that $i(.) \in L_2$, is replaced by equality, and the first equality of (18.5) holds. Then (18.9) holds with equality. In contrast to the proof of (18.9), which is difficult, one can proceed by noting that if $\{F, G, H, J\}$ is a minimal realization, $W'(-s) = W(s)$ implies that $\{-F', H, -G, -J'\}$ is also a minimal realization. This yields existence of $\{F_o, G_o, H_o, J_o\}$ such that

$$\begin{bmatrix} -F_o & G_o \\ -H_o' & J \end{bmatrix} + \begin{bmatrix} -F_o & G_o \\ -H_o' & J \end{bmatrix}' = 0 \tag{18.13}$$

Likewise, if $W'(z^{-1})W(z) = I$, there exists minimal $\{F_0, G_0, H_0, J\}$ with

$$\begin{bmatrix} F_o & G_o \\ H_o' & J \end{bmatrix}' \begin{bmatrix} F_o & G_o \\ H_o' & J \end{bmatrix} = I \tag{18.14}$$

3 NONNEGATIVE REALIZATION

Nonnegative impulse responses arise in a number of applications area, including (linear) compartmental systems [11,12] and charge-routing networks, [13]. A type of generalization arises in the area of hidden Markov models [14]. We pose the basic question in discrete time and for simplicity, but with no loss of generality, for scalar transfer functions. Let $H(z)$ be a rational transfer function with $H(z) = \sum_{k\geq 1} h_k z^{-k}$. Suppose further that the impulse response sequence h_k is nonnegative for all k. Realizations of $H(z)$ are triples $\{F, g, h\}$ for which $H(z) = h'(zI - F)^{-1}g$, and h_k is related to F, g, h by $h_k = h'F^{k-1}g$. We would like the nonnegativity of the impulse response to be reflected in the state-variable realization. Specifically we ask: **when does there exist a nonnegative realization, ie a triple** $A \in R_+^{N\times N}, b \in R_+^N, c \in R_+^N$ **for which**

$$H(z) = c'(zI - A)^{-1}b \tag{18.15}$$

[Here $R_+^N, R_+^{N\times N}$ denote N-vectors and $N \times N$ matrices with all entries nonnegative]. In a number of ways, this is a more difficult question than the ones considered in section 2. It is for example known that: 1. if $H(z) = p(z)/q(z)$ where $p(z)$ and $q(z)$ are coprime polynominals and $q(z)$ has degree n, there may exist no nonnegative realizations with $N = n$, ie no minimal (in the linear systems sense) realization with the nonnegativity property, but there may exist a nonnegative realization with $N > n$, [15]. 2. Some rational $H(z)$ with nonnegative impulse response have no nonnegative realizations of any dimension, [16], eg

$$H(z) = \frac{1}{2}\left[\frac{\frac{1}{2}}{z-\frac{1}{2}} - \frac{z\left(\frac{1}{2}\cos 2\right) - \frac{1}{4}}{z^2 - z\cos 2 + \frac{1}{4}}\right] \tag{18.16}$$

for which

$$h_k = \left(\tfrac{1}{2}\right)^k sin^2 k. \tag{18.17}$$

Evidently, playing around with minimal dimension realizations alone (useful as they are with all their special properties) is unlikely to yield a solution to the problem. Nevertheless, it is possible to reformulate the problem using a minimal realization, [17]

Theorem 18.1 *Let $H(z)$ be a rational transfer function with minimal realization F, g, h of dimension n. Let $\mathcal{R}$ denote the cone spanned by g, Fg, F^2g, ie. the set $\{\sum_{i=0}^{\infty} \alpha_i F^i g | \alpha_i \geq 0\}$. If $H(z)$ has a nonnegative realization, there exists a (finite dimensional) matrix P such that [with $\mathcal{P}$ the cone generated by the columns of P and $\mathcal{P}^*$ the dual cone][1]*

$$\mathcal{R} \subset \mathcal{P} \quad \text{and} \quad \mathcal{FP} \subset \mathcal{P} \quad \text{and} \quad h \in \mathcal{P}^* \tag{18.18}$$

Moreover if for some $n \times N$ matrix P, with $\mathcal{P}$ = cone P, the equations (18.18) hold, there exists a nonnegative realization $\{A, b, c\}$ of $H(z)$ with $A \in R_+^{N\times N}$, and b, $c \in R_+^N$.

Theorem 18.1 in effect replaces the original problem by one concerning the existence of a cone.

[1] Let $\mathcal{X}$ be a cone. Then the dual cone $\mathcal{X}^*$ is defined by $\{y | x'y \geq 0 \ \forall x \in \mathcal{X}\}$

3.1 Simplifying Observations

To make progress in solving the reformulated problem of existence and computation of P, several simplifying observations are helpful. The first is that without loss of generality, one can work with a special minimal realization of $H(z)$. Suppose that with coprime numerator and denominator

$$H(z) = \frac{p_1 z^{n-1} + \cdots + p_n}{z^n + q_1 z^{n-1} + \cdots + q_n}. \tag{18.19}$$

Then one can take

$$F = \begin{bmatrix} 0 & 1 & 0 & \cdots & 0 \\ 0 & 0 & 1 & \cdots & 0 \\ \vdots & \vdots & \vdots & \ddots & \vdots \\ 0 & 0 & 0 & \cdots & 1 \\ -q_n & -q_{n-1} & -q_{n-2} & \cdots & -q_1 \end{bmatrix} \quad g = \begin{bmatrix} h_1 \\ h_2 \\ \vdots \\ h_n \end{bmatrix} \quad h' = \begin{bmatrix} 1 \, 0 \cdots 0 \end{bmatrix}. \tag{18.20}$$

With $h_i = h'F^{i-1}g$ the i-th impulse response coefficient of $H(z)$, and

$$H_n = \begin{bmatrix} h_1 & h_2 & h_3 & \cdots \\ h_2 & h_3 & h_4 & \cdots \\ \vdots & \vdots & \vdots & \ddots \\ h_n & h_{n+1} & h_{n+2} & \cdots \end{bmatrix}. \tag{18.21}$$

it follows that $H(z)$ has a nonnegative realization if and only for some N there exists an $n \times N$ matrix P for which

$$\text{Cone}(H_n) \subset \mathcal{P} \tag{18.22a}$$

$$F\mathcal{P} \subset \mathcal{P} \tag{18.22b}$$

$$\textit{The first row of } P \textit{ has nonnegative entries.} \tag{18.22c}$$

The second observation is

Lemma 18.1 *Let $H(z)$ be a rational n-th order transfer function with nonnegative impulse response. Then $H(z)$ has a nonnegative realization if and only if $H_\alpha(z) = H(\alpha z)$ has a nonnegative realization for any positive α.*

The third observation is linked to the Perron-Frobenius theorem [18]. Any nonnegative matrix has an eigenvalue of maximum modulus which is positive. So if $H(z) = c'(zI - A)^{-1}b$, it is clearly possible for this eigenvalue of A to be a pole of $H(z)$. Because the dimension of A in general may exceed the degree of the denominator of $H(z)$, the poles of $H(z)$ in general will be a subset of the eigenvalue set of A. A nontrivial and highly important result is

Theorem 18.2 *Let $H(z)$ be a rational n-th order transfer function with nonnegative impulse response and suppose it possesses a nonnegative realization $\{A, b, c\}$. Then it also possesses a nonnegative realization of $\{\bar{A}, \bar{b}, \bar{c}\}$ of lesser or equal dimension such that among the eigenvalues of $\bar{A}$ of maximum modulus, which necessarily contain a positive real eigenvalue $\bar{\lambda}, \bar{\lambda}$ is a pole of $H(z)$.*

There is an immediate new necessary condition on H(z) with nonnegative impulse response for it to have a nonnegative realization.

Corollary 18.1 *Let $H(z)$ be a rational transfer function with nonnegative impulse response. A necessary condition for $H(z)$ to have a nonnegative realization is that the poles of $H(z)$ of maximum modulus be a subset of those which are the allowed eigenvalues of maximum modulus of a nonnegative matrix.*

This explains the earlier example with $h_k = (\frac{1}{2})^k sin^2 k$. The poles of $H(z)$ are $\frac{1}{2}$, $\frac{1}{2}exp(\pm 2j)$. If there were a nonnegative realization A, b, c, there would be one with $\frac{1}{2}$ as the maximum modulus of the eigenvalue of A. Then results on nonnegative matrices [18] rule out A having an eigenvalue of $\frac{1}{2}exp(\pm 2j)$, no matter what its dimension.

3.2 Main Result

It turns out that the necessary condition of Corollary 18.1 is almost a sufficient condition for nonnegative realizability [16].

Theorem 18.3 *Let $H(z)$ as given by (18.19) have a nonnegative impulse response. Suppose that*

a) $H(z)$ has one single pole of maximum modulus, which is positive real and may be multiple; or

b) all poles of $H(z)$ of maximum modulus are simple, one is positive real, and all are located at angles corresponding to one or more integer roots of unity; further, if $\bar{\lambda}$ is the magnitude of the maximum modulus poles, $\lim_{k\to\infty} \inf \bar{\lambda}^{-k} h_k > 0$.

Then there exists an $n \times N$ matrix P such that (3.8) holds, and accordingly $H(z)$ has a nonnegative realization.

3.3 Continuous-Time Systems

Consider a transfer function $H(s) = h'(sI - F)^{-1}g$ for which the impulse response $h(t) = h'e^{Ft}g \; for \; t \geq 0$ is nonnegative. The interest is in finding triples A, b, c for which $h(t) = c'e^{At}b$ where all entries of b and c and **off-diagonal** entries of A are nonnegative. The absence of restriction on the diagonal entries of A is standard and motivated in applications. The main result is as follows

Theorem 18.4 *Let* $H(z) = \frac{\sum_{i=1}^{n} p_i s^{n-i}}{s^n + \sum_{i=1}^{n} q_i s^{n-i}}$ *be a rational n-th order transfer function corresponding to an impulse response* $h(\cdot)$*, with* $h(t) \geq 0$ *for all* $t \geq 0$*. Necessary and sufficient conditions for* $H(s)$ *to have a nonnegative realization are that there is a unique (possibly multiple) pole of* $H(s)$ *with maximal real part, and the pole is real; and with* $H(s) = h'(sI - F)^{-1}g$*, there exists* $\lambda > 0$ *such that* $H_{d\lambda}(z) = h'(zI - F - \lambda I)^{-1}g$ *has a nonnegative (discrete-time) impulse response.*

4 CONCLUSION

The main result of this paper is concerned with giving a construction of a nonnegative realization of a transfer function with nonnegative impulse response. As such, it is one result of a number concerned with carrying over an input-output property (ie property of a transfer function matrix or impulse response) to a property of an associated state-variable realization. In contrast to most such results however, realizations which are minimal in the linear systems theory sense may not be relevant.

A number of questions are left hanging. For example, how may the minimal dimension under the nonnegativity constraint, call it N_+, be determined? What is the relation among two nonnegative realizations of dimension N_+? How may the set of all nonnegative realizations of dimension N_+ be characterized?

Acknowledgements

The author wishes to acknowledge the funding of the activities of the Cooperative Research Centre for Robust and Adaptive Systems by the Australian Commonwealth Government under the Cooperative Research Centres Program.

REFERENCES

[1] Kailath, T., "Linear Systems", Englewood Cliffs, N.J: Prentice Hall, Inc., 1978

[2] Ramo, S., Whinnery, J.R., and Van Duzer, T., "Fields and waves in communication electronics", New York:John Wiley, 1965

[3] Beerand, F.P., Johnston, E.R. Jnr., "Mechanics of Materials", New York McGraw Hill, 1992

[4] Penfield, P., "Tellegen's theorem and electrical networks", Cambridge, MA: MIT Press, 1970

[5] Rief, R., "Fundametals of statistical and thermal physics", New York:McGraw Hill, 1965

[6] Bitmead, R.R., and Anderson, B.D.O., "The matrix Cauchy index: properties and applications", *SIAM J. Applied Maths.*, December 1977, pp 655-672.

[7] Newcomb, R.W., "Linear multiport synthesis", New York:McGraw Hill Book Co., 1966

[8] Jones, D.S., "Methods in electromagnetic wave propagation", Oxford University Press, 1979

[9] Vaidyanathan P.P., "Multirate Systems and Filter Banks", Englewood Cliffs, N.J:Prentice Hall, Inc., 1994

[10] Anderson, B.D.O., and Vongpanitlerd, S., "Network Analysis and Synthesis", Englewood Cliffs, N.J:Prentice Hall, Inc., 1973

[11] Maeda, H., Kodama, S., and Kajiya, F., "Compartmental system analysis: Realization of a class of linear systems with physical constraints", *IEEE Transactions on Circuits and Systems*, vol. CAS-24, no. 1, pp. 8-14, 1977

[12] Jacquez, J.A., "Compartmental Analysis in Biology and Medicine", New York: Elsevier, 1972

[13] Gersho, A., and Gopinath, B.B., "Charge-routing networks", *IEEE Trans Circuits and Systems*, vol. CAS-26, 1979, pp. 81-92.

[14] Ito, H., Amari, S., and Kobayashi, "Identifiability of Hidden Markov Information Sources and their minimum degrees of Freedom", *IEEE Transactions on Information Theory*, vol. 38, no. 2, 1992, pp. 324-333.

[15] Farina, L., and Benvenuti, L., "Positive realizations of linear systems", *System and Control Letters*, to appear.

[16] Anderson, B.D.O., Deistler, M., Farina, L,. and Benevenuti, L., "Nonnegative realization of a linear system with nonnegative impulse response", *IEEE Trans Cct and Syst*, to appear

[17] Ohta, Y., Maeda, H., and Kodama, S., "Reachability, observability and realizability of continuous-time positive systems", *SIAM J. Control and Optimization*, vol. 22, no. 2, 1984, pp. 171-180.

[18] Lancaster, P., and Tismenetsky, M., "The Theory of Matrices, Second Edition with Applications," San Diego:Academic Press 1985

19

GENERIC EIGENVALUE ASSIGNABILITY BY REAL MEMORYLESS OUTPUT FEEDBACK MADE SIMPLE

Jan C. Willems

Department of Mathematics,
University of Groningen,
P.O. Box 800, 9700 AV Groningen,
The Netherlands

Dedicated to Tom Kailath on the occasion of his sixtieth birthday.

ABSTRACT

In this paper we will prove that if $n < m*p$, then generically the controlled eigenvalues can be assigned by memoryless output feedback in a real time-invariant linear system with n states, m inputs, and p outputs.

1 INTRODUCTION

It is a pleasure for me to contribute an article in a volume dedicated to Tom Kailath at the occasion of his 60-th birthday. One of the very first system theory seminars which I ever attended – almost 30 years ago – was given by Tom Kailath at MIT. It dealt with Kalman filtering and I distinctly remember admiring his lucid and charming style of presentation and his unequaled ability to explain formal ideas in an informal way.

The topic of my paper is eigenvalue assignment, a subject which is treated in some detail in [4]. This scholarly text also discusses polynomial models in depth. At that time, I looked upon the use of polynomial matrices in linear system theory with some skepticism [11]. However, my recent work has embraced polynomial matrices with the zeal of a convert. In fact, the present paper demonstrates the effectiveness of polynomial representations in dealing with a classical problem in linear system theory: the problem of eigenvalue assignment by memoryless output feedback.

Determining conditions for generic eigenvalue assignability by real memoryless output feedback has been one of the nagging puzzles in linear system theory over the last 25 years. We will not review its history here. Kimura [5] has given a nice account of this problem in the historical session at the 1994 CDC (see also the paper [6]). For an in depth discussion of the status of this problem up to the time of publication, see also the paper by Byrnes [2].

Let n be the number of states, m the number of inputs, and p the number of outputs of a linear time-invariant system with real coefficients. That $n \leq m * p$ is a necessary condition for generic eigenvalue assignability is easy to see by counting the number of equations and unknowns. It has been known since [10] that this condition is not sufficient over $\mathcal{R}$ (although it is sufficient over $\mathcal{C}$ [3]. In [2] some very special cases are given for $n = m * p$ to be a sufficient condition over $\mathcal{R}$: $(m = 1)$, or $(p = 1)$, or $(m = 2)$ and ($p = 2^r - 1$, for some $r \in \mathcal{N}$), or $(p = 2)$ and ($m = 2^r - 1$ for some $r \in \mathcal{N}$) . However, the main result in this area is undoubtedly the result by Alex Wang [8] where he proved, using rather unaccessible mathematics, that $n < m * p$ is a sufficient condition over $\mathcal{R}$: *one additional degree of freedom compared to $\mathcal{C}$ suffices* ! In [7] an elementary proof of this result has been given, exploiting a crucial idea provided again by Alex Wang [9]. By working completely in an (A,B,C)-setting and side-stepping behavioral thinking, the proof in [9], while elementary, turned out to be not particularly transparent. The purpose of the present paper is to provide an elementary and simple proof, based on behavioral thinking.

Throughout the paper we will assume (in order to avoid mathematical technicalities) that all the time functions under consideration are infinitely differentiable. A subset $S \subseteq \mathcal{R}^N$ will be called an *algebraic variety* if there exists a real polynomial in N variables such that S coincides with the zero set of this polynomial. If an algebraic variety S is a strict subset of $\mathcal{R}^N$, then it will be called *proper*. The complement of a proper algebraic variety is called *generic:* it is open, dense, and measure exhausting, i.e., its complement has Lebesgue measure zero.

2 THE SYSTEM AND ITS KERNEL REPRESENTATION

Consider the ubiquitous linear time-invariant system

$$\frac{dx}{dt} = Ax + Bu, \quad y = Cx \tag{19.1}$$

with $x \in \mathcal{R}^n, u \in \mathcal{R}^m, y \in \mathcal{R}^p$, and $(A, B, C) \in \mathcal{R}^{n^2+nm+pn}$. Let $\mu := ENT(\frac{n}{p})$, the smallest integer $\geq \frac{n}{p}$, $p_2 := \mu p - n$, and $p_1 := p - p_2$. When n is divisible by p, then $p_2 = 0$. In this case the vectors and matrices of size p_2 are assumed to be absent. Partition y and C conformably as

$$y = \left[\frac{y_1}{y_2}\right] = \left[\frac{C_1}{C_2}\right] x \tag{19.2}$$

with $y_k \in \mathcal{R}^{p_k}$ and $C_k \in \mathcal{R}^{p_k \times n}, k = 1, 2$.

Let $\mathcal{G} \subset \mathcal{R}^{n^2+nm+pn}$ denote the (generic) subset consisting of those (A, B, C)'s such that the $n \times n$ matrix

$$M(A,B,C) := \begin{bmatrix} C \\ CA \\ \vdots \\ CA^{\mu-2} \\ C_1 A^{\mu-1} \end{bmatrix} \tag{19.3}$$

is invertible. Note that this implies that (19.1) has observability indices

$$(\mu_1, \mu_2, \cdots, \mu_p) \tag{19.4}$$

with, if n is divisible by p, $\mu_1 = \cdots = \mu_p = \mu$, while, if n is not divisible by p, $\mu_1 = \cdots = \mu_{p_1} = \mu$ and $\mu_{p_1+1} = \cdots = \mu_p = \mu - 1$. When it is obvious from the context what (A, B, C) is, we will denote $M(A, B, C)$ simply by M.

It follows from general principles (see [12] for a proof of the analogous discrete-time result – easily be generalized to continuous-time) that the input/output behavior of (19.1) will be governed by a system of differential equations. More precisely, define the *manifest behavior* of (19.1) as

$$\begin{aligned} \mathcal{B} := \{(y,u) \in C^\infty(\mathcal{R}, \mathcal{R}^p \times \mathcal{R}^m) | \ \exists x \in C^\infty(\mathcal{R}, \mathcal{R}^m) \\ \text{such that (19.1) holds } \} \end{aligned} \tag{19.5}$$

Then there exists a polynomial matrix $R \in \mathcal{R}^{p \times (p+m)}$ such that $\mathcal{B}$ consists precisely of the C^∞-solution of

$$R(\frac{d}{dt}) \begin{bmatrix} y \\ u \end{bmatrix} = 0 \tag{19.6}$$

If this is the case we will call (19.6) a *kernel representation* of the manifest behavior of (19.1).

We will now compute such a kernel representation of the manifest behavior of (19.1), assuming $(A, B, C) \in \mathcal{G}$. We thus need to eliminate x from (19.1). Let

$$z := Mx \tag{19.7}$$

and observe, by repeatedly differentiating y using (19.1), that z equals

$$\begin{bmatrix} y \\ \frac{dy}{dt} - CBu \\ \frac{d^2y}{dt} - CB\frac{du}{dt} - CABu \\ \vdots \\ \frac{d^{\mu-2}y}{dt^{\mu-2}} - CB\frac{d^{\mu-3}u}{dt^{\mu-3}} - \cdots - CA^{\mu-3}Bu \\ \frac{d^{\mu-1}y_1}{dt^{\mu-1}} - C_1B\frac{d^{\mu-2}u}{dt^{\mu-2}} - \cdots - C_1A^{\mu-2}Bu \end{bmatrix} \tag{19.8}$$

Now express, using (19.8), that

$$y = CM^{-1}z \tag{19.9a}$$

and that

$$\frac{dz}{dt} = MAM^{-1}z + MBu \tag{19.9b}$$

With z given by (19.8), the system of equations (19.9) consists in total of $(n+p)$ equations in y, u and their derivatives. These equations form the differential equation which we are trying to find. Now substitute the expression for M (19.3) in (19.9). We omit the straightforward details, but it turns that the first n of these are trivial identities. The last p equations lead to the effective relations between y, u and their derivatives. These are of the form

$$R_0w + R_1\frac{dw}{dt} + \cdots + R_{\mu-1}\frac{d^{\mu-1}w}{dt^{\mu-1}} + R_\mu\frac{d^\mu w}{dt^\mu} = 0 \tag{19.10}$$

where

$$w := \begin{bmatrix} y \\ u \end{bmatrix} \tag{19.11}$$

and $R_k \in \mathcal{R}^{p\times(p+m)}, k = 0, 1, \cdots, \mu$. Moreover, $R_{\mu-1}$ and R_μ have the following special form

$$R_\mu = \begin{bmatrix} I_{p_1\times p_1} & O_{p_1\times p_2} & O_{p_1\times m} \\ O_{p_2\times p_1} & O_{p_2\times p_2} & O_{p_2\times m} \end{bmatrix} \tag{19.12}$$

$$R_{\mu-1} = \begin{bmatrix} * & O_{p_1\times p_2} & * \\ * & I_{p_2\times p_2} & O_{p_2\times m} \end{bmatrix} \tag{19.13}$$

The $*$'s in (19.13) and $R_0, \cdots, R_{\mu-2}$ in (19.10) are matrices without any special structure. The total number of free parameters in $R_0, R_1, \ldots, R_{p-1}, R_p$ and hence in the associated differential equation (19.10) is thus $n(m+p)$. By a slight abuse of notation, we will hence write $(R_0, R_1, \cdots, R_\mu) \in \mathcal{R}^{n(m+p)}$. The equation (19.10) is the desired kernel representation referred to earlier.

The above manipulations define a map

$$\begin{aligned} f : \mathcal{G} \subset \mathcal{R}^{n^2+nm+pn} &\to \mathcal{R}^{n(m+p)} \\ (A, B, C) &\stackrel{f}{\longmapsto} (R_0, R_1, \cdots, R_{\mu-1}, R_\mu) \end{aligned} \tag{19.14}$$

f is obviously (see (6)) a rational map and, since $detM(A, B, C) \neq 0$ on $\mathcal{G}$, its denominator does not vanish on $\mathcal{G}$. It is well-known from the work of Wolovich [14] and others that in turn every input/output system (19.10,19.11) such that (19.12,19.13) holds will be the external behavior of a system (19.1) with $(A, B, C) \in \mathcal{G}$.

We will denote the family of linear systems described by

$$R(\frac{d}{dt})w = 0 \tag{19.15}$$

with $R \in \mathcal{R}^{p\times(p+m)}[\xi]$ of the special form (19.10,19.12,19.13) by Σ. We have seen that f is a surjective rational map from $\mathcal{G} \subset \mathcal{R}^{n^2+n(m+p)}$ to $\Sigma \cong \mathcal{R}^{n(m+p)}$.

Note that up to now, we have merely reviewed some simple facts from multivariable linear system theory, as they can be found in Wolovich [14]. Recapitulating, we have seen that considering systems (19.1) with generic observability indices (19.4) amounts to considering systems (19.10, 19.12, 19.13). Actually, persons familiar with the behavioral approach may take (19.10, 19.12, 19.13) as the starting point of the pole placement question, and consider (19.1) merely as the traditional starting point.

3 FEEDBACK AND INTERCONNECTION

Now consider the linear memoryless output feedback law

$$u = Fy \tag{19.16}$$

Applied to (19.1), this yields the closed loop system given by

$$\frac{dx}{dt} = (A + BFC)x \tag{19.17}$$

Its characteristic polynomial is equal to

$$det(I\xi - A - BFC) \tag{19.18}$$

It is straightforward to express this characteristic polynomial directly in terms of the system $\Sigma = f(A, B, C)$. The characteristic polynomial (19.18) equals

$$det\left[\frac{R(\xi)}{-F \mid I_{m\times m}}\right] \tag{19.19}$$

The eigenvalue assignability problem by memoryless output feedback is formulated as follows. Let (19.1) be given and let $d \in \mathcal{R}[\xi]$,

$$d(\xi) = d_0 + d_1\xi + \cdots + d_{n-1}\xi^{n-1} + \xi^n \tag{19.20}$$

be a given monic polynomial. The question is when there exits an $F \in \mathcal{R}^{p\times m}$ such that the characteristic polynomial associated with (19.17) equals d, i.e., such that

$$det(I\xi - A - BFC) = d(\xi) \tag{19.21}$$

Note that we have just shown that this is equivalent to

$$det\left[\frac{R(\xi)}{-F \mid I_{m\times m}}\right] = d(\xi) \tag{19.22}$$

Persons familiar with polynomial representations of linear systems, in particular with the theory of [4], could take (19.22) as the starting point of this paper: (19.6,19.12,19.13) is a canonical representation of a linear system with McMillan degree n and generic observability indices (19.4). The eigenvalue placement question is then formulated by (19.22), and we will prove that for generic elements in $\mathcal{G}$, c.q. in Σ, (19.21), c.q. (19.22), is solvable for all d's if $n < m * p$.

We will now consider control in a behavioral setting, in the spirit of [12, 13]. Let (19.15) denote the *plant* and consider as *controller* the memoryless linear dynamical system

$$Kw = 0 \tag{19.23}$$

with $K \in \mathcal{R}^{m\times(p+m)}$. This yields the *controlled system*

$$\left[\frac{R(\frac{d}{dt})}{K}\right] w = 0 \tag{19.24}$$

Call the real monic polynomial whose roots coincide with

$$det\left[\frac{R(\xi)}{K}\right] \tag{19.25}$$

the *characteristic polynomial* of (19.24). Note that its degree does not exceed n, the McMillan degree of R. We will say that for a plant $R \in \Sigma$ *eigenvalue assignment by real memoryless feedback* is possible if for all monic polynomials $\pi \in \mathcal{R}[\xi]$, with *degree* $(\pi) \leq n$, there exists a K such that (19.24) has characteristic polynomial π.

4 THE MAIN RESULT

The main result of this paper is the following

Theorem 19.1 *Assume that $n < m * p$. Then generically for $R \in \Sigma \cong \mathcal{R}^{n(m+p)}$ eigenvalue assignment by real memoryless feedback is possible.*

Proof: (i) In the first step of the proof we will examine the expansion of (19.25) as a power series in K around a point K_0. Let $R \in \Sigma$ be given. Consider the map h which assigns to K the polynomial (19.25). Let (19.25) be given by

$$\beta_0 + \beta_1 \xi + \ldots + \beta_{n-1} \xi^{n-1} + \beta_n \xi^n \tag{19.26}$$

where we have used the fact, easily proven by examining R, that (19.25) is a polynomial of degree *leq* n. Thus h is a map from $\mathcal{R}^{m(m+p)}$ to $\mathcal{R}^{n+1}$ (associate the coefficients $(\beta_0, \beta_1, \ldots, \beta_{n-1}, \beta_n)$ of the polynomial (19.25) – a polynomial of degree at most n – with a vector in $\mathcal{R}^{n+1}$); h is, in fact, obviously a polynomial map.

Consider the power series expansion of h at the point $K_0 \in \mathcal{R}^{m\times(p+m)}$. There holds

$$h(K_0 + \Delta)(\xi) = det \left[\frac{R(\xi)}{K_0}\right] + \sum_{k=1}^{m} \sum_{\ell=1}^{m+p} \Delta_{k\ell} M_{k\ell}(\xi) + h.o.t. \tag{19.27}$$

where $\Delta_{k\ell}$ denotes the (k, ℓ)-th element of Δ, $M_{k\ell}$ equals $(-1)^{p+k+\ell}$ times the minor obtained by crossing out the $(p+k)$-th row and ℓ-th column of $\left[\frac{R(\xi)}{K_0}\right]$, and $h.o.t.$ means quadratic or higher order terms in the elements of Δ.

We will now show that R has the eigenvalue assignability property if K_0 has the following properties

(a)

$$det \left[\frac{R(\xi)}{K_0}\right] = 0 \tag{19.28}$$

and

(b) the polynomials $M_{k\ell}(\xi)$, $k = 1, \ldots, m; \ell = 1, \ldots, m+p$, span the $(n+1)$-dimensional vector space of real polynomials of degree $\leq n$.

Indeed, by the implicit function theorem, (a) and (b) combined imply (since the linear part of h is surjective at K_0) that the image of h contains an open neighborhood of the origin. However, if the image of h contains the polynomial π, then it also contains $\alpha\pi$ (simply pre-multiply K by the diagonal matrix diag$(\alpha, 1, \cdots, 1)$ and examine (19.24)). Hence (a) and (b) imply that h is surjective indeed.

(ii) In the behavioral literature [13] a controller satisfying (19.28) is called *dependent*, since the plant and the controller have then non-trivial laws in common. In [1] such controllers are called *non-admissible.* In the second step of the proof, we will prove the existence of such a dependent controller.

The row degrees of $R \in \Sigma$ are equal to the observability indices (19.4). Thus if n is divisible by p, there holds $\mu_1 = \cdots = \mu_p = \mu$, while if n is not divisible by p, there holds $\mu_1 = \cdots = \mu_{p_1} = \mu$ and $\mu_{p_1+1} = \cdots = \mu_p = \mu - 1$. In any case, $n < m * p$ implies $\mu_p < m$. We will prove that $\mu_p < m$ implies the existence of a $K_0 \in \mathcal{R}^{m\times(p+m)}$ such that (19.28) holds. Indeed, we can associate with each $R \in \Sigma$ such a K_0 in a canonical way, as follows.

Let the vector polynomial

$$r_p^0 + r_p^1\xi + \cdots + r_p^{\mu_p-1}\xi^{\mu_p-1} + r_p^{\mu_p}\xi^{\mu_p} \tag{19.29}$$

denote the p-th (i.e., the last) row of R. Take

$$K_0 = \begin{bmatrix} r_p^0 \\ r_p^1 \\ \vdots \\ r_p^{\mu_p-1} \\ r_p^{\mu_p} \\ e_{p+\mu_p+1} \\ \vdots \\ e_{m+p-1} \end{bmatrix} \tag{19.30}$$

where e_k equals the row vector $[0, \cdots, 0, 1, 0, \cdots, 0]$ with the 1 in the k-th entry. It is obvious, by examining (19.29) and (19.30), that (19.28) will indeed hold for the K_0 given by (19.30).

(iii) The third part of the proof consists in showing that condition (b) of part (ii) is generically satisfied for the K_0 given by (19.30). In other words, we will prove that $n < m*p$ implies that the K_0 given by (19.30) will yield polynomials $M_{k\ell}(\xi)$ which, generically for Σ, span the polynomials of degree $\leq n$.

In order to see this, observe first that the set of R's in Σ such that the $M_{k\ell}$'s do not span, forms an algebraic variety in $\Sigma \cong \mathcal{R}^{n(p+m)}$. Indeed, the K_0 given by (19.30) is a linear function of R; the coefficients of the $M_{k\ell}$'s are polynomial functions of the coefficients of R; non-spanning thus requires a certain polynomial in the coefficients of R to be zero. The algebraic variety in question is a proper one, for the following R (suggested to me by Joachim

Rosenthal) is an exception

$$\begin{bmatrix} \xi^{\mu_1} & 1 & 0 \cdots 0 & 0 & 0 & \cdots 0\ 0 \cdots 0 \\ 0 & \xi^{\mu_2} & 1 \cdots 0 & 0 & 0 & \cdots 0\ 0 \cdots 0 \\ \vdots & \vdots & \vdots \vdots \vdots \vdots \vdots & \vdots & \vdots & \vdots \vdots \vdots \vdots \vdots \vdots \vdots \\ 0 & 0 & 0 \cdots \xi^{\mu_{p-1}} & 0 & 0 & \cdots 0\ 0 \cdots 1 \\ 0 & 0 & 0 \cdots 0 & \xi^{\mu_p} & \xi^{\mu_p - 1} & \cdots 1\ 0 \cdots 0 \end{bmatrix} \tag{19.31}$$

The corresponding K_0 is

$$[O_{m \times (p-1)}\ I_{m \times m}\ O_{m \times 1}] \tag{19.32}$$

and the set of $M_{k\ell}(\xi)$'s contains the monomials

$$\begin{array}{l} 1, \xi, \cdots, \xi^{\mu_p} \\ \xi^{\mu_1}, \xi^{\mu_1+1}, \cdots, \xi^{\mu_1+\mu_p} \\ \xi^{\mu_1+\mu_2}, \xi^{\mu_1+\mu^2+1}, \cdots, \xi^{\mu_1+\mu_2+\mu_p} \\ \vdots \\ \xi^{\mu_1+\mu_2+\cdots+\mu_{p-1}}, \xi^{\mu_1+\mu_2+\cdots+\mu_{p-1}+1}, \cdots, \xi^{\mu_1+\mu_2+\cdots+\mu_{p-1}+\mu_p} \end{array} \tag{19.33}$$

These obviously span the polynomials of degree $\leq n$. This completes the proof of Theorem 19.1. ∎

Note the surprisingly crucial role played in the proof by the dependent controller (19.30), whose existence was guaranteed by the fact that μ_p is less than m!

5 MEMORYLESS FEEDBACK IN STATE MODELS

Examine again the characteristic polynomial (19.25) and partition K as

$$K = [K_1 \mid K_2] \tag{19.34}$$

with $K_1 \in \mathcal{R}^{m \times p}$ and $K_2 \in \mathcal{R}^{m \times m}$. Now observe that (19.25) has degree exactly n if and only if K_2 is invertible. In the behavioral literature [13], we call a controller (19.23) *regular* if (19.25) has degree n, *singular* (they correspond to non-proper transfer functions) if (19.25) has degree n, and *dependent* if (19.25) is zero. A regular control law (19.34) is obviously equivalent to the feedback control law

$$[K_2^{-1}K_1 \mid I\ ]w = 0 \tag{19.35}$$

i.e.,

$$u = -K_2^{-1}K_1 y \tag{19.36}$$

and lead to polynomials (19.25) which are monic and of degree n. Note that the generic set of R's from Σ for which eigenvalue assignability holds will also have the property that for each real *monic* polynomial π of degree n, there exists a control law as (19.34) , whence, after pre-multiplication with K_2^{-1} we obtain a control law of the type

$$[-F \mid I\ ]w = 0 \tag{19.37}$$

i.e.,

$$u = Fy \tag{19.38}$$

achieving π as characteristic polynomial. By (19.19) this π will also be equal to (19.18).

We are now ready to deliver the *coup de grâce.*

Theorem 19.2 *Assume that $n < m * p$. Then generically for $(A, B, C) \in \mathcal{R}^{n^2+nm+pn}$ there exists, for each real monic polynomial π of degree n, an $F \in \mathcal{R}^{m\times p}$ such that*

$$\pi(\xi) = det(I\xi - A - BFC) \tag{19.39}$$

Proof: Let

$$g : \mathcal{R}^{n(m+p)} \to \mathcal{R} \tag{19.40}$$

be a nonzero polynomial map such that

$$g(R_0, R_1, \cdots, R_{\mu-1}, R_\mu) \neq 0 \tag{19.41}$$

implies that $R \in \Sigma$ has the assignability property. The existence of such a g is the content of theorem 6. Now consider the (A, B, C)'s such that both

$$detM(A, B, C) \neq 0 \tag{19.42}$$

$$g(f(A, B, C)) \neq 0 \tag{19.43}$$

Note that $g \circ f$ is a rational map in (A, B, C). Clear the denominator of $g \circ f$ by multiplying by a sufficiently high power of $det(M(A, B, C)$. Call g' the resulting polynomial in (A, B, C). It follows that those (A, B, C)'s for which

$$(detM(A, B, C))^2 + (g'(A, B, C))^2 \neq 0 \tag{19.44}$$

indeed have the eigenvalue assignability property. Obviously (19.44) defines the complement of an algebraic variety. It is a proper one, since any (A, B, C) such that $f(A, B, C)$ equals (19.31) does not belong to it! ∎

6 CLOSING COMMENT

It is worth contrasting the proof of our result which guarantees generic eigenvalue assignability over $\mathcal{R}$ if $n < m * p$ with the proof of the result [3] which guarantees generic eigenvalue assignability over $\mathcal{C}$ if $n \leq m * p$. Let $\mathcal{F} = \mathcal{R}$ or $\mathcal{C}$. Consider the map h which associates with the $m \times (p+m)$ matrix $K \in \mathcal{F}^{m \times (p+m)}$ the polynomial (19.25) $\in \mathcal{F}[\xi]$. This polynomial is of degree at most n. Let $\nu : \mathcal{F}[\xi] \to \mathcal{F}[\xi]$ denote the map which associates with a polynomial, the monic polynomial with the same roots. Let finally $\mathcal{X} := \nu \circ h$. The problem is to prove the surjectivity of $\mathcal{X}$.

Our proof is based on the fact that if $n < m * p$, then generically the image of h will contain a neighborhood of the origin. This is shown by considering the Taylor series expansion of h around a dependent controller K_0. Using the special structure of h (by *blowing up* the neighborhood of the origin) it follows that h will be surjective if $im(h)$ contains a neighborhood of the origin. Alternatively, we could have shown that, because of the special structure of ν, $\mathcal{X}$ will be surjective if $im(h)$ contains a neighborhood of the origin. Two reasons thus for $\mathcal{X}$ to be surjective.

When $\mathcal{F} = \mathcal{C}$, we can do one degree of freedom better. Expand $\mathcal{X}$ in a Taylor series around any controller K_0 and prove by considering a Taylor series expansion around $K_0 = [0_{m \times p} \mid I_{m \times m}]$ that if $n \leq m * p$, then generically the image of $\mathcal{X}$ will contain an open neighborhood of $\chi(K_0)$. This implies, using the *dominant morphism theorem*, that im $\mathcal{X}$ then contains the complement of a proper algebraic variety, and with some more work we may conclude that $\mathcal{X}$ is surjective.

REFERENCES

[1] R.W. Brockett and C.I. Byrnes, Multivariable Nyquist criteria, root loci, and pole placement, *IEEE Transactions on Automatic Control,* Vol. 26, pp. 271–284, 1981.

[2] C.I. Byrnes, Pole assignment by output feedback, *Three Decades of Mathematical System Theory, A Collection of Surveys at the Occassion of the 50th Birthday of Jan C. Willems*, Lecture Notes in Control and Information Sciences, vol. 135, Springer-Verlag, pp. 31–78, 1989.

[3] R. Hermann and C.F. Martin, Applications of algebraic geometry to system theory, Part I, *IEEE Transactions on Automatic Control,* Vol. 22, pp. 19-25, 1977.

[4] T. Kailath, *Linear Systems*, Prentice Hall, 1980.

[5] H. Kimura, Pole assignment by output feedback: A longstanding open problem, *Proceedings of the 33rd IEEE Conference on Decision and Control,* pp. 2101–2105, 1994.

[6] H. Kimura, From output feedback pole-assignment to robust stabilization, *Journal of the Society of Instrument and Control Engineers of Japan,* Vol. 34, pp.613–620, 1995.

[7] J. Rosenthal, J.M. Schumacher, and J.C. Willems, Generic eigenvalue assignment by memoryless real output feedback, *Systems & Control Letters,* Vol. 26, pp. 253-260, 1995.

[8] X. Wang, Pole placement by static output feedback, *Journal of Mathematical Systems, Estimation, and Control,* Vol. 2, pp. 205–2182, 1992.

[9] X. Wang, Grassmannian, central projection and output feedback pole assignment of linear systems, Preprint, July 1994.

[10] J.C. Willems and W.H. Hesselink, Generic properties of the pole placement problem, *Proceedings of the 7th IFAC Congress,* pp. 1725-1729, 1978.

[11] J.C. Willems, Book review of "*Linear Systems,* by T. Kailath", *Automatica,* Vol. 18, pp. 497–499, 1982.

[12] J.C. Willems, Paradigms and puzzles in the theory of dynamical systems, *IEEE Transactions on Automatic Control,* Vol. 36, pp. 259–294, 1991.

[13] J.C. Willems, On interconnections, control and feedback, Submitted.

[14] W.A. Wolovich, *Linear Multivariable Systems,* Springer Verlag, 1974.

20

FUNDAMENTAL LIMITATIONS OF CONTROL SYSTEM PERFORMANCE

K. J. Åström

Department of Automatic Control
Lund Institute of Technology
Box 118, S-221 00 Lund, Sweden

To Tom Kailath on the occasion of his sixtieth birthday.

ABSTRACT

A typical control design problem has to deal with many different factors such as process dynamics, disturbances, process uncertainties and actuator saturation. There are two principal types of disturbances, load disturbances that drive the system away from its desired behavior and measurement noise that corrupts the information obtained by the sensors. The full design problem is naturally quite complicated. In this paper it is attempted to formulate some problems that capture the essence of the design problem in such a way that we can establish fundamental limitations on achievable performance. Results of this type are a nice complement to computational tools for control system design. They make it possible to quickly explore a design problems for preliminary assessment before attempting to do massive computations.

1 INTRODUCTION

Much research has been devoted to developing methods for designing control systems. A common approach has been to strive for optimality, see [13], and [10]. While such methods indirectly tell what the best performance is they do not give good physical insight into the mechanisms that cause the limitations. There has been relatively little interest to determine the fundamental limitations in control systems. This is an interesting contrast to the situation in the field of information theory which is also addressed in this book. It is interesting to see that the early work in our field paid a lot of attention to these issues, see [9]. The usefulness of investigating fundamental limitations was emphasized very clearly in the 1989 Bode lecture, see [1]. It is also discussed in the papers ([6]) and [12],

In this paper we will develop methods that explicitly gives such limitations. The first result for minimum phase systems is based on a simple investigation of the properties of lead compensation. The other results are based on classical results of Bode, more specifically the relations between the amplitude and the phase of a linear transfer function. Results of a similar type were presented in [1], these results were based on Bode's integral formula. The presentation is restricted to single-input single-output systems.

2 MINIMUM PHASE SYSTEMS

It is often stated in textbooks that arbitrarily high bandwidth can be obtained if the process is minimum phase, i.e. if it has no poles or zeros in the right half plane. This is true if the systems are linear without measurement noise. The situation is, however, very different in the presence of measurement noise, because more measurement noise is injected into the system when the bandwidth is increased. This generates large control signals which will saturate the actuators. Measurement noise and actuator saturations are thus factors that limit the performance for a minimum phase systems. The following calculation gives a quantitative result.

Lead compensation is required to increase the bandwidth of a system. Let us first consider a simple lead compensator with the transfer function.

$$G(s) = \frac{s+a}{s/N+a} \tag{20.1}$$

This compensator has a maximum gain

$$\max_{\omega} |G(i\omega)| = N$$

and it gives a phase lead of at most

$$\varphi = \max_{\omega} \arg G(i\omega) = \arctan \frac{N-1}{2\sqrt{N}}$$

The achievable phase lead increases with N but is limited by 90°. Several compensators can be cascaded to obtain a larger phase lead. If n identical compensators with the transfer function (20.1) are cascaded we find after some calculations that the smallest gain that gives a phase lead φ is given by

$$N_n = \left(1 + 2\tan^2\frac{\varphi}{n} + 2\tan\frac{\varphi}{n}\sqrt{1+\tan^2\frac{\varphi}{n}}\ \right)^n \tag{20.2}$$

Some numerical values of the gain is given in Table 20.1. Notice that it is advantageous to have a large value of n because a given phase lead is then obtained with lower gain. The limiting case as $n \to \infty$ is interesting. We have

$$N_\infty = e^{2\varphi} \tag{20.3}$$

Table 20.1 Maximum of compensators that give a specified phase lead. The table is based on n first order compensators with the transfer function (20.1).

Phase lead	n=2	n=4	n=6	n=8	n=∞
90°	34	25	24	24	23
180°	-	1150	730	630	540
225°	-	14000	4800	3300	2600

This formula is interesting because it gives the smallest gain required to obtain a given phase lead with a compensator of arbitrarily high order. It can be shown that this limit does not depend on the specific form of the transfer function (20.1). Equation (20.3) can be used to determine the largest achievable bandwidth for minimum phase systems.

To illustrate how the results can be used, consider a control system where the inputs and outputs have the range 0 to 10 V. Assume that a digital controller with an AD-converter having 12 bits of resolution is used. If the controller gain is 4096 it then follows that a change of the measured signal of one bit will saturate the output. In this case the maximum gain of the system is thus limited to 4096. It then follows from Equation (20.3) that the phase lead can be at most 238°. If we make the more reasonable assumption that one bit of the measured signal can give a change in the output of the control of at most 10% of its span we find that the maximum lead is limited to 172°. An estimate of the limitations caused by measurement noise can be done in a similar way.

3 PRELIMINARIES

To consider systems where process dynamics imposes limitations we will use the following result which was shown in [9].

Consider a transfer function $G(s)$ with no poles or zeros in the right half plane. Assume that $|\log G(s)|/|s|$ goes to zero as s goes to infinity, then

$$\arg G(i\omega) = \frac{1}{\pi}\int_{-\infty}^{\infty} \frac{d\log|G(iu)|}{du} \log\coth|\frac{u}{2}|du \qquad (20.4)$$

where $u = \log\omega/\omega_0$. It follows from this equation that if the amplitude curve in the Bode diagram is a straight line with slope n the formula reduces to

$$\varphi = n\frac{\pi}{2} \qquad (20.5)$$

This relation is taught in practically all elementary courses in feedback control. In this paper we will show that Equation (20.5) can be used to obtain fundamental limitations on the achievable performance of a control system. Notice also that it is possible to find minimum phase compensators that approximate Equation (20.5) arbitrarily well even for non-integer n.

The Design Inequality

Consider the problem of controlling of a system with the transfer function $G(s)$. Factor the transfer function as

$$G(s) = G_{mp}(s)G_{nmp}(s) \tag{20.6}$$

where G_{mp} is the minimum phase part and G_{nmp} is the nonminimum phase part. Let $\Delta G(s)$ denote the uncertainty in the process transfer function. It is assumed that the factorization is normalized so that $|G_{nmp}(i\omega)| = 1$. We will characterize the achievable bandwidth by the gain crossover frequency ω_{gc}, which is defined as the lowest frequency where the loop transfer function has magnitude one.

Let the minimum phase part be compensated so that

$$\varphi_{mp} = n_{gc}\frac{\pi}{2} \tag{20.7}$$

where n_{gc} is the slope of the compensated minimum phase part of the system. This can always be achieved with arbitrary accuracy by loop shaping.

The idea to compensate a minimum phase system so that slope of the amplitude curve of the loop transfer function is constant over a specified frequency range, was suggested by Bode in his work on feedback amplifier. He coined the term *ideal cut-off characteristics* for such a system.

Assume that the desired phase margin should be at least φ_m, then

$$n_{gc}\frac{\pi}{2} + \arg G_{nmp}(i\omega_{gc}) + \arg \Delta G(i\omega_{gc}) \geq -\pi + \varphi_m \tag{20.8}$$

This equation is a design inequality which gives a limit of the achievable crossover frequency ω_{gc}. It clearly shows the limitations imposed by nonminimum phase elements and process uncertainty.

For minimum phase systems Equation (20.8) reduces to

$$n_{gc} = -2 + \frac{2\varphi_m}{\pi}$$

A phase margin of $\varphi_m = \pi/4$ gives a slope $n_{gc} = -1.5$ but naturally no limitation on the bandwidth. For nonminimum phase systems n_{gc} must be larger to satisfy the inequality.

4 SYSTEMS WITH ZEROS IN THE RIGHT HALF PLANE

We will now discuss limitations imposed by right half plane zeros. We will first consider systems with only one zero in the right half plane. The non-minimum phase part of the plant transfer function then becomes.

$$G_{nmp}(s) = \frac{a-s}{a+s} \tag{20.9}$$

We have

$$\arg G_{nmp}(i\omega) = -2\arctan\frac{\omega}{a}$$

Neglecting process uncertainties it follows from the design inequality, Equation (20.8) that

$$n_{gc}\frac{\pi}{2} - 2\arctan\frac{\omega_{gc}}{a} \geq -\pi + \varphi_m$$

Hence

$$\frac{\omega_{gc}}{a} \leq \tan(\frac{\pi}{2} - \frac{\varphi_m}{2} + n_{gc}\frac{\pi}{4}) \tag{20.10}$$

Assume, for example, that we want a phase margin $\varphi_m = \pi/4$ and that the slope of the compensated minimum phase part is $n_{gc} = -1/2$, we then find that the crossover frequency is limited to

$$\omega_{gc} < a$$

A right half plane zero thus gives an upper bound to the achievable bandwidth. The bandwidth decreases with decreasing frequency of the zero. It is thus more difficult to control systems with slow zeros. A zero at the origin is a limiting case.

The calculations can be extended to the case of several zeros in the right half plane. If there are k distinct real zeros in the right half plane the inequality (20.10) becomes

$$\sum_{i=1}^{k}\frac{\omega_{gc}}{a_i} \leq \tan(\frac{\pi}{2} - \frac{\varphi_m}{2} + n_{gc}\frac{\pi}{4})$$

5 SYSTEMS WITH DEAD TIME

System with dead time will now be investigated. The transfer function for such systems has an essential singularity at infinity. The non-minimum phase part of the transfer function of the process is

$$G_{nmp}(s) = e^{-sL} \tag{20.11}$$

We have

$$\arg G_{nmp}(i\omega) = -\omega L$$

Neglecting process uncertainties it follows from the design inequality, Equation (20.8), that

$$n_{gc}\frac{\pi}{2} - \omega_{gc}L \geq -\pi + \varphi_m$$

Hence

$$\omega_{gc}L \leq \pi - \varphi_m + n_{gc}\frac{\pi}{2} \tag{20.12}$$

Assume that we want a phase margin $\varphi_m = \pi/4$ and a slope at the crossover frequency of $n_{gc} = -1/2$, then

$$\omega_{gc}L \leq \frac{\pi}{2}$$

We thus find that time delays give an upper bound on the achievable bandwidth.

The calculations can be extended to systems with time delay and several zeros in the right half plane.

6 SYSTEMS WITH POLES IN THE RIGHT HALF PLANE

The limitations imposed by right half plane poles will now be investigated. Consider a system with one pole in the right half plane. The non-minimum phase part of the transfer function is thus

$$G_{nmp}(s) = \frac{s+b}{s-b} \tag{20.13}$$

To have a stable closed loop system it follows from Nyquist's stability criterion that the Nyquist curve must make one encirclement of the critical point. The transfer function G_{nmp} has been parameterized to preserve this condition. We have

$$\arg G_{nmp}(i\omega) = -2\arctan\frac{b}{\omega}$$

Neglecting process uncertainties it follows from the design inequality, Equation (20.8), that

$$n_{gc}\frac{\pi}{2} - 2\arctan\frac{b}{\omega_{gc}} \geq -\pi + \varphi_m$$

Hence

$$\omega_{gc} \geq \frac{b}{\tan(\pi/2 - \varphi_m/2 + n_{gc}\pi/4)} \tag{20.14}$$

Assume for example that we want a phase margin $\varphi_m = \pi/4$ and a slope at the crossover frequency of $n_{gc} = -1/2$, the crossover frequency is limited to

$$\omega_{gc} \geq b$$

Unstable poles thus give a lower bound on the crossover frequency. For systems with right half plane poles the bandwidth must thus be sufficiently large. By computing the phase lag of the minimum phase part of the system at ω_{gc}

we can determine the phase lead required to compensate the minimum phase part. Equation (20.3) then gives the gain required to achieve the phase lead. Knowledge of the measurement noise and the saturation levels of the control signal then indicates the feasibility of stabilizing the system.

Notice that systems with fast unstable poles are more difficult to control than systems with slow unstable poles. Compare with unstable zeros where we had the reverse situation.

7 SYSTEMS WITH POLES AND ZEROS IN THE RIGHT HALF PLANE

The calculations can be extended to cases with both poles and zeros in the right half plane. We will give the results for a pole-zero pair. The nonminimum phase part of the transfer function is then

$$G_{nmp}(s) = \frac{(a-s)(s+b)}{(a+s)(s-b)} \tag{20.15}$$

For $a > b$ we have

$$\arg G_{nmp}(i\omega) = -2\arctan\frac{\omega}{a} - 2\arctan\frac{b}{\omega} = -2\arctan\frac{\omega_{gc}/a + b/\omega_{gc}}{1 - b/a}$$

Neglecting process uncertainties it follows from the design inequality, Equation (20.8), that

$$\frac{\omega_{gc}}{a} + \frac{b}{\omega_{gc}} \leq (1 - \frac{b}{a})\alpha$$

where

$$\alpha = \tan(\frac{\pi}{2} - \frac{\varphi_m}{2} + n_{gc}\frac{n_{gc}}{4})$$

The left hand side has its smallest value for

$$\omega_{gc} = \sqrt{ab} \tag{20.16}$$

Hence

$$2\sqrt{\frac{b}{a}} \leq \alpha(1 - \frac{b}{a})$$

This implies

$$\frac{a}{b} \geq 1 + \frac{2 + 2\sqrt{1+\alpha^2}}{\alpha^2} \tag{20.17}$$

Assume for example that we want a phase margin $\varphi_m = \pi/4$ and a slope at the crossover frequency of $n_{gc} = -1/2$, then the inequality becomes

$$a \geq 5.83b$$

It is thus necessary that the unstable zero is faster than the unstable pole and that the ratio between them is sufficiently large. If this is not the case the desired phase margin cannot be obtained. Also notice that the best gain crossover frequency is the geometric mean of the unstable pole and zero.

The X-29

Considerable design effort has been devoted to the design of the flight control system for the X-29 aircraft. One of the design criteria was that the phase margin should be greater than 45° for all flight conditions. At one flight condition the model has the following nonminimum phase component

$$G_{nmp}(s) = \frac{s-26}{s-6}$$

Since $a = 4.33b$, it follows from the discussion given above that a phase margin of 45° cannot be achieved. It is interesting to note that many design methods were used in a futile attempt to reach the design goal. A simple calculation of the type given in this paper would have given much insight.

8 CONCLUSIONS

In this paper we have formulated and solved several problems that make it possible to estimate the limitations on control system performance that are imposed by measurement noise, actuator saturation, and singularities of the transfer function in the right half plane. The results are simple estimates which can complement design calculations. They can also be taught in elementary courses in feedback control.

REFERENCES

[1] B. Stein, "Respect the Unstable", *30th IEEE Conference on Decision and Control*, Honolulu, HI, December 1990.

[2] J. S. Freudenberg and D. P. Looze, "A sensitivity trade-off for plants with time delay", *IEEE Trans. Automatic Control*, **AC-32**:99–104, 1987.

[3] M. M. Seron, and G. C. Goodwin, "Sensitivity Limitations in Nonlinear Feedback Control", *Elsevier Science*, 1995.

[4] J. S. Shamma, "Performance limitations in sensitivity reduction for non-linear plants", *Systems and Control Letters*, **17**:43–47, 1991.

[5] G. C. Goodwin, D. Q. Mayne and J. Shim, "Trade-offs in Linear Filter Design".

[6] J. S. Freudenberg and D. P. Looze, "Right half plane pole and zeros and design tradeoffs in feedback systems", *IEEE Trans. Automatic Control*, **AC-30**:555–565, 1985.

[7] , G. C. Goodwin and M. M. Seron, "Fundamental Design Trade-Offs in Filtering, Prediction and Smoothing".

[8] G. I. Gomez and G. C. Goodwin, "Integral Constraints on Sensitivity Vectors for Multivariable Linear Systems".

[9] H. W. Bode, *Network Analysis and Feedback Amplifier Design*, Van Nostrand, New York, 1945.

[10] S. P. Boyd and C. H. Barratt, *Linear Controller Design – Limits of Performance*, Prentice Hall Inc., Englewood Cliffs, NJ 1991.

[11] J. M. Maciejowski, *Multivariable Feedback Design*, Addison-Wesley, Reading, MA 1989.

[12] R. H. Middleton, "Trade-Offs in Linear Control System Design, *Automatica*, **27**:281–292, 1991.

[13] G. C. Newton, Jr, L. A. Gould and J. F. Kaiser, *Analytical Design of Linear Feedback Controls*, John Wiley & Sons, 1957.

21

LQG CONTROL WITH COMMUNICATION CONSTRAINTS

V. S. Borkar* and Sanjoy K. Mitter**

** Department of Computer Science and Automation, Indian Institute of Science, Bangalore 560012, India*

*** Department of Electrical Engineering and Computer Science and Laboratory for Information and Decision Systems, Massachusetts Institute of Technology, 35–308, Cambridge, MA 02139*

Dedicated to Tom Kailath on the occasion of his sixtieth birthday.

ABSTRACT

The average cost control problem for linear stochastic systems with Gaussian noise and quadratic cost is considered in the presence of communication constraints. The latter take the form of finite alphabet codewords, being transmitted to the controller with ensuing delay and distortion. It is shown that if instead of the state observations an associated "innovations process" is encoded and transmitted, then the separation principle holds, leading to an optimal control linear in state estimate. An associated "off-line" optimization problem for code length selection is formulated. Some possible extensions are also pointed out.

1 INTRODUCTION

Most traditional analyses of control systems presuppose that the observation vector is available in its entirety to the controller at each decision epoch. In many real engineering systems, however, the situation is different. What the controller sees will not often be the original observation vector from the sensor, but a quantized version of it transmitted over a communication channel with accompanying transmission delays and distortion, subject to bit rate constraints. This calls for control systems analysis that explicitly accounts for such communication constraints. This problem has attracted some attention in recent years, see, e.g. [3, 5, 7, 8, 9]. For related work on multirate control of sampled-data systems, see [6] and the references therein. The aim of this work is to show that the classical Linear-Quadratic-Gaussian (LQG) problem does admit a rather clean treatment in this framework, with the proviso that it is not the state or the observation vector that is encoded and transmitted, but an associated process we dub the 'innovations' process by slight abuse of terminology. In fact, a 'separation principle' holds and this will be the main result of this exercise.

There are two key features of our formulation that make this work. The first is the choice of 'innovations process' alluded to above in place of the observation process as the signal to be quantized and encoded. Unlike the latter, the former is an i.i.d. Gaussian sequence with statistics independent of control. This allows us to use a fixed optimal vector quantizer for which extensive analysis is available for the Gaussian case [4]. Secondly, the least squares estimation at the output end of the channel can now be based only on the current channel output and does not have to remember the past outputs, as it ideally should, if the observations were to be encoded directly. This makes the estimation scheme at the controller end completely transparent. These observations will become self-evident as we proceed.

The second key feature is the centroid property of the optimal vector quantizer, which allows us to interpret the quantized random variable as the conditional expectation of the original random variable given an appropriate sub-σ-field. This interpretation nicely fits in with the least squares estimation scheme we use.

The paper is organized as follows. The next section describes the problem formulation in detail. Section 3 derives the optimal controller. Section 4 describes the associated optimal code-length selection problem. Section 5 sketches some possible extensions.

2 PRELIMINARIES

Consider the control system

$$X_{k+1} = AX_k + Bu_k + v_k\,, \quad k \geq 0\,, \tag{21.1}$$

where

i. $\{X_k\}$ is an R^d-valued 'state' process, X_0 prescribed,

ii. $\{u_k\}$ is an R^m-valued control process,

iii. $A \in \mathcal{R}^{d\times d}$, $B \in \mathcal{R}^{d\times m}$,

iv. $\{v_k\}$ is i.i.d. $N(0, Q)$ noise, that is, normally distributed, zero-mean with covariance Q

v. the following 'nonanticipativity' condition holds: $\{v_j, j \geq k\}$ is independent of $\{x_j, u_j, v_{j-1}, j \leq k\}$ for all $k \geq 0$.

Let $G \in \mathcal{R}^{d\times d}$, $F \in \mathcal{R}^{m\times m}$ be prescribed positive semidefinite matrices. Our control problem is to minimize

$$\limsup n \to \infty \frac{1}{n} \sum_{k=0}^{n-1} E\left[X_k^T G X_k + u_k^T F u_k\right]$$

over $\{u_k\}$ as above, subject to the communication mechanism described below. Before getting into the details thereof, we lay down the following assumptions:

A1. The pair (A, B) is controllable.

A2. The pair $(A, G^{\frac{1}{2}})$ is observable.

A3. $||A||_2 \triangleq \lambda_{\max}(A^T A) < 1$.

The above control problem is well-posed under (A1)–(A2) [2, pp. 228–229]. (A3) will be used later. We come now to the encoding and communication mechanism.

Fix an integer $M \geq 1$, the 'code length.' Also let $N \geq 1$ be another integer, the 'communication delay' given by $N = \psi(M)$ for some prescribed nondecreasing map $\psi : \mathcal{N} \to \mathcal{N}$. (Typically, $\psi(n) = [n/r] + 1$ where $[\cdot]$ represents integer part and $r > 0$ is the transmission rate in bits per second.) For $k \geq 0$, let

$$X_{(k+1)N} = A^N X_{kN} + \sum_{i=0}^{N-1} A^{N-i} B u_{kN+i} + \tilde{v}_{k+1},$$

where $\tilde{v}_{k+1} = \zeta_{(k+1)N}$ for

$$\zeta_{kN+i} = \sum_{j=0}^{i-1} A^{i-j-1} v_{kN+j}, \quad 0 < i \leq N.$$

Then $\{\tilde{v}_k\}$ are i.i.d. $N(0, \tilde{Q}_N)$ where

$$\tilde{Q}_i = \sum_{j=0}^{i-1} A^{i-j-1} Q (A^T)^{i-j}, \quad 0 < i \leq N.$$

We call $\{\tilde{v}_k\}$ the innovations process by abuse of terminology.

At time kN, $k \geq 0$, start transmitting M-bit encoding of $\tilde{v}_k$. The transmission is complete at time $(k+1)N$.

Let $\{a_1, \ldots, a_\ell\}$ denote the range of the vector quantizer, assumed to satisfy the usual optimality conditions [4, Section 11.2]. Let $\{A_1, \ldots, A_\ell\}$ denote the finite partition of R^d generated by the vector quantizer, such that A_i gets mapped to a_i, $1 \leq i \leq \ell$. Let ζ_k denote the σ-field generated by the events $\{\tilde{v}_k \in A_i\}$,

$i \leq i \leq \ell$. Then the centroid condition of optimal vector quantizer [4, p. 352] translates into

$$\hat{v}_k = E[\tilde{v}_k/\zeta_k], \quad k \geq 0.$$

Letting $p_i = P(\tilde{v}_k \in A_i)$, $1 \leq i \leq \ell$, it is clear that

$$E[\hat{v}_k] = \sum_i p_i a_i = 0.$$

We assume a memoryless channel that maps a_i to a_j with probability $q(i,j)$, $1 \leq i, j \leq \ell$. Let v_k' be the output of the channel to input $\hat{v_k}$. Then assuming that the channel noise is independent of $\{\tilde{v}_k\}$, the LMS estimate of $\tilde{v}_k$ at time k is $\overline{v}_k = E[\tilde{v}_k/v_k'] = E[\hat{v}_k/v_k']$ calculated as follows using the Bayes rule:

$$\overline{v}_k = \sum_j \sum_i p_i q(i,j) \Big(\sum_s p_s q(s,j) \Big)^{-1} a_i \quad I_{\{v_k' = a_j\}}.$$

Clearly, $E[\overline{v}_k] = 0$ and

$$E \triangleq \text{cov}(\overline{v}_k) = \sum_{i,j,m} p_i p_m q(i,j) q(m,j) \Big(\sum_s p_s q(s,j) \Big)^{-1} a_i a_m^T.$$

The controller thus receives $\overline{v}_k$ at time $(k+1)N$, $k \geq 0$, and has to optimize the control system based on this information. The next section studies this control problem.

3 THE OPTIMAL CONTROLLER

With the aim of formulating a 'separated control problem,' we first study the evolution of $\hat{X}_k$ = the LMS estimate of X_k, $k \leq 0$. At time $(k+1)N$, $\hat{X}_{(k+1)N}$ is obtained as follows:

Step 1. Update $\hat{X}_{kN}$ to

$$\begin{aligned} \tilde{X}_{kN} &= \hat{X}_{kN} + \overline{v}_k \\ &= A^N \tilde{X}_{(k-1)N} + \sum_{i=0}^{N-1} A^{N-i} B u_{(k-1)N+i} + \overline{v}_k. \end{aligned}$$

Step 2. Set $\hat{X}_{(k+1)N} = A^N \tilde{X}_{kN} + \sum_{i=0}^{N-1} A^{N-i} B u_{kN+i}$.
For times $kN + i$, $0 < i < N$, $k \geq 0$, we have

Step 3.

$$\begin{aligned} \hat{X}_{kN+i} &= A^i \hat{X}_{kN} + \sum_{j=0}^{i-1} A^{i-j-1} B u_{kN+j} \\ &\Big(= A\hat{X}_{kN+i-1} + Bu_{kN+i-1} \Big). \end{aligned}$$

Let $e_k = X_k - \hat{X}_k$ denote the estimation error and $R_k = \text{cov}(e_k)$, $k \geq 0$. Then the evolution of $\{R_k\}$ is described by: $R_N = \tilde{Q}_N$ and for $k \geq 1$,

$$R_{kN+i} = \begin{cases} AR_{kN+i-1}A^T + Q\,, & 0 < i < N \\ A^N R^{kN}(A^N)^T + \left(A^N E(A^N)^T - A^N \tilde{Q}_N (A^N)^T\right) + \tilde{Q}_N\,, & i = N \end{cases}$$

In particular, the evolution of $\{R_k\}$ is deterministic and independent of $\{u_k\}$. Combining this with the observation that

$$E\big[X_k^T G X_k\big] = E\big[\hat{X}_k^T G \hat{X}_k\big] + tr\big[GR_k\big]\,, \tag{21.2}$$

we can consider the following 'separated' control problem: Minimize

$$\limsup n \to \infty \frac{1}{n} \sum_{i=0}^{n-1} E\left[\hat{X}_k^T G \hat{X}_k + u_k^T F u_k\right] \tag{21.3}$$

where $\{\hat{X}_k\}$ evolves according to Step 1–Step 3 above. This evolution can be rewritten as

$$\hat{X}_{k+1} = A\hat{X}_K + Bu_k + w_k\,, \quad k \geq 0\,,$$

where $\{w_k\}$ is a zero mean noise sequence given by

$$w_k = \begin{cases} 0 & k \notin \{iN, \quad i \geq 0\} \\ A^N \overline{v}_{i-1} & k = iN, \quad i \leq 0 \end{cases}$$

Now one can mimic the usual arguments for LQG control with minor modifications, to obtain the optimal controller. We sketch them below, closely following the treatment of [2, p. 228–229].

Under our assumptions A1–A2, the unique positive semidefinite solution $\{K_n\}$ to the Riccati equation:

$$K_{k+1} = A^T\left(K_k - K_k B\big(B^T K_k B + F\big)^{-1} B^T K_k\right)A + G, \quad K_0 = 0\,,$$

converges as $k \to \infty$ to the unique positive semidefinite solution K of the algebraic Riccati equation

$$K = A^T\left(K - KB\big(B^T K B + F\big)^{-1} B^T K\right)A + G\,.$$

The optimal value of the n-stage costs [1, p. 130–132]

$$\frac{1}{n} E\left[\sum_{k=0}^{n-1}\left(\hat{X}_k^T G \hat{X}_k + u_k^T F u_k\right)\right]$$

is seen to equal

$$\frac{1}{n}\left(X_0^T K_n X_0 + \sum_{k=0}^{n-1} E[w_k^T K_k w_k]\right)\,,$$

and tends to

$$\lambda = \lim_{n\to\infty} \frac{1}{n}\sum_{k=0}^{n-1} E[w_k^T K_k w_k]$$
$$= \frac{1}{N}\text{trace}\left[(A^T)^N K A^N E\right] .$$

Also, the N-stage optimal feedback policy in its initial stages tends to the stationary feedback policy

$$\mu(x) = -(B^T K B + R)^{-1} B^T K A x . \tag{21.4}$$

Using the definitions of λ, K and $\mu(\cdot)$, it is easy to show that

$$x^T K x = \min_u \left[x^T G x + u^T F x + (Ax + Bu)^T K(Ax + Bu)\right] ,$$

$$x^T K x + N\lambda =$$
$$\min_u \left[x^T G x + u^T F x + E\Big[(Ax + Bu + w_{kN})^T K(Ax + Bu + w_{kN})\Big]\right] ,$$
$$k \geq 0 ,$$

with the minimum in both cases being attained by $u = \mu(x)$. Now one can mimic the standard dynamic programming arguments of [1, p. 191–192] [2, p. 229] to deduce that under arbitrary admissible control $\{u_k\}$, the cost (3) is greater than or equal to λ, with equality if $u_k = \mu(\hat{X}_k)$, $k \geq 0$. We summarize these findings in the following:

Theorem 21.1 *The optimal feedback controller for the separated control problem (and hence for the original control problem) is given by*

$$u_k = \mu(\hat{X}_k) , \quad k \geq 0 ,$$

for $\mu(\cdot)$ defined by (21.4).

4 OPTIMAL CODELENGTH

Consider the matrix equation

$$\overline{R} = A^N \overline{R}(A^N)^T + C(N) \tag{21.5}$$

where $C(N) = A^N E(A^N)^T + \tilde{Q}_N - A^N \tilde{Q}_N (A^N)^T$. Under A3, (21.5) has a unique symmetric positive definite solution $\overline{R}$ given explicitly by

$$\overline{R} = \sum_{i=0}^{\infty} A^{Ni} C(N) (A^T)^{Ni} . \tag{21.6}$$

As $k \to \infty$, $R_{kN} \to \overline{R}$, as can be easily verified.

Given (21.2) and the dynamics of $\{R_k\}$, the total cost for the original control problem is

$$J(N) \triangleq \lambda(N) + \frac{1}{N}\sum_{i=0}^{N-1} \text{trace}\left[GA^i\overline{R}(A^T)^i\right] + \text{trace}[GQ] \,, \tag{21.7}$$

where λ is written as $\lambda(N)$ to make its dependence on N explicit. Recalling that $N = \psi(M)$, the choice of optimal M would be obtained by minimizing $M \to J(\psi(M))$ over $M \in \mathcal{N}$. The expression (21.7) has a complicated dependence on N, but this optimization problem is 'off-line.' Also, the following could be used to advantage for any computational scheme:

i. To go from N to $N+1$, the quantities $A^N K(A^T)^N$, $A^N E(A^T)^N$, $\tilde{Q}_N$ can be updated by pre- and post-multiplying by A, A^T resp. Similarly, $A^N\tilde{Q}_N(A^T)^N$ is updated by pre- and post-multiplying by A^2 and $(A^T)^2$ respectively. This takes care of the updates of $\lambda(N)$ and $C(N)$.

ii. Updating $\overline{R} = \overline{R}(N)$ poses a harder problem. One possible approximation scheme is to use for each N the recursion

$$\tilde{R}_{k+1} = A^N\tilde{R}_k(A^T)^N + C(N) \,, k \geq 1 \,,$$

till a suitable stopping criterion is satisfied and use the resultant matrix $\tilde{R}(N)$ as an approximation for $R(N)$. Repeat the iteration for $N+1$ with $C(N)$ replaced by $C(N+1)$, using $\tilde{R}(N)$ as the initial guess. Our assumption A3 ensures good convergence behavior of the above recursion.

iii. Invoking assumption A3 again, the summations in 21.6 and 21.7 could simply be approximated by a finite summation with a fixed number of terms for moderately large values of N.

In conclusion, observe that in our model, shorter codes correspond to low resolution and low delays, while longer codes mean higher resolution, but longer delays. Thus the above optimization problem captures the tradeoff between delay and accuracy.

5 EXTENSIONS AND OPEN ISSUES

i. Partial observations: Suppose we do not observe $\{X_k\}$, but an accompanying r-dimensional observation process $\{Y_k\}$ described by

$$Y_k = HX_k + \eta_k \,, \quad k \geq 0 \,,$$

where $H \in \mathcal{R}^{d \times r}$ and $\{\eta_k\}$ are i.i.d. $N(0, S)$, independent of $\{v_k\}$. Suppose the pair (A, H) is observable and the pair $(A, Q^{1/2})$ controllable. Then we can carry through the foregoing analysis for this problem with just one change: replace (21.1) by the steady state Kalman filter [4, R. M. Gray].

ii. Coding issues: The preceding section considered optimization over code length, not over codes. Also, the entire analysis ignores the fact that if the encoding is sequentially transmitted and received, each bit carries its own information. The situation is particularly transparent for tree codes [4, Ch. 15]. These are based on successive refinements of partitions, to which we can assign a corresponding increasing family of σ-fields along the lines of ζ (which then corresponds to the largest σ-field in this chain). If the 'centroid rule' is observed at each stage of refinement, the first m bits of an M-bit codeword ($m \leq M$) would correspond to the conditional expectation with respect to the m-th σ-field in the chain. Thus the controller receives a succession of conditional expectations over finer and finer σ-fields during each N-interval. The updates rules of Step 1–Step 3 of Section 3 can be easily modified to accommodate this situation and the rest of the analysis is similar to that above. This would, in fact, seem to make a case for using tree codes in control applications. The situation for other coding schemes, however, is complicated. Also, the problem becomes considerably harder if we consider variable length codes. Finally, we have used a simple model for the channel. More complex situations need to be analyzed.

iii. Distributed control: If several observations are being recorded, encoded and transmitted in a distributed manner, with or without synchronism, the problem would appear to be much more difficult and at the same time, much more interesting for applications. We hope to address this in subsequent work.

Acknowledgements

The authors would like to thank M. Dahleh, J. Tsitsiklis and S. Tatikonda for helpful discussions. This research (V. S. Borkar) supported by a Homi Bhabha Fellowship and by the U.S. Army Research Office under grant number DAAL03-92-G-0115. This research (S. K. Mitter) supported by the Army Research Office under grant number DAAL 03-92-G-0115 (Center for Intelligent Control Systems) and grant number DAAH04-95-1-0103 (Photonic Networks and Data Fusion).

REFERENCES

[1] D. Bertsekas, *Dynamic Programming and Optimal Control*, **Vol. 1**, Athena Scientific, Belmont, Mass., 1995.

[2] D. Bertsekas, *Dynamic Programming and Optimal Control*, **Vol. 2**, Athena Scientific, Belmont, Mass., 1995.

[3] D. F. Delchamps, Stabilizing a linear system with quantized state feedback, *IEEE Trans. Automatic Control*, **35** (1990) 916–924.

[4] A. Gersho and R. M. Gray, *Vector Quantization and Signal Compression*, Kluwer Academic Publ., Boston, 1992.

[5] P. T. Kalamba and S. Hara, Worst-case analysis and design of sampled-data control systems, *IEEE Trans. Automatic Control*, **38** (1993) 1337–1357.

[6] P. Voulgaris, Control of asynchronous sampled-data systems, *IEEE Trans. Automatic Control*, **39** (1994) 1451–1455.

[7] D. Williamson, Finite wordlength design of digital Kalman filters for state estimation, *IEEE Trans. Automatic Control*, **30** (1985) 930–939.

[8] W. S. Wong and R. W. Brockett, Systems with finite communication bandwidth constraints I: state estimation problems, preprint.

[9] W. S. Wong and R. W. Brockett, Systems with finite communication bandwidth constraints II: feedback control problems, preprint.

22

MODELING, IDENTIFICATION AND CONTROL

Michel Gevers

Center for Engineering Systems and Applied Mechanics
Université Catholique de Louvain, Bâtiment Euler
1348 Louvain la Neuve, Belgium

In my PhD dissertation I wrote "Professor Kailath made me work as never before". I haven't stopped since, and I owe much of my dedication to research to this period of work under his supervision.

ABSTRACT

We study the interactions between modeling, identification and control, in the situation where the only purpose of the modeling or identification is the design of a high performance controller. This leads us to suggest that the model building criterion should be determined by the control objective, leading to identification on the basis of closed loop data. We present three different approaches to this 'identification for control' paradigm: a dual control approach, an optimal experiment design approach and a robust control approach. The connections and distinctions between these three viewpoints are discussed, and recent results for each approach are briefly presented.

1 INTRODUCTION

Until about 1990, identification and control were developed as two separate fields of control science. With a few exceptions, there was very little research activity on the interconnection between the two areas. The exceptions (dual control, indirect adaptive control) considered essentially the rather unrealistic situation where the true system is in the model set ($\mathcal{S} \in \mathcal{M}$). The dominant idea was: identify the "best" model, then design the controller on the basis of that best model. "Best" was not "best for control design".

Until the mid-1980's, the concept of identification "design" was essentially non-existent, because the mainstream thinking was that one should identify the 'true system'. Thus, identification was not viewed as an *approximation* theory but as an alternative to modeling from physical principles, with the idea that any decent identification method should have the property that the model converges to this elusive 'true system'. L. Ljung's book [13] presented the first formal identification design concepts. The few available results on identification design for control were again limited to the case $\mathcal{S} \in \mathcal{M}$, i.e. only noise-induced errors were considered. The new theories of robust control were not taken into account in identification design. The interplay between identification design and robust control design was probably first discussed in [3], in the context of adaptive control with unmodelled dynamics ($\mathcal{S} \notin \mathcal{M}$).

The study of the interactions between modeling, identification and control design can be performed at various levels of generality and idealisation. First one should examine whether a model is really necessary for control design, or whether one cannot obtain better performance by the direct tuning of controller parameters towards the minimization of a closed loop performance criterion. We shall return to the idea of direct controller parameter tuning at the end of this paper. Assuming now that one takes a model-based approach to controller design, then the following observations are worthwhile making.

1. Experience shows that extremely simple models often lead to high performance controllers on complex processes. Thus, to compute a high performance controller it is not always necessary to have a very accurate model. The key feature is that the model should capture with high precision the dynamic characteristics that are essential for control.

2. Once a first controller operates on a process, it is typical that large quantities of data continue to be fed to the computer. Why not use these data to compute controllers with higher performance?

3. The best open loop model is not necessarily good for control design. A famous example, produced in his PhD thesis by R. Schrama [14], has illustrated how a 5th order model of an 8th order system, identified using open loop data and validated by commonly used open loop validation criteria, led to a controller that destabilized the 'true' 8th order system, despite the fact that the two (true and model) Nyquist plots were apparently indistuinguishable.

4. Since the model is only a vehicle for control design, it is only natural that the modeling or identification criterion should be a function of the control design objective. One should view the identification and control problem as a *combined optimization problem.* However, except in the simplest of cases, the solution of such problem is presently beyond our computational reach.

To illustrate the first point, let us mention the modeling, identification and control of the Philips Compact Disc (CD) Player. Following the track on a CD involves two control loops: see e.g. [5]. A first permanent magnet/coil system mounted on the radial arm positions the laser spot in the direction orthogonal to the track. A second permanent magnet/coil system controls an objective lens which focuses the laser spot on the disc. The control system therefore consists of a 2-input/2-output system, with the spot position errors (in both radial and focus directions) as the variables to be controlled, and the currents applied to the magnet/coil actuators as the control variables. The modeling of this system using finite element methods or its estimation using spectral analysis techniques would lead to a 2-input/2-output model whose McMillan degree would be of the order of 150. However, by using an identification for control design criterion, a 16-th order model has been identified that leads to excellent control performance: see [5] for details. A comparison between the spectral estimates and the identified models for the 4 input-output channels is presented in Figure 22.1. The spectral estimates have been obtained by taking 100 averages over 409,600 time samples. The parametric models have been identified using 2,000 closed loop data samples.

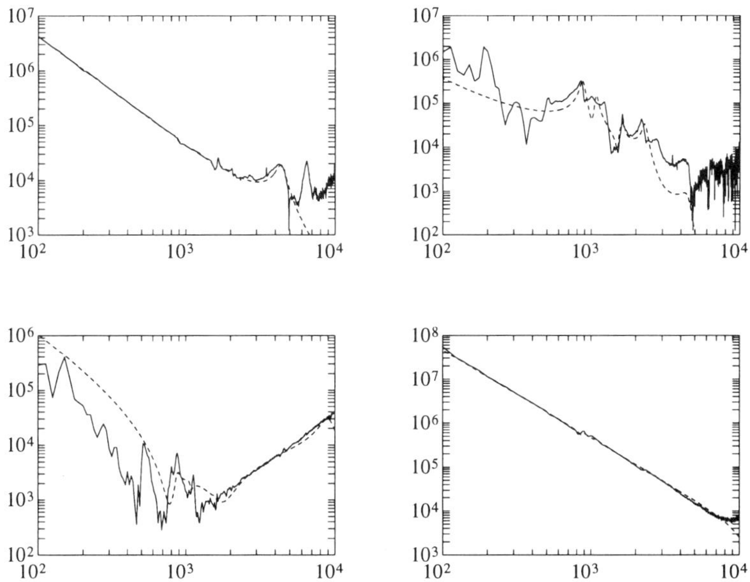

Figure 22.1 Amplitude of spectral estimate (—) and of parametric model (- -).

It follows from our observations that the *modeling* techniques, which typically aim at reproducing the dynamical behavior of the system as accurately as possible on the basis of physical laws, should give way to *identification* techniques, which are data-based, and in which the identification criterion should be tuned towards the control performance objective. This will lead to an approximate (and simplified) model, and hence to a reduced order controller. To return to the example of the CD player, a controller of degree 150, say, based on a full-order model obtained from physical considerations, would be practically useless.[1]

In the remainder of this paper we therefore discuss the concept of control-oriented identification design. The concept of *identification design* was essentially introduced by L. Ljung [13] who laid down some formal design concepts. It concerns the choice of experimental conditions, model set, data filters, criterion, etc so that the approximate model that results from the identification experiment is tuned towards the objective for which it is to be used. However, the contributions to control-oriented identification design were very modest in [13].

Since 1990, there has been a tremendous research activity in control-oriented identification, and it is impossible to cover the whole spectrum of this activity in this single contribution. Rather we shall attempt to present three different conceptual approaches to the problem of *identification for control.*

At the most general (or ideal) level, one can pose the problem as a *dual control* problem, in which identification (or, more precisely, parameter estimation) and control design are posed as a *combined problem.* The solution to this problem, although reasonably easy to formulate, typically leads to intractable computational difficulties. We will explain the basic concepts of dual control and present some recent results in Section 3. We should add that all available results are limited to the case where full-order models are used, i.e. where 'the system is in the model set'.

At the next level of idealisation, one can consider to design the identification such that the ensuing model-based controller performs as close as possible to the 'ideal controller' on the actual closed loop system. By 'ideal controller' is meant the controller that would result from the model-based control design criterion if the true system were known. This problem formulation is often referred to as *optimal experiment design.* We will present some recent results in this direction in Section 4; they are again limited to the case where the 'system is in the model set'.

The next and more realistic approach is to formulate the identification design on the basis of the closeness between the *achieved closed loop system* and the *nominal closed loop sytem* (= identified model with its model-based controller), rather than the unavailable *optimal closed loop system* of the previous approach. This third approach is in line with robust control design thinking. Of course, there is no guarantee that the performance achieved by the model-based de-

[1] Of course, an alternative to this reduced order control-oriented identification is to compute a high-order controller on the basis of an accurate high-order model of the system, and then to perform controller reduction using a closed loop performance criterion.

signed controller will be anywhere near the optimal performance, only that it will be close to the designed performance. In this approach, it is not necessary to restrict oneself to the case where 'the system is in the model set'.

It is now time to introduce some technical information. In Section 2 we set some notations and recall the basics about Least Squares identification. In Section 3 we present the concepts of dual control, as well as some recent results. Section 4 presents some key ideas of optimal experiment design for control. In Section 5 we present the 'robust control' framework, and examine how the identification criterion should be determined by the control design criterion.

2 LEAST SQUARES IDENTIFICATION

We consider that the task is to design a controller for some "true" linear time-invariant scalar system described by

$$\mathcal{S} : y_t = P(q)u_t + H(q)e_t \tag{22.1}$$

where $P(q)$ and $H(q)$ are scalar rational transfer function operators, with $H(q)$ normalized such that $H(\infty) = 1$. Here q^{-1} is the delay operator ($q^{-1}u_t = u_{t-1}$), u_t is the control input signal, y_t is the observed output signal, e_t is white noise of zero mean and variance σ^2, and $v_t \stackrel{\triangle}{=} H(q)e_t$ is the noise acting on y_t.

A controller is to be designed on the basis of a model of the plant identified using a finite set of N input and output data $\{y_t, u_t, t = 1, 2, \ldots, N\}$ collected on the plant. A parametrized model set $\mathcal{M} = \{M(\theta) : \theta \in D_\theta \subset I\!R^d\}$ is used, where D_θ is a set of admissible values and $M(\theta)$ is described by:

$$M(\theta) : y_t = P(q, \theta)u_t + H(q, \theta)e_t. \tag{22.2}$$

If there exists a $\theta_0 \in D_\theta$ such that $P(q) = P(q, \theta_0)$, $H(q) = H(q, \theta_0)$, then we say that 'the system is in the model set': $\mathcal{S} \in \mathcal{M}$.

The data collection can be done in open loop or in closed loop. In the case of closed loop identification, we denote by $C_{id}(q)$ the controller that operates during identification:

$$u_t = C_{id}(q)[r_t - y_t], \tag{22.3}$$

where r_t is the reference excitation signal used during identification.

The Least Squares prediction error method applied to N input-output data delivers an estimate $\hat{\theta}_N$ of θ:

$$\hat{\theta}_N = \arg \min_{\theta \in D_\theta} V_N(\theta) \tag{22.4}$$

where

$$V_N(\theta) = \sum_{t=1}^{N} [\epsilon_t^f(\theta)]^2 \tag{22.5}$$

$$\epsilon_t^f(\theta) = D(q)\epsilon_t(\theta) = \frac{D(q)}{H(q,\theta)}[y_t - P(q,\theta)u_t]. \tag{22.6}$$

Here $\epsilon_t(\theta) \stackrel{\triangle}{=} y_t - \hat{y}_{t|t-1}(\theta)$ is the one-step-ahead prediction error for the model $M(\theta)$, while $D(q)$ is a data filter. In turn, this produces a model:

$$\hat{P}_N = P(e^{j\omega}, \hat{\theta}_N), \quad \hat{H}_N = H(e^{j\omega}, \hat{\theta}_N). \tag{22.7}$$

Under reasonable conditions on the data and the model structure [13], $\hat{\theta}_N$ converges as $N \to \infty$ to

$$\theta^* = \arg \min_{\theta \in D_\theta} \bar{V}(\theta) \tag{22.8}$$

where

$$\bar{V}(\theta) = \lim_{N\to\infty} EV_N(\theta) = E[\epsilon_t^f(\theta)]^2. \tag{22.9}$$

If identification is performed using closed loop data (as is often the case in identification for control), then the asymptotic expression for the cost criterion becomes[2]:

$$\bar{V}(\theta) = \frac{1}{2\pi}\int_{\pi}^{\pi} \left\{ \left| \frac{(P - P(\theta))}{1 + PC_{id}} \right|^2 |C_{id}|^2 \Phi_r + \left| \frac{1 + P(\theta)C_{id}}{1 + PC_{id}} \right|^2 \Phi_v \right\} \frac{|D|^2}{|H(\theta)|^2} d\omega \tag{22.10}$$

This expression gives an implicit characterization of the model to which $P(e^{j\omega}, \hat{\theta}_N)$, $H(e^{j\omega}, \hat{\theta}_N)$ converge if the number of data tends to infinity. Thus, it plays a key role in understanding the *bias error* distribution obtained with closed loop identification. An approximate expression for the *variance error* on the transfer function estimates can also be obtained: see e.g. [13].

3 THE DUAL CONTROL APPROACH

In this section we present the key ideas of *dual control*, as well as some recent developments. Dual control can be seen as one way (perhaps even the optimal way) of solving the joint identification and control design problem when the system is in the model set. Thus, all the ideas and results of this section are limited to the case where $\mathcal{S} \in \mathcal{M}$. Our presentation borrows heavily from the recent PhD thesis of C. Kulcsár [12].

The concept of dual control was introduced by Fel'dbaum [6] who understood that, when one wants to minimize a control performance criterion for a system with unknown parameters, the control has the dual role of maintaining the state close to its desired value while at the same time learning the unknown parameters. These two roles are conflicting.

[2] For reasons of space, we have deleted the ω-dependence in all arguments.

Thus, consider now that the parameter vector θ has a prior probability distribution, say $\Pi(\theta)$, and that the task is to design an optimal control sequence $U_0^N \triangleq \{u_0, u_1, \cdots, u_{N-1}\}$ that minimizes the following cost:

$$J_{0,N}(U_0^{N-1}) = E[\sum_{t=0}^{N-1} c_t(u_t, \theta) \mid I^0], \tag{22.11}$$

where

$$c_t(u_t, \theta) = (y_{t+1} - y_{t+1}^*)^2 + \lambda_t u_t^2, \tag{22.12}$$

with λ_t nonnegative scalars, and where I^0 contains all prior information about the system, i.e. the noise distribution and the prior distribution on θ. The minimum of (22.11) is obtained by solving the following succession of nested optimization problems:

$$\begin{aligned} J_{0,N}^* &= \min_{u_0} E[c_0(u_0, \theta) + \min_{u_1} E[c_1(u_1, \theta) + \dots \\ &+ \min_{u_{N-1}} E[c_{N-1}(u_{N-1}, \theta) \mid I^{N-1}] \dots \mid I^1 \mid I^0] \end{aligned}$$

Define

$$J_{t,N}(u_t) = E[c_t(u_t, \theta) + J_{t+1,N}^* \mid I^t] \tag{22.13}$$

and

$$J_{t,N}^* = \min_{u_t} J_{t,N}(u_t). \tag{22.14}$$

Notice that the 'cost-to-go', $J_{t,N}(u_t)$, can be written as

$$J_{t,N}(u_t) = \int \left[c_t(u_t, \theta) + J_{t+1,N}^*\right] \times \Pi(y \mid \theta, u_t) \, \Pi(\theta \mid I^t) \, dy \, d\theta \tag{22.15}$$

Observe that the distribution of the random vector θ influences the distribution of all future input and output signals, while it is itself influenced by these signals through the parameter estimation procedure. This puts the exact solution to the dual optimal control problem beyond the reach of present-day computer technology except in the simplest of cases.

For this reason, a lot of effort has been spent on computing suboptimal solutions. One particularly simple suboptimal solution is to assume, at time t, that $\hat{\theta}_t$ is exact, i.e. to replace the probability distribution of θ in (22.15) by a probability density centered at $\hat{\theta}_t$ and with zero variance. This idea is called *Certainty Equivalence* (CE) control. It does not take into account the effect of the control on the precision of future estimates of θ.

Interesting progress has been accomplished in the PhD thesis of C. Kulcsár [12]. For Finite Impulse Response (FIR) models, and assuming a Gaussian distribution for all random variables, she observed that the expected values of all future covariances of θ depend on the future only through their dependence on the future control signals. She then proposed a suboptimal solution in which, at time t, the mean of the future $\{\hat{\theta}_k, k \geq t\}$ is frozen at $\hat{\theta}_t$, while the expected values of the covariances $\{\Sigma_k, k \geq t\}$ are calculated exactly. These future covariances depend on all future controls $\{u_t, \cdots, u_{N-1}\}$, that need to be optimized. This suboptimal problem can now be solved using CPU times that are entirely reasonable, yielding solutions that are vastly superior to Certainty Equivalence solutions. Illustrative examples can be found in [12].

4 OPTIMAL IDENTIFICATION DESIGN FOR CONTROL

The optimal identification design is based on the minimization of a quality criterion that compares the *optimal closed loop system* with the *actual closed loop system*: see Figures 22.2 and 22.3. These two loops are assumed driven by the same external signals, the reference r_t with spectrum $\phi_r(\omega)$, and the white noise e_t with variance σ^2. Their outputs are denoted y_t^o and y_t, their inputs u_t^o and u_t, respectively. In the optimal closed loop system of Figure 22.2, the controller C is computed from the true plant $[P, H]$ using the chosen control design criterion: $C = c(P, H)$. In the actual closed loop system of Figure 22.3, the controller $\hat{C}$ is the Certainty Equivalence controller that results from a model $[\hat{P}_N, \hat{H}_N]$ identified using N data: $\hat{C} = \hat{C}_N \stackrel{\triangle}{=} c(\hat{P}_N, \hat{H}_N)$.

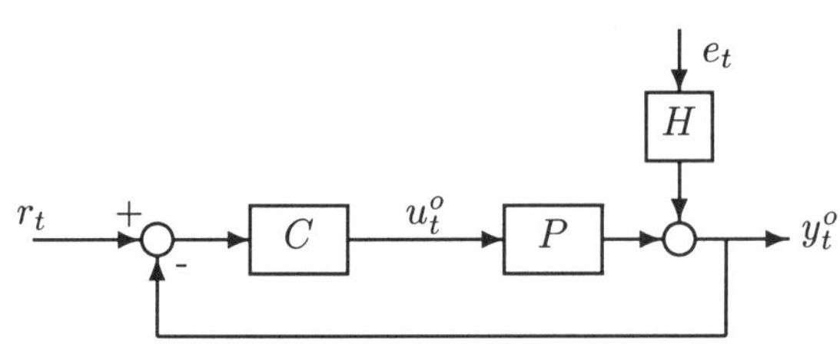

Figure 22.2 Optimal closed loop system.

Our *identification design criterion* will be

$$J_V = E[y_t^o - y_t]^2. \tag{22.16}$$

It is a measure of the degradation that results from using the estimated $\hat{C}_N$ on the plant instead of the optimal C. The use of this measure as an identification design criterion was first proposed in [8].

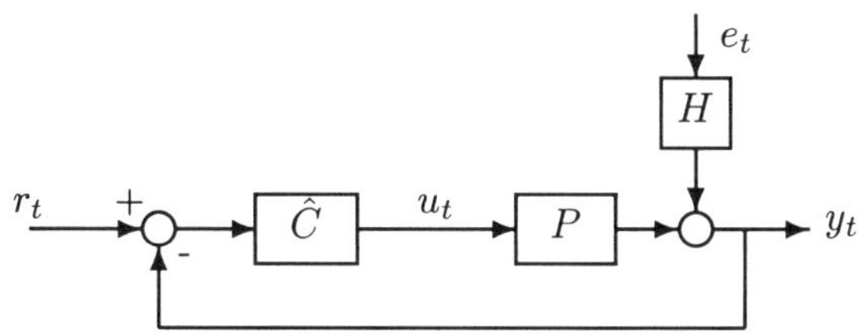

Figure 22.3 Actual closed loop system.

Assuming $\Delta C_N \stackrel{\Delta}{=} \hat{C}_N - C$ to be small and using a Taylor series expansion, one arrives, after some manipulations, at:

$$J_V \cong \frac{1}{2\pi}\int_{-\pi}^{\pi} \frac{|P|^2 \phi_{y^o}}{|1+PC|^2} E|\Delta C_N|^2 d\omega \tag{22.17}$$

Here, the expected value is taken with respect to the probability distribution of the noise during the identification experiment, which produces the random variable $\hat{C}_N = c(\hat{P}_N, \hat{H}_N)$. With ΔC_N small, assume again that we can write :

$$\Delta C_N = [F_1 \;\; F_2] \begin{bmatrix} \Delta P_N \\ \Delta H_N \end{bmatrix}, \tag{22.18}$$

where $\Delta P_N \stackrel{\Delta}{=} P - \hat{P}_N$, $\Delta H_N \stackrel{\Delta}{=} H - \hat{H}_N$, $F_1 = \frac{\partial C}{\partial P}$ and $F_2 = \frac{\partial C}{\partial H}$. Inserting the standard covariance formula for $[\Delta P_N, \Delta H_N]$ [13] in (22.18) yields, after some further manipulations, the following expression for the variance of the controller error at a frequency ω [9]:

$$E|\Delta C_N(e^{j\omega})|^2 \cong \frac{n}{N}|H|^2\{|F_2|^2 + \frac{\sigma^2}{\phi_r}|F_1 + (F_1 G + F_2 H)C_{id}|^2\} \tag{22.19}$$

It follows immediately that the controller variance $E|\Delta C_N(e^{j\omega})|^2$ is minimized, at every frequency, by performing the identification *in closed loop* with an operating controller:

$$C_{id}^{opt}(q) = -\frac{F_1(q)}{F_1(q)P(q) + F_2(q)H(q)}. \tag{22.20}$$

This optimal choice of course also minimizes the criterion J_V.

COMMENTS

- As is typical of optimal experiment design results, this optimal design depends on the unknown system $[P, H]$ and is therefore not feasible.
- It has been shown in [9] that, for Model Reference Control, $C_{id}^{opt}(q) = C(q)$. The same result had been shown for Minimum Variance Control in [8].

- If identification is performed under the ideal closed loop condition with $C_{id} = C_{id}^{opt}(q)$, the control error variance becomes:

$$E|\Delta C_N(e^{j\omega})|^2_{idcl} \cong \frac{n}{N}|H|^2|F_2|^2. \tag{22.21}$$

With open loop identification, we get

$$E|\Delta C_N(e^{j\omega})|^2_{ol} \cong \frac{n}{N}|H|^2\{|F_2|^2 + \frac{\sigma^2|F_1|^2}{\phi_u}\} \tag{22.22}$$

Even though this result is of obvious theoretical interest, it might not appear very useful given that the optimal identification design depends on the unknown system. However, it is shown in [9] that, when $C_{id}^{opt}(q) = C(q)$ (as is the case for Minimum Variance Control and Model Reference Control), an iterative identification and control design leads to a better controller than open loop identification. By iterative design is meant that identification is performed first in open loop for a fraction of the total data collection interval; the model estimated at the end of that interval is used to design a certainty equivalence controller, which is applied to the plant; the identification is continued in closed loop during a second time interval with this controller operating on the plant; at the end of this second interval a new certainty equivalence controller is computed from the present model and applied to the plant again; etc. A simulation example illustrating this iterative scheme is presented in [9].

5 MATCHING IDENTIFICATION AND CONTROL CRITERION

The main drawback of the *optimal identification design for control*, as we have noted, is that the design depends upon the unknown system, because the design criterion J_V is based on a comparison between the *optimal* and the *achieved* closed loop systems of Figures 22.2 and 22.3. An alternative, but suboptimal, formulation of the identification criterion is based on a comparison between the *achieved* closed loop system of Figure 22.3 and the *designed or nominal* closed loop system of Figure 22.4.

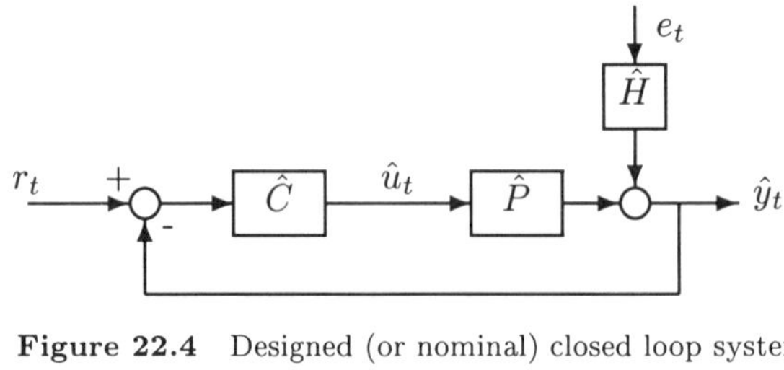

Figure 22.4 Designed (or nominal) closed loop system.

Comparing these two loops is classical in robust control theory; thus, we call this third approach to identification design for control the ***robust control approach.*** However, a key difference (and complication) in our problem formulation with respect to classical robust control thinking is that now both the nominal model $\hat{P}$ and the controller $\hat{C}$ are objects to be designed, as opposed to just $\hat{C}$ in classical robust control design. Thus, one must attempt to perform both the identification design and the control design in such a way that the nominal performance is high and that the two loops of Figures 22.3 and 22.4 are 'close to one another' in a sense to be defined.

If nothing is fixed, then such problem formulation might not make sense. To simplify matters somewhat, we make the following assumptions.

- A control criterion has been selected.
- At some stage of the design, a model structure has been chosen, typically of lower complexity than the 'true system'.

With these assumptions, one can show that every control criterion induces an identification criterion that 'matches' that control criterion. In addition these identification criteria take the form of ***closed loop identification*** criteria. The idea that an identification criterion can be made to match a control criterion was initially advanced for LQG control by Zang et al. [16] and for H_∞ control by Schrama [14]. It was applied to pole placement control and analysed by Åström [1] and Åström and Nilsson [2].

To explain the matching of the control and identification criteria, we take the simplest control design problem, namely the pole placement control problem without disturbances analysed by Åström. Thus, consider the two loops of Figures 22.3 and 22.4, and consider that the control design problem is to design $\hat{C}$ such that the designed closed loop transfer function from r_t to $\hat{y}_t$ is a given reference model, i.e. compute $\hat{C}$ from $\hat{P}$ such that

$$\frac{1}{1+\hat{P}(q)\hat{C}(q)} = S(q), \tag{22.23}$$

where $S(q)$ is some admissible reference model.

It follows from Figures 22.3 and 22.4 that the 'control performance error', defined as the error between the actual and the designed outputs, is given by:

$$y_t - \hat{y}_t = S\left[\frac{P\hat{C}}{1+P\hat{C}}r_t - \frac{\hat{P}\hat{C}}{1+P\hat{C}}r_t\right] = S[y_t - \hat{P}u_t]. \tag{22.24}$$

Equation (22.24) can be seen as an equality between a control performance error on the left hand side (LHS) and a filtered identification error on the right hand side (RHS). Indeed, the RHS is a filtered (by $S(q)$) version of the output error $y_t - P(q,\theta)u_t$, where u_t and y_t are collected on the actual closed loop system of Figure 22.3 with $\hat{C}$ operating. Thus, it appears that the control performance error can be minimized by performing ***identification in closed loop with a data filter*** $S(q)$. However, in the RHS of (22.24) the closed loop signals y and u are functions of the controller, and hence of the model parameter vector θ. In addition, the sensitivity function $S(q)$, which is used as a reference model in the present design, may also be a function of θ as is the case when the model contains nonminimum phase zeros that need to be preserved in the closed loop system. Hence, a more suggestive way to write (22.24) is as follows:

$$y_t - \hat{y}_t = S(q,\theta)[y_t(\theta) - P(q,\theta)u_t(\theta)]. \tag{22.25}$$

Even though the RHS of (22.25) looks like a filtered closed loop prediction error, it cannot be minimized by standard identification techniques, because θ appears everywhere and not just in $P(\theta)$. As a consequence, the approach suggested in all known 'identification for control' schemes is to perform identification and control design steps in an iterative way, whereby the i-th identification step is performed on filtered closed loop data collected on the actual closed loop system with the $(i-1)$-th controller operating in the loop. This corresponds to an i-th identification step in which the following filtered prediction error is minimized with respect to θ:

$$y_t - \hat{y}_t = S(q,\hat{\theta}_{i-1})[y_t(\hat{\theta}_{i-1}) - P(q,\theta)u_t(\hat{\theta}_{i-1})]. \tag{22.26}$$

We refer the reader to [7] and [15] for details on such iterative schemes.

Note: The equality (22.25) between a control error and an apparent 'identification error' has been derived here on the basis of a model reference control desgin scheme. For optimization-based control design criteria, such as e.g. LQG, or H_∞, one does not arrive at an equality such as (22.25), but one can upper-bound the mismatch between the achieved and the designed control criteria by the norm of the RHS of (22.25). Thus, one is also led to iterative identification and control schemes in which the appropriate norm of the RHS of (22.26) is minimized by an identification step.

An interesting question is whether these iterative identification and control schemes converge to the minimum of the achieved cost over the set $\mathcal{C} \triangleq \{\hat{C}(P(\theta)) \ \ \forall\theta \in D_\theta\}$ of all certainty equivalence controllers. For the example given above, this corresponds to asking whether, by successively minimizing over θ the mean square of the prediction errors defined by (22.26), one will converge to the minimum of

$$J(\theta) \triangleq E\{S(q,\theta)[y_t(\theta) - P(q,\theta)u_t(\theta)]\}^2. \tag{22.27}$$

This question has been analyzed in [11], where it has been shown that the answer is in general negative: the iterative identification and control schemes do not generically converge to the minimum of the achieved cost. In fact, it has been shown in [4] that the optimal controller within this reduced order controller set $\mathcal{C}$ is not always the Certainty Equivalence controller of a model $\hat{P}$ that can be obtained as a result of an identification experiment on the true system, whether the data are collected in open loop or in closed loop.

The last observation raises the question whether one could not minimize the criterion (22.27) directly by some optimization method, without resorting to identification. In fact, minimizing (22.27) over all possible θ corresponds to a direct minimization over the controller set $\mathcal{C}$, since every model parameter vector θ defines a controller parameter vector, say $\rho = \rho(\theta)$, via the mapping $C(\rho) = C(P(\theta))$. In particular, returning to the model reference problem above, we note that $y_t - \hat{y}_t$ can also be written as:

$$y_t - \hat{y}_t = \frac{P\hat{C}}{1 + P\hat{C}} r_t - Tr_t = y_t(\rho) - Tr_t. \tag{22.28}$$

where $T = 1 - S$ and is a fixed design quantity. Thus, minimizing $J(\theta)$ is equivalent to minimizing the control criterion $E[y_t(\rho) - Tr_t]^2$ with respect to the controller parameters ρ. Recent work of Hjalmarsson et al. [10] has shown that this is indeed possible using an iterative scheme that avoids any identification step.

6 CONCLUSIONS

We have given some motivation for the idea of performing control-oriented identification. This research activity, though very recent, is now in full bloom. Although the work so far is essentially theoretical, successful industrial applications have already been reported and the rewards in terms of productivity improvements appear to be promising.

The problem of identification design for control, and more generally of the combined design of identification and control, has been addressed from a wide range of angles. In this chapter we have presented three different approaches to the problem. These reflect more the author's perspective on the problem than they cover the very wide range of viewpoints and results that have been presented in the fast growing literature on the subject. We refer to the special issues of journals and the symposia, workshops and numerous invited sessions on this subject for a fuller coverage.

Our presentation has gone from the very idealised 'dual control approach' through the 'optimal design approach' to the more realistic 'robust control approach'. This last approach leads to the much publicised iterative identification and control design schemes. Finally we have raised the question of whether these iterative identification for control approaches and methods should not, in the end, give way to a more direct controller parameter tuning approach, particularly now that such a scheme is available that requires almost no prior knowledge on the real plant.

Acknowledgements[3]

This paper is dedicated to my supervisor Tom Kailath, who not only taught me how to do research, but has continued ever since to take a keen interest in my scientific achievements. Thanks, Tom.

The paper is very much the product of exciting collaborative work and discussions with several colleagues and students whom I am pleased to acknowledge: B.D.O. Anderson, R.R. Bitmead, F. De Bruyne, S. Gunnarsson, H. Hjalmarsson, C. Kulcsar, A. Partanen, P. van den Hof, Z. Zang. In particular, H. Hjalmarsson has made major contributions to these results.

REFERENCES

[1] Åström K.J., "Matching criteria for control and identification", *2nd European Control Conference*, Groningen, Holland, July 1993, pp. 248-251.

[2] Åström K.J. and J. Nilsson, "Analysis of a scheme for iterated identification and control", *Prepr. SYSID'94, 10th IFAC Symp. on System Identification*, Copenhagen, Denmark, July 1994, Vol. 2, pp. 171-176.

[3] Bitmead R.R., M. Gevers and V. Wertz, "Adaptive Optimal Control - The Thinking Man's GPC", Prentice Hall International, London, 1990.

[4] De Bruyne F. and M. Gevers, "Identification for control: can the optimal restricted complexity model always be identified?", *Proc. 33rd IEEE Conf. on Decision and Control*, Orlando, Florida, December 1994, pp. 3912-3917.

[5] de Callafon R.A., P.M.J. Van den Hof and D.K. de Vries, "Control-relevant identification of a compact disc pick-up mechanism", *Proc. 32nd IEEE Conf. on Decision and Control*, San Antonio, TX, December 1993, pp. 2050-2055.

[6] Fel'dbaum A.A., "The theory of dual control", Parts 1-4, *Automation and Remote Control*, 1960-1961.

[7] Gevers M., "Towards a joint design of identification and control?", *Essays on Control: Perspectives in the Theory and its Applications*, Birkhäuser, Boston, 1993, pp. 111-151.

[8] Gevers, M. and L. Ljung, "Optimal experiment designs with respect to the intended model application", *Automatica*, Vol. 22, September 1986, pp. 543-554.

[3]This paper presents research results of the Belgian Programme on Interuniversity Poles of Attraction, initiated by the Belgian State, Prime Minister's Office for Science, Technology and Culture. The scientific responsibility rests with its authors.

[9] Hjalmarsson H., M. Gevers, F. De Bruyne and J. Leblond, "Identification for control: closing the loop gives more accurate controllers", *Proc. 33rd IEEE Conf. on Decision and Control*, Orlando, Florida, December 1994, pp. 4150-4155.

[10] Hjalmarsson H., S. Gunnarsson and M. Gevers, "A convergent iterative restricted complexity control design scheme", *Proc. 33rd IEEE Conf. on Decision and Control*, Orlando, Florida, December 1994, pp. 1735-1740.

[11] Hjalmarsson H., S. Gunnarsson and M. Gevers, "Optimality and sub-optimality of iterative identification and control design schemes", *Proc. American Control Conference*, June 1995, Vol. 4, pp. 2559-2563.

[12] Kulcsár, C, *Planificiation d'expériences et commande duale*, PhD Thesis, Université de Paris-Sud, Centre d'Orsay, 1995.

[13] Ljung, L., *System Identification: Theory for the user*, Prentice Hall, 1987.

[14] Schrama R., "Accurate identification for control: the necessity of an iterative scheme", *IEEE Transactions on Automatic Control*, Vol. 37, No 7, pp. 991-994, July 1992.

[15] van den Hof P.M.J. and R.J.P. Schrama, "Identification and control - closed loop issues", *Prepr. SYSID'94, 10th IFAC Symp. on System Identification*, Copenhagen, Denmark, July 1994, Vol. 2, pp. 1-13.

[16] Zang Z, R.R. Bitmead and M. Gevers, "H_2 iterative model refinement and control robustness enhancement", *Proc. 30th IEEE Conf. on Decision and Control*, Brighton, UK, December 1991, pp. 279-284.

23

IDENTIFICATION AND DIGITAL CONTROL OF VERY FLEXIBLE MECHANICAL SYSTEMS

Ioan D. Landau

Laboratoire d'Automatique (INPG/CNRS), ENSIEG, BP 46, 38402 Saint Martin d'Heres, France

Dedicated to Tom Kailath on the occasion of his sixtieth birthday.

ABSTRACT

A methodology for the identification and robust digital control for very flexible mechanical systems is presented and applied to the control of a 360 degrees flexible robot arm. The importance of the identification in closed loop for improving the performances of the control system is emphasized.

1 INTRODUCTION

It is in 1981 during my stay in Stanford with the group of Professor Thomas Kailath that I discovered the challenge for control raised by very flexible mechanical systems. It is in Stanford that I have seen for the first time very flexible mechanical systems (a flexible transmission and a flexible robot arm).

Few years after my return from Stanford, this area became extremely important for my research activities in System Identification, Adaptive and Robust Digital Control. I have considered these types of systems as a significative benchmark for testing our algorithms.

So my thanks go to Professor Thomas Kailath not only for offering me this opportunity to discover this area but also for encouraging me in many occasions to express my soughts as well as my experience at the frontier between theory and applications.

My objective was to solve the problem of the control of very flexible mechanical systems working from the beginning (identification of the control model) to the end (implementation of the controller) in discrete time. This is a challenge since these systems are of infinite dimension or at least of very large dimension. Note that the usual approach for this problem is to make a very precise continuous time modeling, design the controller in continuous time based on this model and then discretize the controller. My approach will be described next.

The first step was to identify directly from input - output data a finite dimensional discrete time model capturing the significant vibration modes by an appropriate selection of the sampling frequency. Earlier results on identification of such models have been reported in [8]. Further results can be found in [2].

These models are characterized by several very low damped pairs of poles (damping factors < 0.05), by several unstable discrete time zeros (which correspond to truly non minimum phase systems in the case of very flexible arms) and eventually by a small delay of one or two samples. In addition there are important variations of the vibration modes in the presence of load without affecting significantly the damping factors.

The next step was to control such systems showing important variations of their frequency characteristics in the presence of load. The first approach considered was adaptive control. The adaptive control solution considered was an indirect type scheme using a robust parameter estimator (which involves data filtering and data normalization) combined with a control algorithm based on an approximate solution of a pole placement via the use of a "generalized predictive control" strategy [3]. The application of this strategy to flexible transmissions and flexible robot arms has been reported in [7].

However, in the last few years important progress has been made in understanding how to design robust digital controllers and this approach appears to compete with the adaptive control approach. A design method entitled "combined pole placement with sensitivity function shaping" has been developed and successfully applied to the design of robust digital controllers for flexible mechanical systems based on discrete time identified models [4].

Recently the concept of "controller validation" has been introduced [5]. This concept calls for the identification of the closed loop system in order to check the closeness between the designed closed poles computed on the basis of the identified plant model and the achieved closed loop poles.

The use of this concept has shown that the performances of the control loop can be further improved by identifying the plant model in closed loop operation and redesigning the controller on the basis of this new model. In other words, the identification of the plant model in closed loop operation allows to improve the "control" model used for design.

The performance improvement through plant model identification in closed loop operation can be explained by the fact that the effective plant input excitation is in this case much closer to the normal closed loop operation in terms of enhancement of certain frequencies regions critical for control and in terms of magnitude (for open loop identification the level of the excitation is very low and the behavior of the system is influenced by some non-linear effects like friction. This approach joins the strategies calling for iterative identification and control redesign [1].

The present contribution will emphasize how identification in closed loop can improve the synthesis of a robust digital controller for a 360 degrees very flexible robot arm.

2 THE 360 DEGREES FLEXIBLE ARM

The experimental device is a planar mechanism with one degree of freedom used to test the control algorithm for flexible structures which present multiple low damped vibration modes in the band pass of the control.

The flexible arm is constituted of two sheets of aluminum, respectively 1000 mm long and 0.7 mm thick coupled every 100 mm by means of a rigid frame. The free end carries a light source used to measure the orientation of the flexible arm and an electromagnet to fix a load. The other end is directly driven by the main actuating dc motor. It carries the optical sensor and the rotating mirror used for the measure of the tip position of the arm (figure 23.1). The load used for the experiments weighs 240g. The system has a local control loop for the motor axis position control. The digital controller is implemented on a VME system (with OS9 real time operating system) connected to a workstation. The sampling frequency is 20 Hz.

3 IDENTIFICATION OF A DISCRETE TIME MODEL

The flexible arm can be considered as an infinite dimensional system. Previous theoretical analysis using finite elements decomposition and experimental work [8] have shown that such a structure has a very large number of low damped oscillatory modes. However, the energy is essentially located in the first two or three modes. Therefore for control purposes it is enough to control the flexible arm in the range of frequencies covering the main modes. A spectral analysis can easily indicates for a specific application how many modes have to be effectively considered. In the case of the 360 degrees flexible arm, three vibration modes have to be considered.

Once the upper important vibration mode (mode N) being determined, it is possible to define a discrete time control model of the arm by sampling at a frequency f_s such that the Nyquist frequency ($= 0.5 f_s$) lies between the oscillation mode N and the nearest higher mode $N+1$. Of course anti-aliasing filters have to be introduced. After these operations, it has been shown [8,

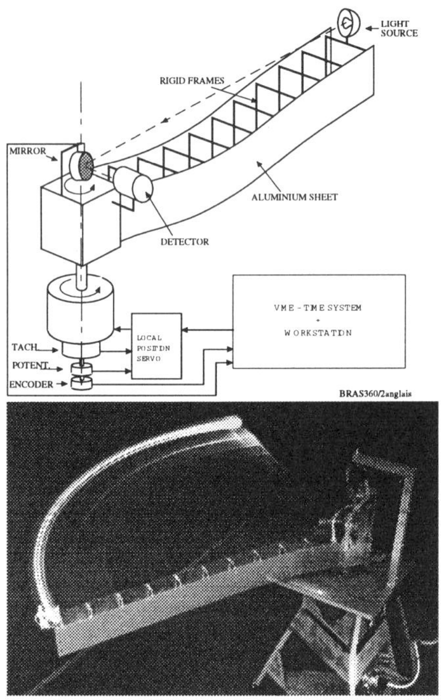

Figure 23.1 Flexible 360° arm

2] that a representative discrete time control model of a flexible arm can be identified from experimental data.

The structure of this model is as follows:

$$H\left(q^{-1}\right)=\frac{q^{-d}B\left(q^{-1}\right)}{A\left(q^{-1}\right)} \tag{23.1}$$

where:

q^{-1} : backward shift operator:

$$y(t-1)=q^{-1}y(t) \tag{23.2}$$

d : integer number of sampling periods on the plant pure time delay

$$A : \; = 1+a_1q^{-1}+\ldots+a_{n_A}q^{-n_A} \tag{23.3}$$

$$B : \; = b_1q^{-1}+\ldots+b_{n_B}q^{-n_B}=q^{-1}B^{\star}\left(q^{-1}\right) \tag{23.4}$$

In eq. 23.3 $n_A = 2N$ where N is the highest vibration mode considered.

Identification is done in open loop using as input a low magnitude P.R.B.S. It was found that prediction error identification methods are suitable for the identification of this type of plant (modified extended least squares, output error with extended prediction model). The identified models are validated by usual whiteness tests on the residual prediction error [2].

The frequency responses of the models identified in open loop for the 360 degrees flexible arm with and without load are shown in figure 23.2 ($n_A = 6$; $n_B = 6$; $d = 0$).

4 CONTROL OF THE 360 DEGREE FLEXIBLE ARM

For briefness, the model obtained by an identification in open-loop is called *open-loop model*, a controller based on this model *open-loop based controller*. Similar definitions are used for the model identified in closed loop (*closed-loop model*) and the corresponding controller (*closed-loop based controller*).

4.1 Controller Design - Open-loop based controller

The digital controller used is of R-S-T form [2] :

$$S(q^{-1})u(t)=T(q^{-1})y^{*}(t+d+1)-R(q^{-1})y(t)$$

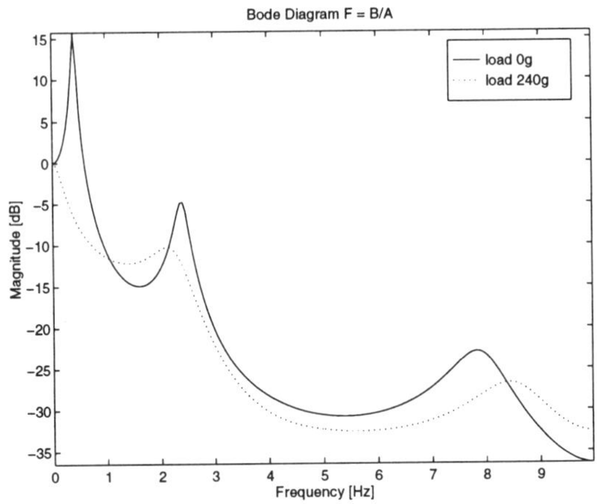

Figure 23.2 Plant models identified in open-loop

where u and y are respectively the plant input and output, y^* defines the desired output trajectory which is generated by a tracking reference model.

The controller design is based on the open-loop model without load. The robustness margins obtained are $\Delta M = -5.17$ dB and $\Delta\tau = 0.119$ s.

4.2 Identification in closed-loop

A new identification is carried out using the open-loop based controller obtained above. For the identification of the model in closed-loop the reference r is left at a constant value while the excitation v (added to the output of the controller) is a PRBS. This assures a better distribution of the spectral power of the plant input than it would be in case of the excitation using the reference input r.

Figure 23.3 shows the model without load identified in closed loop in comparison to the model identified in open loop.

In order to demonstrate the effect of the closed loop identification figure 23.4 shows a map of the closed-loop poles:

- $\times$ computed with the open-loop based controller and the model identified in *open*-loop
- $+$ computed with the open-loop based controller and the model identified in *closed*-loop
- $*$ found by an identification of the entire closed loop

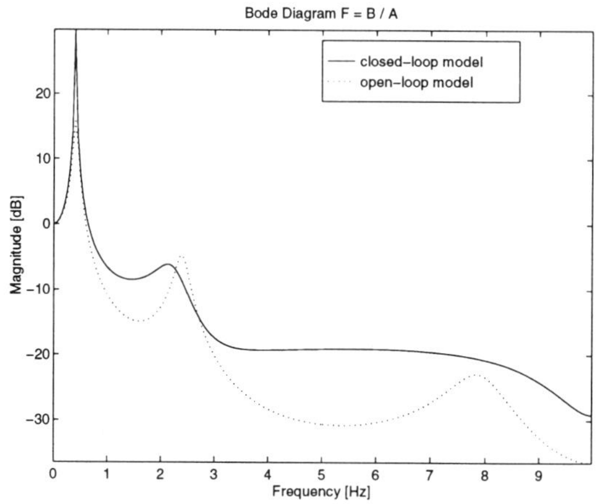

Figure 23.3 Models without load identified in open and closed loop

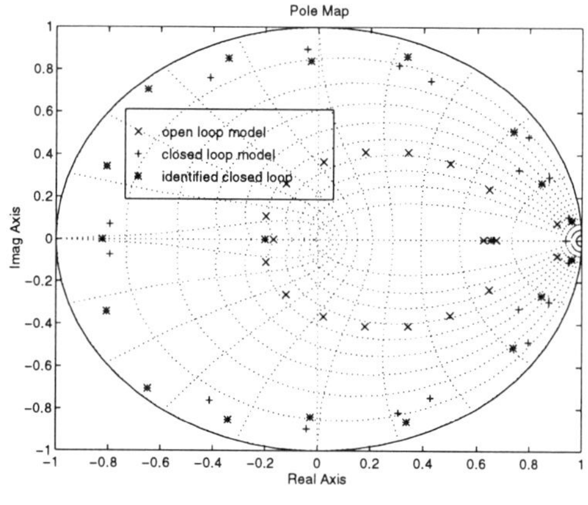

Figure 23.4 Pole Map of the Closed Loop

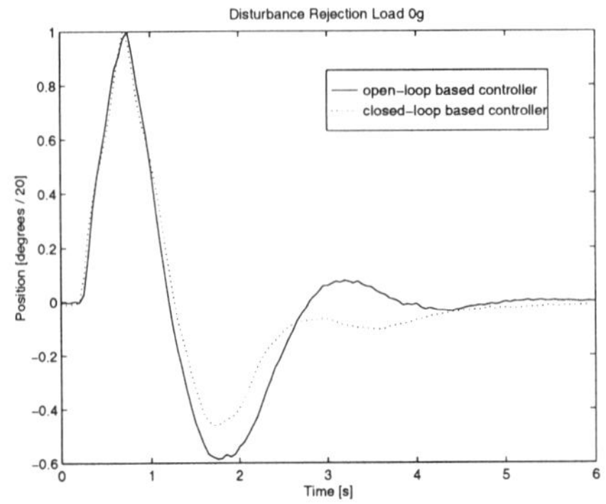

Figure 23.5 Disturbance Rejection without Load

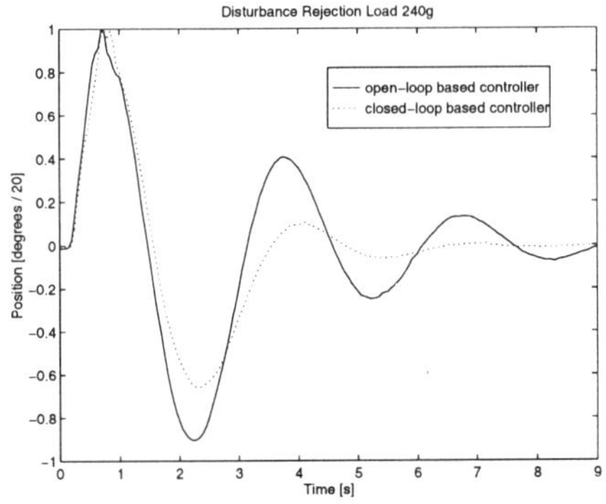

Figure 23.6 Disturbance Rejection with Load

The poles computed with the closed-loop model are much closer to the real poles of the closed loop, especially for the dominant poles of the loop. Thus the closed-loop model gives a much better idea of the behavior of the plant in the closed loop than the model identified in open loop. Identification in closed loop leads also to the improvement of the tracking performances [6].

4.3 Controller re-design (closed loop based controller)

The new controller is computed based on the model identified in closed-loop and starting with the specifications of the open-loop based controller. The robustness margins obtained are $\Delta M = -5.44$ dB et $\Delta\tau = 0.275$ s.

The figures 23.5 and 23.6 show the disturbance rejection behavior of the two controllers for the cases without and with load. The disturbance (of an initial magnitude of about 20°) is normalized to simplify the comparison. The identification in closed loop leads to a clear improvement of the performances of both loaded and unloaded cases.

5 CONCLUSIONS

A methodology dedicated to the identification and robust digital control design for flexible mechanical systems has been briefly presented and illustrated on an example (a 360 degrees flexible arm).

The results obtained in practice show that robust digital controllers designed using discrete time models identified from real data have excellent performances even in the case of important variations of the frequency characteristics of the flexible systems. It was also shown that performances are improved through the redesign of the controller using a model identified in closed loop.

Acknowledgment

The author would like to acknowledge the contribution of J. Langer to this work.

REFERENCES

[1] Gevers, M., "Towards a joint design of identification and control," in H. L. Trentelman and J. C. Willems, editors, *Essays on Control Perspective in the Theory and its Applications*, Birkhauser,Barton, USA, 1993.

[2] Landau, I. D., *System identification and control design*, Prentice-Hall International, Englewood Cliffs, NJ 1990.

[3] Landau, I. D., "Evolution of adaptive control," *Transactions ASME, J. of DSMC*, **115**(2):381-391, June 1993.

[4] Landau, I. D., C. Cyrot and D. Rey, "Robust control design using the combined pole placement/sensitivity function shaping method," In *European Control Conference*, **3**:1693-1697, Groningen, The Netherlands, 1993.

[5] Landau, I. D., "Adaptive control with a french touch – plenary lecture", In *American Control Conference Proceedings*, Seattle, WA 1995.

[6] Landau, I. D., "Closed-loop identification as a design tool for the synthesis of the robust digital controller for a 360° flexible robot arm," Laboratoire d'Automatique de Grenoble, 1995.

[7] M'Saad, M., L. Dugard, and Sh. Hammond, "A suitable generalized predictive adaptive controller case study: Control of a flexible arm," *Automatica*, **29**(3), 1993.

[8] van den Bossche, E., L. Dugard and I. D. Landau, "Modelling and identification of a flexible arm," In *Proceedings of the American Control Conference*, Seattle, WA, 1986.

24

FREQUENCY-SELECTIVE DYNAMIC MODELING: EXPERIENCES IN POWER SYSTEMS AND POWER ELECTRONICS

George C. Verghese

*Laboratory for Electromagnetic and Electronic Systems, &
Department of Electrical Engineering and Computer Science
Massachusetts Institute of Technology
Cambridge MA 02139*

For Professor Thomas Kailath, on his 60th birthday, with admiration, affection, and gratitude.

1 INTRODUCTION

On the occasion of Professor Kailath's sixtieth birthday, I find myself searching out the intellectual threads that connect my doctoral work under him, [1], with my subsequent explorations in areas that seem rather distant from this beginning. I see now that a dominant and recurring theme has been that of frequency-selective dynamic modeling. This paper sketches the development of that theme in the context of my experiences in modeling and control for power systems and power electronics. Along the way, I shall be referring explicitly to the work of three generations of Professor Kailath's academic progeny; each of those I mention is conscious of their link to him, and joins me in this felicitation.

2 INFINITE-FREQUENCY BEHAVIOR

The thesis [1] examined systems of the form

$$E\dot{x}(t) = Ax(t) + Bu(t)\ , \quad y(t) = Cx(t) \tag{24.1}$$

for the case where E is a square but singular matrix, and $sE - A$ is nonsingular as a rational matrix in s. The vector $x(t)$ comprises the internal variables of the system, $u(t)$ denotes a vector of inputs, and $y(t)$ is a vector of outputs. What is perhaps most interesting about these generalized state-space systems (also called descriptor systems, singular systems, differential-algebraic equations or DAEs, ...) is that they can have infinite-frequency modes, corresponding to the infinite zeros of the matrix $sE - A$. (The familiar finite-frequency modes of such a system, manifested as exponential motions when no inputs are present, correspond to the finite zeros of this matrix.) The infinite-frequency modes are displayed in the form of impulsive motions that occur when the system is formed at time $t = 0$ (perhaps by switching together component subsystems), with initial conditions $x(0-)$ that are not constrained by (24.1). Concepts of controllability, observability, realization, equivalence, and so on, can be extended to incorporate the impulsive modes of such systems, [2]. Important contributions by several researchers over the past fifteen years have served to further develop and fill out this picture.

3 SELECTIVE MODAL ANALYSIS (SMA)

Electric power systems are among the largest engineered systems. Individual electric utilities are interconnected so that they can export or import power in accordance with economic or security considerations, and the resulting interconnected power systems can span several countries. The dynamic phenomena of interest can therefore extend across large geographic regions, and detailed analysis of these phenomena begins with very large models.

The primary instance of such long-range dynamics involves electromechanical oscillations in the 0.1 – 2 Hz range, corresponding to oscillatory power flows in the system, and to associated oscillatory deviations of the generator motions from their steady rotations. These so-called *swing modes* can be roughly accounted for by a simple linearized spring-mass model, or linearized swing model, of the form

$$\begin{pmatrix} \dot{\delta}(t) \\ \dot{\omega}(t) \end{pmatrix} = \begin{pmatrix} 0 & I \\ -M^{-1}K & -M^{-1}D \end{pmatrix} \begin{pmatrix} \delta(t) \\ \omega(t) \end{pmatrix} . \tag{24.2}$$

Here δ is the vector of angular deviations of the generator rotors, measured relative to reference axes that turn at constant synchronous speed (corresponding to the nominally 60Hz voltage waveforms that the generators induce); ω is the vector of rotor speed deviations; M is a diagonal matrix of (normalized) rotational inertias; K is a positive semi-definite matrix derived from the admittance matrix of the transmission network, but also dependent on the operating point; and D is a positive-definite damping matrix whose entries are unfortunately not obtainable in any direct way from knowledge of the system. The order of the swing model is evidently twice the number of generators. The number of swing modes — each associated with a conjugate pair of complex eigenvalues — is one less than the number of generators; in addition, the model has a pair of real eigenvalues, one of them at the origin, and the other negative.

The model (24.2) captures with reasonable accuracy the swing-mode frequencies (given by the imaginary parts of the complex eigenvalues) and mode shapes (described by the associated right eigenvectors). However, the damping factors associated with these modes are not accurately represented. Since knowledge of the mode damping is critical to assessing and controlling the swing modes, we look for better models than (24.2). More detailed linearized models can indeed be derived, in the standard state-space form

$$\dot{x}(t) = Ax(t) \ , \tag{24.3}$$

by assembling detailed models of the individual generators, and taking account of their interconnection with each other and the system loads via the transmission network. In practice, a linearized model of a large power system would be left in DAE form, as in (24.1), because one can then take computational advantage of the sparse matrices that occur in this form. No matter which representation is chosen for the detailed model, we are still left with the fact that its dynamic order may now be 12 – 15 times the number of generators in the system. The order of this model can thus be very large.

Soon after arriving at MIT's Electric Power Systems Engineering Laboratory in 1979 (with the ink of Professor Kailath's signature not quite dry on my thesis!), I became involved in the problem of efficiently computing just a few selected modes of a large system such as (24.3). This task was the focus of the first doctoral thesis that I supervised (jointly with Professor Fred Schweppe). Ignacio Pérez-Arriaga's thesis on Selective Modal Analysis (SMA), [3], and extensions of SMA in subsequent doctoral theses supervised by Professor Pérez-Arriaga, [4], [5], [6], have led to the implementation of SMA in a production-grade software package. This package is used by several electric utilities for small-signal stability analysis, and for the siting and tuning of power system stabilizers.

A contribution of SMA that is even more widely used in the power systems community (and perhaps worthy of examination in a variety of other modal analysis contexts) is the notion of *participation factors*. We proposed the use of these numbers to measure the relevance of individual state variables in selected modes, at least from the point of view of reduced-order eigenanalysis. However, reported applications of participation factors as an exploratory or heuristic tool for model partitioning and reduction range well beyond this.

The participation factor p_{ij} for the ith state variable of (24.3) in the jth mode of the system is defined to be the product of the ith entries of this mode's left and right eigenvectors, w_j and v_j respectively (with the normalization $w_j^T v_j = 1$); thus $p_{ij} = w_{ij} v_{ij}$. This quantity is actually the sensitivity of the eigenvalue λ_j to damping of the ith state variable, and is invariant under scaling of the state variables. Furthermore, the sum of participation factors in a subsystem is invariant to changes of coordinates (similarity transformations) in the subsystem model. An examination of the participation matrix $P = [p_{ij}]$ associated with a linearized physical model in the form (24.3) often reveals (after appropriate reordering of state variables and modes) a nearly block-diagonal structure, indicating natural associations between state variable subsets and modal subsets.

At the core of SMA is an iteration aimed at computing a selected eigenvalue λ_o of the model (24.3). Given approximate knowledge of the participation factors of the state variables in the mode associated with λ_o, we can reorder and partition the state variables in (24.3) so as to obtain a description of the form

$$\begin{pmatrix} \dot{x}_r(t) \\ \dot{x}_z(t) \end{pmatrix} = \begin{pmatrix} A_1 & A_2 \\ A_3 & A_4 \end{pmatrix} \begin{pmatrix} x_r(t) \\ x_z(t) \end{pmatrix} , \tag{24.4}$$

where x_r comprises the relevant variables, i.e., those that have high participation factors in the mode of interest, and x_z comprises the less-relevant variables. For the electromechanical or swing modes of a power system, experience with scores of models of various sizes has shown that the most relevant variables are indeed the generator angle perturbations δ and velocity perturbations ω, as found in the linearized swing model (24.2). Also relevant are the rotor flux perturbations (E_q' in conventional notation) on the generators. More specifically, particular swing modes are primarily associated with the δ, ω and E_q' of particular groups of generators. The lower-frequency swing modes generally involve participation by generators spread out over more extensive geographical regions of the power system.

Now, to understand the origins of the basic SMA scheme for iterative computation of λ_o, assume that just this mode is excited in the system (24.4). Then, given an estimate $\lambda_o^{(k)}$ of λ_o at the kth iteration, we proceed by approximating $\dot{x}_z(t)$ on the left side of (24.4) by $\lambda_o^{(k)} x_z(t)$. The second row of (24.4) then becomes algebraic, and may be solved to yield an approximate expression for the less-relevant vector $x_z(t)$. Substitution of the result in the first row then yields the reduced-order system

$$\dot{x}_r(t) = [\, A_1 + A_2(\lambda_o^{(k)} I - A_4)^{-1} A_3 \,]\, x_r(t) \ . \tag{24.5}$$

Eigenanalysis of the reduced matrix in (24.5) now yields the next estimate $\lambda_o^{(k+1)}$ of λ_o. It turns out that this iteration converges locally to λ_o at a linear rate given by the ratio of the total participation of x_r in the mode to the total participation of x_z in the mode. (The relationship of SMA to singular perturbations emerges when we take $\lambda_o^{(k)} = 0$ as our estimate of a *slow* mode.)

The thesis [3] also extends this basic iterative SMA scheme to the multi-mode case, and develops decentralized implementations. Various other extensions and refinements have subsequently been implemented — to obtain quadratic rather than linear convergence, to handle large models through multi-area decomposition, to enable efficient computation of generalized modal sensitivities required for controller siting and tuning, and so on.

4 SYNCHRONIC MODAL EQUIVALENCING (SME)

A *dynamic equivalent* in power systems terminology is a reduced-order dynamic model for a power system. Since dynamic phenomena of interest often span large regions of a power system, the problem of model reduction assumes great importance. The state-of-the-art in dynamic equivalencing for transient stability studies — which are studies of the system's response to significant disturbances, such as shorts on transmission lines — is built on the notion of *coherency*. Two generators are (approximately) coherent for a class of disturbances if their rotor motions are (approximately) equal whenever the system is subjected to such a disturbance. A dynamic equivalent can then be constructed by replacing each group of coherent generators with a single equivalent generator. The actual construction of the equivalent generator can be somewhat cumbersome, [7], but several approaches have been developed for efficiently recognizing coherent generators. One such approach is described next, as it forms the starting point for our recent work.

Earlier studies have established that a linearized swing model of the form (24.2) usually suffices to recognize coherent groups of generators. For such a model, a convincing explanation of why (approximate) coherency is to be expected emerges from the *weak-coupling/slow-coherency* framework set out in [8]. In this framework, the power system is modeled as comprising distinct groups of generators that are strongly interconnected within each group, but interconnected across groups through vanishingly weak ties. Under these circumstances, a singular perturbation analysis shows that the modes of the linearized swing model may be classified as either slow or fast modes, with the fast modes being local to the individual groups, and the slow modes being manifested in the motions of the groups relative to each other. Furthermore, when only the slow inter-group modes are excited, the generators within each group move coherently (in the limit of vanishingly weak inter-group couplings). To apply the slow-coherency framework to dynamic equivalencing, one examines the slowest modes of a linearized swing model in order to try and recognize groups of generators that are approximately coherent in these modes. Following this, the equivalencing procedure of [7] is generally applied.

In recent work, [9], [10], we have built on the slow-coherency framework to arrive at a distinct and promising approach to dynamic equivalencing. Rather than looking for coherent groups of generators, we look for *synchronic* groups, defined as groups whose generator motions are linearly related. Furthermore, instead of restricting our attention to synchronic motions in the slowest modes, we permit alternative selections of modes, and provide an approach to picking these modes so as to yield well-defined inter-group and intra-group behavior. These relaxations of the demands of slow-coherency can be important in practical settings, where the assumption of vanishingly weak interconnections is violated.

Finally, our equivalenced system is obtained by: retaining one of the synchronic groups (or a union of synchronic groups) in full nonlinear detail as a *study group*; retaining, again in full nonlinear detail, one representative or basis generator from each synchronic group external to the study group; and replacing every other generator external to the study group by a nondynamic linear circuit containing a dependent current source that is driven by the motions of the basis generators. The current source gains are computed so as to preserve some or all of the inter-group modes. Experiments with intermediate-sized models have yielded excellent results. For example, a test case involving 23 generators with 476 state variables was reduced to a dynamic equivalent comprising a study group of 6 generators and 4 external basis generators; simulations with the reduced model ran up to five times faster than with the full model, and matched the responses of the full model very closely. Tests on a much larger model are planned for the near future.

5 DYNAMICS OF HARMONIC COMPONENTS

The voltages and currents in a power system are periodic in steady state, and typically sinusoidal, although there are important situations in which higher harmonics are important as well. Similarly, in power electronic circuits, [11], [12], voltages and currents are generally periodic (often dc or sinusoidal), perhaps with some higher-frequency switching ripple superimposed on them. For dynamic analysis and control design in such situations, the dynamics of interest are those that describe deviations from periodic behavior. What we need are *dynamic models for the Fourier coefficients* of waveforms of interest. This idea goes back to classical averaging theory, but the presentation and development in [13], [14], [15] make the application to power systems and power electronics quite transparent; each of these references opens up a novel and very fruitful modeling avenue for the associated application area.

To develop the desired dynamic models, we first note that a waveform $x(\tau)$ can be represented on the interval $(t-T, t]$ by a Fourier series of the form

$$x(\tau) = \mathcal{R}e\left\{\sum_{k\geq 0} X_k(t)e^{jk\omega_s\tau}\right\} , \tag{24.6}$$

where $\mathcal{R}e$ denotes the real part, $\omega_s = 2\pi/T$, $\tau \in (t-T, t]$, and $X_k(t)$ are the complex Fourier coefficients, which we shall also refer to as *phasors*. These

Fourier coefficients are functions of time, since the interval under consideration slides as a function of time. The kth coefficient, or k-phasor, at time t is determined by the following *averaging* operation:

$$X_k(t) = \frac{c}{T}\int_{t-T}^{t} x(\tau)e^{-jk\omega_s\tau}d\tau = \langle x\rangle_k(t) \ , \qquad (24.7)$$

where $c = 1$ for $k = 0$, and $c = 2$ for $k > 0$. The notation $\langle x\rangle_k(t)$ is used to denote the averaging operation in (24.7).

Our objective is to obtain a dynamic model for the dominant phasors, as the window of length T slides over the waveforms of interest. More specifically, beginning with a state-space model that governs the variables of interest, we wish to obtain — with appropriate assumptions and approximations — a state-space model in which the phasors are the state variables. A key fact for our development is that the *derivative* of the kth phasor is given by the following expression (we drop the time argument t here and in other equations below, to avoid notational clutter):

$$\frac{dX_k}{dt} = \left\langle \frac{d}{dt}x \right\rangle_k - jk\omega_s X_k \ . \qquad (24.8)$$

This formula is easily verified by differentiating the definition in (24.7).

Consider, for illustration, the application treated in [15], namely the dynamic modeling of a thyristor-controlled series capacitor (TCSC). This is a power electronic circuit that is inserted in series with a transmission line, and intended to function as a variable capacitance. The TCSC comprises the parallel combination of two branches: one branch contains a capacitor, and the other branch contains an inductor in series with a controlled switch (made up of anti-paralleled thyristors). The switch determines the fraction of each 60Hz-cycle for which the inductor will appear in parallel with the capacitor, thereby modulating the effective capacitance, in some sense.

The capacitor voltage v and inductor current i are natural state variables for the TCSC circuit. The corresponding state-space model (for an appropriate choice of polarities) is then easily seen to be

$$\begin{aligned} C\frac{dv}{dt} &= i_\ell - i \\ L\frac{di}{dt} &= qv \end{aligned} \qquad (24.9)$$

where i_ℓ is the quasi-sinusoidal transmission line current that passes through the TCSC, and q is a switching function that denotes the switch status: $q = 1$ when the switch is closed, placing the inductor in parallel with the capacitor; and $q = 0$ when the switch is open. External control signals and the system state determine which value q takes. Although this model captures the instantaneous behavior of the circuit, it operates at too fine a time-scale to be efficient for simulation or tractable for control design.

We now develop a model for the dynamics of the 1-phasors V_1 and I_1, respectively associated with the variables v and i. Assume for simplicity that the line current i_ℓ is essentially sinusoidal, with corresponding 1-phasor I_ℓ. The dynamics of V_1 and I_1 are then obtained by applying the averaging operation in (24.7), for $k = 1$, to both sides of the state-space model (24.9), followed by application of the differentiation formula in (24.8). The resulting model is:

$$\begin{aligned} C\frac{dV_1}{dt} &= I_\ell - I_1 - j\omega_s CV_1 \\ L\frac{dI_1}{dt} &= \langle qv \rangle_1 - j\omega_s LI_1 \end{aligned} \tag{24.10}$$

If $\langle qv \rangle_1$ depended only on the instantaneous values of V_1, I_1, I_ℓ, and the external control variables, then (24.10) would already be in state-space form, and ready to use. However, $\langle qv \rangle_1$ is not so accommodating, and some approximations are needed before (24.10) can be converted to a state-space description. Fortunately, excellent approximations, of varying degrees of refinement, are possible. The use of the resulting state-space model for the phasors, as opposed to the state-space model (24.9) for the instantaneous variables, results in improvements by more than an order of magnitude in simulation times, and exposes several new control possibilities.

Acknowledgements

My father-in-law and Professor Kailath have the same birthday (a coincidence that isn't!). I am grateful for their attention and generosity as mentors, and for the inspiration that their lives and work are to me.

MIT's Laboratory for Electromagnetic and Electronic Systems (which absorbed the Electric Power Systems Engineering Laboratory that I first joined) has been a most congenial home for research of the sort described in this paper. I am indebted to my colleagues, students, and research sponsors there.

REFERENCES

[1] G.C. Verghese, *Infinite-Frequency Behaviour in Generalized Dynamical Systems*, Ph.D. Thesis, Electrical Engineering Department, Stanford University, December 1978.

[2] G.C. Verghese, B.C. Levy and T. Kailath, "A generalized state-space for singular systems," *IEEE Trans. Automatic Control*, AC-26, 4, 811–831, August 1981.

[3] I.J. Pérez-Arriaga, *Selective Modal Analysis, with Applications to Electric Power Systems*, Ph.D. Thesis, EECS Department, Massachusetts Institute of Technology, May 1981.

[4] F.L. Pagola, I.J. Pérez-Arriaga and G.C. Verghese, "On sensitivities, residues and participations: Applications to oscillatory stability analysis and control," *IEEE Trans. Power Systems*, 4, 278-285, February 1989.

[5] I.J. Pérez-Arriaga, G.C. Verghese, F.L. Pagola, J.L. Sancha and F.C. Schweppe, "Developments in Selective Modal Analysis of small-signal stability in electric power systems," *Automatica*, Vol. 26, No 2, pp. 215–231, 1990.

[6] L. Rouco and I.J. Pérez-Arriaga, "Multi-area analysis of small signal stability in large electric power systems by SMA," *IEEE Trans. Power Systems*, Vol. 8, No. 3, pp. 1257–1265, August 1993.

[7] A.J. Germond and R. Podmore, "Dynamic aggregation of generating unit models," *IEEE Trans. Power Apparatus and Systems*, Vol. PAS-97, No. 4, pp. 1060–1069, July/August 1978.

[8] J.H. Chow, *Time-Scale Modeling of Dynamic Networks with Applications to Power Systems*, Lecture Notes in Control and Information Sciences, Vol. 46, Springer Verlag, 1982.

[9] G.N. Ramaswamy, *Modal Structures and Model Reduction, with Application to Power System Equivalencing*, Ph.D. Thesis, EECS Department, Massachusetts Institute of Technology, June 1995.

[10] G.N. Ramaswamy, C. Evrard, G.C. Verghese, O. Fillâtre and B.C. Lesieutre, "Extensions, simplifications, and tests of Synchronic Modal Equivalencing," IEEE Summer Power Meeting, Denver, July 1996, and to appear in *IEEE Trans. Power Systems.*

[11] J.G. Kassakian, M.F. Schlecht and G.C. Verghese, *Principles of Power Electronics*, Addison-Wesley, 1991.

[12] G.C. Verghese, "Dynamic Modeling and Control in Power Electronics," in *The Controls Handbook*, W.S. Levine (Ed.), CRC Press, 1996.

[13] S.R. Sanders, J.M. Noworolski, X.Z. Liu and G.C. Verghese, "Generalized averaging method for power conversion circuits," *IEEE Trans. Power Electronics*, Special Issue on Modeling, 6, 251–259, April 1991.

[14] C.L. DeMarco and G.C. Verghese, "Bringing phasor dynamics into the power system load flow," *North American Power Symposium*, Howard University, Washington, D.C., October 1993.

[15] P. Mattavelli, G.C. Verghese and A.M. Stankovic, "Phasor dynamics of thyristor-controlled series capacitor systems," IEEE Summer Power Meeting, Denver, July 1996, and to appear in *IEEE Trans. Power Systems.*

PART IV

NONLINEAR SYSTEMS AND CONTROL

25

ON THE NONLINEAR STANDARD H^∞ PROBLEM

Ciprian Foias*, Caixing Gu, and Allen Tannenbaum*****

** Department of Mathematics, Indiana University, Bloomington, Indiana 47405*

*** Department of Mathematics, University of California, Irvine, CA 92717*

**** Department of Electrical Engineering, University of Minnesota, Minneapolis, MN 55455*

This paper is dedicated with affection and admiration to Professor Thomas Kailath on the occasion of his 60th birthday.

ABSTRACT

In this note, using a power series approach [15, 16, 19] we describe a design procedure applicable to analytic nonlinear plants. Our technique is a generalization of the linear H^∞ theory. We can now use this theory to solve the full standard problem in robust control theory in the nonlinear framework.

1 INTRODUCTION

In this note, we extend our work on finding a suitable, implementable nonlinear extension of the linear H^∞ design methodology to the full standard problem. In what follows, we will just consider discrete-time systems, even though the techniques described below carry over to the continuous-time setting as well.

Our approach is valid for systems described by analytic input/output operators. As in [17, 18, 16, 15], our technique involves the expression of each n-linear term of a suitable Taylor expansion of the given operator as an equivalent linear operator acting on a certain associated tensor space which allowed us to iteratively apply the classical commutant lifting theorem in designing a compensator. (Our class of operators includes Volterra series [9].)

More precisely, in our approach we are reduced to applying the classical (linear) commutant lifting theorem to an H^2-space defined on some D^n (where D denotes the unit disc). Now when one applies the classical result to D^n

($n \geq 2$), even though time-invariance is preserved (that is, commutation with the appropriate shift), causality may be lost. Indeed, for systems described by analytic functions on the disc D (these correspond to stable, discrete-time, 1-D systems), time-invariance (that is, commutation with the unilateral shift) implies causality. For analytic functions on the $n-$disc ($n > 1$), this is not necessarily the case. For dynamical system control design and for any physical application, this is of course major drawback for such an approach.

Hence for a dilation result in $H^2(D^n)$ we need to include the causality constraint explicitly in the set-up of the dilation problem. It is precisely this problem which motivated the mathematical operator-theoretic work of [19] and [15] which incorporated Arveson theory [1] into the dilation, commutant lifting framework. In our paper [16], we show how these ideas lead to an explicit solution of a nonlinear extension of the weighted sensitivity minimization problem in linear H^∞ control.

While, the general method explicated in this note is based on a causal extension of the commutant lifting theorem, for the purposes of the operators and spaces which appear in control we will give a direct simple method for finding the optimal causal compensators. In fact, we will show that *the computation of an optimal causal nonlinear compensator may be reduced to a classical linear dilation problem whose solution is given by the Commutant Lifting Theorem.*

We now briefly summarize the contents of this paper. In Section 2, we briefly review the theoretical frameowrk of the Causal Commutant Lifting Theorem. In Section 3, we state and prove a version of the Causal Commutant Lifting Theorem which is most suitable for nonlinear control applications. In Section 4, we work out the theory in $H^2(D^n)$ and then in Sections 5 and 6, we apply our methods to the nonlinear extension of the H^∞ standard problem.

2 PRELIMINARIES ON THE CAUSAL COMMUTANT LIFTING THEOREM

In order to make the presentation as self-contained as possible, we recall in this section the classical Classical Commutant Lifting Theorem [28], [13] as well as the causal version [17], [15].

We let S and U denote isometries on the complex separable Hilbert spaces $\mathcal{G}$ and $\mathcal{K}$, respectively. Let $\mathcal{H} \subset \mathcal{K}$ denote a U^* invariant (closed) subspace. Let $P : \mathcal{K} \to \mathcal{H}$ denote orthogonal projection, and set $T = PU|\mathcal{H}$. For the convenience of the reader, we state the classical Commutant Lifting Theorem [28], [13]:

Theorem 25.1 *Let $A : \mathcal{G} \to \mathcal{H}$ be a bounded linear operator such that $AS = TA$. Then there exists a bounded linear operator $B : \mathcal{G} \to \mathcal{K}$ such that*

$$\begin{aligned} BS &= UB, \\ A &= PB, \\ \|A\| &= \|B\|. \end{aligned}$$

Such an operator B is called an *intertwining dilation* (or *lifting*) of A.

We now define *causality* in this framework. Roughly, causality means that for a given input/output system the past output is independent of the future inputs. This may be given precise mathematical formulation in terms of a family of projections which we shall now do. See also [1] and [11].

Let $P_j,\ j \geq 1$ be a sequence of orthogonal projections on $\mathcal{G}$ satisfying the following conditions:

$$P_1 \leq P_2 \leq \dots \tag{25.1}$$

$$P_j \leq I - S^j S^{*j} \quad j = 1, 2, \dots \tag{25.2}$$

$$P_{j+1} S(I - P_j) = 0 \quad j = 1, 2, \dots . \tag{25.3}$$

We call such a family of projections a *S-causal structure* on $\mathcal{G}$. Note that $I - S^j S^{*j}$, $j \geq 1$ is clearly the "maximal" S-causal structure on $\mathcal{G}$. When no specification of the S-causal structure on $\mathcal{G}$ is made, we will automatically consider that the S-causal structure in $\mathcal{G}$ is the maximal one.

Let $B : \mathcal{G} \to \mathcal{K}$ intertwine S with U, that is

$$UB = BS. \tag{25.4}$$

Note that this implies $U^j B = BS^j$ $(j \geq 1)$, hence $U^j U^{*j} BS^j S^{*j} = BS^j S^{*j}$, and so

$$(I - U^j U^{*j})B = (I - U^j U^{*j})B(I - S^j S^{*j}) \quad j = 1, 2, \dots . \tag{25.5}$$

We now make the following key definition:

Definition 1. Let $\mathcal{G}$ be endowed with a S-causal structure P_j , $j \geq 1$ and $\mathcal{K}$ with a U-causal structure Q_j , $j \geq 1$. An operator B satisfying (25.4) is called *causal* if

$$Q_j B = Q_j B P_j \quad j \geq 1 \tag{25.6}$$

or equivalently,

$$(I - P_j) B^* = (I - P_j) B^* (I - Q_j) \quad j \geq 1. \tag{25.7}$$

Note by (25.5) that B is always causal with respect to the maximal causal structures in $\mathcal{G}$ and $\mathcal{K}$. In what follows the sequence $P_1, P_2, \ldots$ will be fixed and the causality for intertwining dilations B will always be defined relative to this causal structure in $\mathcal{G}$ and the maximal U-causal structure in $\mathcal{K}$.

We now set

$$\|A\|_c := \inf\{\|B\| : B \text{ is a causal intertwining dilation of } A\}. \tag{25.8}$$

By using the weak operator topology, we can easily prove that the infimum in (25.8) is actually a minimum.

Finally, let $\hat{S}$ denote the minimal unitary extension of S on $\hat{\mathcal{G}} \supset \mathcal{G}$, and

$$\mathcal{G}_c := \overline{\left(\cup_{j=0}^{\infty} \hat{S}^{*j} \mathcal{G}_j\right)} \subset \hat{\mathcal{G}}, \quad S_c := \hat{S}|\mathcal{G}_c. \tag{25.9}$$

We now recall the following result from [15]:

Theorem 25.2 (Causal Commutant Lifting Theorem)

1. *If T is invertible, define*
$$A_c \hat{g} := T^{-j} A g_j, \tag{25.10}$$
for $\hat{g} = \hat{S}^{*j} g_j$, where $g_j \in \mathcal{G}_j$ ($j = 0, 1, \ldots$). Then the causal commutant lifting problem is solvable if and only if A_c is bounded. In this case
$$A_c S_c = T A_c, \quad \|A\|_c = \|A_c\|.$$

2. *In general (i.e., we do not assume T is invertible), the causal commutant lifting problem is solvable if and only if there exists a linear, bounded operator*
$$A' : \mathcal{G}_c \to \mathcal{H}, \quad A' S_c = T A', \quad A'|\mathcal{G} = A. \tag{25.11}$$
If such an operator A' exists, then
$$\|A\|_c = \min\{\|A'\| : A' \text{ as in (25.11)}\}. \tag{25.12}$$

As we will see in the next section, the standard problem in robust control theory has a special structure which allows a direction construction of an operator A' as above. For completeness, we will give a direct proof based on [15] for the reduction of the "causal" standard problem to the "classical" standard problem.

3 CONTROL VERSION OF CAUSAL COMMUTANT LIFTING THEOREM

For the standard problem in robust control theory (see our discussion in Section 5 below), we may extract the following mathematical set-up. We are given complex separable Hilbert spaces $\mathcal{E}_1, \mathcal{E}_2, \mathcal{F}_1, \mathcal{F}_2$ equipped with the unilateral shifts $S_{\mathcal{E}_1}, S_{\mathcal{E}_2}, S_{\mathcal{F}_1}, S_{\mathcal{F}_2}$, respectively. Let $\Theta_1 : \mathcal{E}_1 \to \mathcal{F}_1$ and $\Theta_2 : \mathcal{F}_2 \to \mathcal{E}_2$ be operators intertwining $S_{\mathcal{E}_1}$ with $S_{\mathcal{F}_1}$ (i.e., $\Theta_1 S_{\mathcal{E}_1} = S_{\mathcal{F}_1}\Theta_1$), and intertwining $S_{\mathcal{E}_2}$ with $S_{\mathcal{F}_2}$, respectively. We let $U_{\mathcal{E}_1}$, be the minimal unitary dilation of $S_{\mathcal{E}_1}$ on $\mathcal{K}_{\mathcal{E}_1}$, and similarly for $U_{\mathcal{E}_2}$ on $\mathcal{K}_{\mathcal{E}_2}$, $U_{\mathcal{F}_1}$ on $\mathcal{K}_{\mathcal{F}_1}$, and $U_{\mathcal{F}_2}$ on $\mathcal{K}_{\mathcal{F}_2}$.

We let the sequence $P_{\mathcal{E}_2}^{(n)}, n \geq 1$ resp. $P_{\mathcal{F}_2}^{(n)}, n \geq 1$ define the maximal $S_{\mathcal{E}_2}$-causal structure on $\mathcal{E}_2$, resp. maximal $S_{\mathcal{F}_2}$-maximal causal structure on $\mathcal{F}_2$. Moreover, the $S_{\mathcal{E}_1}$-causal structure on $\mathcal{E}_1$ is defined by a general sequence of projections $P_1^{(n)}$, $n \geq 0$ satisfying the causal structure conditions given by (25.1), (25.2), (25.3), and similarly the $S_{\mathcal{F}_1}$-causal structure on $\mathcal{F}_1$ is defined by a sequence of projections $Q_1^{(n)}$, $n \geq 0$ satisfying those conditions as well. We assume that the two input/output operators Θ_1, Θ_2, are causal with respect to the above structures. In Section 4 below, we will give precise definitions of the sequences $P_1^{(n)}$, $Q_1^{(n)}$ for the relevant spaces appearing in the nonlinear H^∞ control problem.

We let $W : \mathcal{E}_1 \to \mathcal{E}_2$ denote a causal operator intertwining $S_{\mathcal{E}_1}$ with $S_{\mathcal{E}_2}$. Thus causality for W means that

$$P_{\mathcal{E}_2}^{(n)} W P_1^{(n)} = P_{\mathcal{E}_2}^{(n)} W, \;\; \forall n \geq 0.$$

Finally, $Q : \mathcal{F}_1 \to \mathcal{F}_2$ will denote a causal operator intertwining $S_{\mathcal{F}_1}$ with $S_{\mathcal{F}_2}$. Define now

$$\mathcal{E}_1^{(n)} := (I - P_1^{(n)})\mathcal{E}_1, \;\; \forall n \geq 0.$$

Since by the causality of W we have

$$(I - S_{\mathcal{E}_2}^n S_{\mathcal{E}_2}^{*n}) W | \mathcal{E}_1^{(n)} = 0,$$

it follows that

$$W_n := S_{\mathcal{E}_2}^{*n} W | \mathcal{E}_1^{(n)},$$

is also uniquely defined by

$$S_{\mathcal{E}_2}^n W_n := W | \mathcal{E}_1^{(n)}, \;\; W_n : \mathcal{E}_1^{(n)} \to \mathcal{E}_2.$$

Clearly,

$$\begin{aligned} W_n S_{\mathcal{E}_1} | \mathcal{E}_1^{(n)} &= S_{\mathcal{E}_2} W_n, \\ W_n &= W_{n+1} S_{\mathcal{E}_1} | S_{\mathcal{E}_1}^{(n)}. \end{aligned}$$

Now let

$$\mathcal{E}_1^{(c)} = \overline{\mathcal{E}_1^{(co)}}$$

where

$$\mathcal{E}_1^{(co)} := \bigcup_{j=0}^{\infty} U_{\mathcal{E}_1}^{*j} \mathcal{E}_1^{(j)} \subset \mathcal{K}_{\mathcal{E}_1}, \;\; S_{\mathcal{E}_1}^{(c)} := U_{\mathcal{E}_1} | \mathcal{E}_1^{(c)}.$$

Finally, we define $W_c : \mathcal{E}_1^{(co)} \to \mathcal{E}_2$, by

$$W_c g := W_n g_n,$$

for $g = U_{\mathcal{E}_1}^{*n} g_n, \;\; g_n \in \mathcal{E}_1^{(n)}, \;\; n \geq 0$. Similarly the same construction is valid for $(W - \Theta_2 Q \Theta_1)_c$.

Note that moreover we can make a similar construction on the spaces $\mathcal{F}_1, \mathcal{F}_2$. In particular, for a causal $Q : \mathcal{F}_1 \to \mathcal{F}_2$, such that $QS_{\mathcal{F}_1} = S_{\mathcal{F}_2} Q$, we can define $Q_c : \mathcal{F}_1^{(co)} \to \mathcal{F}_2$, where

$$\mathcal{F}_1^{(co)} := \bigcup_{j=0}^{\infty} U_{\mathcal{F}_1}^{*j} \mathcal{F}_1^{(j)} \;, \text{ where } \mathcal{F}_1^{(j)} = (I - Q_1^{(j)}) \mathcal{F}_1 \;,\; j \geq 1 \;.$$

Next, it is easy to see both W_c and Q_c extend by continuity to the closure $\mathcal{E}_1^{(c)}$, respectively $\mathcal{F}_1^{(c)} = \overline{\mathcal{F}_1^{(co)}}$. Clearly, we also have

$$\|W_c\| = \|W\|, \;\; W_c | \mathcal{E}_1 = W, \;\; W_c S_{\mathcal{E}_1}^{(c)} = S_{\mathcal{E}_2} W_c,$$

and

$$\|W - \Theta_2 Q \Theta_1\| = \|(W - \Theta_2 Q \Theta_1)_c\|. \tag{25.13}$$

Now set

$$\mu(W, \Theta_1, \Theta_2) := \inf\{\|W - \Theta_2 Q \Theta_1\| : QS_{\mathcal{F}_1} = S_{\mathcal{F}_2} Q\}.$$

This corresponds to the *classical standard control problem.* We also set

$$\mu_c(W, \Theta_1, \Theta_2) := \inf\{\|W - \Theta_2 Q \Theta_1\| : Q \text{ causal}, \; QS_{\mathcal{F}_1} = S_{\mathcal{F}_2} Q\}.$$

This is the *causal standard control problem.*

Let $\hat{\Theta}_1 : \mathcal{K}_{\mathcal{E}_1} \to \mathcal{K}_{\mathcal{F}_1}$ denote the extension of the $\Theta_1 : \mathcal{E}_1 \to \mathcal{F}_1$, that is uniquely defined by

$$\hat{\Theta}_1 U_{\mathcal{E}_1}^{*n} e_1 = U_{\mathcal{F}_1}^{*n} \Theta_1 e_1, \;\; \forall e_1 \in \mathcal{E}_1.$$

Note that $\hat{\Theta}_1 U_{\mathcal{E}_1} = U_{\mathcal{F}_1} \hat{\Theta}_1$ and $\hat{\Theta}_1 U_{\mathcal{E}_1}^* = U_{\mathcal{F}_1}^* \hat{\Theta}_1$.

Lemma 25.1 *Notation as above. Then*

$$(\Theta_2 Q \Theta_1)_c = \Theta_2 Q_c \hat{\Theta}_1 | \mathcal{E}_1^c.$$

Proof. First note that

$$\begin{aligned}\Theta_2 Q(\Theta_1|\mathcal{E}_1^{(n)}) &= \Theta_2(Q|\mathcal{F}_1^{(n)})(\Theta_1|\mathcal{E}_1^{(n)})\\ &= \Theta_2 S_{\mathcal{F}_2}^n Q_n \Theta_1|\mathcal{E}_1^{(n)}\\ &= S_{\mathcal{E}_2}^n \Theta_2 Q_n \Theta_1|\mathcal{E}_1^{(n)}.\end{aligned}$$

Therefore,

$$(\Theta_2 Q \Theta_1)_n = \Theta_2 Q_n \Theta_1.$$

Let $g = U_{\mathcal{E}_1}^{*n} g_n$, $g_n \in \mathcal{E}_1^{(n)}$. Then

$$\begin{aligned}(\Theta_2 Q\Theta_1)_c g &= (\Theta_2 Q\Theta_1)_n g_n\\ &= \Theta_2 Q_n \Theta_1 g_n.\end{aligned}$$

On the other hand,

$$\begin{aligned}\hat{\Theta}_1 g &= \hat{\Theta}_1 U_{\mathcal{E}_1}^{*n} g_n\\ &= U_{\mathcal{F}_1}^{*n}\hat{\Theta}_1 g_n.\end{aligned}$$

(Note that $\Theta_1 g_n \in \mathcal{F}_1^{(n)}$.) Hence

$$Q_c\hat{\Theta}_1 g = Q_c U_{\mathcal{F}_1}^{*n}\Theta_1 g_n = Q_n\Theta_1 g_n.$$

It follows,

$$\Theta_2 Q_c\hat{\Theta}_1 g = \Theta_2 Q_n \Theta_1 g_n = (\Theta_2 Q\Theta_1)_c g,$$

that is

$$(\Theta_2 Q\Theta_1)_c = \Theta_2 Q_c \hat{\Theta}_1|\mathcal{E}_1^{(c)},$$

as required. □

As a consquence of Lemma 1, we have that for any causal Q,

$$\begin{aligned}\|W - \Theta_2 Q\Theta_1\| &= \|(W - \Theta_2 Q\Theta_1)_c\|\\ &= \|W_c - \Theta_2 Q_c\hat{\Theta}_1|\mathcal{E}_1^{(c)}\| \leq\\ &\leq \mu(W_c, \hat{\Theta}_1|\mathcal{E}_1^{(c)}, \Theta_2)\\ &= \|W_c - \Theta_2 Q_{opt}^{(c)}(\hat{\Theta}_1|\mathcal{E}_1^{(c)})\| =: \mu\end{aligned}$$

for some $Q_{opt}^{(c)} : \mathcal{F}_1^{(c)} \to \mathcal{F}_2$. (Note that $\hat{\Theta}_1\mathcal{E}_1^{(c)} \subset \mathcal{F}_1^{(c)}$.) and so,

$$\mu_c := \mu_c(W, \Theta_1, \Theta_2) \leq \mu$$

We should remark that μ is the optimal performance for the standard problem associated to $W_c, \Theta_2, \hat{\Theta}_1|\mathcal{E}_1^{(c)}$.

We now claim that $\mu \le \mu_c$. Indeed, set

$$\mathcal{F}_1^{(-n)} := U_{\mathcal{F}_1}^{*n}\mathcal{F}_1^{(n)},\ n \ge 0,$$

so that $U_{\mathcal{F}_1}^n \mathcal{F}_1^{(-n)} = \mathcal{F}_1^{(n)}$. Define

$$Q := Q_{opt}^{(c)}|\mathcal{F}_1.$$

Clearly,

$$\begin{aligned} Q\mathcal{F}_1^{(n)} &= Q_{opt}^{(c)} U_{\mathcal{F}_1}^n \mathcal{F}_1^{(-n)} \\ &= S_{\mathcal{F}_2}^n Q_{opt}^{(c)} \mathcal{F}_1^{(-n)} \\ &\subset S_{\mathcal{F}_2}^n Q_{opt}^{(c)} \mathcal{F}_1^{(c)} \\ &\subset S_{\mathcal{F}_2}^n \mathcal{F}_2, \end{aligned}$$

that is $P_{\mathcal{F}_2}^{(n)} Q(I - Q_1^{(n)}) = 0, n \ge 1$, and Q is causal. Now since $Q_c = Q_{opt}^{(c)}$ we have $(W - \Theta_2 Q \Theta_1)_c = W_c - \Theta_2 Q_{opt}^{(c)} \hat{\Theta}_1 | \mathcal{E}_1^{(c)}$ and thus

$$\mu = \|(W_c - \Theta_2 Q_{opt}^{(c)} \hat{\Theta}_1 | \mathcal{E}_1^{(c)})\| = \|W - \Theta_2 Q \Theta_1\| \le \mu_c$$

which proves the claim.

We can now summarize the above discussion with the following result:

Theorem 25.3 *Notation as above.*

1. $\mu_c(W, \Theta_1, \Theta_2) = \mu(W_c, \hat{\Theta}_1 | \mathcal{E}_1^{(c)}, \Theta_2)$.
2. Q_{opt} *is a causal optimal solution, i.e.,*

$$\mu_c(W, \Theta_1, \Theta_2) = \|W - \Theta_1 Q_{opt} \Theta_2\|$$

if and only if $Q_{opt,c}$ *is such that*

$$\mu(W_c, \hat{\Theta}_1 | \mathcal{E}_1^{(c)}, \Theta_2) = \|W_c - \Theta_2 Q_{opt,c} \hat{\Theta}_1 | \mathcal{E}_1^{(c)}\|.$$

Finally, let us recall how the classical standard problem can be solved using the Commutant Lifting Theorem. First recall that in the control theory setting one can assume that Θ_2 and $\hat{\Theta}_1^*$ are isometries. Under these supplementary assumptions, set

$$\begin{aligned} \mathcal{H}_1 &:= \mathcal{E}_1^{(c)} \ominus (\hat{\Theta}_1 | \mathcal{E}_1^{(c)})^* \mathcal{E}_1^{(c)}, \\ \mathcal{H}_2 &:= \mathcal{E}_2 \ominus \Theta_2 \mathcal{F}_2. \end{aligned}$$

Let $P : \mathcal{E}_2 \to \mathcal{H}_2$ denote orthogonal projection. Then we define the operator

$$\Lambda = \Lambda(W_c, \hat{\Theta}_1 | \mathcal{E}_1^{(c)}, \Theta_2) : \mathcal{H}_1 \to \mathcal{H}_2, \tag{25.14}$$

by

$$\Lambda h := PW_c h, \quad h \in \mathcal{H}_1. \tag{25.15}$$

Then using the Commutant Lifting Theorem, one may show (see [28], [13]) that

$$\|\Lambda\| = \mu(W_c, \hat{\Theta}_1 | \mathcal{E}_1^{(c)}, \Theta_2).$$

Thus from the above theorem, we have the following result:

Corollary 25.1 *Notation as above. Then*

$$\mu_c(W, \Theta_1, \Theta_2) = \|\Lambda(W_c, \hat{\Theta}_1 | \mathcal{E}_1^{(c)}, \Theta_2)\|.$$

4 CAUSALITY IN $H^2(D^N)$

In this section, we specialize the discussion on causality of the previous sections to the Hardy spaces which appear in the nonlinear control problem that we wish to study.

We first define the class of nonlinear input/output operators in which we will be interested. In order to do this, we will first need to discuss a few standard results about analytic mappings on Hilbert spaces. See [3], [4], [17], [18], [24] and the references therein for complete details.

Let $\mathcal{G}$ and $\mathcal{H}$ denote complex separable Hilbert spaces. Set

$$B_{r_o}(\mathcal{G}) := \{g \in \mathcal{G} : \|g\| < r_o\}$$

(the open ball of radius r_o in $\mathcal{G}$ about the origin). Then we say that a mapping $\phi : B_{r_o}(\mathcal{G}) \to \mathcal{H}$ is *analytic* if the complex function $(z_1, \ldots, z_n) \mapsto \langle \phi(z_1 g_1 + \ldots + z_n g_n), h \rangle$ is analytic in a neighborhood of $(1, 1, \ldots, 1) \in \mathbf{C}^n$ as a function of the complex variables $z_1, \ldots, z_n$ for all $g_1, \ldots, g_n \in \mathcal{G}$ such that $\|g_1 + \ldots + g_n\| < r_o$, for all $h \in \mathcal{H}$, and for all $n > 0$.

We will now assume that $\phi(0) = 0$. It is easy to see that if $\phi : B_{r_o}(\mathcal{G}) \to \mathcal{H}$ is analytic, then ϕ admits a convergent Taylor series expansion ([24], page 97), i.e.,

$$\phi(g) = \phi_1(g) + \phi_2(g, g) + \cdots + \phi_n(g, \cdots, g) + \cdots$$

where $\phi_n : \mathcal{G} \times \cdots \times \mathcal{G} \to \mathcal{H}$ is an n-linear map. Clearly, without loss of generality we may assume that the n-linear map $(g_1, \cdots, g_n) \to \phi(g_1, \cdots, g_n)$ is symmetric in the arguments $g_1, \cdots, g_n$. This assumption will be made throughout this paper for the various analytic maps which we consider. For ϕ a Volterra series, ϕ_n is basically the nth-Volterra kernel.

Now set

$$\hat{\phi}_n(g_1 \otimes \cdots \otimes g_n) := \phi_n(g_1, \cdots, g_n).$$

Then $\hat{\phi}_n$ extends in a unique manner to a dense set of $\mathcal{G}^{\otimes n} := \mathcal{G} \otimes \ldots \otimes \mathcal{G}$ (tensor product taken n times). Notice by $\mathcal{G}^{\otimes n}$ we mean the Hilbert space completion of the algebraic tensor product of the $\mathcal{G}$'s. Clearly if $\hat{\phi}_n$ has finite norm on this dense set, then $\hat{\phi}_n$ extends by continuity to a bounded linear operator $\hat{\phi}_n : \mathcal{G}^{\otimes n} \to \mathcal{H}$. By abuse of notation, we will set $\phi_n := \hat{\phi}_n$. (Recall that an n-linear map on $G \times G \times \cdots \times G$ (product taken n times) becomes linear on the tensor product $\mathcal{G}^{\otimes n}$. For details about the construction of the tensor product, see [2], pages 24-27.)

We now recall the following standard definitions:

Definitions 2.

(i) Notation as above. By a *majorizing sequence* for the analytic map ϕ, we mean a positive sequence of numbers α_n $n = 1, 2, \ldots$ such that $\|\phi_n\| < \alpha_n$ for $n \geq 1$. Suppose that $\rho := \limsup {\alpha_n}^{1/n} < \infty$. Then it is completely standard that the Taylor series expansion of ϕ converges at least on the ball $B_r(\mathcal{G})$ of radius $r = 1/\rho$ ([24], page 97).

(ii) If ϕ admits a majorizing sequence as in (i), then we will say that ϕ is *majorizable.*

Let $H^2(D^n, \mathbf{C}^k)$ denote the standard Hardy space of $\mathbf{C}^k$–valued analytic functions on the n–disc D^n (D denotes the unit disc) with square integrable boundary values. We denote the shift on $H^2(D^n, \mathbf{C}^k)$ by $S_{(n)}$. Note that $S_{(n)}$ is defined by multiplication by the function $(z_1 \cdots z_n)$. (By abuse of notation, we will denote the shift on $H^2(D^n, \mathbf{C}^k)$ by $S_{(n)}$ for any k.)

We now consider an analytic map ϕ with $\mathcal{G} = H^2(D, \mathbf{C}^k) =: H_k$, and $\mathcal{H} = H^2(D^n, \mathbf{C}^m) =: H_m$. Note that

$$H_k \otimes \cdots \otimes H_k = (H_k)^{\otimes n} \cong H^2(D^n, \mathbf{C}^K) \text{ with } K = k^n \tag{25.16}$$

where we map $1 \otimes \cdots \otimes z \otimes \cdots \otimes 1$ (z in the i-th place) to z_i, $i = 1, \cdots, n$.

We will identify ϕ_n as a bounded linear map from $H^2(D^n, \mathbf{C}^K) \to H^2(D, \mathbf{C}^m)$ via the canonical isomorphism (25.16). Then we say that ϕ is *time-invariant* if

$$\phi_n S_{(n)} = S_{(1)} \phi_n, \quad \forall n \geq 1. \tag{25.17}$$

(We will also say each ϕ_n is *time-invariant.*) Equivalently, this means that $S_{(1)}\phi = \phi \circ S_{(1)}$ on some open ball about the origin in which ϕ is defined.

We say that $\phi_n : H^2(D^n, \mathbf{C}^K) \to H_m$ is *causal* if for $F(z_1, \ldots, z_n) \in H^2(D^n, \mathbf{C}^K)$,

$$F(z_1, \ldots, z_n) = \sum_{i_1, \ldots, i_n \geq 0} F_{i_1, \ldots, i_n} z_1^{i_1} \cdots z_n^{i_n}, \quad \phi_n(F)(z) := \sum_{p \geq 0} f_p z^p,$$

each f_p only depends on

$$\{F_{i_1, \ldots, i_n} : 0 \leq i_1, \ldots, i_n \leq p\}.$$

This means that for

$$F(z_1, \ldots, z_n) = \sum_{\max\{i_1, \ldots, i_n\} \geq p} F_{i_1, \ldots, i_n} z_1^{i_1} \cdots z_n^{i_n},$$

we have that

$$(I - S_{(1)}^p S_{(1)}^{*p}) \phi_n(F(z_1, \ldots, z_n)) = 0. \tag{25.18}$$

So in $\mathcal{E} = (H^2(\mathbf{D}^n, \mathbf{C}^K)$ the $S_{(n)}$-causal structure which we will consider is always defined, for $j = 1, 2, \ldots$, by

$$P_{\mathcal{E}}^{(j)} F = \sum_{\max\{i_1, \ldots, i_n\} < j} F_{i_1, \ldots, i_n} z_1^{i_1} \ldots z_n^{i_n}$$

where $F = \sum F_{i_1, \ldots, i_n} z_1^{i_1} \ldots z_n^{i_n} \in H^2(\mathbf{D}^n, \mathbf{C}^K)$. Note that in the case $n = 1$ (and only in this case) the causal structure is precisely the "maximal" one.

In Section 6 we will also make use of the obvious fact that if Q and R are causal, then so is $Q \otimes R$.

5 STANDARD NONLINEAR PROBLEM

We will now describe the physical control problem in which we are interested. First, we will need to consider the precise kind of input/output operator we will be considering. As above, H_k denotes the standard Hardy space of $\mathbf{C}^k$-valued functions on the unit disc. We now make the following definition.

Then we say an analytic input/output operator $\phi : H_k \to H_m$ is *admissible* if it is causal, time-invariant, majorizable, and $\phi(0) = 0$. We denote

$$\mathcal{C}_l := \{\text{space of admissible operators}\}.$$

Since the theory we are considering is local, the notion of admissibility is sufficient for all of the applications we have in mind. Again by abuse of notation, $\mathcal{C}_l$ will denote the set of admissible operators for any k and m.

We now begin to formulate our control problem. Referring to Figure 25.1, G represents the generalized plant which we assume is modelled by an admissible operator, and K the compensator. Let $\mathcal{F}(G, K)$ denote the input/output operator from w to z. Then we want to minimize the "size" of $\mathcal{F}(G, K)$ over all inputs of bounded energy (of fixed given bound) in the sense which will be given below.

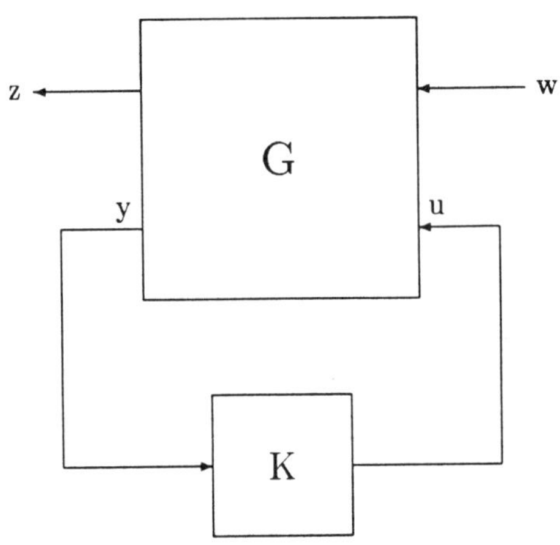

Figure 25.1 Standard Feedback Configuration

One can show that K stabilizes the closed loop if and only if

$$C = Q \circ (I - P \circ Q)^{-1}, \tag{25.19}$$

for some admissible operator Q. We will call such a Q, a *compensating parameter*. Then via this parametrization, we have

$$\mathcal{F}(G, K) = W - P \circ Q \circ R,$$

for admissible operators W, P, R which depend only on the generalized plant G. We now will say in what sense we wish to minimize the size of the operator $W - P \circ Q \circ R$ taken over all Q.

We follow here our convention that for given $\phi \in \mathcal{C}_l$, ϕ_n will denote the bounded linear map on the space $(H^2_k)^{\otimes n} \cong H^2(D^n, \mathbf{C}^K)$ (with $K = k^n$) associated to the n-linear part of ϕ which we also denote by ϕ_n (and which we always assume without loss of generality is symmetric in its arguments). The context will always make the meaning of ϕ_n clear.

We can now state the following definitions:

Definitions 3.

(i) For $W, P, R \in \mathcal{C}_l$, set

$$\mathcal{T}_{(W,P,R)}(Q)(\rho) := \mathcal{T}(Q)(\rho) := \sum_{n=1} \rho^n \|(W - P \circ Q \circ R)_n\|$$

for all $\rho > 0$ such that the sum converges.

(ii) We write $\mathcal{T}(Q) \preceq \mathcal{T}(\tilde{Q})$, if there exists a $\rho_o > 0$ such that $\mathcal{T}(Q)(\rho) \leq \mathcal{T}(\tilde{Q})(\rho)$ for all $\rho \in [0, \rho_o]$. If $\mathcal{T}(Q) \preceq \mathcal{T}(\tilde{Q})$ and $\mathcal{T}(\tilde{Q}) \preceq \mathcal{T}(Q)$, we write $\mathcal{T}(Q) \cong \mathcal{T}(\tilde{Q})$. This means that $\mathcal{T}(Q)(\rho) = \mathcal{T}(\tilde{Q})(\rho)$ for all $\rho > 0$ sufficiently small, i.e., $\mathcal{T}(Q)$ and $\mathcal{T}(\tilde{Q})$ are equal as germs of functions.

(iii) If $\mathcal{T}(Q) \preceq \mathcal{T}(\tilde{Q})$, but $\mathcal{T}(\tilde{Q}) \not\preceq \mathcal{T}(Q)$, we will say that Q *ameliorates* $\tilde{Q}$. Note that this means $\mathcal{T}(Q)(\rho) < \mathcal{T}(\tilde{Q})(\rho)$ for all $\rho > 0$ sufficiently small.

Now with Definitions 3, we can define a notion of "optimality:"

Definitions 4.

(i) $Q_o \in \mathcal{C}_l$ is called *optimal* if $\mathcal{T}(Q_o) \preceq \mathcal{T}(Q)$ for all $Q \in \mathcal{C}_l$.

(ii) We say $Q \in \mathcal{C}_l$ is *optimal with respect to its n-th term* Q_n, if for every n-linear $\hat{Q}_n \in \mathcal{C}_l$, we have

$$\mathcal{T}(Q_1 + \ldots + Q_{n-1} + Q_n + Q_{n+1} \ldots) \preceq \mathcal{T}(Q_1 + \ldots + Q_{n-1} + \hat{Q}_n + Q_{n+1} + \ldots).$$

If $Q \in \mathcal{C}_l$ is optimal with respect to all of its terms, then we say that it is *partially optimal.*

6 ITERATED CAUSAL LIFTING PROCEDURE FOR THE STANDARD PROBLEM

In this section, we discuss a construction from which we will derive partially optimal compensators relative to the closed loop operator $\mathcal{T}$ given in Definitions 3 above. As before, we are given the admissible operators W, P, R. We always suppose that P_1 (the linear part of P) is an isometry, and that $\hat{R}_1$ is a co-isometry.

Using the notation of Section 4, (for each $n = 1, 2, \ldots$) we take $\mathcal{E}_1 := H^2(D^n, \mathbf{C}^{K_1})$, $\mathcal{E}_2 := H^2(D, \mathbf{C}^{k_2})$, $\mathcal{F}_1 := H^2(D^n, \mathbf{C}^{M_1})$, and $\mathcal{F}_2 := H^2(D, \mathbf{C}^{m_2})$, endowed with the causal structures as defined at the end of Section 4. Here $K_1 = k_1^n$, $M_1 = m_1^n$.

We begin by noting the following key relationship:

$$(P \circ Q \circ R)_n = \sum_{1 \leq k \leq n} \sum_{i_1 + \cdots + i_k = n} P_k(Q_{i_1}(R^{\otimes i_1}), \ldots, Q_{i_k}(R^{\otimes i_k})).$$

Note that we may in term write that

$$Q_j(R^{\otimes j}) = \sum_{k_1, \ldots, k_j} Q_j(R_{k_1} \otimes \ldots \otimes R_{k_j}).$$

Thus we see that

$$(W - P \circ Q \circ R)_n = \hat{W}_n - P_1 Q_n(R_1^{\otimes n}),$$

where

$$\hat{W}_n = W_n + A(Q_1, \ldots, Q_{n-1}),$$

and $A(Q_1, \ldots, Q_{n-1})$ is an explicitly computable function of $Q_1, \ldots, Q_{n-1}$.

We are now ready to formulate the *iterative causal commutant lifting procedure.* From the classical Commutant Lifting Theorem, we may choose Q_1 causal such that

$$\|W_1 - P_1 Q_1 R_1\| = \|\Lambda(W_1, R_1, P_1)\|.$$

Now given this Q_1, using Theorem 4, we choose a causal Q_2 such that

$$\begin{aligned} &\|W_2 - P_1 Q_1 R_2 - P_2(Q_1 R_1 \otimes Q_1 R_1) - P_1 Q_2(R_1 \otimes R_1)\| \\ &= \|\Lambda((W_2 - P_1 Q_1 R_2 - P_2(Q_1 R_1 \otimes Q_1 R_1)_c, \widehat{R_1 \otimes R_1} | \mathcal{E}_1^{(c)}, P_1)\|. \end{aligned}$$

Inductively, given causal $Q_1, \ldots, Q_{n-1}$, we may choose Q_n causal such that

$$\|\hat{W}_n - P_1 Q_n(R_1^{\otimes n})\| = \|\Lambda((\hat{W}_n)_c, \widehat{R_1^{\otimes n}} | \mathcal{E}_1^{(c)}, P_1)\|. \tag{25.20}$$

Note that in each step of the procedure, the new "weight" $\hat{W}_n$ is determined by $W_n, P_1, R_1^{\otimes n}$, and the optimal causal parameters chosen. Thus, the iterative commutant lifting procedure is determined by the operator $\Lambda((\hat{W}_n)_c, \widehat{R_1^{\otimes n}} | \mathcal{E}_1^{(c)}, P_1)$, and so may be reduced to a classical dilation problem. The following facts can be proven just as in [17] and [18] to which we refer the reader for the proofs. (See in particular [18], pages 849-853 .) First the causal iterative commutant lifting procedure converges:

Proposition 25.1 *With the above notation, let* $Q^{(1)} := Q_1 + Q_2 + \cdots$. *Then* $Q^{(1)} \in \mathcal{C}_l$.

Next given any $Q \in \mathcal{C}_l$, we can apply the causal iterative commutant lifting procedure to $W - P \circ Q \circ R$. Now set

$$\mathcal{T}_C(q)(\rho) := \sum_{n=1} \rho^n \|\Lambda((\hat{W}_n)_c, \widehat{R_1^{\otimes n}} | \mathcal{E}_1^{(c)}, P_1)\|.$$

Then we have,

Proposition 25.2 *Given $Q \in \mathcal{C}_l$, there exists $\tilde{Q} \in \mathcal{C}_l$, such that $\mathcal{T}(\tilde{Q}) \equiv \mathcal{T}_C(Q)$. Moreover $\tilde{Q}$ may be derived from the causal iterated commutant lifting procedure.*

Moreover, as in [18] we have the following results:

Proposition 25.3 *Q is partially optimal if and only if $\mathcal{T}(Q) \cong \mathcal{T}_C(Q)$.*

Theorem 25.4 *For given P, R and W as above, any $Q \in \mathcal{C}_l$ is either partially optimal or can be ameliorated by a partially optimal compensating parameter.*

Remarks.

1. As in [18], we may prove that optimal compensating parameters always exist. In general however, we have no contructive procedure to find them.
2. For the spaces $\mathcal{E}_1 := H^2(D^n, \mathbf{C}^{k_1})$, $\mathcal{E}_2 := H^2(D, \mathbf{C}^{k_2})$, $\mathcal{F}_1 := H^2(D^n, \mathbf{C}^{k_3})$, and $\mathcal{F}_2 := H^2(D, \mathbf{C}^{k_4})$, Theorem 3 may be expressed as a *Reduction Theorem* (via the Fourier representation) exactly as in [16]. Thus the iterative causal procedure can be reduced to an iterative application of the classical Commutant Lifting Theorem. See [16] for details.

Acknowledgements

This work was supported in part by grants from the Research Fund of Indiana University, by the National Science Foundation DMS-8811084 and ECS-9122106, by the Air Force Office of Scientific Research F49620-94-1-0058DEF, by the Army Research Office DAAH04-94-G-0054 and DAAH04-93-G-0332.

REFERENCES

[1] W. Arveson, "Interpolation problems in nest algebras," *J. Functional Analysis* **20** (1975), 208-233.

[2] M. Atiyah and I. MacDonald, *Introduction to Commutative Algebra*, Addison-Wesley, New York, 1969.

[3] J. Ball, C. Foias, J. W. Helton, and A. Tannenbaum, "On a local nonlinear commutant lifting theorem," *Indiana J. Mathematics* **36** (1987), pp. 693–709.

[4] J. Ball, C. Foias, J. W. Helton, and A. Tannenbaum, "Nonlinear interpolation theory in H^∞," in *Modelling, Robustness, and Sensitivity in Control Systems* (edited by R. Curtain), NATO-ASI Series, Springer-Verlag, New York, 1987.

[5] J. Ball, C. Foias, J. W. Helton, and A. Tannenbaum, "A Poincaré-Dulac approach to a nonlinear Beurling-Lax-Halmos theorem," *Journal of Math. Anal. and Applications* **139** (1989), 496-514.

[6] J. Ball and J. W. Helton, "Sensitivity bandwidth optimization for nonlinear feedback systems," Technical Report, Department of Mathematics, University of California at San Diego, 1988.

[7] J. Ball and J. W. Helton, "H^∞ control for nonlinear plants: connections with differential games," IEEE Conf. Dec. and Control, Tampa, FL, 1989, 956-962.

[8] H. Bercovici, C. Foias, and A. Tannenbaum, "On skew Toeplitz operators. I," *Operator Theory: Advances and Applications* **29** (1988), pp. 21–44.

[9] S. Boyd and L. Chua, "Fading memory and the problem of approximating nonlinear operators with Volterra series," *IEEE Trans. Circuits and Systems* **CAS-32** (1985), pp. 1150–1161.

[10] J. Doyle, B. Francis, and A. Tannenbaum, *Feedback Control Theory*, MacMillan, New York, 1991.

[11] A. Feintuch and B. A. Francis, "Uniformly optimal control of linear systems," *Automatica* **21** (1986), 563-574.

[12] C. Foias, "Contractive intertwining dilations and waves in layered media," *Proc. International Congress Mathematicians*, **2**, Helsinki (1978), 605-613.

[13] C. Foias and A. Frazho, *The Commutant Lifting Approach to Interpolation Problems*, Birkhauser-Verlag, Boston, 1990.

[14] C. Foias and A. Frazho, "Commutant lifting and simultaneous H^∞ and L^2 suboptimization," *SIAM J. Math. Anal.* **23** (1992), 984-994.

[15] C. Foias, C. Gu, and A. Tannenbaum, "Intertwining dilations, intertwining extensions and causality," *Acta Sci. Math. (Szeged)* **57** (1993), 101-123.

[16] C. Foias, C. Gu, and A. Tannenbaum, "Nonlinear H^∞ optimization: A causal power series approach," *SIAM J. Control and Optimization* **33** (1995), 185-207.

[17] C. Foias and A. Tannenbaum, "Iterated cummutant lifting for systems with rational symbol," *Operator Theory: Advances and Applications* **41** (1989), 255-277.

[18] C. Foias and A. Tannenbaum, "Weighted optimization theory for nonlinear systems," *SIAM J. on Control and Optimization* **27** (1989), 842-860.

[19] C. Foias and A. Tannenbaum, "Causality in commutant lifting theory," *Journal of Functional Analysis* **118** (1993), 407-441.

[20] C. Foias, A. Tannenbaum, and G. Zames, "Some explicit formulae for the singular values of certain Hankel operators with factorizable symbol," *SIAM J. Math. Analysis* **19** (1988), 1081-1089.

[21] B. Francis, *A Course in H^∞ Control Theory*, McGraw-Hill, New York, 1981.

[22] B. Francis and A. Tannenbaum, "Generalized interpolation theory in control," *Mathematical Intelligencer* **10** (1988), 48-43.

[23] C. Gu, "Eliminating the genericity conditions in the skew Toeplitz operator algorithm for H^∞ optimization," *SIAM J. Math. Analysis* **23** (1992).

[24] E. Hille and R. Phillips, *Functional Analysis and Semigroups*, AMS Colloquium Publications, vol. 31, Providence, Rhode Island, 1957.

[25] J. W. Helton, "Broadbanding: gain equalization directly from data," *IEEE Trans. Circuits and Systems* **CAS-28** (1981), 1125-1137.

[26] S. Parrott, "On the quotient norm and the Sz. Nagy-Foias lifting theorem," *J. Funct. Anal.* **30** (1970), 311-328.

[27] D. Sarason, "Generalized interpolation in H^∞," *Trans. AMS* **127** (1967), 179-203.

[28] B. Sz.-Nagy and C. Foias, *Harmonic Analysis of Operators on Hilbert Space*, North-Holland Publishing Company, Amsterdam, 1970.

[29] N. J. Young, "An algorithm for the super-optimal sensitivity-minimising controller," *Proc. Workshop on New Perspectives in Industrail Control System Design Using H_∞ Methods,* Oxford University Press, London, 1976.

[30] G. Zames, "Feedback and optimal sensitivity: model reference transformations, multiplicative seminorms, and approximate inverses," *IEEE Trans. Auto. Control* **AC-26** (1981), pp. 301–320.

26

SOME ASPECTS OF NONLINEAR BLACK-BOX MODELING IN SYSTEM IDENTIFICATION

Lennart Ljung

Dept of Electrical Engineering
Linköping University, Sweden
email: ljung@isy.liu.se

Dedicated with affection to Tom Kailath – Remembering happy days at Stanford.

1 INTRODUCTION

The key problem in system identification is to find a suitable model structure, within which a good model is to be found. Fitting a model within a given structure (parameter estimation) is in most cases a lesser problem. A basic rule in estimation is *not to estimate what you already know.* In other words, one should utilize prior knowledge and physical insight about the system when selecting the model structure. It is customary to distinguish between three levels of prior knowledge, which have been color-coded as follows.

- White Box models: This is the case when a model is perfectly known; it has been possible to construct it entirely from prior knowledge and physical insight.
- Grey Box models: This is the case when some physical insight is available, but several parameters remain to be determined from observed data. It is useful to consider two subcases:
  - Physical modeling: A model structure can be constructed on physical grounds, which has a certain number of parameters to be estimated from data. This could, e.g., be a state space model of given order and structure.
  - Semi-physical modeling: Physical insight is used to suggest certain nonlinear combinations of measured data signal. These new signals are then subjected to model structures of black box character.

- Black Box models: No physical insight is available or used, but the chosen model structure belongs to families that are known to have good flexibility and have been "successful in the past".

A nonlinear black box structure for a dynamical system is a model structure that is prepared to describe virtually any nonlinear dynamics. There has been considerable recent interest in this area with structures based on neural networks, radial basis networks, wavelet networks, hinging hyperplanes, as well as wavelet transform based methods and models based on fuzzy sets and fuzzy rules. This paper describes the common framework for these approaches.

It is pointed out that the nonlinear structures can be seen as a concatenation of a mapping from observed data to a regression vector and a nonlinear mapping from the regressor space to the output space. These mappings are discussed separately. The latter mapping is usually formed as a basis function expansion. The basis functions are typically formed from one simple scalar function which is modified in terms of scale and location. The expansion from the scalar argument to the regressor space is achieved by a radial or a ridge type approach.

Basic techniques for estimating the parameters in the structures are criterion minimization, as well as two step procedures, where first the relevant basis functions are determined, using data, and then a linear least squares step to determine the coordinates of the function approximation. A particular problem is to deal with the large number of potentially necessary parameters. This is handled by making the number of "used" parameters considerably less than the number of "offered" parameters, by regularization, shrinking, pruning or regressor selection.

A more comprehensive treatment is given in [8] and [4].

2 SYSTEM IDENTIFICATION

System Identification is the art and methodology of building mathematical models of dynamical systems based on input-output data. See among many references [5], [6], and [9].

We denote the output of the dynamical system at time t by $y(t)$ and the input by $u(t)$. The data are assumed to be collected in discrete time. At time t we thus have available the data set

$$Z^t = \{y(1), u(1), \dots, y(t), u(t)\} \tag{26.1}$$

A *model* of a dynamical system can be seen as a mapping from past data Z^{t-1} to the next output $y(t)$ (a *predictor model*):

$$\hat{y}(t) = \hat{g}(Z^{t-1}) \tag{26.2}$$

We put a "hat" on y to emphasize that the assigned value is a prediction rather than a measured, "correct" value for $y(t)$.

The problem is to use the information in a data record Z^N to find a mapping $\hat{g}_N$ that gives good predictions in (26.2).

3 NON-LINEAR BLACK BOX MODELS

In this section we shall describe the basic ideas behind model structures that have the capability to cover any non-linear mapping from past data to the predicted value of $y(t)$. A model structure is a parameterized mapping of the kind (26.2):

$$\hat{y}(t|\theta) = g(\theta, Z^{t-1}) \tag{26.3}$$

The parameter θ is a vector of coefficients that are to be chosen with the help of the data. We shall consequently allow quite general non-linear mappings g. This section will deal with some general principles for how to construct such mappings.

Now, the model structure family (26.3) is really too general, and it turns out to be useful to write g as a concatenation of two mappings: one that takes the increasing number of past observations Z^{t-1} and maps them into a finite dimensional vector $\varphi(t)$ of fixed dimension and one that takes this vector to the space of the outputs:

$$\hat{y}(t|\theta) = g(\theta, Z^{t-1}) = g(\varphi(t), \theta) \quad \text{where} \quad \varphi(t) = \varphi(Z^{t-1}) \tag{26.4}$$

Let the dimension of φ be d. We shall call this vector the *regression vector* and its components will be referred to as the *regressors*.

The choice of the non-linear mapping in (26.3) has thus been reduced to two partial problems for dynamical systems:

1. How to choose the non-linear mapping $g(\varphi)$ from the regressor space to the output space (*i.e.*, from R^d to R^p).
2. How to choose the regressors $\varphi(t)$ from past inputs and outputs.

The second problem is the same for all dynamical systems, and it turns out that the most useful choices of regression vectors are to let them contain past inputs and outputs, and possibly also past predicted/simulated outputs. The basic choice is thus

$$\varphi(t) = [y(t-1), \ldots, y(t-n), u(t-1), \ldots, u(t-m)] \tag{26.5}$$

More sophisticated variants are obtained by letting φ contain also $\hat{y}(t-k|\theta)$. Then (some of) the regressors will also depend on the parameter vector, which lead to so called *recurrent networks* and more complicated algorithms.

In case $n = 0$ in the above expression, so that the regression vector only contains $u(t-k)$ and possibly predicted/simulated outputs (from u) $\hat{y}(t|\theta)$, we talk about *output error models.*

Function Expansions and Basis Functions

The non-linear mapping $g(\varphi, \theta)$ is from R^d to R^p for any given θ. At this point it does not matter how the regression vector φ is constructed. It is just a vector that lives in R^d.

It is natural to think of the parameterized function family as function expansions:

$$g(\varphi, \theta) = \sum \theta(k) g_k(\varphi) \tag{26.6}$$

where g_k are the *basis functions* and the coefficients $\theta(k)$ are the "coordinates" of g in the chosen basis.

Now, the only remaining question is: How to choose the basis functions g_k? Depending on the support of g_k (*i.e.* the area in R^d for which $g_k(\varphi)$ is (practically) non-zero) we shall distinguish between three types of basis functions: *Global, Ridge-type, and Local.* A typical and classical global basis function expansion would then be the Taylor series, or polynomial expansion, where g_k would contain multinomials in the components of φ of total degree k. Fourier series are also relevant examples. We shall however not discuss global basis functions here any further. Experience has indicated that they are inferior to the semi-local and local ones in typical practical applications.

Local Basis Functions Local basis functions have their support only in some neighborhood of a given point. Think (in the case of p=1) of the indicator function for the unit cube:

$$\kappa(\varphi) = 1 \text{ if } |\varphi_k| \leq 1 \ \forall k\,, \text{ and } 0 \text{ otherwise} \tag{26.7}$$

By scaling the cube and placing it at different locations we obtain the functions

$$g_k(\varphi) = \kappa(\beta_k(\varphi - \gamma_k)) \tag{26.8}$$

By allowing β to be a matrix we may also reshape the cube to be any parallelepiped. The parameters β are thus *scaling* or *dilation* parameters while γ determine *location* or *translation*.

This choice of g_k in (26.6) gives functions that are piecewise constant over areas in R^d that can be chosen arbitrarily small by proper choice of the scaling parameters, and placed anywhere using the location parameters. Expansions like (26.6) will thus be able to approximate any function by a function that is piecewise constant over arbitrarily small regions in the φ-space. It should be fairly obvious that will allow as to approximate any reasonable function arbitrarily well. It is also reasonable that the same will be true for any other localized function, such as the Gaussian bell function: $\kappa(\varphi) = exp(-|\varphi|^2)$

Expanding a scalar function the the regressor space In the discussion above, κ was a function from R^d to R. It is quite common to construct such a function from a scalar function $\sigma(x)$ from R to R, by expanding it. Two typical ways to do that are the *radial* and the *ridge* approaches.

Radial Basis Functions In the radial approach we have

$$\kappa(\varphi) = \sigma(\|\varphi\|_\beta^2) \tag{26.9}$$

Here the quadratic norm β will automatically also act as the scaling parameters. β could be a full (positive semi-definite, symmetric) matrix, or a scaled version of the identity matrix.

Ridge-type Basis Functions A useful alternative is to let the basis functions be local in one direction of the φ-space and global in the others. This is achieved quite analogously to (26.9) as follows.

$$g_k(\varphi) = \sigma(\beta_k^T(\varphi - \gamma_k)) \tag{26.10}$$

Here β_k is a d-dimensional vector. Note the difference with (26.8)! The scalar product $\beta_k^T\varphi$ is constant in the subspace of R^d that is perpendicular to the scaling vector β_k. Hence the function $g_k(\varphi)$ varies like σ in a direction parallel to β_k and is constant across this direction. This motivates the term *semi-global* or *ridge-type* for this choice of functions.

Connection to "Named Structures"

Here we briefly review some popular structures, other structures related to interpolation techniques are discussed in [8, 4].

Wavelets The local approach corresponding to (26.6,26.8) has direct connections to wavelet networks and wavelet transforms. The exact relationships are discussed in [8]. Loosely, we note that via the dilation parameters in β_k we can work with different scales simultaneously to pick up both local and not-so-local variations. With appropriate translations and dilations of a single suitably chosen function κ (the "mother wavelet"), we can make the expansion (26.6) orthonormal. The typical choices is to have $\beta_k = 2^k$ and $\gamma_j = j$, and work with doubly indexed expansions in (26.6). This is discussed extensively in [4].

Wavelet and Radial Basis Networks. The choice of the Gaussian bell function as the basic function without any orthogonalization is found in both wavelet networks, [10] and radial basis neural networks [7].

Neural Networks The ridge choice (26.10) with

$$\sigma(x) = \frac{1}{1+e^{-x}}$$

gives a much-used neural network structure, viz. the *one hidden layer feedforward sigmoidal net.*

Hinging Hyperplanes If instead of using the sigmoid σ function we choose "V-shaped" functions (in the form of a higher-dimensional "open book") Breiman's *hinging hyperplane* structure is obtained, [2]. Hinging hyperplanes model structures [2] have the form

$$g(x) \;=\; \frac{1}{2}[(\beta^+ + \beta^-)x + \gamma^+ + \gamma^-] \pm \frac{1}{2}|(\beta^+ - \beta^-)x + \gamma^+ - \gamma^-|\,.$$

Thus a hinge is the superposition of a linear map and a semi-global function. Therefore, we consider *hinge* functions as semi-global or ridge-type, though it is not in strict accordance with our definition.

Nearest Neighbors or Interpolation By selecting κ as in (26.7) and the location and scale vector γ_k, β_k in the structure (26.8), such that exactly one observation falls into each "cube", the nearest neighbor model is obtained: just load the input-output record into a table, and, for a given φ, pick the pair $(\widehat{y}, \widehat{\varphi})$ for $\widehat{\varphi}$ closest to the given φ, $\widehat{y}$ is the desired output estimate. If one replaces (26.7) by a smoother function and allow some overlapping of the basis functions, we get interpolation type techniques such as kernel estimators.

Fuzzy Models Also so called *fuzzy models* based on fuzzy set membership belong to the model structures of the class (26.6). The basis functions g_k then are constructed from the fuzzy set membership functions and the inference rules. The exact relationship is described in [8].

4 ESTIMATING NON-LINEAR BLACK BOX MODELS

The predictor $\hat{y}(t|\theta) = g(\varphi(t), \theta)$ is a well defined function of past data and the parameters θ. The parameters are made up of coordinates in the expansion (26.6), and from location and scale parameters in the different basis functions.

A very general approach to estimate the parameter θ is by minimizing a criterion of fit. This will be described in some detail in Section 5. For Neural Network applications these are also the typical estimation algorithms used, often complemented with *regularization*, which means that a term is added to the criterion (26.11), that penalizes the norm of θ. This will reduce the variance of the model, in that "spurious" parameters are not allowed to take on large, and mostly random values. See e.g. [8].

For wavelet applications it is common to distinguish between those parameters that enter linearly in $\hat{y}(t|\theta)$ (i.e. the coordinates in the function expansion) and those that enter non-linearly (i.e. the location and scale parameters). Often the latter are seeded to fixed values and the coordinates are estimated by the linear least squares method. Basis functions that give a small contribution to the fit (corresponding to non-useful values of the scale and location parameters) can then be trimmed away ("pruning" or "shrinking").

5 GENERAL PARAMETER ESTIMATION TECHNIQUES

In this section we shall deal with issues that are independent of model structure. Principles and algorithms for fitting models to data, as well as the general properties of the estimated models are all model-structure independent and equally well applicable to, say, linear ARMAX models and Neural Network models.

It suggests itself that the basic least-squares like approach is a natural approach, even when the predictor $\hat{y}(t|\theta)$ is a more general function of θ:

$$\hat{\theta}_N = \arg\min_{\theta} V_N(\theta, Z^N) \tag{26.11}$$

where

$$V_N(\theta, Z^N) = \frac{1}{N}\sum_{t=1}^{N} \|y(t) - \hat{y}(t|\theta)\|^2 \tag{26.12}$$

This procedure is natural and pragmatic – we can think of it as "curve-fitting" between $y(t)$ and $\hat{y}(t|\theta)$. It also has several statistical and information theoretic interpretations. Most importantly, if the noise source in the system is supposed to be a Gaussian sequence of independent random variables $\{e(t)\}$ then (26.11) becomes the Maximum Likelihood estimate (MLE),

It is generally agreed upon that the best way to minimize (26.12) is a *damped Gauss-Newton scheme*, [3]. By this is meant that the estimates iteratively updated as

$$\hat{\theta}_N^{(i+1)} = \hat{\theta}_N^{(i)} - \mu_i \left[\frac{1}{N}\sum_{t=1}^{N} \psi(t, \hat{\theta}_N^{(i)})\psi^T(t, \hat{\theta}_N^{(i)})\right]^{-1} \times \left[\frac{1}{N}\sum_{t=1}^{N}(y(t) - \hat{y}(t|\hat{\theta}_N^{(1)}))\psi(t, \hat{\theta}_N^{(i)})\right] \tag{26.13}$$

where

$$\psi(t, \theta) = \frac{\partial}{\partial\theta}\hat{y}(t|\theta) \tag{26.14}$$

The step size μ is chosen so that the criterion decreases at each iteration. Often a simple search in terms of μ is used for this. If the indicated inverse is ill conditioned, it is customary to add a multiple of the identity matrix to it. This is known as the *Levenberg-Marquard* technique.

It is also quite useful work with a modified criterion

$$W_N(\theta, Z^N) = V_N(Z^N) + \delta\|\theta\|^2 \tag{26.15}$$

with V_N defined by (26.12). This is known as *regularization.* It may be noted that stopping the iterations (i) in (26.13) before the minimum has been reached has the same effect as regularization. See, e.g., [8].

Measured of model fit Some quite general expressions for the expected model fit, that are independent of the model structure, can be developed.

Let us measure the (average) fit between any model (26.3) and the true system as

$$\bar{V}(\theta) = E|y(t) - \hat{y}(t|\theta)|^2 \tag{26.16}$$

Here expectation E is over the data properties (i.e. expectation over "Z^∞" with the notation (26.1)).

Before we continue, let us note the very important aspect that the fit $\bar{V}$ will depend, not only on the model and the true system, *but also on data properties*, like input spectra, possible feedback, etc. We shall say that the fit depends on the *experimental conditions.*

The estimated model parameter $\hat{\theta}_N$ is a random variable, because it is constructed from observed data, that can be described as random variables. To evaluate the model fit, we then take the expectation of $\bar{V}(\hat{\theta}_N)$ with respect to the estimation data. That gives our measure

$$F_N = E\bar{V}(\hat{\theta}_N) \tag{26.17}$$

The rather remarkable fact is that if F_N is evaluated for data with the same properties as those of the estimation data, then, asymptotically in N, (see, e.g., [5], Chapter 16)

$$F_N \approx \bar{V}(\theta^*)(1 + \frac{dim\ \theta}{N}) \tag{26.18}$$

Here θ^* is the value that minimizes the expected value of the criterion (26.12). The notation $dim\ \theta$ means the number of estimated parameters. The result also assumes that the model structure is successful in the sense that $\varepsilon(t)$ is approximately white noise.

It is quite important to note that the number $dim\ \theta$ in (26.18) will be changed to *the number of eigenvalues of* $\bar{V}''(\theta)$ (the Hessian of $\bar{V}$) *that are larger than* δ in case the regularized loss function (26.15) is minimized to determine the estimate. We can think of this number as the *efficient number of parameters.* In a sense, we are "offering" more parameters in the structure, than are actually "used" by the data in the resulting model.

Despite the reservations about the formal validity of (26.18), it carries a most important conceptual message: If a model is evaluated on a data set with the same properties as the estimation data, then *the fit will not depend on the data properties*, and it will depend on the model structure *only in terms of the number of parameters used and of the best fit offered within the structure.*

The expression (26.18) clearly shows the trade off between *variance* and *bias.* The more parameters used by the structure (corresponding to a higher dimension of θ and/or a lower value of the regularization parameter δ) the higher the variance term, but at the same the lower the fit $\bar{V}(\theta^*)$. The trade off is thus to increase the efficient number of parameters only to that point that the improvement of fit per parameter exceeds $\bar{V}(\theta^*)/N$. This can be achieved by estimating F_N in (26.17) by evaluating the loss function at $\hat{\theta}_N$ for a *validation data set*. It can also be achieved by Akaike (or Akaike-like) procedures, [1], balancing the variance term in (26.18) against the fit improvement.

The expression can be rewritten as follows. Let $\hat{y}_0(t|t-1)$ denote the "true" one step ahead prediction of $y(t)$, and let

$$W(\theta) = E|\hat{y}_0(t|t-1) - \hat{y}(t|\theta)|^2 \tag{26.19}$$

and let

$$\lambda = E|y(t) - \hat{y}_0(t|t-1)|^2 \tag{26.20}$$

Then λ is the *innovations* variance, i.e., that part of $y(t)$ that cannot be predicted from the past. Moreover $W(\theta^*)$ is the *bias error*, i.e. the discrepancy between the true predictor and the best one available in the model structure. Under the same assumptions as above, (26.18) can be rewritten as

$$F_N \approx \lambda + W(\theta^*) + \lambda \frac{dim\ \theta}{N} \tag{26.21}$$

The three terms constituting the model error then have the following interpretations

- λ is the unavoidable error, stemming from the fact that the output cannot be exactly predicted, even with perfect system knowledge.
- $W(\theta^*)$ is the bias error. It depends on the model structure, and on the experimental conditions. It will typically decrease as $dim\ \theta$ increases.
- The last term is the *variance error*. It is proportional to the (efficient) number of estimated parameters and inversely proportional to the number of data points. It does not depend on the particular model structure or the experimental conditions.

6 CONCLUSIONS

Non-linear black box models for regression in general and system identification in particular has been widely discussed over the past decade. Some approaches like Artifical Neural Networks and (Neuro-)Fuzzy Modeling, and also to some extent the wavelet-based approaches have typically been introduced and described out of their regression-modeling context. This has lead to some confusion about the nature of these ideas.

In this contribution we have stressed that these approaches indeed are "just" corresponding to special choices of model structures in an otherwise well known and classical statistical framework. The well known principle of *parsimony* – to keep the effective number of estimated parameters small – is an important factor in the algorithms. It takes, however, quite different shapes in the various suggested schemes: explicit regularization (a pull towards the origin), implicit regularization (stopping the iterations before the objective function has been minimized), pruning and shrinking (cutting away parameters – terms in the expansion (26.6) – that contribute little to the fit) etc. All of these are measures to eliminate parameters whose contributions to the fit is less than their adverse variance effect.

REFERENCES

[1] H. Akaike. A new look at the statistical model identification. *IEEE Transactions on Automatic Control*, AC-19:716–723, 1974.

[2] L. Breiman. Hinging hyperplanes for regression, classification and function approximation. *IEEE Trans. Info. Theory*, 39:999–1013, 1993.

[3] J. E. Dennis and R. B. Schnabel. *Numerical methods for unconstrained optimization and nonlinear equations.* Prentice-Hall, 1983.

[4] A. Juditsky, H. Hjalmarsson, A. Benveniste, B. Deylon, L. Ljung, J. Sjöberg, and Q. Zhang. Nonlinear black-box modeling in system identification: Mathematical foundations. *Automatica*, 31, 1995.

[5] L. Ljung. *System Identification - Theory for the User.* Prentice-Hall, Englewood Cliffs, N.J., 1987.

[6] L. Ljung and T. Glad. *Modeling of Dynamic Systems.* Prentice Hall, Englewood Cliffs, 1994.

[7] T. Poggio and F. Girosi. Networks for approximation and learning. *Proc. of the IEEE*, 78:1481–1497, 1990.

[8] J. Sjöberg, Q. Zhang, L. Ljung, A. Benveniste, B. Deylon, P.Y. Glorennec, H. Hjalmarsson, and A. Juditsky. Nonlinear black-box modeling in system identification: A unified overview. *Automatica*, 31, 1995.

[9] T. Söderström and P. Stoica. *System Identification.* Prentice-Hall Int., London, 1989.

[10] Q. Zhang and A. Benveniste. Wavelet networks. *IEEE Trans Neural Networks*, 3:889–898, 1992.

27

A NEW CLASS OF ADAPTIVE NONLINEAR SYSTEMS

Petar Kokotović* and Miroslav Krstić**

** Department of Electrical and Computer Engineering*
University of California, Santa Barbara
*** Department of Mechanical Engineering*
University of Maryland

Dedicated with esteem to Tom Kailath, on the occasion of his sixtieth birthday

1 INTRODUCTION

Adaptive control is one of only a few research areas in which Tom Kailath has not been directly involved. This wouldn't speak well for adaptive control if it weren't for Tom's fascination with estimation. His contributions to nonlinear estimation and filtering are discussed elsewhere in this volume. Although many sharp teeth were broken on that hard nut, for important special classes of nonlinear systems a discernible progress has recently been made.

For adaptive systems, stronger parameter estimators and stronger controllers have been developed. They owe much to the idea of backstepping with which our short story begins. After a simple but dramatic example showing that the popular "certainty equivalence" concept is a misfit in the nonlinear world, we present in Section 4 a class of new controllers with strong boundedness properties. These controllers accept parameter estimates from identifiers in Section 5 which employ standard estimation algorithms, including Tom's favorite least squares [1, 2]. Married with our controllers, these identifiers constitute a family of new adaptive nonlinear systems [3]. In this way, Tom's estimation results have indirectly contributed to the development of adaptive control.

2 BACKSTEPPING

We introduce the idea of backstepping by carrying out a *nonadaptive* design for the system

$$\dot{x}_1 = x_2 + \varphi(x_1)^{\mathrm{T}}\theta\,, \qquad \varphi(0) = 0 \tag{27.1}$$

$$\dot{x}_2 = u \tag{27.2}$$

where θ is a *known* parameter vector and $\varphi(x_1)$ is a smooth nonlinear function. Our goal is to stabilize the equilibrium $x_1 = 0$, $x_2 = -\varphi(0)^{\mathrm{T}}\theta = 0$. Backstepping design is recursive. First, the state x_2 is treated as a **virtual control** for the x_1-equation (27.1), and a **stabilizing function**

$$\alpha_1(x_1) \;=\; -c_1 x_1 - \varphi(x_1)^{\mathrm{T}}\theta\,, \qquad c_1 > 0 \tag{27.3}$$

is designed to stabilize (27.1) assuming that $x_2 = \alpha_1(x_1)$ can be implemented. Since this is not the case, we define

$$z_1 = x_1 \tag{27.4}$$
$$z_2 = x_2 - \alpha_1(x_1)\,, \tag{27.5}$$

where z_2 is an error variable expressing the fact that x_2 is not the true control. Differentiating z_1 and z_2 with respect to time, the complete system (27.1), (27.2) is expressed in the error coordinates (27.4), (27.5):

$$\dot z_1 = \dot x_1 = x_2 + \varphi^{\mathrm{T}}\theta = z_2 + \alpha_1 + \varphi^{\mathrm{T}}\theta = -c_1 z_1 + z_2 \tag{27.6}$$
$$\dot z_2 = \dot x_2 - \dot\alpha_1 = u - \frac{\partial\alpha_1}{\partial x_1}\dot x_1 = u - \frac{\partial\alpha_1}{\partial x_1}\left(x_2 + \varphi^{\mathrm{T}}\theta\right)\,. \tag{27.7}$$

It is important to observe that the time derivative $\dot\alpha_1$ is implemented analytically, without a differentiator. For the system (27.6)–(27.7) we now design a control law $u = \alpha_2(x_1, x_2)$ to render the time derivative of a Lyapunov function negative definite. It turns out that the design can be completed with the simplest Lyapunov function

$$V(x_1, x_2) = \frac{1}{2}z_1^2 + \frac{1}{2}z_2^2\,. \tag{27.8}$$

Its derivative for (27.6), (27.7) is

$$\begin{aligned} \dot V &= z_1\left(-c_1 z_1 + z_2\right) + z_2\left[u - \frac{\partial\alpha_1}{\partial x_1}\left(x_2 + \varphi^{\mathrm{T}}\theta\right)\right] \\ &= -c_1 z_1^2 + z_2\left[u + z_1 - \frac{\partial\alpha_1}{\partial x_1}\left(x_2 + \varphi^{\mathrm{T}}\theta\right)\right]\,. \end{aligned} \tag{27.9}$$

An obvious way to achieve negativity of $\dot V$ is to employ u to make the bracketed expression equal to $-c_2 z_2$ with $c_2 > 0$, namely,

$$u = \alpha_2(x_1, x_2) = -c_2 z_2 - z_1 + \frac{\partial\alpha_1}{\partial x_1}\left(x_2 + \varphi^{\mathrm{T}}\theta\right)\,. \tag{27.10}$$

This control may not be the best choice because it cancels some terms which may contribute to the negativity of $\dot V$. Backstepping design offers enough flexibility to avoid cancellation. However, for the sake of clarity, we will assume that none of the nonlinearities is useful, so that they all need to be cancelled as in the control law (27.10). This control law yields

$$\dot{V} = -c_1 z_1^2 - c_2 z_2^2 \,, \tag{27.11}$$

which means that the equilibrium $z = 0$ is globally asymptotically stable. In view of (27.4), (27.5), the same is true about $x = 0$. The resulting closed-loop system in the z-coordinates is linear:

$$\begin{bmatrix} \dot{z}_1 \\ \dot{z}_2 \end{bmatrix} = \begin{bmatrix} -c_1 & 1 \\ -1 & -c_2 \end{bmatrix} \begin{bmatrix} z_1 \\ z_2 \end{bmatrix} . \tag{27.12}$$

We present examples of adaptive nonlinear designs and include summaries of design procedures. A complete theory of this approach can be found in the recently published book by Krstić, Kanellakopoulos, and Kokotović [3].

3 WEAKNESS OF CERTAINTY EQUIVALENCE

The so called "certainty equivalence principle," so pervasive in the linear theory, is, in general, not applicable to nonlinear systems. To illustrate this, let us consider the system

$$\dot{x} \;=\; -x + x^3 \tilde{\theta} \,, \tag{27.13}$$

where $\tilde{\theta} = \theta - \hat{\theta}$ is the parameter estimation error. This systems is obtained by applying the certainty equivalence controller $u = -x - x^3\hat{\theta}$ to the scalar system $\dot{x} = u + x^3\theta$, where θ is the unknown constant parameter and $\hat{\theta}$ is its estimate. Most common parameter estimators generate bounded estimates $\hat{\theta}(t)$ with convergence rates not faster than exponential. Assuming that such an estimator is applied, we let $\tilde{\theta}(t) = \mathrm{e}^{-t}$, which, upon substitution in (27.13), gives

$$\dot{x} \;=\; -x + x^3 \mathrm{e}^{-t} \,. \tag{27.14}$$

Does this connection of a certainty equivalence controller and an exponentially converging estimator produce a stable feedback system? No! For initial conditions $|x_0| > \sqrt{\frac{3}{2}}$, the solutions of system (27.14) escape to infinity in finite time:

$$x(t) \to \infty \quad \text{as} \quad t \to \frac{1}{3} \ln \frac{x_0^2}{x_0^2 - 3/2} \,. \tag{27.15}$$

It is clear that even the best exponentially converging estimator cannot prevent this catastrophic instability. From this reason, we need stronger controllers which prevent unbounded behavior caused by parameter estimation error $\tilde{\theta}$.

4 CONTROLLER DESIGN

We now proceed to design a controller stronger than any "certainty equivalence" controller. The key ingredient of this design is "nonlinear damping," thanks to which the controller acquires desired boundedness properties.

4.1 Backstepping with Nonlinear Damping

We strengthen the controller for the preceding example, $u = -x - \varphi(x)\hat{\theta}$, with a **nonlinear damping** term $-\varphi(x)^2 x$, that is, $u = -x - \varphi(x)\hat{\theta} - \varphi(x)^2 x$. With this stronger controller, the closed-loop system is

$$\dot{x} \;=\; -x - \varphi(x)^2 x + \varphi(x)\tilde{\theta}\,. \tag{27.16}$$

To see that x is bounded whenever $\tilde{\theta}$ is, we consider the Lyapunov function $V = \frac{1}{2}x^2$. Its derivative along the solutions of (27.16) is

$$\begin{aligned}\dot{V} &= -x^2 - \varphi(x)^2 x^2 + x\varphi(x)\tilde{\theta} \\ &= -x^2 - \left[\varphi(x)x - \frac{1}{2}\tilde{\theta}\right]^2 + \frac{1}{4}\tilde{\theta}^2 \\ &\le -x^2 + \frac{1}{4}\tilde{\theta}^2\,. \end{aligned}\tag{27.17}$$

This inequality shows that $|x(t)|$ will not grow larger than $\frac{1}{2}|\tilde{\theta}(t)|$, because then $\dot{V}$ becomes negative and $V = \frac{1}{2}x^2$ decreases. Thus, thanks to the nonlinear damping, the boundedness of $\tilde{\theta}(t)$ guarantees that $x(t)$ is bounded.

To show how nonlinear damping is incorporated into a backstepping design, we consider the system

$$\begin{aligned}\dot{x}_1 &= x_2 + \varphi(x_1)^{\mathrm{T}}\theta \\ \dot{x}_2 &= u\,.\end{aligned}\tag{27.18}$$

Viewing x_2 as a control input, we first design a control law $\alpha_1(x_1, \hat{\theta})$ to guarantee that the state x_1 in $\dot{x}_1 = x_2 + \varphi(x_1)^{\mathrm{T}}\theta$ is bounded whenever $\tilde{\theta}$ is bounded. In the first stabilizing function we include a nonlinear damping term[1] $-\kappa_1|\varphi(x_1)|^2 x_1$:

$$\alpha_1(x_1, \hat{\theta}) = -c_1 x_1 - \varphi(x_1)^{\mathrm{T}}\hat{\theta} - \kappa_1|\varphi(x_1)|^2 x_1, \quad c_1, \kappa_1 > 0\,. \tag{27.19}$$

Then we define the error variable $z_2 = x_2 - \alpha_1(x_1, \hat{\theta})$, and for uniformity denote $z_1 = x_1$. The first equation is now

$$\dot{z}_1 = -c_1 z_1 - \kappa_1|\varphi|^2 z_1 + \varphi^{\mathrm{T}}\tilde{\theta} + z_2\,. \tag{27.20}$$

[1]The Euclidian norm of a vector v is denoted as $|v| = \sqrt{v^{\mathrm{T}}v}$.

If z_2 were zero, the Lyapunov function $V_1 = \frac{1}{2}z_1^2$ would have the derivative

$$\dot{V}_1 = -c_1 z_1^2 - \kappa_1|\varphi|^2 z_1^2 + z_1\varphi^{\mathrm{T}}\tilde{\theta} = -c_1 z_1^2 - \kappa_1\left|\varphi z_1 - \frac{1}{2\kappa_1}\tilde{\theta}\right|^2 + \frac{1}{4\kappa_1}|\tilde{\theta}|^2$$
$$\leq -c_1 z_1^2 + \frac{1}{4\kappa_1}|\tilde{\theta}|^2\,, \tag{27.21}$$

so that z_1 would be bounded whenever $\tilde{\theta}$ is bounded. With $z_2 \neq 0$ we have

$$\dot{V}_1 \leq -c_1 z_1^2 + \frac{1}{4\kappa_1}|\tilde{\theta}|^2 + z_1 z_2\,. \tag{27.22}$$

Differentiating $x_2 = z_2 + \alpha_1(x_1, \hat{\theta})$, the second equation in (27.18) yields

$$\dot{z}_2 = \dot{x}_2 - \dot{\alpha}_1 = u - \frac{\partial\alpha_1}{\partial x_1}\left(x_2 + \varphi^{\mathrm{T}}\theta\right) - \frac{\partial\alpha_1}{\partial\hat{\theta}}\dot{\hat{\theta}}\,. \tag{27.23}$$

The derivative of the Lyapunov function

$$V_2 = V_1 + \frac{1}{2}z_2^2 = \frac{1}{2}|z|^2 \tag{27.24}$$

along the solutions of (27.20) and (27.23) is

$$\dot{V}_2 \leq -c_1 z_1^2 + \frac{1}{4\kappa_1}|\tilde{\theta}|^2 + z_1 z_2 + z_2\left[u - \frac{\partial\alpha_1}{\partial x_1}\left(x_2 + \varphi^{\mathrm{T}}\theta\right) - \frac{\partial\alpha_1}{\partial\hat{\theta}}\dot{\hat{\theta}}\right]$$
$$\leq -c_1 z_1^2 + \frac{1}{4\kappa_1}|\tilde{\theta}|^2$$
$$+z_2\left[u + z_1 - \frac{\partial\alpha_1}{\partial x_1}\left(x_2 + \varphi^{\mathrm{T}}\hat{\theta}\right) - \left(\frac{\partial\alpha_1}{\partial x_1}\varphi^{\mathrm{T}}\tilde{\theta} + \frac{\partial\alpha_1}{\partial\hat{\theta}}\dot{\hat{\theta}}\right)\right]\,. \tag{27.25}$$

We note that now, in addition to the $\tilde{\theta}$-dependent disturbance term $\frac{\partial\alpha_1}{\partial x_1}\varphi^{\mathrm{T}}\tilde{\theta}$, we also have a $\dot{\hat{\theta}}$-dependent disturbance $\frac{\partial\alpha_1}{\partial\hat{\theta}}\dot{\hat{\theta}}$. No such term appeared in the scalar system (27.16). We now use nonlinear damping terms $-\kappa_2\left|\frac{\partial\alpha_1}{\partial x_1}\varphi\right|^2 z_2$ and $-g_2\left|\frac{\partial\alpha_1}{\partial\hat{\theta}}^{\mathrm{T}}\right|^2 z_2$ to counteract the effects of both $\tilde{\theta}$ and $\dot{\hat{\theta}}$:

$$u = -z_1 - c_2 z_2 - \kappa_2\left|\frac{\partial\alpha_1}{\partial x_1}\varphi\right|^2 z_2 - g_2\left|\frac{\partial\alpha_1}{\partial\hat{\theta}}^{\mathrm{T}}\right|^2 z_2 + \frac{\partial\alpha_1}{\partial x_1}\left(x_2 + \varphi^{\mathrm{T}}\hat{\theta}\right), \tag{27.26}$$

where $c_2, \kappa_2, g_2 > 0$. Upon completing the squares as in (27.20), we get

$$\dot{V}_2 \leq -c_1 z_1^2 - c_2 z_2^2 + \left(\frac{1}{4\kappa_1} + \frac{1}{4\kappa_2}\right)|\tilde{\theta}|^2 + \frac{1}{4g_2}|\dot{\hat{\theta}}|^2\,, \tag{27.27}$$

which means that the state of the error system

$$\dot{z} = \begin{bmatrix} -c_1 - \kappa_2 |\varphi|^2 & 1 \\ -1 & -c_2 - \kappa_2 \left|\frac{\partial \alpha_1}{\partial x_1}\varphi\right|^2 - g_2 \left|\frac{\partial \alpha_1}{\partial \hat{\theta}}^{\mathrm{T}}\right|^2 \end{bmatrix} z$$

$$+ \begin{bmatrix} \varphi^{\mathrm{T}} \\ -\frac{\partial \alpha_1}{\partial x_1}\varphi^{\mathrm{T}} \end{bmatrix} \tilde{\theta} + \begin{bmatrix} 0 \\ -\frac{\partial \alpha_1}{\partial \hat{\theta}} \end{bmatrix} \dot{\hat{\theta}} \tag{27.28}$$

is bounded whenever the disturbance inputs $\tilde{\theta}$ and $\dot{\hat{\theta}}$ are bounded. Moreover, since V_2 is quadratic in z, see (27.24), we can use (27.27) to show that the boundedness of z is guaranteed also when $\dot{\hat{\theta}}$ is square-integrable but not bounded. As we shall see in Section 5, this property is crucial for adaptive control with passive identifiers where $\dot{\hat{\theta}}$ cannot be a priori bounded.

A systematic backstepping design with nonlinear damping has been developed for the class of nonlinear systems transformable into the **parametric strict-feedback form**:

$$\begin{aligned} \dot{x}_1 &= x_2 + \varphi_1(x_1)^{\mathrm{T}}\theta \\ \dot{x}_2 &= x_3 + \varphi_2(x_1, x_2)^{\mathrm{T}}\theta \\ &\vdots \\ \dot{x}_{n-1} &= x_n + \varphi_{n-1}(x_1, \dots, x_{n-1})^{\mathrm{T}}\theta \\ \dot{x}_n &= \beta(x)u + \varphi_n(x)^{\mathrm{T}}\theta \\ y &= x_1 \end{aligned} \tag{27.29}$$

where β and

$$F(x) = [\varphi_1(x_1),\ \varphi_2(x_1, x_2),\ \cdots, \varphi_n(x)] \tag{27.30}$$

are smooth nonlinear functions, and $\beta(x) \neq 0$, $\forall x \in \mathbb{R}^n$.

In the recursive design, summarized in (27.31)–(27.35) below, the output $y = x_1$ of the system (27.29) is forced to asymptotically track the reference output $y_{\mathrm{r}}(t)$ whose first n derivatives are assumed to be known, bounded, and piecewise continuous.

The recursive controller design for (27.29) is given by:

$$z_i = x_i - y_{\mathrm{r}}^{(i-1)} - \alpha_{i-1} \tag{27.31}$$

$$\alpha_i(\bar{x}_i, \hat{\theta}, \bar{y}_{\mathrm{r}}^{(i-1)}) = -z_{i-1} - c_i z_i - w_i^{\mathrm{T}}\hat{\theta} + \sum_{k=1}^{i-1}\left(\frac{\partial \alpha_{i-1}}{\partial x_k} x_{k+1} + \frac{\partial \alpha_{i-1}}{\partial y_{\mathrm{r}}^{(k-1)}} y_{\mathrm{r}}^{(k)}\right) - s_i z_i \tag{27.32}$$

$$w_i(\bar{x}_i, \hat{\theta}, \bar{y}_{\mathrm{r}}^{(i-2)}) = \varphi_i - \sum_{k=1}^{i-1} \frac{\partial \alpha_{i-1}}{\partial x_k}\varphi_k \tag{27.33}$$

$$s_i(\bar{x}_i, \hat{\theta}, \bar{y}_{\mathrm{r}}^{(i-2)}) = \kappa_i |w_i|^2 + g_i \left|\frac{\partial \alpha_{i-1}}{\partial \hat{\theta}}^{\mathrm{T}}\right|^2 \tag{27.34}$$

$$i = 1, \dots, n$$

$$\bar{x}_i = (x_1, \dots, x_i), \qquad \bar{y}_{\mathrm{r}}^{(i)} = (y_{\mathrm{r}}, \dot{y}_{\mathrm{r}}, \dots, y_{\mathrm{r}}^{(i)})$$

The adaptive control law is:

$$u = \frac{1}{\beta(x)} \left[\alpha_n(x, \hat{\theta}, \bar{y}_{\mathrm{r}}^{(n-1)}) + y_{\mathrm{r}}^{(n)} \right] \tag{27.35}$$

This design endows the resulting adaptive system with the properties described in the next subsection.

4.2 Boundedness Properties

The system governing the error states z is

$$\dot{z} \;=\; A_z(z, \hat{\theta}, t)z + W(z, \hat{\theta}, t)^{\mathrm{T}} \tilde{\theta} + Q(z, \hat{\theta}, t)^{\mathrm{T}} \dot{\hat{\theta}} \tag{27.36}$$

where A_z, W, Q are:

$$A_z(z, \hat{\theta}, t) = \begin{bmatrix} -c_1 - s_1 & 1 & 0 \cdots & 0 \\ -1 & -c_2 - s_2 & 1 \ \ddots & \vdots \\ 0 & -1 & \ddots \ \ddots & 0 \\ \vdots & \ddots & \ddots \ \ddots & 1 \\ 0 & \cdots & 0 \ -1 & -c_n - s_n \end{bmatrix}$$

$$W(z, \hat{\theta}, t)^{\mathrm{T}} = \begin{bmatrix} w_1^{\mathrm{T}} \\ w_2^{\mathrm{T}} \\ \vdots \\ w_n^{\mathrm{T}} \end{bmatrix}, \qquad Q(z, \hat{\theta}, t)^{\mathrm{T}} = \begin{bmatrix} 0 \\ -\frac{\partial \alpha_1}{\partial \hat{\theta}} \\ \vdots \\ -\frac{\partial \alpha_{n-1}}{\partial \hat{\theta}} \end{bmatrix}. \tag{27.37}$$

Using a Lyapunov argument we can show that the controller guarantees:

$$\text{If } \tilde{\theta} \in \mathcal{L}_\infty \text{ and } \dot{\hat{\theta}} \in \mathcal{L}_2 \text{ or } \mathcal{L}_\infty \quad \text{then} \quad x \in \mathcal{L}_\infty$$

Thus x is bounded whenever $\tilde{\theta}$ is bounded and $\dot{\hat{\theta}}$ is either bounded or square-integrable. It follows that we need identifiers which guarantee these properties independent of the controller.

In addition to the boundedness of $x(t)$, our goal is to achieve asymptotic tracking, that is, to regulate $z(t)$ to zero. With z and $\hat{\theta}$ bounded, it is not hard to prove that $z(t) \to 0$ provided

$$W(z(t), \hat{\theta}(t), t)^{\mathrm{T}} \tilde{\theta}(t) \to 0 \quad \text{and} \quad \dot{\hat{\theta}}(t) \to 0.$$

Let us factor the regressor matrix W, using (27.36), (27.33) and (27.30) as

$$W(z,\hat{\theta},t)^{\mathrm{T}} = \begin{bmatrix} 1 & 0 & \cdots & 0 \\ -\frac{\partial\alpha_1}{\partial x_1} & 1 & \ddots & \vdots \\ \vdots & \ddots & \ddots & 0 \\ -\frac{\partial\alpha_{n-1}}{\partial x_1} & \cdots & -\frac{\partial\alpha_{n-1}}{\partial x_{n-1}} & 1 \end{bmatrix} F(x)^{\mathrm{T}} \triangleq N(z,\hat{\theta},t)F(x)^{\mathrm{T}}. \quad (27.38)$$

Since $N(z,\hat{\theta},t)$ is invertible, the tracking condition $W(z(t),\hat{\theta}(t),t)^{\mathrm{T}}\tilde{\theta}(t) \to 0$ becomes $F(x(t))^{\mathrm{T}}\tilde{\theta}(t) \to 0$.

5 DESIGN OF IDENTIFIERS

We now develop identifiers for the parametric model

$$\dot{x} = f(x,u) + F(x,u)^{\mathrm{T}}\theta. \quad (27.39)$$

The parametric strict-feedback system (27.29) is a special case of this model with $F(x,u)$ given by (27.30) and $f(x,u) = [x_2, \ \ldots, \ x_n, \ \beta_0(x)u]^{\mathrm{T}}$.
From Section 4 we know that the identifiers to be designed must possess the following boundedness properties:

(i) $\tilde{\theta} \in \mathcal{L}_\infty$ and $\dot{\hat{\theta}} \in \mathcal{L}_2$ or $\mathcal{L}_\infty$,

(ii) if $x \in \mathcal{L}_\infty$ then $F(x(t))^{\mathrm{T}}\tilde{\theta}(t) \to 0$ and $\dot{\hat{\theta}}(t) \to 0$.

We present two types of such identifiers: **the passive identifier** and **the swapping identifier**.

5.1 Passive identifiers

For the parametric model (27.39) we implement the "observer"

$$\dot{\hat{x}} = \left[A_0 - \lambda F(x,u)^{\mathrm{T}} F(x,u) P\right](\hat{x} - x) + f(x,u) + F(x,u)^{\mathrm{T}}\hat{\theta}, \quad (27.40)$$

where $\lambda > 0$ and A_0 is an arbitrary constant matrix such that

$$PA_0 + A_0^{\mathrm{T}}P = -I, \quad P = P^{\mathrm{T}} > 0. \quad (27.41)$$

By direct substitution it can be seen that the observer error

$$\epsilon = x - \hat{x} \quad (27.42)$$

is governed by

$$\dot{\epsilon} = \left[A_0 - \lambda F(x,u)^{\mathrm{T}} F(x,u) P\right]\epsilon + F(x,u)^{\mathrm{T}}\tilde{\theta}. \quad (27.43)$$

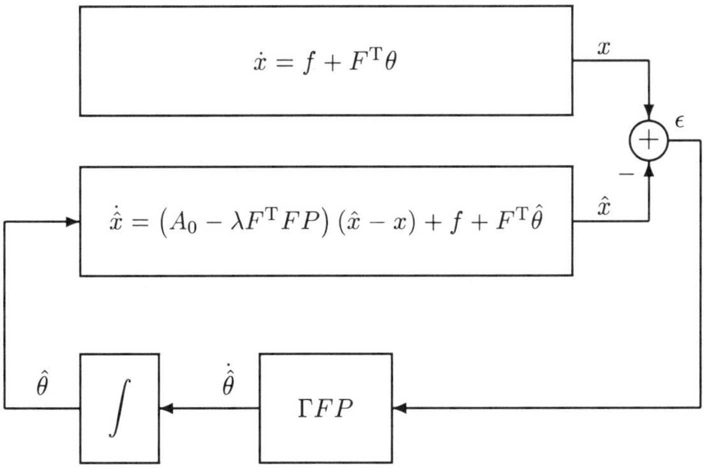

Figure 27.1 The passive identifier.

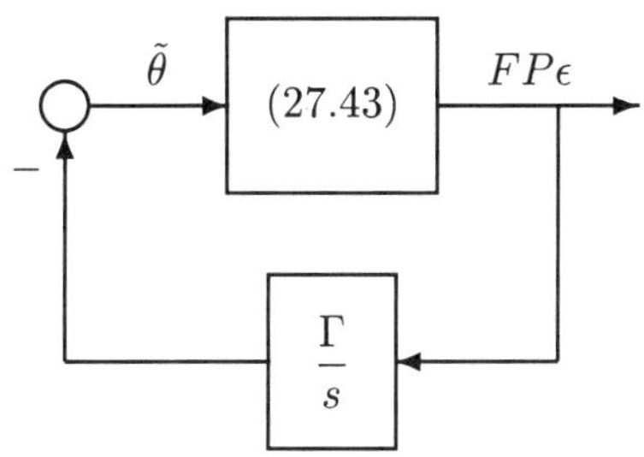

Figure 27.2 Feedback connection of the strictly passive system (27.43) with the passive system Γ/s.

The observer error system (27.43) has a *strict passivity* property from the input $\tilde{\theta}$ to the output $F(x,u)P\epsilon$. A standard result of passivity theory is that the equilibrium $\tilde{\theta} = 0,\ \epsilon = 0$ of the negative feedback connection of one strictly passive and one passive system is globally stable. Using integral feedback such a connection can be formed as in Figure 27.2. This suggests the following parameter update law

$$\dot{\hat{\theta}} = \Gamma F(x,u)P\epsilon\,, \qquad \Gamma = \Gamma^{\mathrm{T}} > 0\,. \tag{27.44}$$

To analyze the stability properties of the passive identifier we use the Lyapunov function

$$V = \tilde{\theta}^{\mathrm{T}}\Gamma^{-1}\tilde{\theta} + \epsilon^{\mathrm{T}}P\epsilon\,. \tag{27.45}$$

Its derivative can be shown to satisfy

$$\dot{V} \leq -\epsilon^{\mathrm{T}}\epsilon - \frac{\lambda}{\bar{\lambda}(\Gamma)^2}|\dot{\hat{\theta}}|^2 . \tag{27.46}$$

This guarantees the boundedness of $\tilde{\theta}$ and ϵ, even when $\lambda = 0$. However, $\dot{\hat{\theta}}$ cannot be shown to be bounded (unless x and u are known to be bounded). Instead, for the passive identifier one can show that $\dot{\hat{\theta}}$ is square integrable. For this we must use $\lambda > 0$, that is, we rely on the nonlinear damping term $-\lambda F(x,u)^{\mathrm{T}}F(x,u)P$ in the observer. The boundedness of $\tilde{\theta}$ and the square-integrability of $\dot{\hat{\theta}}$ imply that x is bounded.

To prove the tracking, we need to show that the identifier guarantees that $F(x(t))^{\mathrm{T}}\tilde{\theta}(t) \to 0$ and $\dot{\hat{\theta}}(t) \to 0$ whenever x is bounded. Both properties follow from standard adaptive control arguments [3]. The latter property can easily be shown to follow from the square-integrability of $\dot{\hat{\theta}}$. The regulation of $F(x)^{\mathrm{T}}\tilde{\theta}$ to zero follows upon showing that both $\epsilon(t)$ and $\dot{\epsilon}(t)$ converge to zero. While the convergence of $\epsilon(t)$ follows by deducing its square-integrability from (27.46), the convergence of $\dot{\epsilon}(t)$ follows from the fact that its integral, $\int_0^\infty \dot{\epsilon}(\tau)d\tau = \epsilon(\infty) - \epsilon(0) = -\epsilon(0)$, exists.

5.2 Swapping identifiers

For the parametric model (27.39) we implement two filters,

$$\dot{\Omega}^{\mathrm{T}} = \left[A_0 - \lambda F(x,u)^{\mathrm{T}}F(x,u)P\right]\Omega^{\mathrm{T}} + F(x,u)^{\mathrm{T}} \tag{27.47}$$

$$\dot{\Omega}_0 = \left[A_0 - \lambda F(x,u)^{\mathrm{T}}F(x,u)P\right](\Omega_0 - x) - f(x,u)\,, \tag{27.48}$$

where $\lambda \geq 0$ and A_0 is as defined in (27.41). The **estimation error**,

$$\epsilon = x + \Omega_0 - \Omega^{\mathrm{T}}\hat{\theta}\,. \tag{27.49}$$

can be written in the form

$$\epsilon = \Omega^{\mathrm{T}}\tilde{\theta} + \tilde{\epsilon} \tag{27.50}$$

where $\tilde{\epsilon} \triangleq x + \Omega_0 - \Omega^{\mathrm{T}}\theta$ decays exponentially because it is governed by

$$\dot{\tilde{\epsilon}} = \left[A_0 - \lambda F(x,u)^{\mathrm{T}}F(x,u)P\right]\tilde{\epsilon}\,. \tag{27.51}$$

The filters (27.47) and (27.48) have converted the dynamic model (27.39) into the linear static parametric model (27.50) to which we can apply standard estimation algorithms. As our update law we will employ either the gradient

$$\dot{\hat{\theta}} = \Gamma\frac{\Omega\epsilon}{1 + \nu\mathrm{tr}\{\Omega^{\mathrm{T}}\Omega\}}\,, \qquad \begin{array}{l}\Gamma = \Gamma^{\mathrm{T}} > 0\\ \nu \geq 0\end{array} \tag{27.52}$$

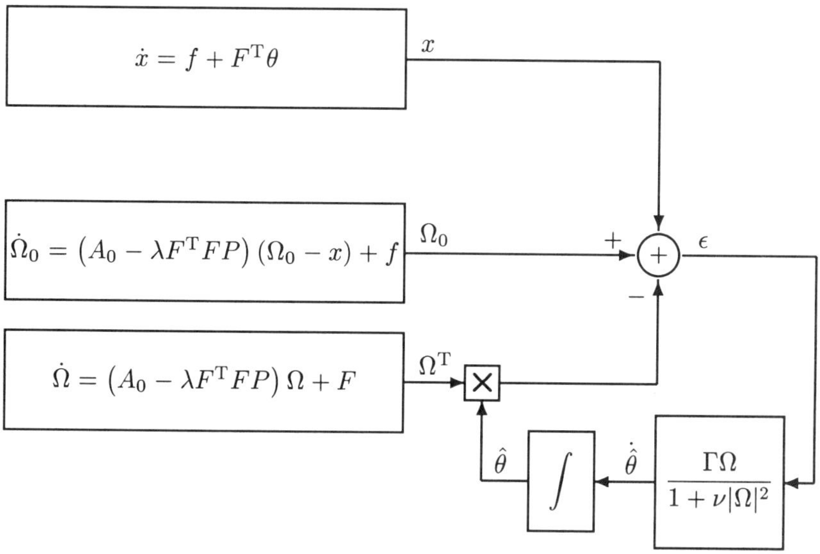

Figure 27.3 The swapping identifier.

or the least squares algorithm

$$\begin{aligned} \dot{\hat{\theta}} &= \Gamma \frac{\Omega\epsilon}{1+\nu \operatorname{tr}\{\Omega^{\mathrm{T}}\Omega\}} \\ \dot{\Gamma} &= -\Gamma \frac{\Omega\Omega^{\mathrm{T}}}{1+\nu \operatorname{tr}\{\Omega^{\mathrm{T}}\Omega\}}\Gamma\,, \qquad \Gamma(0) = \Gamma(0)^{\mathrm{T}} > 0 \\ & \qquad\qquad\qquad\qquad\qquad\quad \nu \geq 0\,. \end{aligned} \tag{27.53}$$

By allowing $\nu = 0$, we encompass unnormalized gradient and least-squares. The complete swapping identifier is shown in Figure 27.3.

The update law normalization, $\nu > 0$, and the nonlinear damping, $\lambda > 0$, are two different means for slowing down the identifier in order to guarantee the boundedness and square-integrability of $\dot{\hat{\theta}}$.

For the gradient update law (27.52), the identifier properties (boundedness of $\tilde{\theta}$ and $\hat{\theta}$ and regulation of $F(x)\tilde{\theta}$ and $\dot{\hat{\theta}}$) are established via the Lyapunov function

$$V = \frac{1}{2}\tilde{\theta}^{\mathrm{T}}\Gamma^{-1}\tilde{\theta} + \tilde{\epsilon}P\tilde{\epsilon} \tag{27.54}$$

whose derivative is

$$\dot{V} \leq -\frac{3}{4}\frac{\epsilon^{\mathrm{T}}\epsilon}{1+\nu \operatorname{tr}\{\Omega^{\mathrm{T}}\Omega\}}\,. \tag{27.55}$$

The Lyapunov function for the least-squares update law (27.53) is $V = \tilde{\theta}^{\mathrm{T}}\Gamma(t)^{-1}\tilde{\theta} + \tilde{\epsilon}P\tilde{\epsilon}$.

6 CONCLUSION

A connection of either of the above two identifiers with a controller designed in Section 4 constitutes an adaptive nonlinear system with desirable global boundedness properties of all the signals and the tracking property that the error between the output and a reference input converges to zero.

Acknowledgements

This work was supported in part by the National Science Foundation under Grant ECS-9203491 and in part by the Air Force Office of Scientific Research under Grant F-49620-92-J-0495.

REFERENCES

[1] T. Kailath, Ed., *Linear Least-Squares Estimation*, Stroudsburg, PA: Dowden, Hutchinson & Ross, 1977.

[2] T. Kailath, *Lectures on Wiener and Kalman Filtering*, New York : Springer-Verlag, 1981.

[3] M. Krstić, I. Kanellakopoulos, and P. V. Kokotović, *Nonlinear and Adaptive Control Design*, New York, NY: Wiley, 1995.

28

NONLINEAR CONTROL OF CONSTRAINED DYNAMIC SYSTEMS

David Quinn Mayne and Warren R. Schroeder

Department of Electrical and Computer Engineering
University of California
Davis, California 95616

Dedicated with affection and admiration to Professor Kailath on the occasion of his 60th birthday.

ABSTRACT

Linear system theory has reached an impressive maturity, with elegant solutions to a wide range of problems, and associated design procedures. Recent research in nonlinear system theory has seen the emergence of important techniques such as feedback linearization and backstepping, as well as interesting extensions of H_∞ theory to nonlinear systems. These impressive advances underline our relative inability to deal with hard constraints. In this paper we examine design of controllers for linear systems subject to constraints on states and controls. A major motivation is the presence of hard constraints in most applications; efficient handling of these constraints requires nonlinear controllers even if the system being controlled is linear.

1 INTRODUCTION

Constraints on controls and states arise in most control applications; actuators saturate and safe operation requires limitations on states such as velocity and acceleration in mechanical systems, and temperature and pressure in chemical processes. Efficient handling of constraints requires nonlinear control even if the system being controlled is linear. A well known example is a zero velocity error controller; linear control enforces the integral of the error in response to a step input to be zero. Acceleration constraints cause the integral of error, over the interval $[0, t_1]$ where t_1 is the first time at which the error is zero, to be negative and irreducible. Hence, small settling time necessitates large overshoot, and small overshoot necessitates large settling time, an impasse which cannot be overcome with linear control. This classic problem has been solved, in practice, by ad-hoc methods which have little theoretical justification. This is a criticism, not of practice since sheer necessity has enforced the development of ad-hoc solutions, but of theory which has been slow to address the problem of constrained control.

Sheer necessity was also the driving force in the development of model predictive control in the petrochemical industry where steady state optimization forces the operating point to lie on or near the boundary of the set of admissible states and controls. Model predictive control obtains a feedback control for constrained systems, not by solving the Hamilton-Jacobi-Bellman partial differential equation (generally impossible) but by repeatedly solving online an *open-loop* optimal control problem for the current state (a considerably simpler task) and applying the minimizing control for a short time before repeating the process. In essence, the value function $V^0(x)$ is determined for the sequence of states $\{x[k] := x(k\Delta)\}$ actually encountered rather than for each state in the state space; an impossible problem is replaced by a soluble problem. The price is the complexity of the on-line controller (which has to have the capability of solving open-loop optimal control problems); this restricts its use to 'slow' plants. An interesting feature of this research was its development, independently by industry, to address an important problem. Theory has, in this endeavour, been more helpful: an adequate stability theory, to which Kailath [1] was one of the early contributors, exists [2, 3, 4, 5], and robustness and implementation problems have been partially addressed.

The work of Gilbert and Tan [6] is applicable to the control of 'fast' constrained systems in that the controller is much more easily implemented. Gilbert and Tan introduce the useful concept of *maximal output admissible set* associated with a linear controller $u = Kx$ which is stabilizing in the absence of constraints. If the initial state x lies in this output admissible set, the controller $u = Kx$ steers this state to the origin exponentially fast along a trajectory which satisfies the control and state constraints. They show how this set may be characterized, and computed, and employ it to construct a nonlinear controller with a very specific structure; the controller has the form $u = K_{i(x)}x$ where $i(x)$ is the smallest integer i in $\{1 \dots N\}$ such that x lies in the output admissible set of $u = K_i x$ and $\{K_i \mid i = 1, \dots, N\}$ is a sequence of linear, stabilizing controllers, ranging from 'aggressive' to 'conservative'. This ingenious and useful extension to linear control is easily implementable.

Other relevant research includes the work of Teel [7] and Sussmann et al [8] which establishes the existence of stabilizing controllers, composed of linear and saturation functions, for linear constrained systems, and the recent work by Shamma [9] on the construction of nonlinear ℓ_1-optimal control laws for linear discrete time systems with additive disturbances.

In this paper, we take a different route. We seek a finite dimensional approximation to the value function; the approximation is, itself, a value function for a new problem. The associated controller for the approximate problem also stabilizes the original system. The value function for the modified problem tends, in an appropriate sense, to the value function of the original problem, as an approximation parameter converges to zero, but close approximation is not required. The potential advantage of this approach is that optimality may be approached as far as desired; the price is controller complexity. The fact that the modified problem has a value function which can be finitely parameterized is crucially important. The results presented here are based on research reported in [10, 11, 12, 13].

We define the modified problem in §2, and show how it may be exactly solved. A robust version is discussed in §3. In §4, we show how the value function for the new problem may be employed to derive two classes of controllers, both relatively easily implemented, at least when compared with model predictive control.

Notation: Given any set X, $\text{int}(X)$ denotes its interior, and $\partial(X)$ its boundary. Subscripts denote elements of a sequence; superscripts denote components of a vector. For any integer j, Σ_j is the convex set defined by:

$$\Sigma_j = \{\mu \in \mathbb{R}^j \mid \mu^i \geq 0, \sum_{i=0}^{j} \mu^i = 1\}$$

If $v \in \mathbb{R}^p$ and $w \in \mathbb{R}^q$, are column vectors, then (v, w) denotes the column vector $(v^T, w^T)^T \in \mathbb{R}^{p+q}$. For any two subsets A and B of $\mathbb{R}^n$, $A+B := \{a+b \mid a \in A,\ b \in B\}$ denotes set addition, and $A - B := \{a - b \mid a \in A,\ b \in B\}$.

2 THE PROBLEM

Consider, initially, the problem of determining a minimum time state feedback controller for the system which is described by

$$\dot{x} = Ax + Bu$$

and is subject to 'hard' control and state constraints

$$u(t) \in \Omega,\ x(t) \in E,$$

where $\Omega \subset \mathbb{R}^m$ is convex and compact and E is closed and convex; each set contains the origin in its interior. We make the following assumptions:

A1: The pair (A, B) is controllable.

A2: The matrix B has maximal rank m.

Let $V_c^0 : \mathbb{R}^n \to \mathbb{R}$ denote the value function for this problem. The value function is determined by its level sets $X_\tau^c, \tau \in \mathbb{R}_+$, where

$$X_\tau^c := \{x \mid V_c^0(x) \leq \tau\}$$

For the minimum time problem, the boundaries of these level sets are isochrones. It is not generally possible to parameterize the value function V_c^0 or its level sets finitely. As a first step to obtaining a finite parameterization consider the discrete-time version of the minimum time problem. The system being controlled is now

$$x[k+1] = Fx[k] + Gu[k]$$

where $x[k] := x(k\Delta)$, $u[k] := u(k\Delta)$, $F := \exp(A\Delta)$ and $G := \int_0^\Delta \exp(As)Bds$ where Δ is the sampling period. Let V^0 denote the value function for the discrete time, minimum time problem, and X_k, $k \in I\!N$ the associated level sets defined now by:

$$X_k := \{x \mid V^0(x) \leq k\}$$

The fact that k is integer valued is important: $V^0(x)$ has the constant value k in $\text{int}(X_k \setminus X_{k-1})$, a significant simplification. The value function is fully defined in any compact set (in the interior of X_∞) by a *finite* number of its level sets. Despite this simplification, it is still not possible, in general, to parameterize finitely the value function or its level sets. The latter can, in principle, be generated recursively by the following algorithm which is an alternative to the dynamic programming recursion:

Algorithm 1

Data: $X_0 = \{0\}$.
Step 1: Set k=0.
Step 2: Compute the set $X_{k+1} = \{F^{-1}X_k - F^{-1}G\Omega\} \cap E$.
Step 3: Set $k = k + 1$. If $k = N$, stop. Else, go to Step 2.

We can obtain finitely parameterized level sets by requiring the constraint sets Ω and E to be polyhedrons. Therefore, we add the following assumptions

A3: Ω is a polytope in $I\!R^m$, $0 \in \text{int}(\Omega)$.

A4: E is a polyhedron in $I\!R^n$, $0 \in \text{int}(E)$.

A polytope is the convex hull of a finite number of points, and is, therefore, compact and convex. A polytope X may be specified by a set of linear inequalities. A polyhedron is a convex set specified by a set of linear inequalities, and is not necessarily compact; a compact polyhedron is a polytope. The sum of two polytopes, the intersection of two polytopes, and the intersection of a polytope and a polyhedron, are all polytopes. If A is a matrix and X is a polytope, then AX is a polytope.

We need, in the sequel, the concept of viability [14] or control invariance:

Definition 28.1 *A set $X \subset E$ is said to be* control invariant *for the discrete time system (F, G, Ω, E) if, for every $x \in X$, there exists a $u \in \Omega$ such that $Fx + Gu \in X$*

Given any initial state in a control invariant set X, there exists an admissible control which maintains the state in X. The significance of the assumptions A3 and A4 is that they lead to a finite parameterization of the sets X_k as shown by the result:

Proposition 28.1 *(i) Algorithm 1 is well defined and generates a non-decreasing sequence $\{X_k\}$ of control invariant polytopes for the system (F, G, Ω, E), (ii) for all k, all $x \in X_k$, there exists a $u \in \Omega$ such that $Fx + Gu \in X_{k-1}$, (iii) for each k, X_k is a level set for the minimum time problem, i.e. X_k is the set of states which can be steered to the origin along an admissible trajectory in no more than k steps.*

The essential point to observe is that the polytope property is invariant under set addition, multiplication (of a set) by a matrix, and set intersection, so that Step 2 of Algorithm 1 generates a polytope if X_k is a polytope. It follows that every element of the sequence $\{X_k\}$ generated by the algorithm is a polytope. Since each polytope X_k is finitely parameterized by its vertices, or, equivalently, by its dual representation as a set of linear inequalities, we have achieved our objective of obtaining a problem whose solution can be finitely parameterized and relatively easily computed.

Since the class of piecewise constant controls is a subset of the set of measurable controls, $X_k \subset X^c_{k\Delta}$. Moreover, since the vertices of X_k correspond to bang-bang controls, each vertex of X_k lies on the boundary of $X^c_{k\Delta}$. In this sense, X_k is an approximation to $X^c_{k\Delta}$. By dividing each interval $[k\Delta, (k+1)\Delta]$ into M subintervals, and allowing the control to have independent values in each subinterval, the number of vertices of X_k can be increased, and, hence, the approximation of X_k to $X^c_{k\Delta}$ improved.

3 ROBUST CONTROL

Minimum time control is notoriously sensitive. However, robustness may be achieved by modifying the above procedure. Consider the perturbed system

$$\dot{x} = Ax + Bu + w$$

where $w(t) \in W_c$. We assume

A5: W_c is compact, convex subset of $\mathbb{R}^n$, $0 \in \text{int}(W_c)$.

The corresponding perturbed discrete time system is

$$x[k+1] = Fx[k] + Gu[k] + w[k]$$

where $w[k] \in W$. To maintain finite parameterization, and to ensure the variation in $x(k\Delta)$ due to $w(\cdot)$ does not exceed the variation in $x[k]$ due to $\{w[k]\}$, we assume

A6: W is a polytope in $\mathbb{R}^m$, $0 \in \text{int}(W)$ satisfying

$$\{x^{u_0,w}(\Delta; 0, 0) \mid w(t) \in W_c, t \in [0, \Delta]\} \subset W$$

where $x^{u,w}(t; x_0, t_0)$ denotes the solution at t of $\dot{x} = Ax + Bu + w$ due to control $u(\cdot)$ and disturbance $w(\cdot)$ when the initial state is x_0 at time t_0, and u_0 is the zero control ($u_0(t) \equiv 0$). We require the following [14]

Definition 28.2 *The set $X \subset E$ is said to be* robust control invariant *for the discrete time system (F, G, Ω, W, E) if, for every $x \in X$, there exists a $u \in \Omega$ such that $Fx + Gu + W \subset X$.*

If X is robust control-invariant for (F, G, Ω, W, E), there exists a control law $h : X \to \Omega$ such that $x^{h,w}[k; x, 0] \in X$, for all $x \in X$, all $k \geq 0$, and every admissible disturbance sequence.

Because of the disturbance, we can no longer control to the origin but merely to a robust control invariant set X_0 whose determination will be discussed later. The value function, given the robust control invariant terminal set X_0, is the minimum time required to steer a state x to X_0 over all admissible disturbance sequences ($V^0(x) = \inf_u \sup_w \{k \mid x[k] \in X_0, x[s] \in E, 0 \leq s \leq k\}$ where $x[s] = x^{u,w}[s; x, 0]$ and $x \mapsto u(x)$ and $x \mapsto w(x)$ are state feedback controllers satisfying their respective constraints). Consider the following extension of Algorithm 1:

Algorithm 2

Data:	Polytope X_0.
Step 1:	Set k=0.
Step 2:	Compute $X'_k := \{x \in X_k \mid x + W \subset X_k\}$.
Step 3:	Compute the set $X_{k+1} = \{F^{-1}X'_k - F^{-1}G\Omega\} \cap E$.
Step 3:	Set $k = k + 1$. If $k = N$, stop. Else, go to Step 2.

Theorem 28.1 *Suppose X_0 is a robust control invariant polytope. Then, (i) Algorithm 2 is well defined and generates a non-decreasing sequence $\{X_k\}$ of robust control invariant polytopes for the system (F, G, Ω, W, E), (ii) for all k, all $x \in X_k$, there exists a $u \in \Omega$ such that $Fx + Gu + W \subset X_{k-1}$, and (iii) X_k is the set of states that can be steered to the set X_0 in no more than k steps, despite the disturbance $\{w[k]\}$.*

That $\{X_k\}$ is a sequence of polytopes follows from the observation that the polytope property is also invariant under the contraction operation in Step 2; it is this contraction operation which ensures robustness. The set X_0 may be constructed using the following result:

Theorem 28.2 *Suppose the controller $u = Kx$ steers any state of the unconstrained discrete time system to the origin in $s \leq n$ steps. Let $F_K := F + GK$. Then the set*

$$X_0 := W + F_K W + \ldots + F_K^{s-1} W$$

is a robust-control-invariant set for (F, G, Ω, W, E) if

$$X_0 \subset E \text{ and } KX_0 \subset \Omega$$

Note that X_0 converges to the set $\{0\}$ as $\max\{\|w\| \mid w \in W\}$ converges to zero.

4 NONLINEAR CONTROLLERS

The information encoded in the sets $\{X_k\}$, which are the level sets of the value function for the minimum time (algorithm 1) and robust minimum time (algorithm 2) problems, is sufficient to construct a controller. We present two alternatives below for the robust case.

4.1 One step model predictive controller

The controller determines, firstly, the minimum integer k such that the current state $x \in X_k \setminus X_{k-1}$, secondly, determines the control

$$u(x) = \arg\min\{\|u\| \mid u \in \Omega,\ Fx + Gu \in X'_{k-1}\}$$

and, thirdly, applies this (constant) control (to the discrete or continuous time system) for time Δ, after which the process is repeated.

4.2 Simplex controller

This makes use of the following result:

Lemma 28.1 *Suppose, for $i = 0, 1, \ldots s$, that control $u_i(\cdot)$ steers the continuous time system from state x_i to state z_i in time T along a trajectory that satisfies both state and control constraints, and that $x \in S_1$, where $x_1 \ldots x_s$ and $z_1 \ldots z_s$ are affinely independent and simplices S_1 and S_2 are defined by:*

$$\begin{aligned} S_1 &:= \text{co}\,\{x_i \mid i \in \{1 \ldots s\}\} \\ S_2 &:= \text{co}\,\{z_i \mid i \in \{1 \ldots s\}\}. \end{aligned}$$

Then $x = \sum_{i=0}^{s} \mu^i x_i$ for some $\mu \in \Sigma_s$ and the control $u(\cdot) = \sum_{i=0}^{s} \mu^i u_i(\cdot)$ steers the system from state $x \in S_1$ to the state $z = \sum_{i=0}^{s} \mu^i z_i \in S_2$ in time T along a trajectory that satisfies both state and control constraints.

Each set polytope X_k is decomposed into a sequence $\{X_{kj} \mid j \in \{1 \ldots j_k\}\}$ of possibly overlapping simplices whose union covers $X_k \setminus X_{k-1}$. The simplex controller determines the simplex X_{kj} in which the current state x lies (where k is, as before, the least integer such that $x \in X_k \setminus X_{k-1}$), determines $\mu \in \Sigma_n$ such that $x = \sum_{i=0}^{n} \mu^i x_{kji}$, sets $u(x) = \sum_{i=0}^{n} \mu^i u_{kji}$, and applies the constant control $u(x)$ to the discrete or continuous time system for time Δ. Here x_{kji}, $i = \{0 \ldots n\}$ are the vertices of X_{kj}, and, for each i, u_{kji} steers the discrete time system from x_{kji} to X_{k-1}; these controls are obtained as a byproduct of algorithm 2. By Lemma 1, control $u(x)$ steers either system from state $x \in X_{km} \subset X_k$ to the set X_{k-1}.

4.3 Properties of the two controllers

Theorem 28.3 *(i) The robust one step model predictive and simplex controllers steer the perturbed discrete time system from any initial state $x \in X_j \subset X_N$ to X_0 in j steps, and then maintains the state in X_0, satisfying, for all k, the state and control constraints, and, (ii) there exists a $\delta > 0$ such that the controllers steer the perturbed continuous time system from any initial state $x \in X_j \subset X_N$ to $X_0 + N(0, \delta)$ in time $j\Delta$, and then maintains the state in $X_0 + N(0, \delta)$, satisfying, for all t, the control constraints, and, for all sampling instants $k\Delta$, the state constraint.*

Acknowledgements

This work was supported by the National Science Foundation under grant ECS-93-12922.

REFERENCES

[1] W. H. Kwon, A. M. Bruckstein, and T. Kailath. Stabilizing state-feedback design via the moving horizon method. *International Journal of Control*, 37(3):631–643, 1983.

[2] S.S. Keerthi and E.G. Gilbert. Optimal, infinite horizon feedback laws for a general class of constrained discrete time systems. *Journal of Optimization Theory and Applications*, 57:265–293, 1988.

[3] D. Q. Mayne and H. Michalska. Receding horizon control of non-linear systems. *IEEE Transactions on Automatic Control*, 35(5):814–824, 1990.

[4] H. Michalska and D. Q. Mayne. Robust receding horizon control of constrained nonlinear systems. *IEEE Transactions on Automatic Control*, 38:1623–1632, 1993.

[5] J.B. Rawlings and K.R. Muske. The stability of constrained receding horizon control. *IEEE Transactions on Automatic Control*, AC-38:1512–1516, 1993.

[6] E. G. Gilbert and K. T. Tan. Linear systems with state and control constraints: the theory and application of maximal output admissible sets. *IEEE Transactions on Automatic Control*, AC-36:1008–1020, 1991.

[7] A. R. Teel. Global stabilization and restricted tracking for multiple integrators with bounded controls. *Systems & Control Letters*, 18:165–171, 1992.

[8] H. J. Sussmann, E. D. Sontag, and Y. Yang. A general result on the stabilization of linear systems using bounded controls. *IEEE Transactions on Automatic Control*, 39(12):2411–2425, December 1994.

[9] J. S. Shamma. Construction of nonlinear feedback for ℓ_1–optimal control. In *Proceedings of the 33rd IEEE Conference on Decision and Control*, pages 40–45, Lake Buena Vista, Florida, December 1994.

[10] D. Q. Mayne and W. R. Schroeder. Nonlinear control of constrained dynamic systems. *International Journal of Control*, 60:1035–1043, 1994.

[11] D. Q. Mayne and W. R. Schroeder. Nonlinear control of constrained linear systems: Regulation and tracking. In *Proceedings 33rd IEEE Conference on Decision and Control*, pages 2370–2375, Lake Buena Vista, Florida, December 1994.

[12] D. Q. Mayne and W. S. Schroeder. Nonlinear control of linear systems with state and control constraints. In *Proceedings of the IFAC symposium on nonlinear control systems design*, Lake Tahoe, California, June 1995.

[13] D. Q. Mayne and W. S. Schroeder. Robust nonlinear control of constrained linear systems. Technical Report UCD-ECE-SCR-95/3, College of Engineering, University of California, Davis, July 1995.

[14] J. P. Aubin. *Viability theory*. Systems & control: foundations & applications. Birkhauser, Boston, Basel, Berlin, 1991.

29

A SYSTEM THEORY FOR PRODUCTION LINES

D. Jacobs*, C.-T. Kuo**, J.-T. Lim**, and S.M. Meerkov**

*Ford Motor Company, EFHD, Rawsonville, Box 922
Ypsilanti, MI 48197

**Department of Electrical Engineering and Computer Science
University of Michigan, Ann Arbor, MI 48109-2122

Dedicated to Tom Kailath, with respect and admiration.

ABSTRACT

A system-theoretic approach to serial production lines is described. The approach is based on the notion of Improvability. A production system is called improvable under constraints if limited resources involved in its operation can be redistributed so that a performance index is improved. Unconstrained improvability is a problem of determining the operation the improvement of which leads to the largest improvement of the production system as a whole. In this work, criteria for improvability, both constrained and unconstrained, are derived. These criteria, referred to as Indicators of Improvability, can be applied using the off-line and, more importantly, real-time data. Based on the indicators obtained, production line personnel could design and implement various continuous improvement projects. As an illustration, an application of this approach to an automotive component plant is briefly reviewed.

1 INTRODUCTION AND PROBLEM FORMULATION

1.1 Manufacturing Considerations

Production system is a set of machines and material handling devices designed to manufacture a product or a component of a finished product. Production systems play an important role in any manufacturing organization. No manufacturing enterprise can be successful without efficient production lines. Unfortunately, however, not much rigorous engineering knowledge is available for their analysis and design. As a result, the field is advanced by manufacturing gurus (Deming, Taguchi, Goldratt) and by discoveries of practitioners (Henry Ford's serial assembly line or Taichi Ohno's just-in-time operation). The lack of precise engineering knowledge may be responsible for the fact that in many practical situations manufacturing systems produce at a rate of about a half of their nominal production rate. The main reason for this is the machines breakdowns. A breakdown of a machine leads to a starvation of the downstream machine and to a blockage of the upstream machine. The material handling system, through its buffers, is supposed to attenuate these perturbations. However, when the knowledge of the laws that govern the behavior (statics and dynamics) of a production system is missing, design of an efficient material handling system becomes a matter of intuition and luck. Quite often, one runs out of luck, and as a result a production system capable of producing, say, 10 parts/min produces at the rate of 5 parts/min. It is reasonable, therefore, to expect that the development of rigorous engineering methods for production systems analysis and design could facilitate economic gains in many manufacturing facilities.

1.2 System-Theoretic Considerations

The properties of traditional plants considered within the scope of Systems and Control are well understood and widely utilized. The notions of steady states and their stability, properties of controllability and observability – all are practical tools for analysis and design of electrical, mechanical, chemical, and other engineering systems. On the other hand, system-theoretic properties of production systems are virtually unknown. In this paper we describe and analyze one of such properties – the property of Improvability. Roughly speaking, a production system is improvable under constraints if limited resources involved in its operation can be re-distributed so that a performance index is improved. Unconstrained improvability has as its goal the determination of a machine which is the main impediment in the operation of the system as a whole, with a subsequent assignment of additional resources to this machine. In both of these situations, we derive the so-called Indicators of Improvability. Based on either off-line or, more importantly, on-line data, these indicators permit the production line personnel to design and put in place various continuous improvement projects.

Improvability is somewhat similar to optimality. The emphasis, however, is different. In the optimality, one is interested in finding conditions for the best possible system performance. Given the lack of precise information available on the factory floor, this may be both unnecessary and impossible. Just determining whether the system can be improved and indicating steps leading to an improvement may be sufficient. This information is provided by the Indicators of Improvability, the main outcome of this work.

1.3 Improvability under Constraints

Consider a production system with machines $m_i, i = 1, ..., M$, and buffers $b_j, j = 1, ..., B$, connecting the machines according to a technologically defined topology. Assume that the machines are subject to random breakdowns and let $p_i, i = 1, ..., M$, be the average production rate of m_i in isolation (i.e. when the machine operates alone and parts to work on are available at any point in time). Let $N_i, i = 1, ..., B$, be the capacity of b_i. The average production rate of the system is, then, a function of these parameters:

$$PR = PR(p_1, ..., p_M, N_1, ..., N_B).$$

Constraints on the resources available are introduced as follows:

$$\sum_{i=1}^{B} N_i = N^*, \qquad \prod_{i=1}^{M} p_i = p^*.$$

For the obvious reasons, the first of these relationships is referred to as the Work-in-Process (WIP) constraint. The second can be called the Workforce (WF) constraint. Indeed, if it is assumed that the isolation production rate of a machine depends on the personnel assigned to this operation (both machine operators and skilled labor for repairs and maintenance) and the total number of people assigned to the production system as a whole is fixed, the effect of the workforce constraint can be conceptualized in the form given above. Note that a formalization in the form of a summation, rather than a multiplication, is also possible. Since in some cases p_i's represent a probability (see Section 2 below), the product form seems more natural.

Definition 29.1 *A production system is called* **improvable with respect to WIP** *if there exists a sequence $N_1^*, ..., N_B^*$ such that $\sum_{i=1}^{B} N_i^* = N^*$ and*

$$PR(p_1, ..., p_M, N_1^*, ..., N_B^*) > PR(p_1, ..., p_M, N_1, ..., N_B).$$

Definition 29.2 *A production system is called* **improvable with respect to WF** *if there exists a sequence $p_1^*, ..., p_M^*$ such that $\prod_{i=1}^{M} p_i^* = p^*$ and*

$$PR(p_1^*, ..., p_M^*, N_1, ..., N_B) > PR(p_1, ..., p_M, N_1, ..., N_B).$$

Definition 29.3 *A production system is called* **improvable with respect to WIP & WF simultaneously** *if there exist sequences $N_1^*, ..., N_B^*$ and $p_1^*, ..., p_M^*$ such that $\sum_{i=1}^{B} N_i^* = N^*$, $\prod_{i=1}^{M} p_i^* = p^*$, and*

$$PR(p_1^*, ..., p_M^*, N_1^*, ..., N_B^*) > PR(p_1, ..., p_M, N_1, ..., N_B).$$

The first problem addressed in this paper is: Given a production system, find Indicators of Improvability under constraints that can be implemented using real-time data.

1.4 Unconstrained Improvability

When a system is no longer improvable under constraints or when resource re-allocations cannot be carried out, further improvements require additional resources, i.e., relaxation of the constraints. The question arises: Which of the p_i's and N_i's should be increased so that the most benefits are obtained for the system as a whole? To formalize this question, introduce

Definition 29.4 *Machine i is the* **bottleneck machine** *if*

$$\frac{\partial PR(p_1,...,p_M,N_1,...,N_B)}{\partial p_i} > \frac{\partial PR(p_1,...,p_M,N_1,...,N_B)}{\partial p_j}, \forall j \neq i.$$

Definition 29.5 *Buffer i is the* **bottleneck buffer** *if $\forall j \neq i$*

$$PR(p_1,...,p_M,N_1,...,N_i+1,...,N_B) > PR(p_1,...,p_M,N_1,...,N_j+1,...,N_B).$$

It should be pointed out that the worst machine in isolation and the smallest buffer are not necessarily the bottlenecks (see Section 4). Therefore, the second problem addressed in this work is: Given a production system, identify the bottleneck machines and buffers, again using real-time data.

1.5 Potency of the Material Handling System

As was mentioned above, production systems consist of machines and buffers or, more generally, material handling system (MHS). For a fixed cycle time (i.e., the time necessary to carry out an operation), the potency of the machines is characterized by their reliability. This is measured by the average up- and down-time. The potency of the MHS in most cases is not monitored and, moreover, it is not clear how this potency could be measured. Below we introduce a possible characterization.

Definition 29.6 *MHS is* **weakly potent** *if the machine with the smallest isolation production rate is the bottleneck; otherwise, MHS is* **not potent**. *MHS is* **potent** *if it is weakly potent and, in addition, production rate of the system is sufficiently close to that of the slowest machine. MHS is* **strongly potent** *is it is potent and this performance is achieved using the smallest possible total buffer capacity N^*.*

Based on the bottleneck analysis described in Section 4, this paper gives a tool for a characterization of the potency of material handling systems.

Improvability theory developed to-date provides solutions to the problems mentioned above. The goal of this paper is to describe some of the solutions and to outline an application at an automotive component plant.

The remainder of this paper is organized as follows: In Section 2, mathematical preliminaries are described. Sections 3 and 4 are devoted to the constrained and unconstrained improvability, respectively. The results of an application are reviewed in Section 5. The conclusions are formulated in Section 6. Due to space limitations, no proofs are included; they can be found in [1] - [3].

2 PRELIMINARIES

2.1 Model

Although the problems formulated above are of importance for many production systems, results obtained to-date address, mostly, the simplest but important class – serial production lines. Many models for such lines have been described in the literature (see, for instance, [4] - [5]). The following model is considered throughout this work:

(i) The system consists of M machines arranged serially, and $M-1$ buffers separating each consecutive pair of machines.

(ii) The machines have identical cycle time T_c. The time axis is slotted with the slot duration T_c. Machines begin operating at the beginning of each time slot.

(iii) Each buffer is characterized by its capacity, $N_i < \infty, \;\; 1 \leq i \leq M-1$.

(iv) Machine i is starved during a time slot if buffer $i-1$ is empty at the beginning of the time slot. Machine 1 is never starved.

(v) Machine i is blocked during a time slot if buffer i has N_i parts at the beginning of the time slot and machine $i+1$ fails to take a part during the time slot. Machine M is never blocked.

(vi) Machine i, being neither blocked nor starved during a time slot, produces a part with probability p_i and fails to do so with probability $q_i = 1 - p_i$. Parameter p_i is referred to as the **production rate** of machine i **in isolation.**

2.2 Production Rate Estimate

Production rate of the line (i) - (vi) cannot be evaluated in the closed form. Therefore, below we give a recursive procedure that is proven to converge to a small neighborhood of the sought production rate.

Introduce

$$\tilde{p}_i^f = \text{Prob}(m_i \text{ produces a part during a time slot } | \text{ it is not blocked at the beginning of this time slot}),$$

$$\tilde{p}_i^b = \text{Prob}(m_i \text{ produces a part during a time slot } | \text{ it is not starved at the beginning of this time slot}).$$

Since m_M is never blocked and m_1 is never starved,

$$PR(p_1, ..., p_M, N_1, ..., N_{M-1}) = \tilde{p}_M^f = \tilde{p}_1^b.$$

It turns out that $\tilde{p}_M^f$ and $\tilde{p}_1^b$ can be evaluated using the following recursive procedure:

$$\begin{aligned} p_i^b(s+1) &= p_i[1 - Q(p_{i+1}^b(s+1), p_i^f(s), N_i)], \quad 1 \le i \le M-1, \\ p_i^f(s+1) &= p_i[1 - Q(p_{i-1}^f(s+1), p_i^b(s+1), N_{i-1})], \quad 2 \le i \le M, \\ p_1^f(s) &= p_1, \qquad p_M^b(s) = p_M, \qquad s = 1, 2, 3, ..., \end{aligned} \tag{29.1}$$

with initial conditions

$$p_i^f(0) = p_i, \qquad i = 1, ..., M,$$

where

$$Q(x, y, N) = \begin{cases} \frac{(1-x)(1-\alpha)}{1-\frac{x}{y}\alpha^N} & x \neq y \\ \frac{1-x}{N+1-x} & x = y \end{cases}, \qquad \alpha = \frac{x(1-y)}{y(1-x)}.$$

Lemma 29.1 *Recursive procedure (2.1) is convergent and, therefore, the following limits exist:*

$$p_i^f = \lim_{s\to\infty} p_i^f(s), \qquad p_i^b = \lim_{s\to\infty} p_i^b(s), \qquad i = 1, ..., M.$$

In addition,

$$p_1^b = p_M^f.$$

Proof: Is based on a monotonicity argument and can be found in [1].

Define the production rate estimate as follows:

$$PR_{est}(p_1, ..., p_M, N_1, ..., N_{M-1}) = p_M^f = p_1^b.$$

We show below that this estimate is close to the real production rate $\tilde{p}_M^f = \tilde{p}_1^b$. Consider the joint steady state probability $X_{i,...,j}(h_i, ..., h_j)$ that the consecutive buffers $i, i+1, ..., j$, $1 \le i < j \le M-1$, contain $h_i, h_{i+1}, ..., h_j$ parts respectively. In general, one cannot expect that these joint probabilities are close to the product of their marginal distributions, i.e. $X_{i,...,j}(h_i, ..., h_j) \neq X_i(h_i)X_{i+1,...,j}(h_{i+1}, ..., h_j)$, where $X_i(h_i)$ is the probability that the i-th buffer contains h_i parts. It turns out, however, that for certain values of $h_i, h_{i+1}, ..., h_j$, related to blockages and starvations, they are indeed close. Specifically, define

$$\begin{aligned} \delta_{i,j}(b) &= |X_{i,...,j}(0, b, N_{i+2}, ..., N_j) - X_i(0)X_{i+1,...,j}(b, N_{i+2}, ..., N_j)|, \\ \delta^{i,j}(a) &= |X_{i,...,j}(a, N_{i+1}, ..., N_j) - X_i(a)X_{i+1,...,j}(N_{i+1}, ..., N_j)|, \\ \delta &= \max_{i,j} \max_{a,b} \{\delta_{i,j}(b), \delta^{i,j}(a)\}. \end{aligned} \tag{29.2}$$

Then, as it follows from extensive numerical experimentation, δ is always small ($\delta << 1$.) An illustration is given in Table 29.1 for several lines with $N_i = 3$, $i = 1, 2, 3$. At present, we do not have an analytical proof that $\delta << 1$, although we believe that such a proof is possible. Therefore, we formulate this as

Numerical Fact 2.1: *For serial production lines defined by (i)-(vi),*

$$\delta << 1.$$

Table 29.1 Behavior of δ and estimation error

p_1	p_2	p_3	p_4	δ	Error
0.80	0.80	0.80	0.80	0.0073	0.0008
0.70	0.80	0.70	0.80	0.0233	0.0047
0.70	0.90	0.70	0.90	0.0568	0.0144
0.60	0.99	0.99	0.60	0.1181	0.0283
0.99	0.60	0.60	0.99	0.0083	$2*10^{-7}$

It should be pointed out that δ seem to be small only in the so-called "open" serial lines, such as defined by (i) - (vi). In the so-called "closed" lines, where the first machine could be starved and the last blocked, δ may be of the order 1.

Theorem 29.1 *For serial production lines (i) - (vi),*

$$|\tilde{p}_i^f - p_i^f| \approx |\tilde{p}_i^b - p_i^b| \approx \mathcal{O}(\delta), \quad i = 1, ..., M.$$

Therefore,

$$Error = |PR_{est} - PR| \approx \mathcal{O}(\delta).$$

Proof: Is based on a perturbation theory for Markov chains and is given in [1].

The analysis of the improvability property described in this work is carried out in terms of the PR_{est}.

3 IMPROVABILITY UNDER CONSTRAINTS

3.1 Improvability with Respect to WF

Theorem 29.2 *Serial production line (i)-(vi) is unimprovable with respect to WF if and only if*

$$p_i^f = p_{i+1}^b, \quad i = 1, ..., M-1. \tag{29.3}$$

Proof: Is based on analysis of the equilibria of the recursive procedure (2.1); see [2] for details.

Improvability Indicator 3.1: *Condition (3.1) is satisfied and, therefore, the line is unimprovable with respect to WF if each buffer is, on the average, close to being half full in the following sense:*

$$\mathrm{E}[h_i] = \frac{N_i}{2} \frac{N_i + 1}{N_i + 1 - p_i^f} + \mathcal{O}(\delta) \approx \frac{N_i}{2}, \quad i = 1, ..., M-1, \tag{29.4}$$

where h_i is the steady state occupancy of the i-th buffer and $E[\cdot]$ denotes the expectation.

Proof: See [2].

An interpretation of this Improvability Indicator can be given as follows: Buffer b_i protects m_i from blockages and m_{i+1} from starvations. From the point of view of m_i, buffer b_i should be empty; from the point of view of m_{i+1}, buffer b_i should be full. The compromise is – buffer half full. Note that although the speeds of the machines affect this relationship (through p_i^f), this effect is insignificant for any reasonable value of N_i's.

3.2 Improvability with Respect to WF and WIP Simultaneously

Theorem 29.3 *Let N^* be an integer multiple of $M-1$. Then serial production line (i)-(vi) is unimprovable with respect to WF & WIP simultaneously if and only if (3.1) takes place and, in addition,*

$$p_i^f = p_i^b, \quad i = 2, ..., M-1. \tag{29.5}$$

Proof: Is based on analysis of the equilibria of the recursive procedure (2.1); see [2] for details.

To formulate a real-time based improvability indicator that follows from (3.3), introduce the probabilities of **manufacturing blockage** and **manufacturing starvation**:

$$mb_i = \text{Prob}\ (\{m_i \text{ is up during a time slot}\} \cap \{b_i \text{ is full at the beginning of this slot}\} \cap \{m_{i+1} \text{ fails to take a part from } b_i \text{ at the beginning of this slot}\}),$$

$$ms_i = \text{Prob}\ (\{m_i \text{ is up during a time slot}\} \cap \{b_{i-1} \text{ is empty at the beginning of this slot}\}).$$

Improvability Indicator 3.2: *Conditions (3.1) and (3.3) are satisfied and, therefore, the line is unimprovable with respect to WF and WIP simultaneously if each buffer is half full in the sense of (3.2) and, in addition, manufacturing blockage and manufacturing starvation of each machine are close to each other in the following sense:*

$$|mb_i - ms_i| \approx \mathcal{O}(\delta), \quad i = 2, ..., M-1. \tag{29.6}$$

Proof: See [2].

An interpretation of (3.4) is as follows: Buffers b_i and b_{i+1} serve to protect m_i from starvation and blockages, respectively. If m_i is more often blocked than starved, b_{i-1} can be reduced and b_i increased until the compromise is reached – blockages equal starvations.

Note, that using the recursive procedure (2.1), mb_i and ms_i can be evaluated as follows:

$$\begin{aligned} mb_i &= p_i Q(p_{i+1}^b, p_i^f, N_i), \qquad i = 1, ..., M-1, \\ ms_i &= p_i Q(p_{i-1}^f, p_i^b, N_{i-1}), \qquad i = 2, ..., M. \end{aligned} \tag{29.7}$$

3.3 Improvability with Respect to WIP

Theorem 29.4 *Serial production line (i)-(vi) is unimprovable with respect to WIP if and only if the quantity*

$$\min_{i=1,\ldots,M} p_i \left(\min \left\{ \frac{p_i^f}{p_i^b}, \frac{p_i^b}{p_i^f} \right\} \right) \tag{29.8}$$

is maximized over all sequences $N_1, ..., N_{M-1}$ such that $\sum_{i=1}^{M-1} N_i = N^*$.

Proof: Is based on analysis of the equilibria of the recursive procedure (2.1); see [2] for details.

Unfortunately, we do not have either a good interpretation or a real-time based reformulation of (3.6). Therefore, based on the insight provided by (3.4), we formulate

Improvability Indicator 3.3: *Serial production line (i) - (vi) is unimprovable with respect to WIP if*

$$\max_{i=2,\dots,M-1} \frac{1}{p_i^2}|mb_i - ms_i| \tag{29.9}$$

is minimized over all sequences N_i, $i = 1, \dots, M-1$, *such that* $\sum_{i=1}^{M-1} N_i = N^*$.

It turns out that the performance of the system designed according to (3.6) and (3.7) is practically the same. Indeed, the following has been arrived at through discrete event simulations:

Numerical Fact 3.1: *Let* $N_1^*, \dots, N_{M-1}^*$ *be the sequence which maximizes (3.6). Let* $N_1^{**}, \dots, N_{M-1}^{**}$ *be the sequence which minimizes (3.7). Then*

$$|PR_{est}(p_1, .., p_M, N_1^*, .., N_{M-1}^*) - PR_{est}(p_1, .., p_M, N_1^{**}, .., N_{M-1}^{**})| =: \delta_A << 1.$$

An illustration of this statement for $N^* = 10$ is given in Table 29.2.

Buffer capacity assignments obtained according to Improvability Indicator 3.3 has been compared with that obtained using the Goldratt's Theory of Constraints [6]. According to this theory, all available buffer capacity should be allocated to protect the slowest machine. The result of this comparison is given in Table 29.3. Thus, due to its qualitative nature, Improvability Theory approach results in a more efficient buffer allocation.

Table 29.2 Illustration of Numerical Fact 3.1

p_1	p_2	p_3	p_4	PR_{est}^*	δ_A
0.80	0.80	0.80	0.80	0.7211	0
0.70	0.80	0.70	0.80	0.6496	0
0.70	0.90	0.70	0.90	0.6795	0.0016
0.60	0.99	0.99	0.60	0.6000	0.0024
0.99	0.60	0.60	0.99	0.5679	0

4 UNCONSTRAINED IMPROVABILITY

4.1 Bottlenecks in Unimprovable Systems

Contrary to conventional wisdom, a machine with the smallest isolation production rate is not necessarily the bottleneck in the sense of Definition 29.4.

Table 29.3 Performance comparison between Improvability Theory and Theory of Constraints for the system with $M = 11, p_i = 0.8,\ i \neq 6, p_6 = 0.6, N^* = 60$

Results	**Improvability theory solution**	**Theory of Constraints solution**	
Buffers distribution	$N_i = 6,\ i \neq 6, 10,$ $N_6 = 7,\ N_{10} = 5$	$N_5 = 51,$* $N_i = 1,\ i \neq 5$	$N_5 = N_6 = 26,$** $N_i = 1,\ i \neq 5, 6$
PR_{est}	0.599	0.427(29% less)	0.49(18% less)

*: Allocate the buffering capacity in front of the bottleneck only
**: Allocate the buffering capacity on both sides of the bottleneck

Table 29.4 Bottleneck example

i	1	2	3	4
p_i	0.80	0.83	0.77	0.80
N_i	2	2	5	—
$\frac{\partial PR_{est}}{\partial p_i}$	0.369	0.452	0.443	0.022

An illustration is given in Table 29.4 where the fastest, rather than the slowest, machine is the bottleneck

The situation, however, is different in unimprovable lines:

Theorem 29.5 *In serial production lines (i) - (vi) unimprovable with respect to WF or WF and WIP simultaneously, the following relationship holds:*

$$p_i \frac{\partial PR_{est}}{\partial p_i} = const, \; i = 1, ..., M. \tag{29.10}$$

Proof: Follows from the algebraic properties of the $PR(p_1, ..., p_M, N_1, ...N_{M-1})$ when (3.1) is satisfied; see [2] for details.

Improvability Indicator 4.1: *A line unimprovable with respect to WF or WIP and WF simultaneously always has as its bottleneck the machine with the smallest isolation production rate.*

Proof: Follows immediately from (4.1).

The situation with bottleneck buffers in unimprovable systems is similar to the above. Due to the space limitations, we omit it from this manuscript; pertinent results can be found in [2].

4.2 Bottlenecks in Improvable Systems

Theorem 29.6 *Consider a serial production line (i) - (vi) with $M = 2$. If $mb_1 > ms_2$, then*

$$\frac{\partial PR_{est}}{\partial p_1} < \frac{\partial PR_{est}}{\partial p_2},$$

i.e. m_2 is the bottleneck. If $mb_1 < ms_2$,

$$\frac{\partial PR_{est}}{\partial p_1} > \frac{\partial PR_{est}}{\partial p_2},$$

i.e. m_1 is the bottleneck.

Proof: Is based on the properties of function $Q(x, y, N)$ defined in (2.1); see [3] for details.

At present, we do not have a proof of a generalization of this Theorem for an arbitrary M. However, based on extensive numerical investigation, the following rule is shown to take place:

Improvability Indicator 4.2: *If $mb_i > ms_{i+1}$, the bottleneck machine is downstream of m_i. If $mb_i < ms_{i+1}$, the bottleneck machine is upstream of m_i. If, according to this rule, the system has more than one bottleneck, the primary one is the machine with the largest severity defined as*

$$S_i = (mb_{i-1} + ms_{i+1}) - (mb_i + ms_i), \quad i \in [\text{set of bottlenecks}]$$

$$(\text{if } i = 1, S_1 = ms_2 - bm_1; \text{ if } i = M, S_M = mb_{M-1} - ms_M).$$

This rule is illustrated in Figures 29.1 - 29.3. The numbers in the circles and rectangles represent the p_i's and N_i's, respectively. The ms_i's and the mb_i's are shown below each machine. The last row shows the $\frac{\partial PR_{est}}{\partial p_i}$ evaluated numerically using discrete event simulations.

In Figures 29.1 - 29.3, the arrows are obtained by placing a line within the sign of inequality between mb_i and ms_{i+1}. According to Improvability Indicator 4.2, the machine which has the arrow pointed towards it from both sides (or one side for the first and the last machine) is the bottleneck. Thus, in Figures 29.1 and 29.2, m_1 and m_3 are the bottlenecks, respectively. In Figure 29.3, there are three bottlenecks: m_1, m_3, and m_9. The severity of each is 0.12, 0.31, and 0.48, respectively. Thus, the primary bottleneck is m_9. The last row of numbers in Figures 29.1 - 29.3 support these conclusions. Note that we obtained these conclusions without using either p_i's, or N_i's, or even PR. The only data necessary for this analysis is mb_i's and ms_i's, which are often available in real time. Thus, this approach allows one to find not only stationary bottlenecks but also bottlenecks changing in time.

Improvability Indicator 4.2 can be used to characterize the potency of a MHS. Indeed, the MHS if Figure 29.1 is not potent; that of Figure 29.2 is potent; and that of Figure 3 is weakly potent. Due to space limitation, we omit the analysis of strong potency. This can be found in [3].

The results described in Section 3 and 4 form a foundation of the Improvability Theory for serial production lines. A number of applications of this approach has been carried out during the last 3 years. One of these applications is briefly described below.

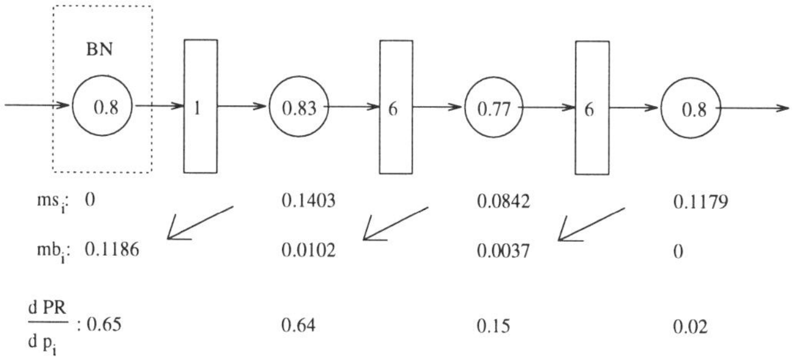

Figure 29.1 Improvability Indicator 4.2 for system with $M = 4$ and m_1 as a bottleneck

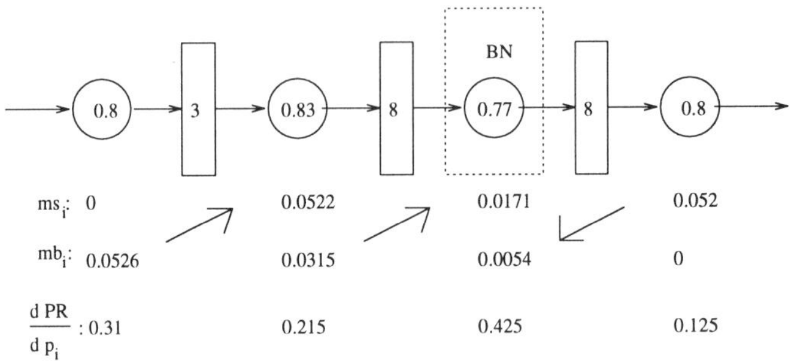

Figure 29.2 Improvability Indicator 4.2 for system with $M = 4$ and m_3 as a bottleneck

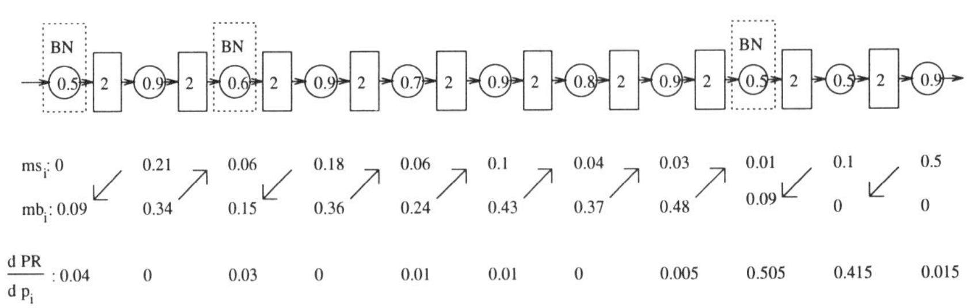

Figure 29.3 Improvability Indicator 4.2 for system with $M = 11$ and multiple bottlenecks

5 AN APPLICATION AT AN AUTOMOTIVE COMPONENT PLANT

5.1 System Description

The production line considered is shown in Figure 29.4. The material handling system is structured as a set of palletized circular conveyers connecting each pair of consecutive operations and providing buffering capabilities (Figure 29.5).

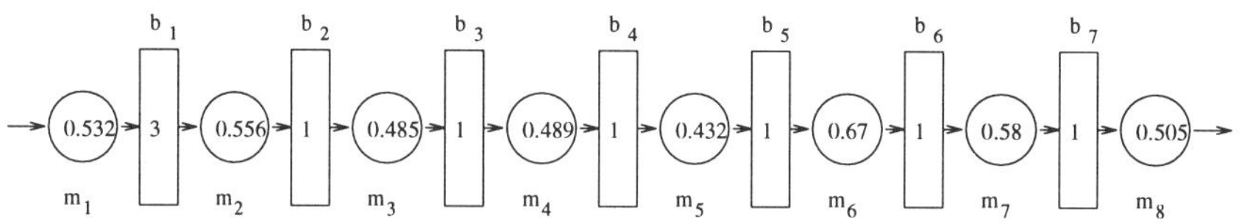

Figure 29.4 Serial production line

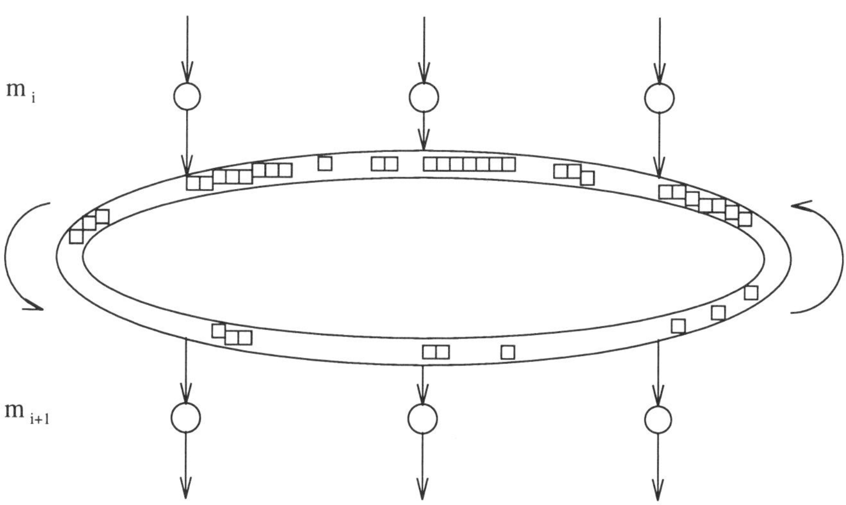

Figure 29.5 Material handling

Parameters of the machine and buffers identified are shown in Figure 29.4 (see [7] for details). Based on these parameters and using the recursive procedure (2.1), the production rate of the line has been evaluated to be 0.2068 which is less than 50% of the production rate of the slowest machine (m_5). Thus, the system incurs substantial losses, due to a relative inefficacy of the material handling system. The goal of this continuous improvement project was to recover these losses by modifying the MHS.

5.2 Analysis

Table 29.5 shows the values of mb_i's and ms_i's calculated using (3.5). From these data and Improvability Indicator 4.2, it follows that m_5 is the bottleneck. To protect the bottleneck, increase b_4 and b_5 to $N_4 = N_5 = 5$. The resulting system has the production rate of 0.2683 and the new bottlenecks, as it follows from the data shown in Table 29.6, are m_3 and m_7 with the severities $S_3 = 0.2523$ and $S_7 = 0.1091$. Protecting the primary bottleneck, we increase b_2 and b_3 to $N_2 = N_3 = 5$. Resulting system has $PR = 0.3249$ with the bottleneck at m_8 (Table 29.7). Increasing now b_7 to $N_7 = 5$, we, finally, obtain $PR = 0.3882$ and the primary bottleneck at m_5 (Table 29.8). Since this production rate is within 10% of the slowest machine in isolation and this machine is the primary bottleneck, we conclude that the MHS arrived at is potent.

Table 29.5 Starvation and blockage of the system without the MHS improvement

i	1	2	3	4	5	6	7	8
ms_i	0	0.0042	0.0815	0.1494	0.1737	0.3978	0.3207	0.2982
mb_i	0.3252	0.3477	0.2365	0.1912	0.0862	0.161	0.1175	0

Bottleneck: m_5; Production rate = 0.2068

Table 29.6 Starvation and blockage of the system with $N_4 = N_5 = 5$

i	1	2	3	4	5	6	7	8
ms_i	0	0.01	0.1083	0.2173	0.1604	0.2511	0.2159	0.2367
mb_i	0.2634	0.2825	0.1392	0.0057	0.0048	0.2408	0.1525	0

Bottlenecks: m_3 ($S_3 = 0.2523$), m_7 ($S_7 = 0.1091$); Production rate = 0.2683

Table 29.7 Starvation and blockage of the system with $N_2 = N_3 = N_4 = N_5 = 5$

i	1	2	3	4	5	6	7	8
ms_i	0	0.0192	0.0025	0.0065	0.0057	0.0238	0.1032	0.1801
mb_i	0.2071	0.2194	0.1584	0.1597	0.1027	0.3331	0.1847	0

Bottleneck: m_8; Production rate = 0.3249

Table 29.8 Starvation and blockage of the system with $N_2 = N_3 = N_4 = N_5 = N_7 = 5$

i	1	2	3	4	5	6	7	8
ms_i	0	0.0369	0.0102	0.0225	0.023	0.1156	0.181	0.1168
mb_i	0.1438	0.1402	0.0884	0.0821	0.0219	0.2008	0.0157	0

Bottlenecks: m_5 ($S_5 = 0.1528$), m_7 ($S_7 = 0.1209$); Production rate = 0.3882

5.3 Implementation

A version of the system designed above has been implemented on the factory floor. A new robotic MHS has been installed to accommodate the buffering recommendations developed. At present, the line exhibits a satisfactory performance. More details can be found in [7].

6 CONCLUSIONS

The results reported here contribute to the system theory of production lines. Specifically, the Indicators of Improvability derived could be viewed as characteristics of the statics of a well designed production system. The properties of dynamics of such systems, however, remain completely unexplored.

REFERENCES

[1] D. Jacobs and S.M. Meerkov, "A system-theoretic property of serial production lines: Improvability," *Control Group Report No. CGR-93-01*, University of Michigan, Ann Arbor, MI , Jan. 1993. Also to appear in *Int. J. Syst. Sci*, 1995.

[2] D. Jacobs and S.M. Meerkov, "On the process of continuous improvement in production systems: Fundamental properties and guidelines," *Control*

Group Report No. CGR-93-10, University of Michigan, Ann Arbor, MI, July, 1993. Also to appear in *Math. Prob. in Eng.*, 1995.

[3] C.-T. Kuo, J.-T. Lim and S.M. Meerkov, "A system-theoretic characterization of bottlenecks in serial production lines," *Control Group Report No. CGR-95-05*, University of Michigan, Ann Arbor, MI, June,1995.

[4] J.A. Buzacott and J.G. Shanthikumar, *Stochastic Models of Manufacturing Systems*, Prentice Hall, N.Y., 1993.

[5] S.B. Gershwin, *Manufacturing Systems Engineering*, Prentice Hall, 1994.

[6] E. Goldratt and J Cox, *The Goal*, North River Press, N.Y., 1986.

[7] C.-T. Kuo, J.-T. Lim and S.M. Meerkov, "Improvability analysis of a machining transfer line: An application," *Control Group Report No. CGR-94-08*, University of Michigan, Ann Arbor, MI, Apr. 1994.

30

SIMULATED ANNEALING APPROACH TO GROUP TECHNOLOGY

Shashi M. Sharma and N. Viswanadham*

Department of Computer Science and Automation,
Indian Institute of Science, Bangalore 560 012, India
Email : shashi@mitl.soft.net

** Department of Computer Science and Automation*
Indian Institute of Science, Bangalore 560 012, India
Email : vishu@csa.iisc.ernet.in

Dedicated with admiration to Professor Thomas Kailath on the occasion of his sixtieth birthday.

ABSTRACT

In this paper, we use the simulated annealing approach to solve the group technology (GT) problem. It is well known that this problem belongs to the class of NP complete nonlinear integer programming problems. Several heuristic approaches have been reported in the literature but these do not work for the large scale problems. We report in this paper, a simulated annealing based approach that is very effective for solving large scale GT problems in terms of CPU time and the quality of the solution.

1 INTRODUCTION

Group technology (GT) problem involves clustering of machines that are dedicated to the manufacture of part families in one physical location to form the cells. GT is the central concept behind cellular manufacturing systems. The benefits of cellular manufacturing systems includes:

- Reduced set-up time and work in process inventory.
- Increased design and manufacturing efficiency.
- Reduced non-value adding flow of parts resulting in improvements of manufacturing lead time and increased throughput.

In order to successfully implement GT, the problem of machine grouping has to be solved together with the machine cell layout problem. In the machine grouping problem, one has to determine the composition of the machine cells and corresponding part families. In the facility layout problem, on the other hand one determines the arrangement of the machine cells in the factory such that the transportation time, i.e. the time spent by the material handling systems in moving parts across the machine is minimized. In this paper, we consider the GT problem only.

The concept of GT was introduced by Mitrafanov in late sixties [10]. The part family concept was introduced and extended by Burbridge [3]. McAuley [8] introduced the concept of identifying the machine cells by using single linkage clustering analysis. This method consists of constructing the similarity coefficient for each machine/machine pair, and then grouping together the pairs which are close in the similarity coefficient metric. This algorithm is essentially a greedy algorithm and gives only suboptimal solutions. Rajgopalan and Batra [12] formulated the family formation as a graph theoretic model and applied graph partitioning technique to form clusters. Constraints on utilization of machine cells and the machine cell size were also incorporated in their model.

The technique of generalized family formation was developed by McCormick et. al. [9] using a heuristic bond energy algorithm. This method was improved greatly by Gongware and Ham by eliminating tedious manual sorting procedure. King [5], King and Nakornchai developed [6] the rank-order clustering approach to GT. Another approach to generalized part family was developed by Ballakur and Steudel [2]. They attempt to minimize the intercellular movement of parts and maximize machine utilization under various constraints.

Recently, there has been lot of work done on the GT problem using neural network techniques. Moon and Chi [11] modeled a generalized part family problem as a constraint satisfaction network. Their method however, does not work well even for medium sized problems.

In this paper, we develop simulated annealing algorithm for generalized family problem which will take into account various technological and feasibility constraints. Our approach to the GT problem is to formulate it as an integer quadratic optimization problem. The objectives are to minimize the intercellular movement and hence the material handling cost and also to distribute the workload uniformly among all the machine cells.

The rest of the paper is organized as follows: In section 2, we formulate the generalized part family problem as integer quadratic programming. In section 3, we introduce simulated annealing algorithm and techniques to generate a feasible solution. In section 4, we apply simulated annealing method to various cell formation examples from the literature. In the last section, we discuss our computational experience and compare our method with various techniques available in the literature.

2 PROBLEM FORMULATION

Consider a manufacturing system consisting of n machines and producing m parts. The objective is to group n machines in to a distinct number, say B cells with the constraint that each machine resides in exactly one cell.

Let x_{ik} be a binary valued variable which takes a value 1 if i'th machine is in k'th cell and 0 otherwise. Let p_q be the production requirement of part q in the given period. The machines requirement for the various parts is represented by the $A = [a_{iq}]$ matrix, where a_{iq} is the processing time for part q on the i'th machine. If machine i is not required for processing part q, a_{iq} is assumed to be zero. Let T_i be the total available time on machine i in a given planning horizon.

One has to satisfy several engineering constraints while performing the cell grouping. Sometimes it may be necessary to have two machines of different type in the cell to enable smooth production of the required parts and yet other times it is necessary that two different machines are not placed in the same cell. To take care of such constraints, we define the relationship indicator matrix $R = [r_{ij}]$. If $r_{ij} = 1$ then the machines i and j are to be placed in the same cell and $r_{ij} = 0$ indicates that their is no special adjacency relationship between machines i and j. On the other hand, $r_{ij} = -1$ indicates that machines i and j are not to be placed in the same cell. In our discussion, we will limit ourselves to two values 0 and -1 of r_{ij}. The case $r_{ij} = 1$ can be converted to the standard formulation by treating the machine i and j as a composite machine.

We define $W = [w_{ij}]$, as a $n \times n$, symmetric matrix which is derived from the A matrix as follows

$$w_{ij} = \begin{cases} \sum_{q=1}^{m} (a_{iq} \wedge a_{jq}) p_q & i \neq j \\ 0 & i = j \end{cases} \tag{30.1}$$

wherer $\wedge$ denote the boolean AND operation defined by

$$a_{iq} \wedge a_{jq} = \begin{cases} 1 & \text{if } a_{iq} \neq 0 \text{ and } a_{jq} \neq 0 \\ 0 & \text{otherwise} \end{cases}$$

Thus w_{ij} represents the cumulative part flow between machines i and j. Naturally to minimize the inter cellular movement, the pair of machines with high value of w_{ij} are to be grouped in the same cell. One can define w_{ij} in many other ways as appropriate to the problem. The only restriction being that the resultant W matrix should be symmetric, in which case, the resultant cell formation will be the one which will have minimal interaction among the cells in terms of the w_{ij} metric.

We further define $V = [v_{iq}]$ the $(m \times n)$ matrix and $M = [m_{ji}]$ as an $(B \times m)$ average cell load matrix with components

$$v_{iq} = \frac{a_{iq} \times p_q}{T_i}$$

and

$$m_{jq} = \frac{\sum_{i=1}^{n} x_{ij} v_{iq}}{\sum_{i=1}^{n} x_{ij}}$$

The element v_{iq} is the fraction of the workload induced by part q on machine i, m_{jq} is the total load in cell j induced by part q and $\sum_{i=1}^{n} x_{ij}$ is the total number of machines in cell j.

With the above notation, we define the objective function Φ_1 as

$$\Phi_1(X) = \sum_{i=1}^{n} \sum_{j>i}^{n} \sum_{k=1}^{B} x_{ik}(1 - x_{jk}) w_{ij} \tag{30.2}$$

Notice that the term $x_{ik}(1 - x_{jk})$ is 1, if and only if the machine i is in k'th cell and the machine j is not. Hence, the objective function which is the sum of the above term multiplied with w_{ij}, over all the cells and machine represents the total inter-cellular movement of the parts.

If we consider $i = 1, \cdots, n$, as the nodes and w_{ij} as the weight of the edge between machine i and j of a graph, then the minimization of Φ_1 is reduced to B graph partitioning, which is known to be NP complete.

In order to ensure that the average workload in each cell is distributed uniformly, we define the objective function Φ_2 as

$$\Phi_2(X) = \sum_{i=1}^{n} \sum_{j=1}^{B} x_{ij} \sum_{q=1}^{m} (v_{iq} - m_{jq})^2 \tag{30.3}$$

The term $x_{ij} \sum_{q=1}^{m} (w_{iq} - m_{jq})^2$ in $\Phi_2(X)$ gives the difference between the the total load on machine i in cell j induced by all the parts, from the mean load of cell j. The objective function $\Phi_2(X)$ is the sum total of all such variations for all the cells and machines. Hence the minimization of $\Phi_2(X)$ will result in load leveling among all machine cells variation among machines.

Now we formulate a number of optimization problems essentially minimizing $\Phi_1(X) + \gamma\Phi_2(X)$ subject to certain feasibility and engineering constraints. The constant $\gamma > 0$ is used to scale the two objective functions appropriately. We consider fixed cell size and bounded cell size problem below

Case 1 Fixed Cell Size Problem

In this formulation the sizes of the cells(number of machines in the cell) are fixed. Let c_i be the size of the i'th cell. Then the optimization problem

$$\text{Minimize } \Phi_1(X) + \gamma\Phi_2(X)$$

subject to

$$x_{ik} = \{0, 1\}, \;\; \forall i = 1, \cdots n; \; k = 1, \cdots B \tag{30.4}$$

$$\sum_{k=1}^{B} x_{ik} = 1 \qquad \text{for } i = 1, \cdots, n \tag{30.5}$$

$$\sum_{i=1}^{n} x_{ik} = c_i \qquad \text{for } k = 1, \cdots, B \tag{30.6}$$

$$r_{ij} x_{ik} x_{jk} \leq 0 \qquad \forall i \neq j \tag{30.7}$$

gives the solution to the GT problem. The constraint (30.5) ensures that the machine i resides exactly in one cell. The second constraint (30.6) ensures that there are exactly c_k machines in k'th cell. The third constraints (30.7) imposes relationship constraint among machines i and j as dictated by R.

Case 2 Bounded Cell Size Problem

The cell size in this case is not fixed but has a lower bound $\bar{c_k}$ and upper bound $\bar{C}_k$. The GT problem can be formulated as a minimization problem:

$$\text{Minimize } \Phi_1(X) + \gamma\Phi_2(X)$$

subject to

$$x_{ik} = \{0, 1\}, \ \forall i = 1, \cdots n; \ k = 1, \cdots B \tag{30.8}$$

$$\sum_{k=1}^{B} x_{ik} = 1 \qquad \text{for } i = 1, \cdots, n \tag{30.9}$$

$$\bar{c_k} \leq \sum_{i=1}^{n} x_{ik} \leq \bar{C_i} \qquad \text{for } k = 1, \cdots, B \tag{30.10}$$

$$r_{ij} x_{ik} x_{jk} \leq 0 \qquad \forall i \neq j \tag{30.11}$$

The constraint (30.9) and (30.11) are similar to case 1. The constraint (30.10) in this case enforces the upper and lower bounds on the cells sizes.

In both the cases, it is assumed that the cell size is known *a priori*. The number of cells is usually not known in advance, however case 2 can be extended to general GT formulation by setting $B = m$, $\bar{c_i} = 0$ and $\bar{C_i} = m$. The value of m_{jq} is assumed to be zero if the number of machines in the cell j is zero. The number of variables in this formulation explodes to m^2.

The parts which are processed in a cell are grouped together as a single part family. The part q belongs to part family j which maximizes $\sum_i v_{iq} x_{ij}$ The exceptional machines can also be identified by using a threshold ϵ

$$EM = \{i : \sum_{j} a_{ij} > \epsilon \, \forall j \in PF_i\} \tag{30.12}$$

This information is used to generate efficient cell formation by machine duplication strategy or part contract strategy.

3 SIMULATED ANNEALING ALGORITHM AND GROUP TECHNOLOGY

A simulated annealing algorithm, has two kind of parameters

- Problem specific
- Generic

The problem specific parameters are

1. *State:* The state of the system is represented by the binary valued variables $x_{ij} : i = 1, \cdots, n; j = 1 \cdots, B$
2. *Neighborhood of the state:* The neighborhood of state S is the set of states in which the system can move from S with non-zero probability. For our problem the neighborhood of state S is the set $\mathcal{S}$ of state which are generated by swapping the cell of two machine at random or picking up one machine at random and putting it to some other cell. This is explained in detail in the next section.
3. *Cost of the state:* The cost of the state is the measure of the goodness of the state. We have already defined an appropriate cost function for our problem in the previous section.
4. *Initial State:* The initial state is chosen picking up a random state from the set of all possible feasible states.

The generic parameters of the simulated annealing are

- Initial value of the temperature T_0.
- Decreasing factor α .
- Stopping criterion.
- Length of the iteration L, at each value of the temperature.

For this problem, we use three different values for T_0, L and the linear cooling schedule parameter α.

3.1 Fast Annealing Algorithm

The key to successfully solve the optimization problem lies in its ability generate the next state and calculating the Δ function efficiently. These two issues are interconnected and the computationally efficient ways of performing these operations are considered below.

Generation of the Next State

The key issue here is the generation of the next state which is feasible, i.e, constraints on the cell size are satisfied. If we are not careful the annealing algorithm may settle down to a low cost but infeasible state. There are essentially two strategies to generate the next state depending upon the constraints of the cell size.

- **Rule 1** Pick two machines at random and swap their locations.
- **Rule 2** Pick at random a machine i from the set of machines and a cell j from the set of cells. Now transfer the machine i to the cell j provided that such a transfer will not lead to an infeasible state in terms of cell size constraints, i.e.
  (a) The number of machines in the cell where machine i is currently located , is strictly greater than the minimum number of machines permitted in that cell.
  (b) The number of machines in the cell j is strictly less than the maximum number of machines permitted in the cell j.

For Case 1, rule 1 is employed and for Case 2 we employ the combination of rule 1 and rule 2 to generate the next state. Its easy to see that both strategies generate the next state which satisfies the constraints on the size of the machine cells. The relationship constraints are maintained by checking explicitly for the feasibility. If the generated state does not satisfy the relationship constraint R the corresponding generate procedure is repeated again until a feasible state is produced a valid state. This causes additional burden on the annealing algorithm but our computational experience shows that if the relationship matrix is sparse then the additional time spent on the checking the relationship does not appreciably effect the efficiency of the annealing algorithm.

Implementation of Rule 1 and Rule 2 requires order $O(1)$ time. However they use data structure of size $O(b)$ and $O(n+b)$ respectively. The additional check of the relationship is done in of the order $O(n)$.

Calculation of Δ

The cost efficiency of the SA algorithm can be significantly increased by efficient calculation of Δ, the difference in the cost after the transition. A naive way of calculating Δ by calculating the cost function will run in order $O(n^2B+n^2mB)$ times. However the special structure of the generate function will enable us to calculate Δ in order $O(n+m^2)$ time. Since Δ is the sum of two terms Δ_1 and Δ_2, the difference in $\Phi_1(X)$ and $\Phi_2(X)$ after the transition, we will now consider the computation of Δ_1 and Δ_2 for the two cases.

Calculation of Δ_1

In general the computation of Δ_1 will depend upon the perturbation scheme used to generate the next state. The formulae for the two different schemes used here are

- For Rule 1

  Suppose machines p in cell k and q in cell l are swapped, then Δ is computed using

$$\Delta_1 = \sum_{i=1}^{n} x_{ip}w_{ik} - \sum_{i \neq l} x_{iq}w_{ik} + \sum_{i=1}^{n} x_{iq}w_{ip} - \sum_{i \neq k} x_{ip}w_{il} \tag{30.13}$$

- For Rule 2

  Let p be the machine and q be the bin chosen and the location of the machine p is cell k prior to the generate call. Then Δ is given by

$$\Delta_1 = \sum_{i=1}^{n} x_{ip}w_{ik} - \sum_{i \neq n} x_{iq}w_{ik} \tag{30.14}$$

Calculation of Δ_2

Computation of Δ_2 is more involved than Δ_1. In order to speed up the computation we maintain two additional data structures in form of matrix M^t and an array C^t at t'th iteration of the annealing algorithm where M^t is the matrix of m_{ij}'s and C^t contains the number of machines in each cells at t'th iteration. The procedures for computation of Δ_1 for Rule 1 and Rule 2 are:

- For Rule 1

  Let machine p in cell k and machine q in cell l are swapped at the t'th iteration of the annealing algorithm then Δ_2 is given by

$$\begin{aligned}\Delta_2 = & \sum_{j=1}^{m}(w_{pj} - m_{lj}^{t+1})^2 + \sum_{j=1}^{m}(w_{qj} - m_{kj}^{t+1})^2 \\ & - \sum_{j=1}^{m}(w_{pj} - m_{kj}^{t})^2 + \sum_{j=1}^{m}(w_{qj} - m_{lj}^{t})^2 \\ & + \sum_{i \neq p,q}^{n}\sum_{j=1}^{m}\left\{(w_{ij} - m_{kj}^{t+1})^2 - (w_{ij} - m_{kj}^{t})^2\right\} \\ & + \sum_{i \neq p,q}^{n}\sum_{j=1}^{m}\left\{(w_{ij} - m_{lj}^{t+1})^2 - (w_{ij} - m_{lj}^{t})^2\right\}\end{aligned} \tag{30.15}$$

The matrix M and array C is updated after the iteration only if the generated state is accepted as the new state, by the following formulae

$$m_{kj}^{t+1} = m_{qj}^{t} + \frac{w_{lj} - w_{qj}}{c_k^{t+1}}$$

$$m_{lj}^{t+1} = m_{lj}^{t} + \frac{w_{qj} - w_{lj}}{c_l^{t+1}} \tag{30.16}$$

$$c_{k,l}^{t+1} = c_{k,l}^{t} \tag{30.17}$$

It is evident from the above formulae that the matrix M can be updated in the order $O(m)$ times and the array C is the order of $O(1)$ times. Thus the full computation of Δ_2 will be the order of $O(mn)$ times.

- For Rule 2.

  Let machine p which was in cell k is moved to the cell l then Δ_2 is given as

$$\Delta_2 = \sum_{j=1}^{m}(w_{pj} - m_{lj}^{t+1})^2 - \sum_{j=1}^{m}(w_{pj} - m_{qj}^{t})^2$$

$$+ \sum_{i \neq p,q}^{n}\sum_{j=1}^{m}\left\{(w_{ij} - m_{kj}^{t+1})^2 - (w_{ij} - m_{kj}^{t})^2\right\}$$

$$+ \sum_{i \neq p,q}^{n}\sum_{j=1}^{m}\left\{(w_{ij} - m_{lj}^{t+1})^2 - (w_{ij} - m_{lj}^{t})^2\right\} \tag{30.18}$$

(30.19)

The data structures M and C are updated as follows:

$$m_{qj}^{t+1} = m_{qj}^{t} - \frac{w_{pj}}{c_k^{t+1}}$$

$$m_{lj}^{t+1} = m_{lj}^{t} + \frac{w_{lj}}{c_l^{t+1}}$$

$$c_k^{t+1} = c_k^{t} - 1$$

$$c_l^{t+1} = c_l^{t} + 1 \tag{30.20}$$

As in the case of Δ_1 the total computation involved in calculating Δ_2 is of the order $O(mn)$

4 COMPUTATIONAL RESULTS

In order to demonstrate the efficiency of the method developed here, we solved a large number of problems with the number of machines ranging from 5–20 and parts ranging from 8–25. Favorable results were achieved in comparison with the methods described in [7] and [13]. The details can be obtained from the authors.

5 CONCLUSION

In this paper, we developed a methodology to solve the GT problem by using simulated annealing algorithm. We modeled the GT problem as quadratic integer optimization problem with linear inequality constraints. We have shown that the formulation is equivalent to multi-graph partitioning problem with constraints. We have presented an efficient method to calculate Δ, the difference in the cost function, to accelerate the annealing algorithm. We report that the annealing algorithm gives superior result even for the faster cooling schedule.

Our method can be easily extended to accommodate all the other variants of GT available in the literature. Our method is superior when applied to some large scale problems, however the preprocessing takes time of the order $O(Bn^2)$. It may be possible to obtain further computational saving by using by using special data structure for speeding up the preprocessing.

Acknowledgements

This work was supported by the Department of Science and Technology (India) and the Office of Naval Research under the project "Modelling Analysis and Control of Discrete Event Systems."

REFERENCES

[1] Aarts, E. and Korst, J., "Simulated annealing and boltzmann machine: A stochastic approach to combinatorial optimization with neural computing", John Wiley, New York, 1989.

[2] Ballakur, A. and H. J. Steudel, H.J., "A within-cell utilization based heuristic for designing cellular manufacturing systems", International Journal of Production Research 25, 1987, pp. 5, 639–665.

[3] Burbridge, J. L., "Production flow analysis", The Production Engineer 42, 1963, pp. 11, 742.

[4] CollinsR, E. and Stratmann, K. H., "Simulated annealing: An annotated bibliography", American Journal of Mathematical and Management Sciences 8, pp. 3, 209–207.

[5] King, J. R., "Machine-component grouping in production flow analysis: An approach using a rank order clustering algorithm", International Journal of Production Research 18, 1980, 213–219.

[6] King, J. R. and Nakornchai, V., " Machine-component group formation in group technology: Review and extension", International Journal of Production Research 20, 1982, pp. 2, 117–133.

[7] K. Ravi Kumar and Anthony Vannelli, *Efficient algortihms for grouping component-processor families*, Tech. Report RC 10636 (# 47695) 7/20/84, IBM Research Divison, 1984.

[8] J. McAuley, *Machine grouping for efficient production*, The Production Engineer **51** (1972), no. 2, 53–57.

[9] W. T. McCormick, P. J. Schweitzer, and T. W. White, *Problem decomposition andd data reorganozation by a clustering technique*, Operation Research **20** (1972), no. 992–1009.

[10] S. P. Mitrafanov, *Group technology in industry*, vol. 1 and 2, USSR: Mashinosroienie, 1983.

[11] Y. B. Moon and S. C. Chi, *Generalized part family formation using neural network techniques*, Journal of Manufacturing Systems **11**, no. 3, 149–159.

[12] R. Rajgopalan and J. L. Batra, *Design of cellular production sytems: A graph theoretic approach*, International Journal of Production Research **13**, no. 6, 567–579.

[13] V. Venugopal and T. T. Narendran, *Cell formation in manufacturing systems through simulated annealing: An experimental evaluation*, European Journal of Operation Research **63** (1992), 409–422.

PART V

SIGNAL PROCESSING

31

SPEECH RECOGNITION BY MACHINES

Bishnu S. Atal

Bell Laboratories
Murray Hill, NJ U.S.A.

Dedicated to Tom Kailath on the occasion of his sixtieth birthday.

ABSTRACT

Voice communication with machines is no longer a dream but a reality. The tremendous progress that has been accomplished has come about as a result of solving successfully some of the fundamental problems caused by the immense variability present in the speech signal. This paper will discuss important issues in automatic recognition of speech, the major advances made in this field, the current state of the technology, and future developments.

1 INTRODUCTION

Speech is a natural form of communication for humans and thus the problem of understanding or "recognizing" speech by machines has challenged scientists for many years. We do not yet understand in any detail how humans understand speech. But considerable advances in automatic speech recognition and understanding by machines have taken place [1]-[2].

Research in automatic speech recognition (ASR) since the 1970s has produced solutions for increasingly difficult tasks, from the correct recognition of a few isolated words from a single speaker to recognition of fluent speech from virtually any speaker. Progress in automatic recognition of speech is continuing and the research frontiers are shifting towards the solution of an even harder problem – unconstrained dialogue with machines. The availability of high-speed processors and high density memories at reasonable cost in digital computers, along with large databases of recorded speech, has made it possible to develop sophisticated signal representations, pattern-matching techniques, and language models in support of automatic speech recognition.

2 COPING WITH ACOUSTIC VARIABILITY

Speech is the acoustic form of language. Speech recognition is essentially a process of recognizing acoustic patterns of the spoken language. Human communication by voice appears to be so simple that we often forget how variable these acoustic patterns are. Vast differences occur in the spoken utterance dependent on context, speaking style, speaker, dialect, speaking environment, microphone characteristics, etc. The major obstacle to achieving high accuracy in speech recognition is the large variability present in the speech signal, with only a small part that is important for carrying the linguistic information. A large part of this variability is due to various redundancies introduced by the human speech production process to achieve reliable speech communication in noisy and reverberant acoustic environments. Other sources of variability are introduced by differences in the vocal systems of speakers, differences in speaking rates, and the influence of neighboring sounds on the acoustic realization of a particular sound due to sluggish articulatory movements. Automatic methods of speech recognition must be able to handle this large variability in a proper manner. We illustrate here a few examples of the variability inherent in the speech signal.

Consider a simple case of the same word spoken by the same speaker on two different occasions. Acoustic realizations of the two utterances are in general not identical due to variations in the speaking rate or the speaking style. The speech recognition procedure must be able to compare the two utterances and conclude that they are same. Variations in speaking rate cause nonlinear distortions of the time axis of speech patterns. Linear scaling of the time axis is generally not sufficient to cope with speaking-rate variability. Nonlinear time normalization using dynamic programming is necessary to achieve time alignment between unknown and reference utterances.

The same word spoken by two different speakers will in general have different acoustic characteristics, due to the differences in their vocal tracts and speaking styles. A speaker-dependent recognizer uses the utterance of a single speaker to learn the speech patterns of that speaker. In contrast, a speaker-independent recognizer is trained on speech from many speakers and is used to recognize speech from speakers that may be outside of the training population.

Recognition of continuous speech introduces additional problems. In isolated words or speech where words are separated by distinct pauses, the beginnings and ends of words and clearly marked. In continuous speech, word boundaries are blurred and words evolve smoothly in time with no acoustic separation.

Automatic methods of segmenting continuous speech into words therefore had to be devised. Machine recognition of continuous speech with a large vocabulary requires that syntactic and semantic constraints be incorporated in the recognition process.

The progress in automatic speech recognition has come about as a result of solving successfully problems created by the large variability present in speech signals. Figure 31.1 shows some of the major advances that were made during the past twenty years and were important in achieving this progress. Dynamic time warping (DTW) was the most important step taken in early 1970s to handle variations in speaking rates [3]. Clustering of speech patterns into multiple templates [4] for each word made it possible to recognize words spoken by any speaker (speaker-independent speech recognition). Recognition of individual words from a string of connected words required development of level-building ASR algorithms [5]. The introduction of Hidden Markov Models (HMM) techniques [6] in early 1980s was a giant step that set the stage for big success in ASR. Recognition of continuous fluent speech was made possible by breaking words into phone-like subword units, use of bigram and higher-order language models, and the development of efficient beam search techniques [7]. Finally, introduction of concept spotting [8] and robust parsing techniques moved ASR from simple recognition of words to understanding meaning conveyed by a group of words.

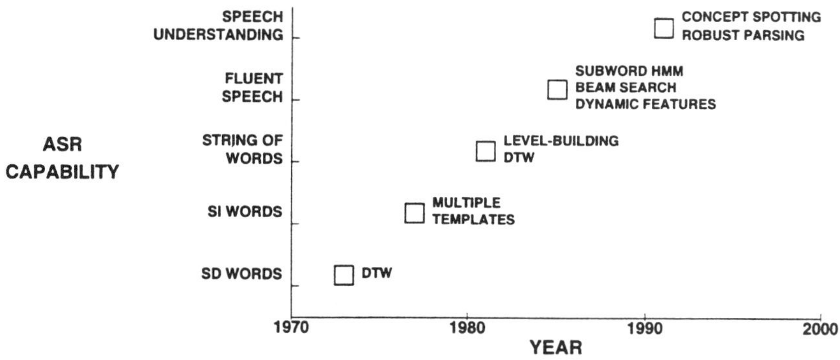

Figure 31.1 Major advances in automatic speech recognition from 1970 onwards that made it possible to achieve steady progress, from the simple problem of speaker-dependent recognition of isolated words to the more difficult problem of recognizing spontaneous speech encountered in dialogues.

3 AUTOMATIC SPEECH RECOGNITION PROCESS

There are at least three principal approaches to speech recognition: The first is based on statistical techniques of pattern recognition [6] that utilizes a training set of speech data to learn important information about the speech signal. The second approach, commonly known as acoustic-phonetic approach, uses knowledge of the relationship between acoustic and phonetic structures of the language. The third approach uses artificial neural networks. Best results in speech recognition have so far been achieved by using the statistical pattern recognition approach supplemented by the knowledge of acoustic-phonetic relationships in speech.

The basic steps of the pattern recognition approach are illustrated in the block diagram of Figure 31.2. The speech signal is analyzed to provide a parametric representation at the acoustic level. These parameters ("features") are then compared to a stored set of patterns derived from a large collection of speech utterances from many speakers using a HMM-based training procedure. This comparison provides a set of scores representing the similarity between the unknown pattern and each of the stored patterns. The last step augments these scores with other knowledge about the speech utterance, such as the language, the context, and semantics, to yield the best recognition results.

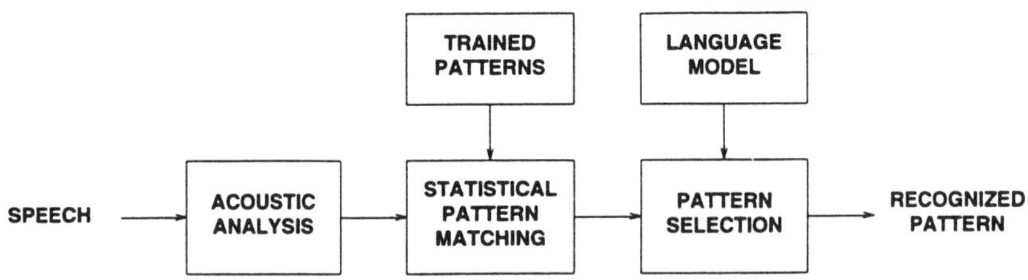

Figure 31.2 A block diagram showing the basic steps of the pattern recognition approach.

3.1 Acoustic Features of Speech

The selection of proper acoustic or spectral features is crucial for achieving high performance in speech recognition. The short-time spectral envelope of speech, obtained either by filtering or linear prediction analysis, is still considered to be the most effective representation for speech recognition, especially if rendered on a critical-band ("Bark") scale. The spectra are computed sequentially in time at intervals of 10 to 20 ms and are usually converted into cepstral coefficients. The cepstrum is defined as the inverse Fourier transform of the logarithm of the power spectrum.

The cepstral coefficients are instantaneous (static) features. One of the most important advances in the acoustic representation of speech has been the introduction of dynamic features [9], such as first- and second-order derivatives of the cepstrum. The static and dynamic features are generally combined to form a larger feature set; a smaller set can then be obtained by proper "pruning" from the larger set.

3.2 Training and Pattern Matching

Because of its statistical nature and its simple algorithmic structure for handling the large variability in speech signals, the HMM approach has found widespread use for automatic speech recognition. Most of the successful systems today are based on this approach. An HMM representation can be used to model a sound, a word, a phrase, or a long utterance. During the training phase, the HMMs are trained from an ensemble of observation vectors coming from spoken utterances and stored for each of the basic pattern to be recognized. During recognition, the unknown acoustic patterns are compared with a set of stored reference patterns ("templates") established from the training data to provide a set of similarity scores between the test and reference patterns.

3.3 Pattern Selection

In the recognition phase, the utterance is decoded by determining the optimal sequence of HMM states and the corresponding speech units based on the observed sequence of acoustic feature vectors in the utterance. Search procedures based on dynamic programming methods are used to find the sequence of states with the maximum likelihood. Additional information based on the syntax and semantics of the source language is included in the recognition process to produce admissible outputs.

4 CURRENT CAPABILITIES

The performance of ASR systems continues to improve steadily. Figure 31.3 shows the word error rate for various test materials and the steady decrease in the error rate achieved since 1980. The performance of the current ASR systems degrades considerably in the presence of noise, reverberation, or distortion and for conversational speech.

There are many factors that influence the performance of automatic speech recognition systems. The most important of these are the size of the vocabulary and the speaking styles. Figure 31.4 shows examples of ASR tasks that can

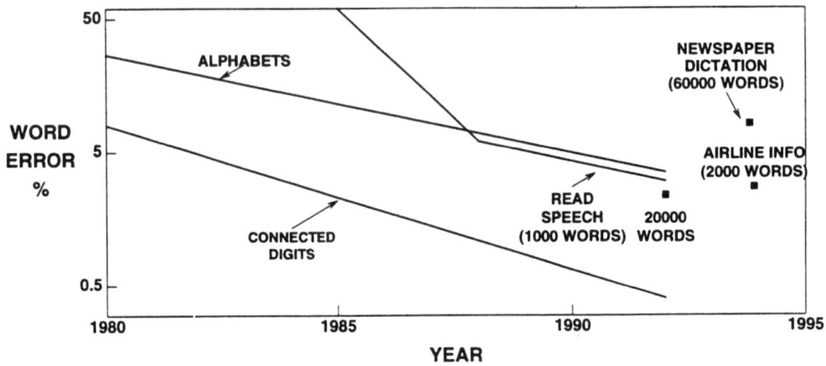

Figure 31.3 Reduction in the word error rate for different automatic speech recognition tasks between 1980 and 1995.

be handled by automatic methods for different vocabulary sizes and speaking styles. Generally, the number of confused words increases with the vocabulary size. Current systems can properly recognize a vocabulary of as many as a several thousand words, while the speaking style can vary over a wide range, from distinct isolated words to spontaneous running speech.

Examples of speech recognition applications that can be handled by the current technology are shown on the left side of the diagonal line in Figure 31.4. These include recognition of voice commands for computers, names, digit strings, and keyword spotting. The items on the right of the diagonal line in Figure 31.4 are examples of the speech recognition tasks that work well in laboratory environments but which need more research to become useful for real-world applications. Automatic recognition of fluent speech with a large vocabulary is not feasible unless constraints on syntax and semantics are introduced. The present capability in handling natural languages and in following a dialogue is limited because we do not understand how to model the variety of expressions that natural languages use to convey concepts and meanings.

5 CHALLENGING ISSUES IN SPEECH RESEARCH

For speech technology to be used widely, it is necessary that the major road-blocks faced by the current technology be removed. Some of the key issues that pose major challenges in speech research research are listed below:

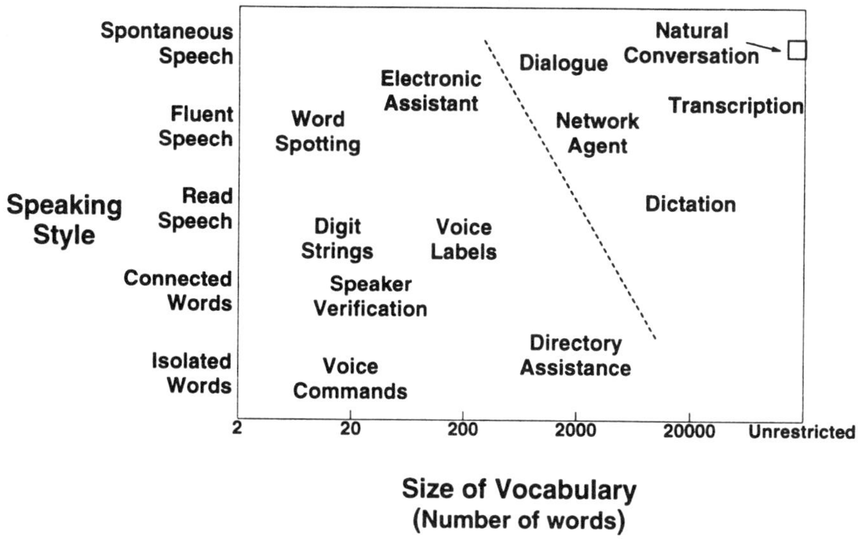

Figure 31.4 Different speech recognition tasks shown in a space of two dimensions: speaking style and size of vocabulary. Tasks that can be handled by current technology are shown to the left of the diagonal line. Items to the right of the line need more research to bring performance to a useful level.

- *Robust performance.* Can the recognizer work well for different speakers and in the presence of the noise, reverberation, and spectral distortion that are often present in real communication channels?

- *Automatic learning of new words and sounds.* In real applications, the users will often speak words or sounds that are not in the vocabulary of the recognizer. Can it learn to recognize such new words or sounds automatically?

- *Grammar for spoken language.* The grammar for spoken language is quite different from that which is used in carefully constructed written text. How does the system learn this grammar?

- *Flexibility.* Unless it is flexible, speech technology will have limited applications. What restrictions are there on the vocabulary? Can it handle spontaneous speech and natural spoken language?

A number of methods have been proposed to deal with the problem of robustness. The proposed methods include signal enhancement, noise compensation, spectral equalization, robust distortion measures, and novel speech representations. These methods provide partial answers valid for specific situations, but do not provide a satisfactory answer to the problem. Clean, carefully articulated, fluent speech is highly redundant, with the signal carrying significantly more information than is necessary to recognize words with high accuracy. However, the challenge is to realize the highest possible accuracy when the signal is corrupted with noise or other distortions and part of the information is lost.

6 FUTURE DEVELOPMENTS

The advances in digital technology is rapidly changing the fabric of telecommunications and the way we access information. A new mode of interacting with computers through voice is emerging. When combined with video, the voice mode offers an easy natural communication interface with computers. Speech recognition technology is a key component of such an interface. Human speech communication is a complex process and it will require scientific understanding of many other issues beyond acoustics and pattern recognition to mimic this process in computers. Speech science is expanding its frontiers to answer the basic question of how we put words together to express ideas in the spoken language.

REFERENCES

[1] Rabiner, L. and Juang, B.-H., *Fundamentals of speech recognition*, Prentice-Hall, Englewood Cliffs, NJ, 1993.

[2] Makhoul, J. and Schwartz, R., "State of the art in continuous speech recognition", in *Voice communication between humans and machines*, D. Roe and J. Wilpon (eds.), National Academy Press, Washington, pp. 165-198, 1994.

[3] Itakura, F., "Minimum prediction residual principle applied to speech recognition", *IEEE Transactions on Acoustics, Speech, and Signal Processing*, **23**:57-72, 1975.

[4] Rabiner, L., Levinson, S., Rosenberg, A., Wilpon, J., "Speaker independent recognition of isolated words using clustering techniques", *IEEE Transactions on Acoustics, Speech, and Signal Processing*, **27**:336-349, 1979.

[5] Meyers, C. and Rabiner, L., "A level-building dynamic time warping algorithm for connected word recognition", *IEEE Transactions on Acoustics, Speech, and Signal Processing*, **29**:284-297, 1981.

[6] Bahl, L., Jelenik, F., and Mercer, R., "A maximum likelihood approach to continuous speech recognition", *IEEE Transactions on Patt. Anal. Machine Intell.*, **5**:179-190, 1983.

[7] Chow, Y., et al., "BYBLOS: The BBN continuous speech recognition system", *Proc. IEEE-ICASSP*, Dallas, TX, pp. 89-92, 1987.

[8] Pieraccini, R. and Levin, E., "Stochastic representation of semantic structure for speech understanding", *Speech Communications*, **11**:283-288, 1992.

[9] Furui, S., "Speaker-independent isolated word recognition using dynamic features of speech spectrum", *IEEE Transactions on Acoustics, Speech, and Signal Processing*, **34**:52-59, 1986.

32

NOVEL HOS-BASED APPROACH TO STOCHASTIC PREDICTION AND ESTIMATION

Tariq S. Durrani and A. Rahim Leyman

Signal Processing Division,
Department of Electronic and Electrical Engineering,
UNIVERSITY OF STRATHCLYDE, Glasgow G1 1XW,
Tel:041-5524400 (ext.2686), Fax:041-5522487,
email:rahim@spd.eee.strathclyde.ac.uk

Dedicated to Tom Kailath with affection and profound esteem.

ABSTRACT

This Letter proposes an approach for optimum state prediction and estimation in the presence of additive non-Gaussian measurement and observation noises. The proposed method exploits higher order statistics of the state variables and searches for the optimum estimate by minimizing the cumulant power of the prediction/estimation error. Theoretical results are presented and computer simulations are included to support the theory and to provide comparisons with Kalman filtering method.

1 INTRODUCTION

In this Letter, a novel approach is presented for state estimation and prediction when the state and observation models are subjected to non-Gaussian noise. The derivation of the proposed method relies on the third order statistics of the data, and employs a recently proposed performance criterion[1][2] to provide optimal solutions. Since cumulant statistics are involved, Gaussian noise effects are automatically suppressed. An easily implementable iterative algorithm is provided to achieve an optimum solution. In the sequel, state estimation/prediction procedures are developed based on a single input single output (SISO) dynamic system to illustrate the approach. The performance of the proposed algorithm is verified by using computer simulations, and when compared with the Kalman Filter, the proposed method are shown to provide improved estimation/prediction solutions.

2 STATE MODEL

Consider a SISO dynamical system whose state is described by the following equations:

$$x(k+1) = Ax(k) + Bu(k) \tag{32.1}$$

Observation Model

$$z(k+1) = Mx(k+1) + v(k+1) \tag{32.2}$$

The following assumptions are taken to be valid: $x(k)$, system state scalar at discrete time k, a zero mean non-Gaussian random sequence; A, state transition constant; $u(k)$, is a measurement noise, assumed to be non-Gaussian whose the third order cumulant is defined as $\mathbf{C}_u(\tau_1, \tau_2) \triangleq E\{u(k)u(k+\tau_1)u(k+\tau_2)\} \neq 0$; B, state constant which relates the external input to the state $x(k)$; $z(k)$, scalar observation or measurement, a zero mean non-Gaussian random sequence; M, a constant describing the relationship between the state scalar, $x(k)$ and the measurement scalar; $v(k)$, scalar observation noise, a zero mean asymmetric, non-Gaussian noise where the third order cumulant matrix is defined as $\mathbf{C}_v(\tau_1, \tau_2) \triangleq E\{v(k)v(k+\tau_1)v(k+\tau_2)\}$.

As in the Kalman filter case, we begin by making a prediction estimate as follows:

$$\hat{x}(k+1|k) = A\hat{x}(k|k) \tag{32.3}$$

On the arrival of a new observation, $z(k+1)$, we seek to make a correction to the state estimation according to the following function:

$$\hat{x}(k+1|k+1) \triangleq w_1\hat{x}(k+1|k) + w_2 z(k+1) \tag{32.4}$$

where w_1 and w_2 are the weights to be determined to make the correction to the state estimate. Equation (32.4) represents a new correction procedure which is a linear combination of both the weighted previous estimate and the weighted new observation. We seek to determine these weights to form a new optimum for the state estimate, $\hat{x}(k+1|k+1)$, by minimising a cost function related to the prediction error which is given by

$$e(k+1|k+1) \triangleq \hat{x}(k+1|k+1) - x(k+1) \tag{32.5}$$

By substituting Eqs.(32.1) and (32.2) into Eq.(32.5) we get the following

$$e(k+1|k+1) = w_1\hat{x}(k+1|k) - x(k+1)\left[1 - w_2 M\right] + w_2 v(k+1) \tag{32.6}$$

Now we define the following constraint

$$w_1 \triangleq 1 - w_2 M \tag{32.7}$$

This constraint is a key requirement for solving the prediction problem as shown below. Substituting Eq.(32.7) into Eq.(32.6) leads to the following expression:

$$e(k+1|k+1) = w_1 e(k+1|k) + w_2 v(k+1) \tag{32.8}$$

Eq.(8) states that the estimation error is a weighted combination of the prediction error $\{e(k+1|k)\}$ and the observation noise $\{v(k+1)\}$. Using Eq.(32.8) we now proceed to determine the weights w_1 and w_2 by minimising the cumulant power of the estimation error. The third order cumulant of $e(k+1|k+1)$ is defined as

$$\mathbf{C}_e(\tau_1, \tau_2; k+1|k+1) \triangleq E\{e(k+1|k+1)e(k+\tau_1+1|k+1)e(k+\tau_2+1|k+1)\} \tag{32.9}$$

For notational simplicity we write $\mathbf{C}_e(\tau_1, \tau_2; k+1|k+1)$ as $\mathbf{C}_e(k+1|k+1)$ where the cumulant lag values τ_1 and τ_2 have been suppressed. Throughout, we assume that for some support domain $D(\tau_1 \tau_2 \in D)$, the third order cumulant of the estimation error, Eq.(32.9), is non-zero. Therefore from Eqs.(32.8) and (32.9) we get the following

$$\mathbf{C}_e(k+1|k+1) = w_1^3 \mathbf{C}_e(k+1|k) + w_2^3 \mathbf{C}_v(k+1) \tag{32.10}$$

with the cross products being zero due to orthogonality. Defining the third order cumulant power[1][2] of Eq.(32.10)as

$$J_e(k+1|k+1) \triangleq \sum_{\tau_1} \sum_{\tau_2} |\mathbf{C}_e(k+1|k+1)|^2 \tag{32.11}$$

we obtain from Eq.(32.10)

$$J_e(k+1|k+1) = w_1^6 J_e(k+1|k) + 2w_1^3 w_2^3 J_{ev} + w_2^6 J_v(k+1) \tag{32.12}$$

where we have defined J_{ev} as

$$J_{ev} \triangleq \sum_{\tau_1} \sum_{\tau_2} [\mathbf{C}_e(k+1|k) \bullet \mathbf{C}_v^*(k+1)] \quad \text{and}$$
$$J_v \triangleq \sum_{\tau_1} \sum_{\tau_2} |\mathbf{C}_e(k+1|k)|^2 \tag{32.13}$$

where $*$ denotes conjugate transpose. To solve this optimisation problem we append the constraint of Eq.(32.8) to the cost function of Eq.(32.12) as

$$J_e(k+1|k+1) = w_1^6 J_e(k+1|k) + 2w_1^3 w_2^3 J_{ev} + w_2^6 J_v(k+1) + \lambda\{w_1 + w_2 M - 1\} \tag{32.14}$$

where λ is a scalar Lagrange multiplier. If we define the weights as $w_1 = \sin^2\theta$ and $w_2 = \frac{\cos^2\theta}{M}$ and $a = \tan^2\theta$, it can be easily shown (details see [4]) that the optimum weights, w_1 and w_2 can be obtained by setting $\frac{\partial J_e}{\partial w_1}$ and $\frac{\partial J_e}{\partial w_2}$ to zero and solve the following polynomial function

$$J_e(k+1|k)M^5a^5 - J_{ev}M^2a^3 + J_{ev}M^2a^2 - \frac{Jv(k+1)}{M} = 0 \qquad (32.15)$$

From Eq.(32.15) we can then determine the optimum weights, w_1 and w_2 and the whole iterative procedure of our proposed state estimation algorithm can be summarised as follows :

INITIALIZATION

- Set $\hat{x}(k|k)$
- Set $\mathbf{C}_e(k|k)$, $\mathbf{C}_u(k)$ and $\mathbf{C}_v(k)$

PREDICTION

- **Prediction** :$\hat{x}(k+1|k) = A\hat{x}(k|k)$
- Using Eqs.(32.1),(32.3) and (32.5) to obtain $\mathbf{C}_e(k+1|k) = A^3\mathbf{C}_e(k|k) - B^3\mathbf{C}_u(k)$
- $J_e(k+1|k) = \sum_{\tau_1}\sum_{\tau_2}|\mathbf{C}_e(k+1|k+1)|^2$
- $J_{ev} = \sum_{\tau_1}\sum_{\tau_2}[\mathbf{C}_e(k+1|k) \bullet \mathbf{C}_v^*(k+1)]$
- $J_v(k+1) = \sum_{\tau_1}\sum_{\tau_2}|\mathbf{C}_v(k+1)|^2$

UPDATE

- Solve Eq.(32.15)
- Obtain w_1 and w_2
- **New Estimate** $\hat{x}(k+1|k+1) = w_1\hat{x}(k+1|k) + w_2 z(k+1)$
- **Third Order Cumulant of Prediction error at correction step** $\mathbf{C}_e(k+1|k+1) = w_1^3\mathbf{C}_e(k+1|k) + w_2^3\mathbf{C}_v(k+1)$

The above iterative procedure will converge and give an optimal estimate of $\hat{x}(k+1|k+1)$ and with a minimised third order cumulant of the prediction error, $e(k+1|k+1)$. Note that the whole framework is an extension of the approach for the derivation of the classical Kalman Filter, in that, the state estimates for the Kalman Filter are given as

$$\begin{aligned}\hat{x}(k+1|k+1) &= \hat{x}(k+1|k) + K_{k+1}[z(k+1) - M\hat{x}(k+1|k)] \\ &= \hat{x}(k+1|k)[1 - K_{k+1}M] + K_{k+1}z(k+1) \qquad (32.16)\end{aligned}$$

and

$$P(k+1|k+1) = [1 - K_{k+1}M]\, P(k+1|k) \qquad (32.17)$$

where $P(k+1|k+1)$ is the estimation error at step $(k+1)$ variance and K_{k+1} is the associated Kalman Gain. By comparing Eqs.(32.4) and (32.15), it is clear that $w_1 = 1 - K_{k+1}M$ and $w_2 = K_{k+1}$. To prove that w_2 is actually the Kalman Gain, we compute the covariance of Eq.(32.8) with the adjoint constraint of Eq.(32.7), i.e.

$$P(k+1|k+1) = w_1^2 P(k+1|k) + w_2^2 R + \lambda \{w_1 + w_2 M - 1\} \qquad (32.18)$$

where we have defined $R \triangleq E\{v^2(k+1)\}$ as the variance of the observation noise. Setting $\frac{\partial P(k+1|k+1)}{\partial w_1}$ and $\frac{\partial P(k+1|k+1)}{\partial w_2}$ to zero and satisfying the constraint equation, yields

$$w_1 = \frac{R}{M^2 P(k+1|k) + \mathbf{R}} \; \textit{and} \; w_2 = \frac{M P(k+1|k)}{M^2 P(k+1|k) + R} \qquad (32.19)$$

However from the classical Kalman filter, for a SISO dynamic system we have

$$K_{k+1} = \frac{M P(k+1|k)}{M^2 P(k+1|k) + R} \qquad (32.20)$$

hence obviously $w_2 = K_{k+1}$. Therefore, the constraint of Eq.(32.7) ensures that our methodology in deriving the new optimal estimate of $\hat{x}(k+1|k+1)$ follows closely the Kalman filter development. As in the latter case, we would expect the results to be recursive in nature.

3 SIMULATION EXPERIMENTS

Computer simulations were carried out to ascertain the performance of the proposed method, as compared to that of the Kalman filter. We setup a simple first order process as follows:

$$x(k) = Ax(k-1) + Bu(k) \quad \text{Process Model} \qquad (32.21)$$

$$z(k) = Mx(k) + v(k) \quad \text{Observation} \qquad (32.22)$$

The mean value was subtracted from $u(k)$ and $v(k)$ and a total of 256 samples were used. The third order cumulants were estimated upto lag value of 4. For all the state prediction plots shown in the experiment, the solid line represents the actual state, $x(k)$, the dash line denotes the classical Kalman filter algorithm and the dotted line represents the proposed algorithm.

Example 1: The first example is to verify that for Gaussian $u(n)$ and $w(n)$, the Kalman filter provides optimal tracking of $\{x(n)\}$ and proposed algorithm gives incorrect estimates. For the system realisation in the simulation we set $A = 0.98$, $B = 1$ and $M = 1$. $u(n)$ is Gaussian process with zero mean and $\sigma_u^2 = 1$ and $R = 3$. We perturb the propose algorithm with a small amount of non-Gaussian $v(n)$, i.e. the cumulant power , $J_v = 0.0001$. It is seen from Fig. 32.1 that the Kalman filter tracks the actual process neatly while the proposed algorithm is unable to follow $x(n)$.

Example 2: In this example, $u(k)$ and $v(k)$ are exponentially distributed with measures,$\sigma_u^2 = 0.8$ and $J_u = 1.52$ with $J_v = 0.98$. The system parameters are $A = 0.98$, $B = 1$ and $M = 1$. Fig. 32.2 shows the performance of the KF and the porposed method. It is obvious that the proposed method tracks the process much more closely as compared to the KF.

4 CONCLUSION

We have presented a new approach to state prediction and estimation by employing third order statistics for the case of systems driven by non-Gaussian sequences. The proposed filter represents a time-varying recursive filter and therefore it is well adapted to non-stationary non-Gaussian signalling environment. Simulations were carried out to illustrate the performance and accuracy of the proposed filter for exponentially distributed input and noise processes. Extensions to multiple input and output and the steady state properties will be presented in another communication

Acknowledgements

This work was supported by US Navy under contract N00014-93-1-0055.

REFERENCES

[1] T.S.Durrani, A.R.Leyman, "New Algorithms for DOA Estimation using Higher Order Statistics", Proc. IEEE,*Intl.Conf.Acoust. Speech and Sig. Proc.*, pp.596-600, Minneapolis, Minnesota, April 1993.

[2] A.R.Leyman & T.S.Durrani, "HOS-based Direction of Arrival Estimation", *IEE Electronic Letters*,,pp.540-541 Vol.30,No.7, 31st March 1994.

[3] T.Kailath, "Linear System", *Prentice Hall Inc*, 1980, New Jersey.

[4] A.R.Leyman, "Applications of Higher Order Statistics to Array and Signal Processing.", *Ph.D Thesis*, UNIVERSITY OF STRATHCLYDE, 1994.

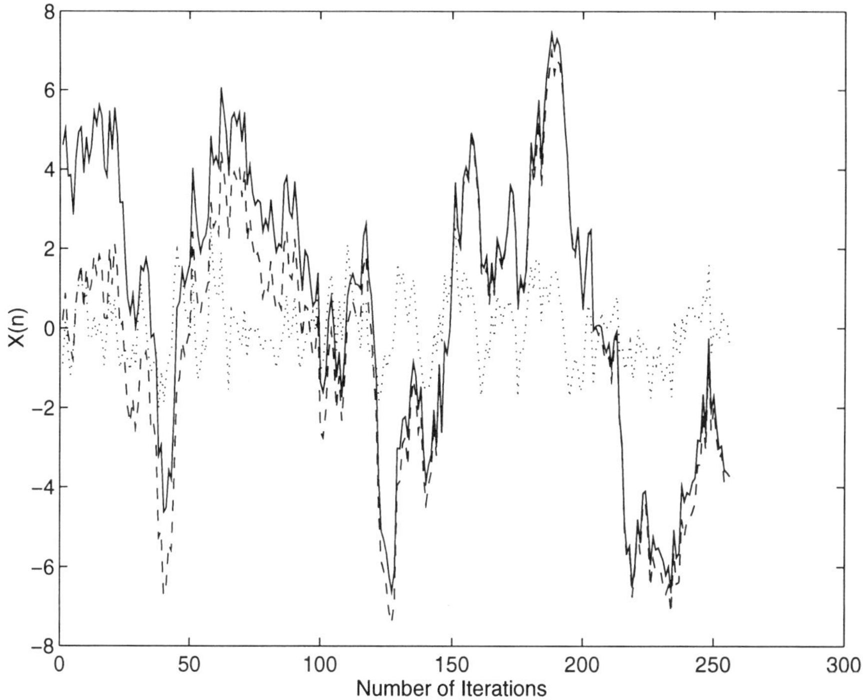

Figure 32.1 First Order process with Gaussian excitation and noise with $A = 0.98$, $B = 1$ and $M = 1$. Solid line represents the actual values of $\{x(n)\}$; the dash line denotes the Kalman filter estimates. And the dotted line represents the estimates from the proposed method.

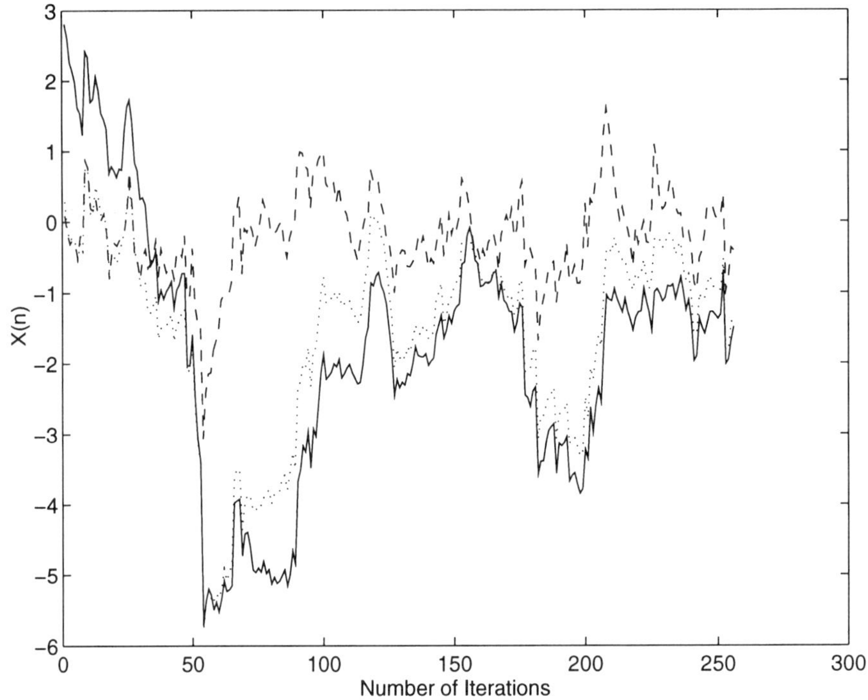

Figure 32.2 $u(k)$ and $v(k)$ are exponentially distributed with $A = 0.98$, $B = 1$ and $M = 1$.

The sizes of $\mathbf{W}_a$ and $\mathbf{R}_a$ are 2F x P and 2F x 3.

Let us further define $\mathbf{M} = [\mathbf{T}_a|\mathbf{R}_a]$ and $\mathbf{S} = \begin{bmatrix} \mathbf{E}_a \\ \hline \mathbf{S}_a \end{bmatrix}$, then the measurement matrix can be expressed by $\mathbf{W} = \mathbf{W}_a = \mathbf{MS}$. This factorization suggests the matrix $\mathbf{W}$ has a generic rank of 4.

The multi-target case has been treated in several prior works[2],[1]: *Given a sequence of 2-D video images of multiple moving targets, the problem is to track the 3-D motion of the targets and reconstruct their 3-D shapes.* For the multi-object MSE problem, a new challenge is to cluster the FPs of (closely moving) objects so that they are classified into correct objects.

Let us for simplicity concentrate on two targets, A and B, with their shapes denoted by $\mathbf{S}_a$ and $\mathbf{S}_b$. Same as the single-target case, the measurement matrices for A and B, can be derived as: $\mathbf{W}_a = \mathbf{R}_a\mathbf{S}_a + \mathbf{T}_a\mathbf{E}_a$ and $\mathbf{W}_b = \mathbf{R}_b\mathbf{S}_b + \mathbf{T}_b\mathbf{E}_b$ Concatenate these matrices,

$$\mathbf{W} = [\mathbf{W}_a|\mathbf{W}_b] = [\mathbf{T}_a|\mathbf{T}_b|\mathbf{R}_a|\mathbf{R}_b] \left[\begin{array}{c|c} \mathbf{E}_a & 0 \\ \hline 0 & \mathbf{E}_b \\ \hline \mathbf{S}_a & 0 \\ \hline 0 & \mathbf{S}_b \end{array}\right] \equiv \mathbf{MS} \tag{33.4}$$

Let k denote the number of the targets, then the sizes of $\mathbf{W}$, $\mathbf{M}$, and $\mathbf{S}$ are $2F \times (P+Q)$, $2F \times 4k$, and $4k \times (P+Q)$ respectively. In this case, $k = 2$, so that rank of a noise-free measurement matrix W is exactly $4k = 8$.

SVD on Measurement Matrix It is realistic to assume that the measurement matrix $\mathbf{W}$ is corrupted by noise and therefore turns into full rank. The SVD of the measurement matrix is

$$\mathbf{W} = \overline{U}\ \overline{\Sigma}\ \overline{V} = U\Sigma V + U'\Sigma' V'$$

Removing the "noise" singular values Σ', we have

$$\mathbf{W} \approx U\Sigma V = U\Sigma^{1/2}\Sigma^{1/2}V \tag{33.5}$$

2 PRIOR APPROACHES

In real applications, the ideally-ranked measurement matrix $\mathbf{W}$ is often corrupted by various noise sources. So any practical clustering algorithm has to be able to handle a variety of noises, for examples, (1) *perspective distortions*, (2) *tracking errors*, and (3) *occlusions* (occluded FPs can only be very roughly estimated by extrapolation).

Among the prior works, the most relevant methods to this paper are algebraic-based *subspace clustering method*[2], and another based on a shape interaction matrix (by e.g. minimum-spanning-tree solution) [1]. They both claims theoretical rigor under idealistic (para-perspective and noise-free) situations. However, they suffer from the weakness of being numerically sensitive to noise corruptions. Our study suggests that any realistic noise level - conforming to real-world application need - could easily render these (theoretically valid) techniques very inefficient.

Subspace Clustering Approach The *Subspace Rank Properties*[2] includes: (1) *Inclusive Rank Property:* If the column vectors of the matrix W (or, equivalently, V) is correctly grouped (each corresponds to one object), then each group (e.g. V_a, and V_b) spans a *rank-4* space. (2) *Exclusive Rank Property:* Generically, any mixture of (5 or more) columns from two groups or more would cause the rank to exceed 4. (This property may be exploited to prevent over-subscribing alien columns into an object.) However, the numerical behavior involved in *rank* checking is of a serious concern. The theoretically non-zero and zero singular values become almost undistinguishable under practically inevitable noise corruption.

Minimum-Cut Approach Based on Shape Interaction Matrix A notion of shape interaction matrix Q is introduce in [1]:

$$Q = V^T V$$

Ideally, the canonical form of Q (sorted to form correct clusters) would have block-diagonal structure due to the orthogonality between V_i and V_j (i.e. $V_i^T V_j = 0$) when V_i and V_j correspond to different objects. (The orthogonality corresponds to the exclusive rank property.) So Q may be potentially very useful for separating FPs. However, the following properties of Q may prove to be a constraint: (1) **Q is invariant to motion and number of frames.** (Although V is dependent on motion, however the information about motion (hidden in a unitary matrix) is lost in the product $Q = V^T V$, because a unitary matrix times its transpose yields an identity matrix. (2) The orthogonality is very vulnerable to noise corruption. The theoretically non-zero and zero entries in Q become almost undistinguishable under perspective distortion. The histogram in Figure 33.2 shows substantial overlap between off-block-diagonal entries and the block-diagonal entries. To deal with this situation, a suboptimal minimum-spanning-tree search method is proposed[1]. Still, the non-uniqueness of the solution is a very realistic cause for concern, especially when the perspective distortion and NSR are large.

3 PSV CLUSTERING METHOD

The **PSV Clustering Method** for separating multiple targets consists of the following steps:

1. The first 2 (or sometimes 3) PSVs form a $2 \times (N+M)$ matrix

$$\left[\frac{V_1}{V_2}\right] = [\mathbf{v}_1 \ \mathbf{v}_2 \ \cdots \mathbf{v}_{N+M}]$$

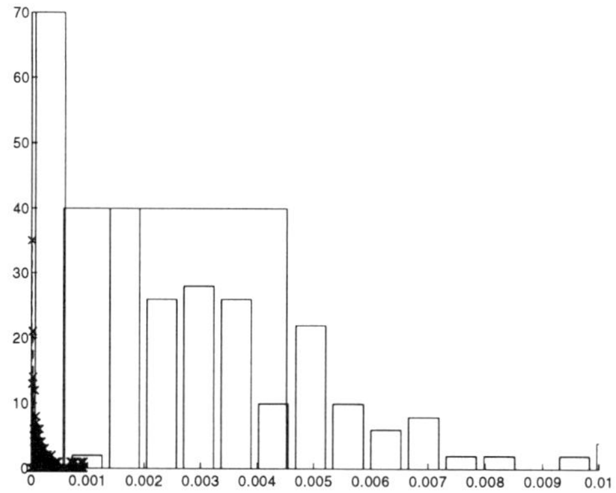

Figure 33.2 The histogram shows substantial overlap between off-block-diagonal Q entries (shown by dense "x" marks) and the block-diagonal entries. Here the perspective distortion is approximately 15% and NSR = 10%. Note that the first 70 (the peak) of the diagonal entries are totally "buried" by the off-diagonal entries ("x"). In general, the separability is easily corrupted by noise.

where the column vector are denoted by $\{ \mathbf{v}_i,\ i = 1, N+M \}$.

2. Classify these column vectors $\{ \mathbf{v}_i,\ i = 1, N+M \}$ into 2 groups (or 3 if 3 targets) by a clustering method, such as VQ, k-mean, or neural network techniques.

3. Once the targets are separated, then the motion and shape of individual targets can be constructed[3].

As to be justified in our discussion, the PSV clustering method provides high noise susceptibility and is readily applicable to degenerate situations. Both SVD and VQ are very mature numerical techniques, so the method approach is computationally and numerically more efficient than previous approaches[1, 2].

Mathematical Foundation The first two PSVs are formed by some linear combinations of the matrix **S**:

$$\left[\frac{V_1}{V_2}\right] = \mathbf{PS} = [\mathbf{P}_a^t|\mathbf{P}_b^t|\mathbf{P}_a^r|\mathbf{P}_b^r] \begin{bmatrix} \mathbf{E}_a & 0 \\ \hline 0 & \mathbf{E}_b \\ \hline \mathbf{S}_a & 0 \\ \hline 0 & \mathbf{S}_b \end{bmatrix} \tag{33.6}$$

$$= [\mathbf{P}_a^t\mathbf{E}_a + \mathbf{P}_a^r\mathbf{S}_a|\mathbf{P}_b^t\mathbf{E}_b + \mathbf{P}_b^r\mathbf{S}_b] \tag{33.7}$$

To illustrate this, let us refer to a simulation example (Figure 33.4), where $\mathbf{P}_a^t$ and $\mathbf{P}_b^t$ determine the centroids of the clusters (see marks "o" and "x" in Figure 33.4) while $\mathbf{S}_a$ and $\mathbf{S}_b$ define the shapes of the clusters. In Figure 33.4 the cube and pyramid shapes are quite vividly displayed.

Empirical studies suggest that the translation-oriented term $\mathbf{P}^t \equiv [\mathbf{P}^t_a|\mathbf{P}^t_b]$ will most likely prevail over the rotation terms $\mathbf{P}^r_a$ and $\mathbf{P}^r_b$ in the first 2 PSVs (3 PSVs for 3 targets, and so on.) This is easily verified by simulations, cf. Section 4. Such a dominance is very noticeable in most practical situations, barring rare degenerate situations. The PSV clustering method hinges upon the following observations:

1. The more dominant is $\mathbf{P^t}$, the better the separability.
2. When $\mathbf{P}^r_a$ and $\mathbf{P}^r_b$ are negligible compared with $\mathbf{P}^t$, then $\{\ \mathbf{v}_i,$ would have formed two highly concentrated clusters.
3. Since all the SVs tend to be equally corrupted by noise, so the principal SV's will remain less affected.
4. The first few PSV's can potentially display (distinctively) the shapes of different objects (See e.g. Figure 33.4.)

Under such a dominance assumption, clusters belonging to different targets will be sufficiently separated. Moreover, the cluster patterns may be described by a closed-form formula derived below.

Derivation of Ideal $\mathbf{P}^t_{ideal}$ To estimate the centroids of the vector clusters of $\{\ \mathbf{v}_i,\ i = 1, N + M\ \}$, it is useful to have a analytical formula describing the behavior of $\mathbf{P}^t$ and the PSVs.

Recall that $\mathbf{T}_a$ and $\mathbf{T}_b$ represents the translational motions of the mass centers of two targets. This also implies that the shape matrices $\mathbf{S}_a$ and $\mathbf{S}_b$ are object-centered. It can be verified that, on the RHS of Eq. 33.4, the (upper) E block and the (lower) S block are orthogonal to each other.

To simplify mathematical derivation further, we shall make an instrumental (though unrealistic) assumption that, on the LHS, the (left) translational T- block and the (right) rotational R-block are also orthogonal. Such an assumption allows us to separately treat the singular value analyses on $[\mathbf{T}_a|\mathbf{T}_b]\left[\begin{array}{c|c}\mathbf{E}_a & 0\\ \hline 0 & \mathbf{E}_b\end{array}\right]$ and $[\mathbf{R}_a|\mathbf{R}_b]\left[\begin{array}{c|c}\mathbf{S}_a & 0\\ \hline 0 & \mathbf{S}_b\end{array}\right]$. By SVD, we have $[\mathbf{T}_a|\mathbf{T}_b]\left[\begin{array}{c|c}\mathbf{E}_a & 0\\ \hline 0 & \mathbf{E}_b\end{array}\right] = \tilde{U}\ \tilde{\Sigma}\ \tilde{V}$ and $\mathbf{P}^t_{ideal}$ may be obtained by solving R

$$\tilde{V} = \mathbf{P}^t_{ideal}\left[\begin{array}{c|c}\mathbf{E}_a & 0\\ \hline 0 & \mathbf{E}_b\end{array}\right]$$

In our simulation example, (Section 4), we have verified $\mathbf{P}^t_{ideal} \approx \mathbf{P}^t$ (with an error around 6%) The comparison of the SVD values also indicates the same. The largest two singular values do coincide closely with the two singular values corresponding to the translation components (with errors around 20% and 33% respectively).

Closed-form Analysis In this section, we assume the no noise nor perspective distortion case. To analyze how the motion trajectories affect the PSVs, we take the (well-known) QR decomposition on $[\mathbf{T}_a|\mathbf{T}_b]$, which results in

$$[\mathbf{T}_a|\mathbf{T}_b]\left[\begin{array}{c|c}\mathbf{E}_a & 0\\ \hline 0 & \mathbf{E}_b\end{array}\right] = \mathbf{X}\left[\begin{array}{c}Y\\ \hline 0\end{array}\right]\left[\begin{array}{c|c}\mathbf{E}_a & 0\\ \hline 0 & \mathbf{E}_b\end{array}\right] \equiv \mathbf{X}\Phi$$

where

$$\Phi = Y\left[\begin{array}{c|c}\mathbf{E}_a & 0\\ \hline 0 & \mathbf{E}_b\end{array}\right]$$

where Y is the upper triangular matrix derived by the QR decomposition. The left factorization of QR decomposition can be dropped from the analysis, since it (being orthonormal) does not affect the singular value analysis.

We can express Y as follows:

$$Y = y_{11}\left[\begin{array}{c|c}1 & \beta\\ \hline 0 & \alpha\end{array}\right]$$

The singular values of Φ can be analytically derived as $y_{11}\sqrt{\lambda_i}$, $i = 1, 2$, where $\lambda_i = M\rho_i$ and

$$\rho_i = 0.5\left[r + (\alpha^2 - \beta^2) \pm \sqrt{[r - (\alpha^2 + \beta^2)]^2 + 4r\beta^2}\right]$$

where N and M are the numbers of FPs of targets A and B, and $r = N/M$. The singular vectors of Φ (before normalization) can be represented by

$$[\quad (1 - \frac{\alpha^2}{\rho_i})\mathbf{E}_a \quad | \quad \beta\mathbf{E}_b \quad] \tag{33.8}$$

Theorem 33.1 (Characteristics of Principal SVs) *The first PSV and the second PSV (assuming the SVD is translation-dominant) have the following characteristics:*

If $\beta < 0$, it usually implie much easier separability. (Details omitted here.) If $\beta \approx 0$ (or in a more general term, if $|\beta| << |\alpha|$), then the singular vectors will be simply (before normalization) $[\ \mathbf{E}_a \mid 0\ ]$ and $[\ 0 \mid \mathbf{E}_b\ ]$, giving rise clear separability. Therefore, we shall focus our discussion only on the harder case when $\beta > 0$, i.e. the two trajectories are somewhat correlated.

1. *$\lambda_i \geq 0$ and $\rho_i \geq 0$*
2. *It can be easily verified that $\rho_1 \geq \alpha^2 \geq \rho_2$. This implies that the centroids of the two object are located at upper-right and lower-right quarter planes. (Note that $1 - \frac{\alpha^2}{\rho_1} > 0$ and $1 - \frac{\alpha^2}{\rho_2} < 0$.)*
3. *The separation between the two targets is proportional to α^2. The larger is the value α the greater the separation.*

4. The separation between the two targets is proportional to difference between ρ_1 and ρ_2, which is greater than $2|\beta|\sqrt{r}$ by inspection. So a minimum distance of the two centroids is guaranteed. Given the same value of α, the larger is β the greater the separation.

Similar formula can be obtained for situations involving three moving targets or more. The detailed derivation is left to the reader.

4 EXPERIMENTAL RESULTS

The it PSV clustering method is applied to synthetic data to show its performance and potential promise for practical applications. Simulation results justify that for the 2-target and 3-target cases, the first and second PSVs can offer sufficient information to clearly determine the corresponding clusters for FPs.

This experimental sequence consists of 15 frames of two moving targets under perspective projection. Other sources of noise (e.g. tracking errors, etc.) are modeled as white noise and is added to the perspective distorted images. Initially, the targets (assumed to be transparent) follow a very similar trajectory on the image plane. During the first 5 frames their images are almost overlapping. Thereafter they start to separate slightly. Figure 33.3 depicts the trajectories of the two objects in frame #1 ~ 3 and #6 ~ 8 respectively. Figure 33.3(b) depicts the depth map of the two targets.

Theory Verification: In our frames 1-15 simulation, we have obtained under translation-only orthographic projection case, $\mathbf{P}^t_{ideal} = \left[\begin{array}{c|c} 0.1865 & 0.1745 \\ \hline -0.1234 & 0.2637 \end{array}\right]$

Translation-only but with perspective projection: $\mathbf{P}^t_{ideal} = \left[\begin{array}{c|c} 0.16 & 0.21 \\ \hline -0.15 & 0.23 \end{array}\right]$ Their full-motion counterpart (computed from $\mathbf{W}$) is $\mathbf{P}^t = \left[\begin{array}{c|c} 0.17 & 0.20 \\ \hline -0.14 & 0.24 \end{array}\right]$ Note that all these three are very close.

Their corresponding singular values also coincide closely. The singular values of idealistic translation-only (under perspective projection) are 27.9 and 1.5, very close to the largest two full-motion singular values: 22.2 and 2.4.

The dominance of $[\mathbf{P}^t_a|\mathbf{P}^t_b]$ over $\mathbf{P}^r_a$ and $\mathbf{P}^r_b$ is also verified by the simulation:

$$[\mathbf{P}^t_a|\mathbf{P}^t_b|\mathbf{P}^r_a|\mathbf{P}^r_b] = \left[\begin{array}{c|c|c|c|c|c|c|c} 0.1720 & 0.2019 & -0.0082 & 0.0038 & -0.0066 & -0.0039 & 0.0153 & 0.0025 \\ -0.1421 & 0.2418 & 0.0341 & 0.0216 & -0.0029 & 0.0125 & 0.0770 & 0.0362 \end{array}\right]$$

Note that $\mathbf{P}^r$ has much smaller entries than $\mathbf{P}^t$. Consequently, the clusters' radii are much smaller than the distance of centroids, cf. Figure 33.4.

Performance Analysis: This section reports the performances of the PSV clustering method based on several metrics.

Figure 33.4 shows the effect on PSVs by different numbers of frames. Our minimum experiment uses **only** 5 frames and the PSV clusters are depicted in

Figure 33.4(a). the performance will in general improve if more frames are used. It can be explained as follows: when the number of frames increases, smaller values of $\mathbf{P}^r$ will result and the PSV clusters become more concentrated, as demonstrated in Figure 33.4(b),(c). Note also that, as theoretically predicted, *the centroids of the two objects are indeed located at upper-right and lower-right quarter planes.*

In our experiment, the tracking errors are modeled by additive white noise with NSR ranging from 0% - 100%. Figure 33.5 reveals a very strong noise susceptibility of the PSV clustering method. The strong vigilance of the PSVs prevents them from being corrupted by white noise as high as 100% NSR. As displayed in Figure 33.5, the clusters are easily separable.

We also experimented on different levels of perspective distortion. Unfortunately, the rank properties collapse badly under nonlinear perspective distortion, which cannot be modeled by white noise. It is important to note that the two kinds of (perspective vs. additive) noises have totally different effects. For a similar amount of deviations, the perspective distortion is by far much more damaging than additive white noises. This result is confirmed by the simulation.

Figure 33.6 exhibits the PSV result when the perspective distortion increases from 10% to approximately 15% and 20% respectively. Figure 33.3(b) depicts their depth maps. In perspective projection, the sizes of a target as seen by the camera becomes larger as it moves closer. This affects the PSV clusters in a curious way. The farther target (cube) yields a more concentrated clustering, since it (1) shows a smaller size and (2) suffers less perspective distortion.

Clearly separable clustering patterns are also exhibited in our 3-target simulation. Figure 33.7 is the PSV plot for 3 targets with similar trajectories on the image plane.

REFERENCES

[1] J. Costerira and T. Kanade. A Multi-body Factorization Method for Motion Analysis. *International Conf. on Computer Vision*, June 1995.

[2] S.Y. Kung and J.S. Taur and M.Y. Chiu. An SVD Approach to Multi-Camer- Multi-Target 3-D Motion-Shape Analysis. *Proceedings, International Conf. on Image Processing*, 573-577, 1994.

[3] C. Tomasi and T. Kanade. Shape and motion from image streams under orthography: a factorization method. *Internation Journal of Computer Vision*, 9(2):330–334, 1992.

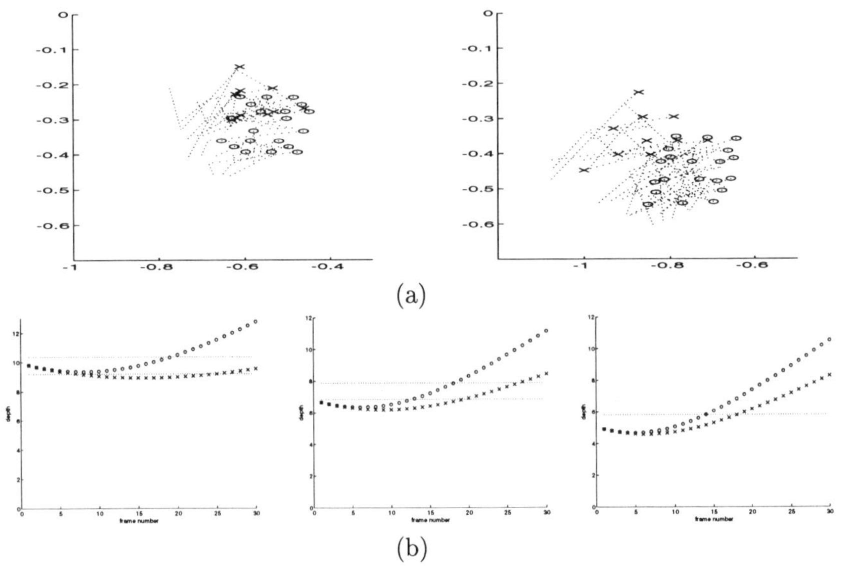

Figure 33.3 (a) Trajectories of the cube and the pyramid, the perspective distortion(P.D.)n is approximately 10%. *Left:* frame number 1 to 3. *Right:* frame number 6 to 8. (b) Depth maps corresponding to 10%, 15%, 20% perspective distortion.

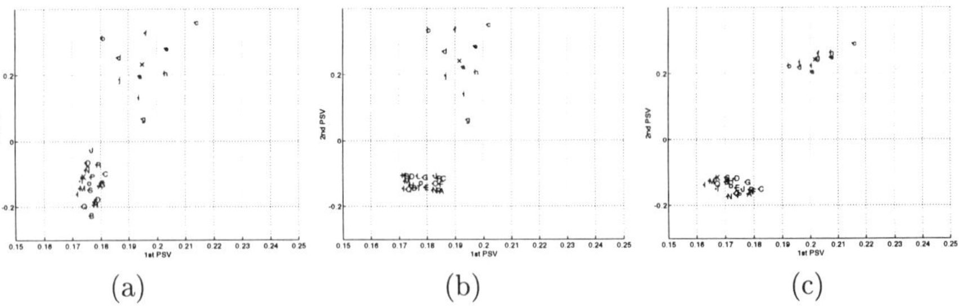

Figure 33.4 Clustering of PSV method for different number of frames (2 targets, (NSR = 10%, P.D. ≈ 10%.) For shape interpretation, see Section 3, Eq. 33.7. (a) frame 6 ~ 10. (b) frame 1 ~ 10. (c) frame 1 ~ 15. The PSVs are approximated by linear combinations of E and S: E serves as separating power and S (shape) is roughly preserved by linear combination.

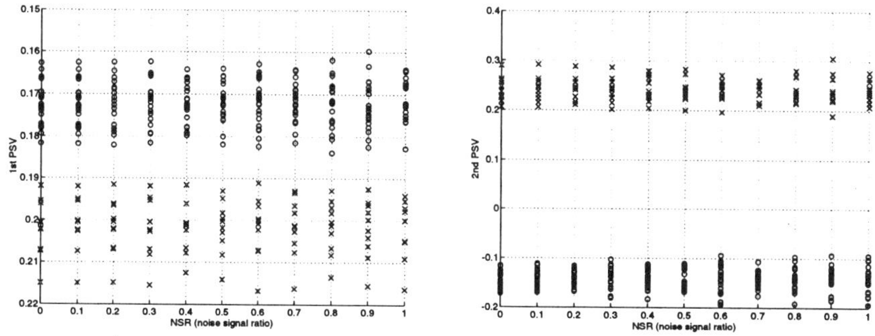

(a) (b)

Figure 33.5 Noise susceptibility of PSV clustering method. (NSR ranges from 0% to 100%.) (a) 1st PSV vs. NSR. (b) 2nd PSV vs. NSR.

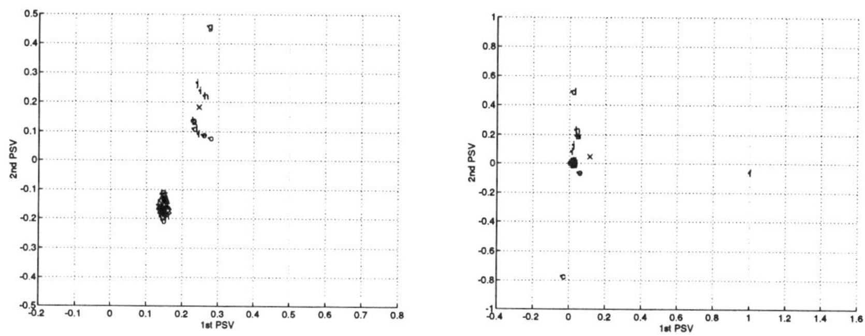

(a) (b)

Figure 33.6 PSV clustering under different levels of perspective distortion. (a) P.D. ≈ 15%: separated clusters. (b) P.D. = ≈ 20%: while the cube is highly concentrated, the pyramid is somewhat spread around.

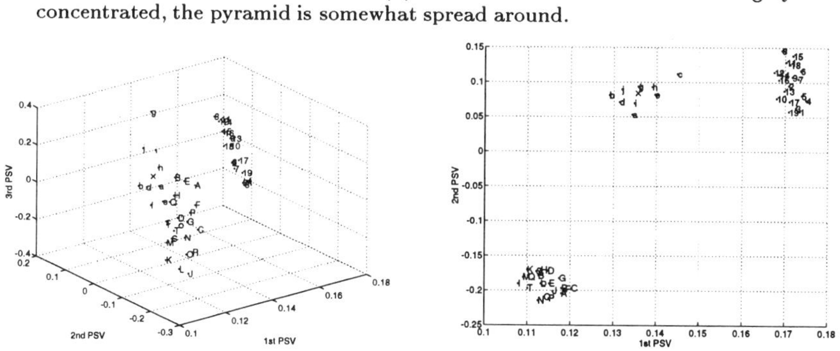

(a) (b)

Figure 33.7 Three-object example. (a) Display of first 3 PSVs. (b) First 2 PSVs.

34

A DETERMINISTIC ANALYSIS FOR LEARNING ALGORITHMS WITH CONSTANT LEARNING RATES

R. Liu*, X. Ling, G. Dong***

** Dept. of Electrical Engr.*
Univ. of Notre Dame
Notre Dame, IN 46556
*** Dept. of Electronic Engr.*
Fudan University
Shanghai, China

This article is dedicated to Professor Kailath on his 60th birthday with affection and best wishes. It is also to cherish long-time friendship and memories.

ABSTRACT

A novel deterministic approach to the convergence analysis of (stochastic) learning algorithms is presented. The link between the two is a new concept of *time-average invariance* (TAI) which is a property of deterministic signals but with applications to stochastic signals. It is then extended to the case of perturbed TAI signals.

1 PRELIMINARY DISCUSSION

Consider an unsupervised learning algorithm (neural network) of the form

$$\frac{d}{dt}x = bf(x, s) \tag{34.1}$$

or its counter part in time-discrete form,

$$x(k+1) = x(k) + bf(x(k), s(k)) \tag{34.2}$$

where $b > 0$ is called the learning rate, $x(t) \in R^n$ and $s(t) \in R^m$ are the state and the input signals respectively. For signal processing purposes, the objective of such learning algorithms is usually to abstract some (statistical) information from the input signals s at a time when the state x converges.

For example, the purpose of Oja's neural network [1], given by

$$x(k+1) = x(k) + b[s(k)s(k)^T x(k) + x(k)x(k)^T s(k)s(k)^T x(k)] \tag{34.3}$$

is to abstract the principal component of the correlation matrix of s. More specifically, it is shown that $x(k) \to e_1$ when $k \to \infty$, where e_1 is the principal component of $R = E\{ss^T\}$.

A fundamental question arises: since Eq. (34.1) or (34.2) is non-autonomous, what do we mean by their equilibrium point x^0; and since x is a stochastic process, in what sense $x(k) \to x^0$?

Ljung [2] provided an answer to the above equations. From Eq. (34.2), an associated ordinary differential equation is first constructed

$$\frac{d}{dt}x = F(x) \tag{34.4}$$

where

$$F(x) = E_s\{f(x,s)\} \tag{34.5}$$

In the expectation $E_s\{\}$, *x is considered as a constant.* Note that $F(x)$ is independent of t (or asymptotically independent of t), if s is stationary (or asymptotically stationary). Under these conditions Eq. (34.4) is autonomous and hence its equilibrium point x^0 is well defined and is given by

$$F(x^0) = 0 \tag{34.6}$$

He then shows that if Eq. (34.4) is asymptotically stable at x^0, then under certain conditions, we have

$$x(k) \to x^0 \textit{with probability one, as } k \to \infty \tag{34.7}$$

where $x(k)$ is the solution of Eq. (34.2). Since the equilibrium points are obtained from an associated *ordinary differential equation*, this approach sometimes is referred to as the *ODE method.* This method is widely used for this type of learning algorithm.

Among these conditions, they include:

(1) $s(\cdot)$ is a sequence of independent random variables

(2) The learning rate is time-varying and $b(k) \to 0$ as $k \to \infty$.

The first condition is technical which may not be necessary. The second condition is fundamental and without which Conclusion (34.7) is not possible.

On the other hand, the second condition is very critical for on-line applications. Because it implies that its capability of learning is diminished with time and hence losing its capability of reacting to new disturbances. In view of this, a learning algorithm with a constant (or a non-diminishing) learning rate is more desirable.

Kushner and Shwartz [3] have investigated the case of constant learning rate for a special case of Eq. (34.2). Using a similar ODE method for equilibrium points, they have shown that

$$x(k) \to x^0 \text{ weakly, as } k \to \infty \tag{34.8}$$

Other authors, see, for example, [4], [5] and [6], have similar results: either the neural network converges with probability one but with diminishing learning rate and hence becomes less robust with time, or it converges in a weaker sense when the learning rate is constant. On the other hand, from many simulation studies, these learning algorithms do converge with a constant learning rate but to a small ϵ-ball, i.e., it does show a "convergence" to a constant with a small variation, typically in its third or fourth significant digit. Furthermore, the condition of statistical independency seems to be not necessary. In fact, many neural networks converge even for (deterministic) periodical inputs. In summary, the current theoretical studies do not indicate the full potential of the ODE method.

In this paper, signals are viewed as deterministic functions of t, but satisfy a property called *time-average invariance.* This is a property of deterministic signals but resembles that of stationary stochastic signals. As such, deterministic-based analysis can be applied to stochastic-like signals. Consequently, the complexity of the convergence analysis is significantly reduced.

2 THE MAIN THEOREMS

In this paper, the analog equation (34.1) is used instead of the equation (34.2). Let a T-time-average operator of a given time-function x be defined by

$$\langle x \rangle_t = \frac{1}{T} \int_t^{t+T} x(\alpha) d\alpha \tag{34.9}$$

Note that the T-time-average of x is a function of t.

Definition 1 A signal s is said to be *time-average invariant* (or TAI), if there exists a T, such that the T-time-average of $h(s)$ is independent of t for any function h, i.e.,

$$\langle h(s) \rangle_{t_1} = \langle h(s) \rangle_{t_2} = \langle h(s) \rangle \text{ for any } h \text{ and any } t_1 \text{ and } t_2$$

The constant T is called the *cycle* of s.

This class of signals includes deterministic periodic signals, as well as realizations from some stochastic process. For example, if s is a stationary and ergodic stochastic process, then its expectation can be approximated by a time-average of its realization if T is large enough, i.e.,

$$\langle s \rangle_t \cong E\{s\} \text{ for all } t \tag{34.10}$$

Under this approximation, such a realization of s can be considered as TAI. Whenever s is TAI, the subscript t is omitted.

Let s be TAI. Take a T-time-average to its right-hand side of Eq. (34.1) with x held as a constant, where T is a cycle of s, i.e.,

$$F(x) = < f(x,s) > \tag{34.11}$$

Note that F is independent of t because s is TAI. As associated equation is then constructed by

$$\frac{d}{dt}x = F(x) \tag{34.12}$$

which is autonomous. Let x^0 be an equilibrium position, i.e., it satisfies

$$F(x^0) = 0 \tag{34.13}$$

Definition 2. A learning algorithm (34.1) is said to be $\epsilon - convergent\ to\ x^0$ if there exists a neighborhood N of x^0, such that $\forall \epsilon > 0$ there exists $B > 0$, so that for any $b \leq B$ any trajectory $x(t)$ issued in N will remain in N, and further

$$\| x(t) - x^0 \| \leq \epsilon \text{ for } t \geq K \tag{34.14}$$

for some finite K.

The following theorem is given in [7].

Let Σ denote the class of signals which are TAI with cycle T, and their T-time-averages are bounded by a constant C.

Theorem 1. Let the following conditions be satisfied:

(C1) $s \in \Sigma$

(C2) System (34.1) is ultimately bounded by X for all $s \in \Sigma$, i.e.,

$$x(t) \in \Omega \text{ as } t \to \infty \text{ for all } s \in \Sigma$$

where $\Omega = \{x | \ \| x \| < X\}$. Furthermore, there exists a set $D \subset \Omega$ such that for every initial condition $x_0 \in D$ and every $s \in \Sigma$, we have $x(t) \in \Omega$ for all t. W.1.g., let D denote the largest such set.

(C3) the function f satisfies a Lipschitz conition in x uniformly in s, i.e., there exists an $L > 0$ such that

$$\| f(x_1,s) - f(x_2,s) \| \leq L \| x_1 - x_2 \|, \ \forall x_1, x_2 \in \Omega, \text{ and } \forall s \tag{34.15}$$

Then the learning algorithm (34.1) is ϵ-convergent to x^0, if its associated equation (34.12) is asymptotically stable at x^0.

In view of Eq. (34.23), if Theorem 1 is valid for cycle T, it is also valid for any cycle $T' < T$.

In [6] and [10], the concept of sweep and block-adaptive updating have been introduced by Kung and Diamantaras for the convergence analysis of temporal neural networks. The current analysis differs from it in that the updating is at every time instant instead of every block-time, and further, Theorem 1 is formally proved.

In this paper, we will show an extention to this theorem. Let s be perturbed by a noise n, i.e.,

$$\tilde{s} = s + n \tag{34.16}$$

where s is TAI and n satisfies the condition that $\| n(t) \| \leq bE$ for some E. $\tilde{s}$ is called a *perturbed TAI signal.*

Theorem 2. Theorem 1 remains valid for its perturbed signal $\tilde{s}$, if condition (C3) is replaced by (C4):
(C4)

$$\| f(x_1, s) - f(x_2, s) \| \leq L \| x_1 - x_2 \|, \ \forall x_1, x_2, \text{ and } s \tag{34.17}$$

and

$$\| f(x, s_1) - f(x, s_2) \| \leq L \| s_1 - s_2 \|, \ \forall s_1, s_2, \text{ and } x \tag{34.18}$$

for some L.

This Theorem shows that Theorem 1 is robust and it can be extended to non-stationery signals. Let us prove Theorem 2.

In order to exploit the TAI property of s, a new variable z is introduced. Let the T-time-average of x be denoted by z, i.e.,

$$z(t) = \langle x \rangle_t = \frac{1}{T} \int_t^{t+T} x(\alpha) d\alpha \tag{34.19}$$

which is called the *forward moving average* of x. It follows that

$$\frac{d}{dt} z(t) = \langle \frac{d}{d\alpha} x(\alpha) \rangle_t \tag{34.20}$$

Lemma 1. Let Conditions (C1), (C2) and (C4) be satisfied and let $\tilde{s}$ be the input. Then, $\forall x_0 \epsilon D, \forall s \epsilon \Sigma$ and $\forall \tau$

$$\| z(\tau) - x(t) \| \leq bA \ \text{ for } \tau \leq t \leq \tau + T \tag{34.21}$$

where

$$A = 2n(LTX + S)e^{bLT} \tag{34.22}$$

and

$$S = \sup_{\tau} \sup_{s \epsilon \Sigma} \int_{\tau}^{\tau+T} \| f(0, \tilde{s}(\alpha) \| d\alpha \tag{34.23}$$

Note that S exists because f satisfies Condition (34.19) and $s\epsilon\Sigma$. The following corollary follows immediately from Eq. (34.22).

Corollary 1.1. For every t,

$$\| z(t) - x(t) \| \to 0 \quad \text{as } b \to 0$$

Now, consider the convergence property of z when s is replaced by the perturbed $\tilde{s}$. For any τ, let

$$e(t) = x(t) - z(\tau) \quad \text{for } \tau \leq t \leq \tau + T \tag{34.24}$$

Substituting this into Eq. (34.21) and with s replaced by $\tilde{s}$, we have

$$\frac{d}{d\tau} z(\tau) = b \langle f(z(\tau) + e(t), \tilde{s}(t)) \rangle_\tau \tag{34.25}$$

This equation defines the dynamics of $z(\tau)$ when perturbed signal $\tilde{s}$ is applied. The relation between Eq. (34.26) and its associated equation (34.12) is now given.

Lemma 2. If the associated equation (34.12) is asymptotically stable at an equilibrium point x^0, then there is a neighborhood N of x^0 such that for every $\epsilon > 0$, there is a B small enough so that for any $b \leq B$, any trajectory $z(t)$ of Eq. (34.26) issued in N remains in N and satisfies the condition

$$\| z(t) - x^0 \| \leq \epsilon \text{ for } t \geq K \tag{34.26}$$

for some finite K.

Proofs of Lemmas 1 and 2 are given in the Appendix.

Proof of Theorem 2. Let the associated equation (34.12) be asymptotically stable at x^0. It follows from Lemma 2 that there exists a neighborhood N of x^0 such that for every $\epsilon > 0$, there is a B_1 small enough so that for any $b \leq B_1$, any trajectory $z(t)$ of Eq. (34.26) issued in N remains in N and satisfies the condition

$$\| z(t) - x^0 \| \leq \frac{\epsilon}{2} \text{ for } t \geq K \tag{34.27}$$

for some K. Following Corollary 1.1, choose B_2 small enough so that $\forall b \leq B_2$

$$\| z(t) - x(t) \| \leq \frac{\epsilon}{2} \text{ for } \forall t \tag{34.28}$$

Now, choose $B = \min(B_1, B_2)$. Then, for any $b \leq B$, we have

$$\| x(t) - x^0 \| \leq \| z(t) - x(t) \| + \| z(t) - x^0 \| \leq \epsilon \text{ for } t \geq K$$

which is Eq. (34.14). Hence the theorem is proved.

3 CONCLUSION

A deterministic approach to the convergence analysis of (stochastic) learning algorithms is presented. The link between the two is a new concept of *time-average invariance* which is a property of deterministic signals but resembles stationary stochastic signals. As such, deterministic-based analysis can be applied to stochastic-like signals. Consequently, the complexity of the convergence analysis is significantly reduced.

The case of constant learning rate b was investigated. It was shown [7] that if the signal s is time-average invariant, then the learning algorithm will converge to an ϵ-ball in a finite time and remain in the ϵ-ball thereafter. The value of ϵ can be made arbitrarily small by choosing b small enough. The location of the ϵ-ball is determined by an associated equation. In this paper, we have shown that the above remains valid even if s is perturbed by a bounded noise.

From the application point of view, the procedure of deterministic method essentially follows the ODE method, but with the following differences:

- The ordinary differential equation is obtained by a time-average, not by an ensemble average.
- The concept of ϵ-convergence is new with which the convergence with constant learning rate can be derived.
- The conditions on the learning algorithm and the input signal are weaker. In particular, they do not require statistical independence on the input signals.

As such, the same ODE approach has been extended to a broader class of algorithms.

Appendix A

In the following proofs, only the case $x_0 \in D$, hence $\| x(t) \| \leq X$, is considered.

Proof of Lemma 1. For any τ, Eq. (34.1), with s replaced by $\tilde{s}$, can be written in an integral form,

$$x(t) - x(\tau) = b \int_{\tau}^{t} f(x(\alpha), \tilde{s}(\alpha)) d\alpha \text{ for } \tau \leq t \leq \tau + T \tag{A1}$$

Applying the Lipschitz condition (34.18) and using Condition (C2) and Eq. (34.24), we have

$$\| x(t) - x(\tau) \| \leq bL \int_{\tau}^{t} \| x(\alpha) - x(\tau) \| d\alpha + b(LTX + S) \tag{A2}$$

It follows from Gronwall inequality [9] that

$$\| x(t) - x(\tau) \| \leq b(LTX + S)e^{bLT} \text{ for } \tau \leq t \leq \tau + T \tag{A3}$$

Therefore,

$$\| x(t) - x(\tau) \| \leq \frac{bA}{2n} \text{ for } \tau \leq t \leq \tau + T \tag{A4}$$

where

$$A = 2n(LTX + S)e^{bLT}$$

which is Eq. (34.23). It follows from Eq. (A4) that for each component i, we have

$$\| x_i(t) - x_i(\tau) \| \leq \frac{bA}{2n} \text{ for } \tau \leq t \leq \tau + T \tag{A5}$$

Choose t_i in $[\tau, \tau + T]$ so that

$$x_i(t_i) = z_i(\tau) \text{ for } i = 1, 2, \cdots, n \tag{A6}$$

Then for $\forall\tau$, we have

$$\| x_i(t) - z_i(\tau) \| \leq \| x_i(t) - x_i(\tau) \| + \| x_i(\tau) \| \leq \frac{bA}{n} \tag{A7}$$

for $i = 1, 2, \cdots, n$. Therefore,

$$\| x(t) - z(\tau) \| \leq bA \text{ for } \tau \leq t \leq \tau + T$$

which is Eq. (34.22). Lemma 1 is proved.

Proof of Lemma 2. Without loss of generality, we assume that the equilibrium point of the associated equation (34.12) is at the origin, i.e., $x^0 = 0$. Since it is asymptotically stable, there exists [8] a Lyapunov function $V(x)$ in a neighborhood of the origin. Without loss of generality, choose the neighborhood $N \subset D$ as follows

$$N = \{x | V(x) \leq v\} \tag{A8}$$

where v is small enough so that in N, V is positive definite and $\dot{V}$ is negative definite. This implies that

$$\dot{V}(x) = b\nabla V(x) \cdot \langle f(x, s) \rangle < 0 \ \ \forall x \in N, \text{ and } x \neq 0 \tag{A9}$$

where ∇V is the gradient of V, and the dot denotes the dot product. Applying the same Lyapunov function V to Eq. (34.26), we have

$$\dot{V}_e(z) = b\nabla V(z) \cdot \langle f(z + e, s + n) \rangle_\tau \ \ \forall z \in N \tag{A10}$$

They can be estimated by

$$\dot{V}_e(z) \leq \dot{V}(z) + b^2 ML(A + E) \tag{A11}$$

where M is the upper bound of $\| \nabla V(z) \| \leq M$ in N. Here we have used the Lipschitz conditions (34.18) and (34.19) and Lemma 1. Choose $\epsilon > 0$ small enough, so that $\delta < v$, where v is given by Eq. (A8) and δ is given by the following equation

$$\delta = \min\{V(z)| \text{ for } \| z \|= \epsilon\} \tag{A12}$$

Hence, the set defined by

$$Z = \{z|\delta \leq V(z) \leq v\} \tag{A13}$$

is non-empty and is contained in N. In $Z, \dot{V}_e$ is negative and bounded away from zero. Hence, B can be chosen small enough so that $\forall b \leq B$

$$\dot{V}_e(z) \leq -\delta_1 < 0 \text{ for } z \in Z, \text{ and for all } \tau \tag{A14}$$

for some δ_1. Since V is positive definite in Z, any solution $z(t)$ of Eq. (34.26) issued in Z, will remain in Z until it reaches the surface $V(z) = \delta$ for some finite time K. Denote $Z_\delta = \{z|V(z) = \delta\}$. Then, $z(K) \in Z_\delta$ and Z_δ is contained in the ϵ-ball. In view of Eq. (A14), $\dot{V}_e(z)$ is negative on Z_δ, and bounded away from zero. Hence,

$$\| z(\tau) \| \leq \epsilon \text{ for } \tau \geq K \tag{A15}$$

which is Eq. (34.27). Hence Lemma 2 is proved.

Acknowledgements

This work is supported in part by a grant from the Office of Naval Research, N00014-91-J-1461.

REFERENCES

[1] E. Oja, "A Simplified Neuron Model as a Principal Component Analyzer", *J. Math. Biology*, Vol. 15, pp. 267-273, 1982.

[2] L. Ljung, "Analysis of Recursive Stochastic Algorithms", *IEEE Trans. on Automatic Control*, pp. 551-575, August 1977.

[3] H. Kushner and A. Shwartz, "Weak Convergence and Asymptotic Properties of Adaptive Filters with Constant Gains", *IEEE Trans. on Information Theory*, pp. 177-182, March 1984.

[4] E. Oja and J. Karhunen, "On Stochastic Approximation of the Eigenvectors and Eigenvalues of the Expectation of a Random Matrix", *J. of Math. Ana. and Appl.*, pp. 69-84, 1985.

[5] T. D. Sanger, "Optimal Unsupervised Learning in a Single-Layer Linear Feedforward Neural Network", *Neural Networks*, Vol. 2, pp. 459-473, 1989.

[6] R. Liu, Y. F. Huang, and Z. T. Ling, "A Novel Approach to the Convergence of Neural Networks for Signal Processing", *IEEE Trans. on Circuits and Systems-I: Fundamental Theory and Applications*, pp. 187-190, March 1995.

[7] W. Hahn, *Theory and Applications of Lyapunov's Direct Method*, Prentice-Hall, 1963.

[8] M. Vidyasagar, *Nonlinear System Analysis*, Prentice-Hall, 1993.

[9] S. Y. Kung, *Digital Neural Networks*, Prentice-Hall, Inc., 1993.

35

TOMOGRAPHIC DECONVOLUTION OF ECHOGRAMS

Enders A. Robinson

Maurice Ewing and J. Lamar Worzel Professor of Geophysics
Columbia University
New York, NY 10027 USA

This paper is dedicated to Thomas Kailath on the occasion of his sixtieth birthday, a scientist with both genius and taste: the genius to produce works of excellence and the taste to imbue these works with the finest impulses of the imagination.

ABSTRACT

This paper gives a new method of deconvolution called tomographic deconvolution. The interior of an object or body may be inaccessible, so the structure of its interior must be found indirectly. Of particular interest are imaging techniques that generate a cross-sectional picture of the unknown interior. To obtain such a picture a non-destructive energy source can be used to probe the body, and the resulting wave motion can be collected at various points on the surface of the body. The determination of the interior of an object from data collected on the surface represents an inverse medium problem. For our purposes, we assume that a source of man-made energy generates a incident signal that penetrates the body. The incident signal is called the source wavelet. The body is bounded by two sides with the source on one of the sides. In the simplest case there are two detectors. The first detector is on the same side as the source and the second detector is on the other side. The first detector records the wave motion reflected by the heterogeneities within the body. This record is called the echogram. The second detector records the wave motion transmitted through the body. This record is called the tomogram. Thus, for a given source signal, two response signals are recorded, namely, the echogram and the tomogram. These two records provide critical information about the internal structure of the medium. Often, the echogram or the tomogram, as the case may be, can be used directly to depict the internal structure of the body. However, in many cases the data is smeared and clouded by two effects. One is the effect produced by the shape of the source wavelet. The other is the effect produced reverberations due to multiple reflections within the body. Often only a hazy picture, if any, can be reconstructed from the observed records. In this paper, a new method, called tomographic deconvolution of the echogram, is proposed. This method makes use of the information in both the echogram and the tomogram. More precisely, the tomogram is used to deconvolve the echogram. This deconvolved signal is readily computed because neither involved statistical estimation methods nor sophisticated numerical approximations are required. Because the proposed deconvolution method does not require knowledge of the source wavelet, the source wavelet does not have to be recorded or otherwise measured. The proposed deconvolution process automatically removes the unknown source wavelet entirely, and in addition clears up a significant component of the reverberations due to internal multiples. The tomographic-deconvolved echogram gives a better picture of the internal structure of the body than either the echogram or the tomogram alone can possibly give.

1 INTRODUCTION

Inverse medium problems are encountered in many areas of engineering, physics, geophysics, and medicine. Examples occur in reflection seismology, in radar, in sonar, and in the medical probing of the human body. Deconvolution and related inversion methods [1, 2, 3] as well as spectral methods [4, 5] play a significant role in obtaining meaningful solutions. In order to process the data digitally, it is necessary to use a discrete approximation to the continuous body. We cannot, may not, or will not open up, mutilate, or destroy the body, but we will probe its interior with non-destructive signals. A standard approach is to represent the body by a discrete layered model. This model represents a useful approximation to handle the propagation of waves through an inhomogenous medium and had found wide use in solving inverse problems involving wave motion. The internal structure of the body is the unknown quantity that we most desire. Our objective is to build a picture of the internal structure from the observed data that we record on the surface of the body. These observed data are the tomogram and the echogram.

The internal structure of an inhomogenous body can be characterized by its impedance function. Let us consider a body where the impedance is a function of the x dimension only. Using this functional dependence, it is convenient to measure on the x axis, not distance, but travel time. In this way, space and time are measured in terms of the same space-time unit. This unit, known as the wave-second, represents the distance that a material wave propagates in one second. This space-time convention is commonplace in astronomy where distance is measured in terms of light-years. A light-year is the distance that a light wave propagates in one year. Commuters also use this space-time convention when they describe their town as being 50 minutes away from New York City. In seismic work the discrete time unit is often chosen to be one millisecond. Consider a material in which the seismic wave velocity is 10,000 feet per second, or, in other words, 10 feet per millisecond. For such a material one millisecond would represent a distance of 10 feet. By using this space-time convention in our figures, we are able to use the same horizontal scale for both the x axis of the body and the time axes of the various signals.

Let the x direction be the horizontal direction, and let the source be on the left side of the body. The source, activated at time zero, gives rise to a right-going plane wave as input into the body. This input is called the "source wavelet". The source wavelet often is not measured and so in such cases would also represent an unknown. The recording of the source wavelet can present extra difficulties or added expenses, so a deconvolution method that does not require knowledge of the source wavelet is at a premium. Such is the deconvolution method proposed in this paper. The source wavelet need not be recorded, measured, or estimated in order to apply this deconvolution method to observational data.

The source wavelet serves only a tool to help us determine the internal structure of the body, and, in the final analysis, the effects of the source wavelet have to be eliminated to obtain the clear picture of the interior. For a sharp energy source, such as an explosion in seismic exploration, a large amount of high amplitude energy is injected into the Earth over a short time span. Such dynamite explosions can be undesirable in certain locations, so alternative sources are used instead. In cases when the medium should be disturbed as little as possible, a low amplitude source is required. By spreading a low amplitude input signal over a comparatively long time span, it is possible to reach a desired total input energy level. The vibroseis source in seismic exploration represents such a case. Another such case is the air-gun source in marine seismic exploration. The air-gun was designed so as to minimize damage to fish and other marine life. Like the internal structure of the body, the source wavelet represents an unknown quantity, but it is one that we want to eliminate. Its presence can only cloud the desired picture of the inside of the body. For the purpose of this paper, we assume the source wavelet is not recorded. Knowledge of the source wavelet is not needed, because the method of tomographic deconvolution proposed here automatically eliminates the effects of the source wavelet, except for a constant scale factor.

2 THE PROPOSED DECONVOLUTION METHOD

There are two outputs that we record on the surface of the body. One output occurs on the right-hand side of the body, and represents the wave motion transmitted through the body. This transmission response to the given input is called the "tomogram" [6,7]. The other output occurs on the left-hand side of the body (namely, the same side as the input), and represents the wave motion reflected by the inhomogeneities within the body. This reflection response to the given input is called the "echogram". Thus the system has one input and two outputs. See Figure 35.1. The tomogram is a right-going plane wave emitted from the right-hand interface. The echogram is a left-going plane wave emitted from the left-hand interface. If the special case when the source wavelet is an impulsive right-going unit spike occurring at time zero, the resulting tomogram and the echogram are each impulse responses.

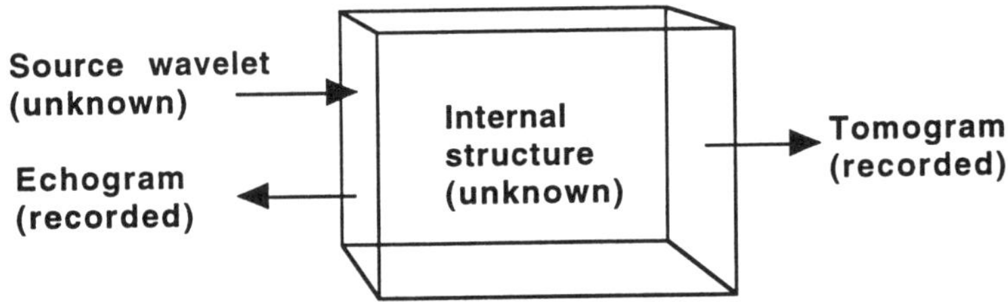

Figure 35.1 The model. A source wavelet of unknown shape is sent into an heterogeneous body with an unknown structure. The recorded transmitted signal is called the tomogram and the recorded reflected signal is called the echogram. The problem is to determine a picture of the internal structure from these two recorded signals.

The recorded tomogram and the recorded echogram are now used in the deconvolution process to yield the signal labeled "deconvolved". In the Figures, both the tomogram and the echogram are plotted from the time of their first breaks. Each is a discrete signal made up of a sequence of amplitude coefficients. In its bare essentials, the proposed deconvolution process can be described as follows. Divide the z-transform of the echogram by the z-transform of the tomogram. The coefficients of the result give the deconvolved signal. This deconvolved signal, so computed, except for a constant scale factor, is the desired approximation to the internal structure of the body. Generally displays are normalized before they are plotted, so the effect of the constant scale factor disappears in the final picture. The deconvolved signal appears under the title "deconvolved" in the figures that follow. See Figure 35.2.

The theory of this deconvolution method is based on the Robinson recursion [2]. In the above paragraph we have given the bare-bones theoretical description of the deconvolution process. In practice, one is confronted with noisy data,

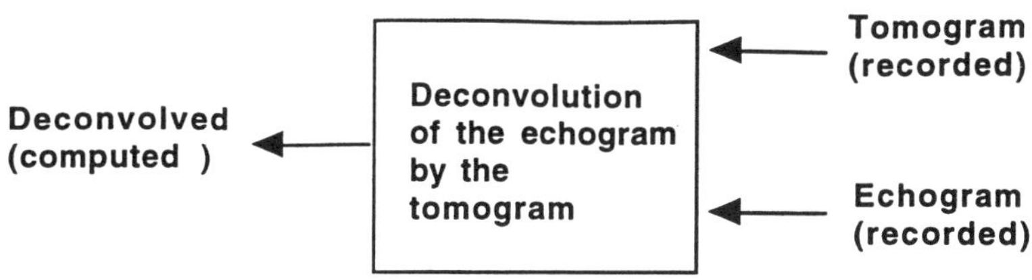

Figure 35.2 The deconvolution process. The recorded echogram is deconvolved by use of the recorded tomogram. The result is the deconvolved signal.

and then a least squares approach [2] can be used to implement the deconvolution process. The proposed tomographic method of deconvolution is more robust than the more precise method known as dynamic deconvolution[2]. For more accurate results in deconvolution, iterative improvement methods can be applied.

The structure of each of the Figures 35.4, 35.5, 35.6, and 35.7 is shown in Figure 35.3.

Because of the space-time convention, the x axis and the time axis shown in the diagrams of each Figure are compatible, so direct comparisons of the diagrams can be made. In particular, one should observe that in all cases the deconvolved diagram give a significantly better picture of the internal structure than either the tomogram or the echogram. This evidence supports the worth of the tomographic deconvolution process proposed. As a final step, the tomographic deconvolved signal can be used to compute an approximation to the acoustic impedance of the body.

Figure 35.4 shows the case of an impulsive source, that is, the source wavelet is a spike. Because the source wavelet is an impulse, the echogram and the tomogram suffer no masking effects due to the source. However, they do suffer masking effects due to reverberations due to internal multiple reflections. The deconvolved record has many of the reverberations removed, and thus gives a clearer picture of the unknown structure.

Figure 35.5 shows the case of an oscillating source wavelet. The internal structure is identical to that of Figure 35.4. The masking effect due to the source

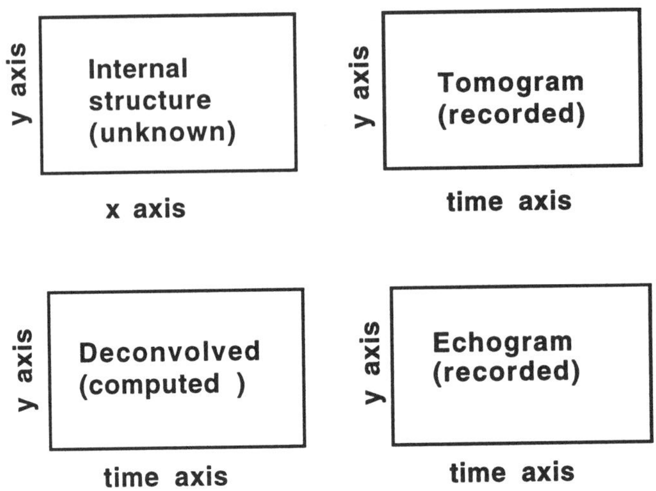

Figure 35.3 An explanation of the figures that follow. The upper left diagram shows the unknown internal structure. The lower left diagram shows the tomographic deconvolved signal. How well the "deconvolved" diagram agrees with the "internal structure" diagram indicates the success of the method. The two right hand diagrams give the recorded information that is used to compute the deconvolved diagram, namely, the "tomogram" and the "echogram". Without tomographic deconvolution, it is apparent that these two right hand pictures do not convey much easily interpretable information.

wavelet together with the masking effects due to reverberations makes both the tomogram and the echogram unintelligible. The deconvolved record completely removes the source wavelet and removes the same amount of reverberations as deconvolved record shown in Figure 35.4. In fact, the deconvolved record in Figure 35.5 is the same as that in Figure 35.4, as the source wavelet, whatever it is, is always completely removed. The removal of the source wavelet alone would justify using the proposed tomographic deconvolution method. The additional effect of removing a significant part of the reverberations makes the proposed method even more worthwhile. A further example of the tomographic deconvolution process is given in Figures 35.6 and 35.7.

3 THE MODEL USED

The stratified model of the interior of the body represents a discrete approximation to a continuous inhomogenous medium. For the purposes of this model, the body is mathematically sliced into many thin layers normal to the x direc-

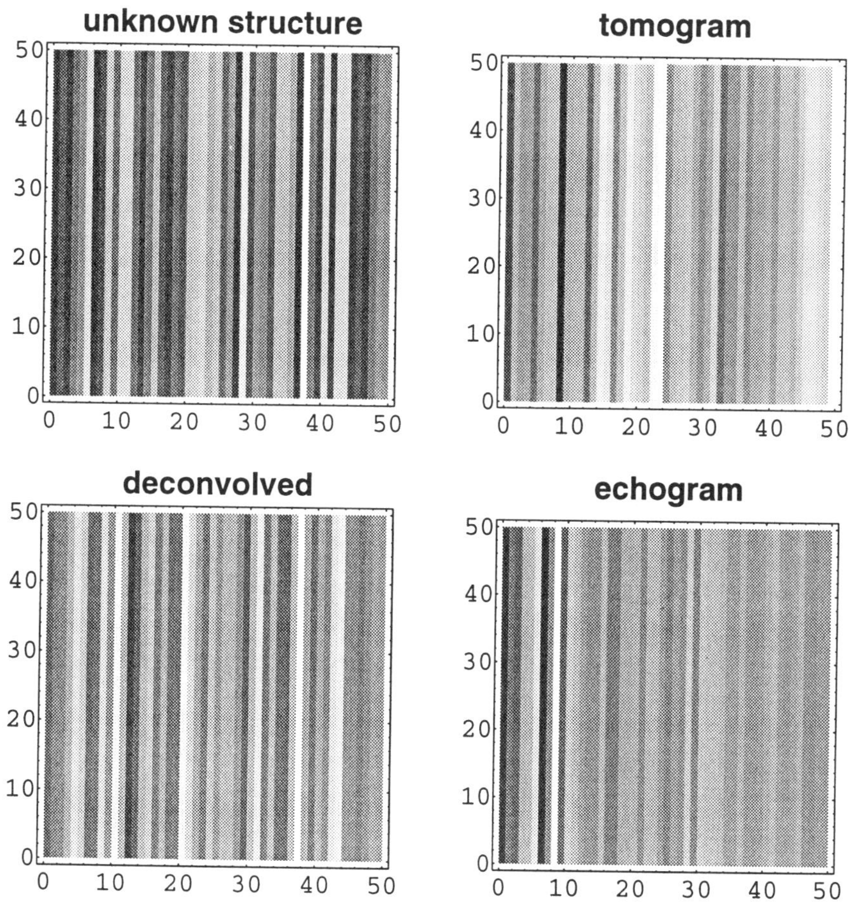

Figure 35.4 Case of impulsive source for a body with a given internal structure. We see that the echogram gives an indication of the internal structure, but it is not nearly as good as the clear picture given by the deconvolved diagram.

tion. This theoretical division of the body into thin layers produces a stratified medium characterized by the interfaces between the layers. An arbitrary time unit is chosen, and the thickness of each layer is chosen as one-half of the time unit. The reason for the choice for the thickness is so that the two-way travel time (forward and back) in each layer is equal to one unit. In the mathematical construction of the stratified model, it is assumed that the impedance is constant within each layer, but the impedance can change from layer to layer.

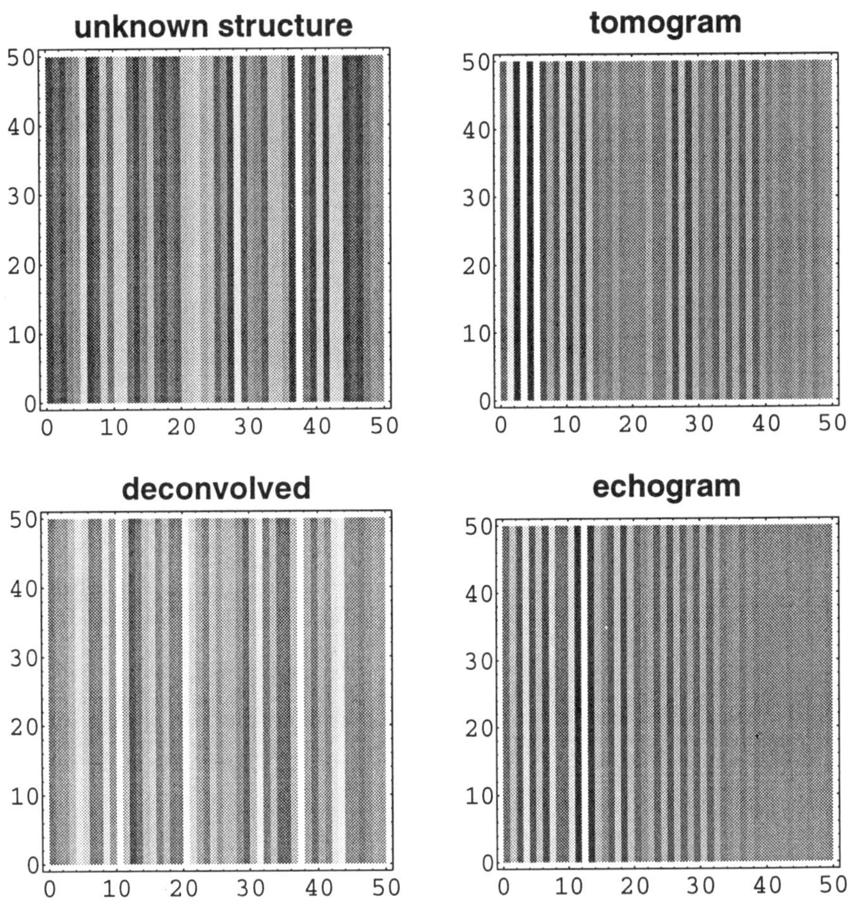

Figure 35.5 Case of an oscillating source wavelet. The body has the same internal structure as that shown in Figure 35.4. Now the echogram and tomogram cannot be interpreted at all, but the deconvolved record yields the same clear approximation of the internal structure as before.

As is well known, an impedance contrast gives rise to a reflection coefficient. Each interface separating two adjacent layers with different impedances has a non-zero reflection coefficient. The greater the impedance contrast between the two adjacent layers, the greater is the magnitude of the reflection coefficient.

As a matter of convenience, let us measure pulse amplitudes in terms of a unit that represents the square root of the energy of the pulse. In this way, the square of the amplitude of a pulse gives the energy of the pulse. Also, we assume that there is no dissipation of kinetic energy into heat. Thus all the source energy imparted into the body can be accounted for, over time, in terms of the resulting wave motion.

By definition, each reflection coefficient is a local reflection coefficient, or a so-called Fresnel reflection coefficient. Such a reflection coefficient characterizes the interface separating two adjacent strata in isolation from all the other strata. The Fresnel reflection coefficient depends only on the impedances of the two strata in question, and not on the impedances of the remaining strata. A pulse normally incident on such an isolated interface is divided into a reflected pulse and a transmitted pulse. Energy is conserved. As a result, the energy incident on an interface is equal to the sum of the energy transmitted through the interface and the energy reflected from the interface. As a consequence, the magnitude of the Fresnel reflection coefficient must less than one. The magnitude of the reflection coefficient does not depend upon the direction in which the pulse travels through the interface, but the sign of the reflection coefficient does. For a given interface, the reflection coefficient for a left-gong pulse is the negative of the reflection coefficient for a right-gong pulse.

The transmission coefficient deals with a pulse traveling across an isolated interface. According to our convention, the pulse amplitudes are measured in terms of the square root of energy. The Fresnel transmission coefficient for a normally incident pulse must be less than one in magnitude. In fact, the transmission coefficient and the reflection coefficient for a given interface satisfy a Pythagorean relationship, namely, the sum of their squares is equal to one. Thus the transmission coefficient is given by the square root of the quantity one minus the reflection coefficient squared. It is convenient to choose the positive square root in both the case of a left-gong incident pulse and the case of a right-gong incident pulse. As a result, the transmission coefficient for an interface is the same for either direction. In summary, for a normally incident pulse encountering an interface, the Fresnel reflection coefficient is a real number between minus one and plus one, and the Fresnel transmission coefficient is a real number between zero and plus one.

A plane wave is the simplest form of wave motion. We consider plane-wave motion normal to the interfaces, that is, waves traveling in the x direction. The sequence of reflection coefficients associated with the entirety of the interfaces represents the internal structure of the body. Thus what we call the "internal structure" is the sequence of Fresnel reflection coefficients for a right-going pulse. Mathematically, the internal structure is represented by a sequence of numbers each less than one in magnitude. This sequence of reflection coefficients, once they are determined, can be used to compute the required impedance function of the body in a straight-forward manner. In the figures, the diagram labeled internal structure depicts the sequence of Fresnel reflection coefficients. The internal structure is the unknown in our problem.

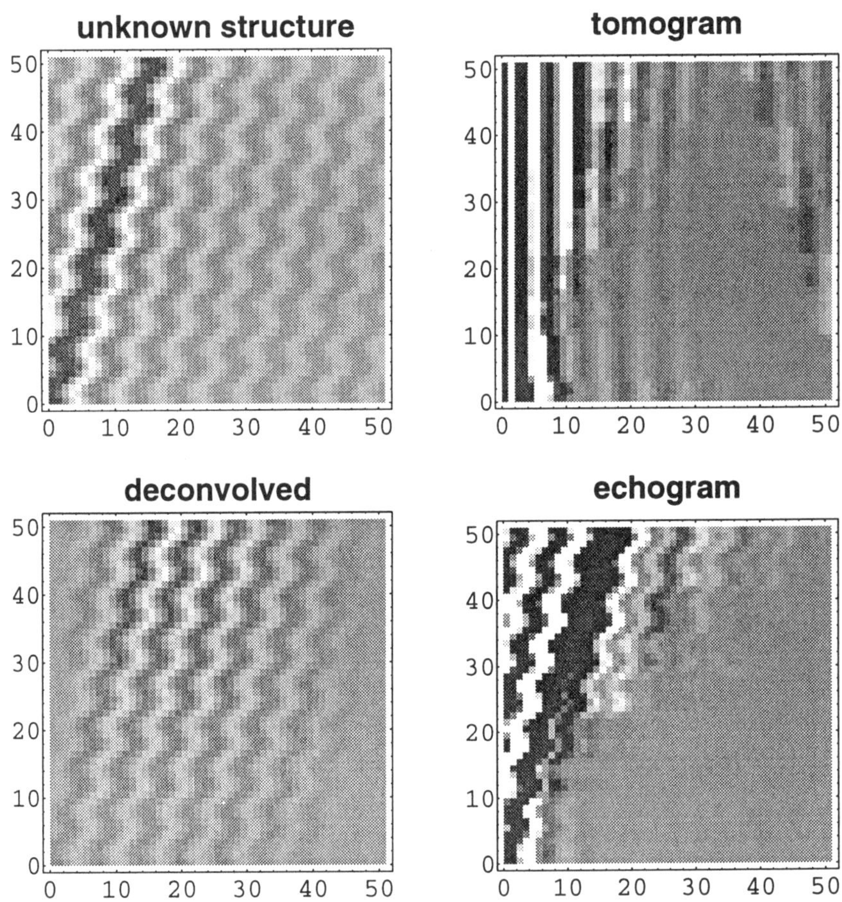

Figure 35.6 Case of impulsive source for a body with a complicated internal structure. Even with no masking effects due to the source wavelet, the echogram suffers from the masking effects due to reverberations. The deconvolution removes much of the reverberating energy, and thus the deconvolved signal gives a clearer picture of the unknown structure than either the echogram or the tomogram can give.

4 CONCLUSION

This paper proposes a method of deconvolution that we call tomographic deconvolution. The problem is to find a picture of the internal structure of an body without cutting into the body or destroying its integrity in any way. A

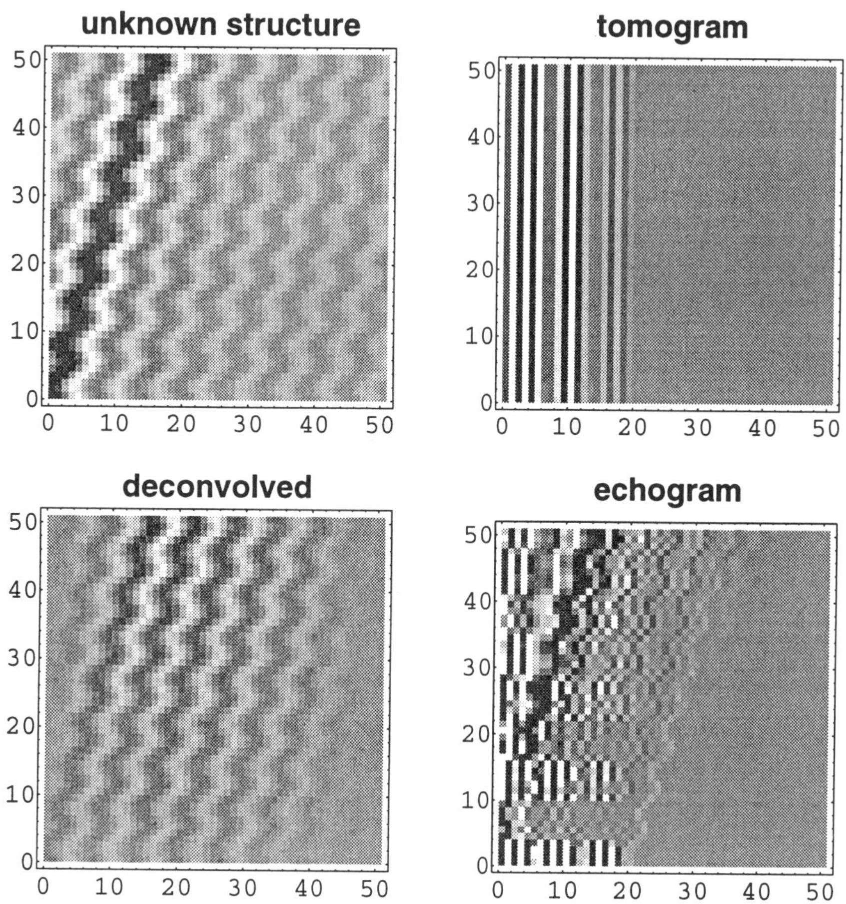

Figure 35.7 Case of an oscillating source wavelet. The body has the identical internal structure as that in Figure 35.6. The echogram and tomogram cannot be interpreted at all, but the deconvolved record yields the same clear approximation of the internal structure as that shown in Figure 35.6. Without any knowledge of the source wavelet, the deconvolution process has completely removed the source wavelet. The removal of the source wavelet is in addition to the removal of much of the reverberating energy due to the multiple reflections occurring within the body.

source wavelet is introduced into the body as a probe. No knowledge of the source wavelet is required for the proposed deconvolution method, so the source wavelet does not have to be measured or recorded. This situation is fortunate because often as a good measurement of a source signal is difficult to capture. The echogram is the reflection response to the source wavelet and the tomogram as the corresponding transmission response. The echogram is measured on the same side of the body as the source, and the tomogram is measured on the opposite side of the body. In the proposed tomographic deconvolution process, the recorded tomogram is used to deconvolve the recorded echogram. The result is a deconvolved signal that produces a reasonably faithful depiction of the interior of the body. In this depiction, the entirety of the disturbing effect caused by the source wavelet is removed, and a significant portion of the disturbing effect caused by the reverberations due to internal multiple reflections is also removed.

REFERENCES

[1] Mendel, J. M., *Optimal Seismic Deconvolution, An Estimation-based Approach,* Academic Press, New York, 1977.

[2] Robinson, E. A., *Seismic Inversion and Deconvolution*, Pergamon Press, New York, 1984.

[3] Gardner, W. A., *Statistical Spectral Analysis, A Nonprobabilistic Theory,* Prentice Hall, 1988.

[4] Marple, S. L, *Digital Spectral Analysis with Applications*, Prentice Hall, 1987.

[5] Robinson, E. A., "Spectral approach to geophysical inversion by Lorentz, Fourier, and Radon transforms," *Proc. IEEE*, **70**:1039, 1982

[6] Lo, T. W., and Inderweisen, P., *Fundamentals of Seismic Tomography,* Society of Exploration Geophysics, Tulsa, Oklahoma, 1994.

[7] Stewart, Robert, *Exploration Seismic Tomography: Fundamentals*, Society of Exploration Geophysics, Tulsa, Oklahoma, 1993.

36

ANTENNA ARRAYS IN MOBILE COMMUNICATIONS

Björn Ottersten

Signal Processing, Department of Signals, Sensors and Systems
Royal Institute of Technology
S–100 44 Stockholm, Sweden

Dedicated to Tom Kailath, my mentor and friend, on his 60th birthday.

ABSTRACT

This paper describes the utilization of antenna arrays at the base stations of mobile communication systems. Multiple antennas can provide a processing gain to increase the base station range and improve coverage. Also, by exploiting the angular selectivity of an antenna array at the base stations of a wireless system, users may be spatially multiplexed to increase system capacity. We address several aspects of the reception and transmission problems that arise when the spatial dimension is considered. Both the forward and reverse channels are discussed and an overview of some interesting research topics in the area is presented.

1 INTRODUCTION

The dramatic expansion of mobile communications over the last years has emphasized the importance of efficient use of frequency bandwidth. There is an increasing demand for capacity in wireless systems which traditionally directly translates into a demand for more bandwidth which is quite limited. Also, the infrastructure investment costs are often a limiting factor when deploying a new system that must have wide area coverage. It is therefore of great interest to increase the range by employing antenna arrays.

Traditional telecommunication schemes multiplex channels in frequency and/or time. However, the spatial dimension is in general used in a very rudimentary fashion by, for example, using certain channels in certain geographical areas (cellular systems!cell planning). By employing an array of antennas, it is possible to multiplex channels in the spatial dimension just as in the frequency and time dimensions. Recently, a more efficient use of the spatial dimension has appeared as a means of increasing the capacity in wireless communication systems without exploiting additional bandwidth [1, 2]. By employing an array of antennas at the base stations of a cellular system when receiving and transmitting over the communication channel, the spatial dimension may be used to separate several users operating in the same channel. This more efficient use of the spatially dimension will be termed Spatial Division Multiple Access (SDMA).

Using an array of antennas is also a way of increasing the gain of the system thereby extending range and improving coverage. Of course, the hardware requirements are more demanding but this permits a sparser infrastructure and will often be more cost effective. In general, increasing the range of cellular systems is of great interest initially, for example, when deploying the new PCS system in the US. However, we may expect that demand for increased system capacity will follow shortly after adequate coverage is achieved.

2 EXPLOITING THE SPATIAL DIMENSION

To achieve increased range in a cellular systems, it may be argued that the mobile to base communication (up link) is the critical link. It is desirable that the mobiles operate at low powers and thus, for acquisition, the base stations must be able to detect weak signals of short duration in a noisy and possibly interfering environment. In the down link (base to mobile communication), increased range may be achieved by for example increasing the transmit power.

When receiving communication signals at an antenna array, the proposed signal processing methods for distinguishing different messages, can be grouped in two main categories; those that exploit array response information and those that do not. Since in general, the array configuration is known, the array response is known (or may be calibrated) for an incident wave from a given location [3]. This assumes that the scenario is well behaved in the sense that the propagation may be modeled by a single, or small number of paths. These methods will be referred to as using *directional information* and include techniques proposed in e.g. [1, 4, 5]. The other class of methods, make little or no assumptions on the array response but rely on other properties for separating the signals. In [4, 6], a reference signal is assumed available which may be correlated with the array output to achieve signal separation. This reference signal may be a known training sequence, a known code sequence [7], or may be generated by feeding back decisions [8]. There are a number of methods that make use of the constant modulus property or finite alphabet of communication signals to separate them [9, 10].

To achieve increased system capacity by employing an array of antennas at the base stations, the frequency reuse distance may be decreased [11, 12] or the frequency channels may be reused with in a cell [1] (or a combination thereof). In both cases, the interference in the system induced by other users is of course increased. In the up-link, this is manifested by the cross-talk problem. Mobiles operating on the same channel (frequency/time slot) with dramatically different signal amplitudes caused by, for example, fading are difficult to separate. It is difficult to adequately suppress the stronger signal when estimating the weaker signal resulting in cross-talk. In some sense the down link problem may be even more sever, especially in frequency division duplex (FDD) systems [1]. The fading caused by local scattering around the mobile (or the base station) is observable in the up-link but unobservable in the down link due to the uncorrelatedness of the fading processes at the different frequencies. The up and down link channels are not reciprocal. The down link problem has received limited attention. In [11] a method is proposed which does not exploit directional information whereas in [1] a model based approach using this information is proposed.

3 A SPATIAL CHANNEL MODEL

In [2, 13] a model of the flat fading due to local scattering is developed taking the spatial dimension into account. The propagation between the mobile and the array is modeled as a superposition of a large number of rays originating from local scatterers in the vicinity of the mobile. We assume independent scattering, an angular distribution of the scatterers which is Gaussian (as seen from the array), and that the relative time delays for different propagation paths are small compared to the inverse of the bandwidth of the communication signal.

Assuming a uniform linear array with element spacing Δ in wavelengths, the signal received at the array may then be modeled as

$$\mathbf{x}(t) = \mathbf{v}s(t) + \mathbf{n}(t) \tag{36.1}$$

$$\mathbf{v} \in N(0, \mathbf{R}(\theta, \sigma)) \tag{36.2}$$

$$\mathbf{R}(\theta, \sigma) \approx p\, \mathbf{a}(\theta)\mathbf{a}^*(\theta) \odot \mathbf{B}(\theta, \sigma) + \sigma_n^2 \mathbf{I} \tag{36.3}$$

$$\mathbf{a}(\theta) = [1, e^{j2\pi\Delta \sin\theta}, \ldots, e^{j2\pi\Delta(m-1)\sin\theta}]^T \tag{36.4}$$

$$\{\mathbf{B}(\theta, \sigma)\}_{kl} = e^{-2[\pi\Delta(k-l)]^2\sigma^2\cos^2\theta} \tag{36.5}$$

where, $\mathbf{x}(t)$, is a complex valued $(m \times 1)$ vector, $s(t)$ complex envelop of the transmitted signal with power p, $\mathbf{n}(t)$ is the additive spatially white noise, $\odot$ denotes element-wise multiplication, and $\mathbf{v}$ is a complex, Gaussian random vector with a distribution function parameterized by the nominal direction to the mobile, θ, and the angular spread (standard deviation), σ, see Figure 36.1.

Equations (36.1-36.5) model the Rayleigh fading of the channel. If a direct path is present giving rise to a Rice distribution of the received amplitude, this may

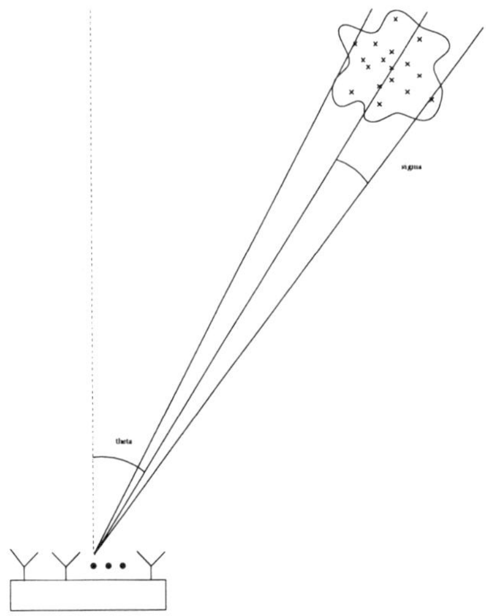

Figure 36.1 Geometry of the model characterizing the local scattering in the vicinity of the mobile.

be modeled by introducing a mean which is a scaled version of $\mathbf{a}(\theta)$ in (36.2). The vector $\mathbf{a}(\theta)$ is often termed the array response vector and represents the array output to a point source from direction θ. Frequency selective fading may be incorporated in the model by adding time delayed versions of the signal with different spatial characteristics. Also, interfering sources on the same frequency channel may easily be added to the model.

Propagation Modeling and Data Experiments

The spatial channel model described above has been validated against experimental data collected by Ericsson Radio Systems. In the field experiments, a transmitter has been placed in urban areas approximately 1km from the receiving array [12]. The data has been processed to gain insight into propagation effects as well as into the behavior of some receiving algorithms. The standard deviation, σ, of the angular distribution is a critical parameter for SDMA systems, [1]. In [2, 13] the angular spread is found to be between two and six degrees in the experiments when the transmitter is placed 1km from the receiving array.

4 SDMA IN THE UP LINK

Here, we will discuss some observations related to the model presented above. Due to the local scattering, the wavefront at the array represented by $\mathbf{v}$ is not planar, i.e.,

$$\mathbf{v} \neq \mathbf{a}(\theta), \quad \text{for any } \theta\,. \tag{36.6}$$

This may be interpreted as spatial diversity, i.e., the correlation between antenna elements decreases with distance, this is seen in the structure of the second moment of $\mathbf{v}$ in (36.3). The flat fading becomes less sever at the array as the diversity increases, i.e., σ increases. Techniques that make no use of directional information, e.g., [6] efficiently exploit this fact and perform better as the angular spread increases. Methods that are based on directional information, $\mathbf{a}(\theta)$, for estimating the signals [4, 14] will in general deteriorate as the angular spread becomes larger. These methods which are related to traditional *beamforming* techniques, are derived from a point source model. This behavior is not surprising since $\mathbf{v}$ will not correspond to an array response vector for any θ.

If the goal is to increase the range of a cellular system, this model error is not critical, however, the situation is quite different when attempting to host multiple mobiles in the same cell on the same frequency channel. This is most noticeable in situations where multiple mobiles are present and the signal amplitudes differ significantly. When estimating the weaker signal, it will be difficult to impose orthogonality to the stronger signal when constrained to the point source model resulting in cross talk between the spatial channels.

Note that the signal at the array (neglecting noise) is still a low rank process lending itself to subspace based estimation techniques. However, the point source array response model is not appropriate.

5 SDMA IN THE DOWN LINK

Note that we may model the down link spatial channel statistics as in (36.1) however, in most current FDD systems the up and down link flat fading may be considered independent. If the main objective is increased range, this does not pose a major problem. However, the unobservable down link channel this one of the main obstacles if the intention is to also increase system capacity. An array could be employed at the mobile site as well, but in many applications this is not considered a feasible solution. Another alternative is to attempt to estimate the channel by employing feedback [15]. This requires a complete redesign of protocols and signaling and is probably only possible in environments which are very slowly varying. This technique may be feasible for *movable* (rather than mobile) systems such as indoor wireless local area networks.

If we are attempting to increase capacity in current FDD systems in the down link, the information gained from the signal separation techniques in the up link, can not be used directly. Since the channels are not reciprocal, it is not possible to reuse an "optimal" weight vector obtained from receive data, in the transmit mode. One must at least attempt to "transform" the weights to the

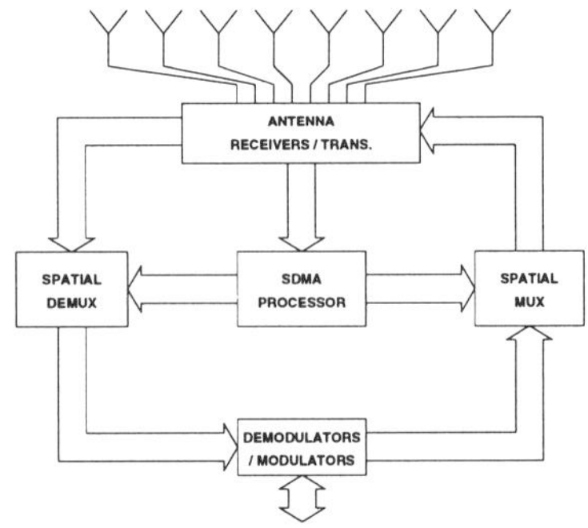

Figure 36.2 Possible configuration of an SDMA system.

transmit frequency. However, this is not a well conditioned problem unless an array model is introduced. When using an array model to transform weight vectors, directional information is exploited. It should be noted that in [11], a transmit scheme is proposed which does not use directional information. The down link scheme is based on statistical information estimated in the up link to take into account the unobservable fading. However, the frequency duplex distance is not compensated for causing the system to degrade in the presence of strong direct paths.

In time division duplex (TDD) systems, the up and down link channels can be considered reciprocal if there is limited movement between receive and transmit. Up link channel information may then be used to achieve spatially selective transmission and thus increasing capacity [16]. However, efficient use of the spatial dimension in current FDD cellular systems requires the use of directional information. Array response modeling is feasible for medium to large size cells with high placement of the base station antennas avoiding near field scattering.

6 SDMA SYSTEM

There are two main approaches for increasing capacity with antenna arrays. The frequency reuse distance may be decreased or multiple mobiles may be allocated to the same cell (or some combination of the above). In [1] it is shown that when directional information can be used, multiple mobiles per cell is a more efficient way of increasing capacity. Figure 36.2 displays the general structure of a SDMA system. There are several advantages with this approach to increasing capacity. To fully exploit decreased reuse distance, one must suppress signals to mobiles in other cells by forming nulls in the down link transmit pattern. This is very difficult even in a synchronous TDMA system because of propagation delays. With a small reuse distance, the desired and interfering signals fade independently at the mobile causing problems. Also, allocating mobiles to a frequency/spatial channel is easier when treated locally within a cell. By reducing the reuse distance, capacity is maximized when all frequencies are used in all cells whereas in the other scheme, at least in theory, capacity is hardware limited.

7 SUMMARY

Providing adequate coverage and sufficient capacity are two challenging problems for wireless communication systems. Antenna arrays at the base stations of cellular systems can increase range compared to current systems. The capacity problem can be significantly mitigated by ***spatial division multiple access*** *(SDMA)* techniques. SDMA supports multiple connections on a single conventional channel, based on user localization steered reception and transmission schemes, and therefore offers capacity increases over current wireless system implementations.

REFERENCES

[1] P. Zetterberg and B. Ottersten, The spectrum efficiency of a basestation antenna array system for spatially selective transmission, *To appear in IEEE Transactions on Vehicular Technology.*

[2] P. Zetterberg, Mobile communication with base station antenna arrays: Propagation modeling and system capacity, Technical Report TRITA-SB-9502, Signals, Sensors & Systems, February 1995, Licentiate Thesis.

[3] D.H. Johnson and D.E. Dudgeon, *Array Signal Processing – Concepts and Techniques*, Prentice-Hall, Englewood Cliffs, NJ, 1993.

[4] S. Andersson, M. Millnert, M. Viberg, and B. Wahlberg, "An Adaptive Array for Mobile Communication Systems", *IEEE Trans. on Veh. Tec.*, **40**(1):230–236, 1991.

[5] S.C. Swales, M.A. Beach, D.J. Edwards, and J.P. McGeehan, The performance enhancement of multibeam adaptive base-station antennas for cellular land mobile radio systems, *IEEE Trans. Vehicular Technology*, **39**:56–67, Feb. 1990.

[6] J.H. Winters, Optimum combining in digital mobile radio with cochannel interference, *IEEE Trans. Vehicular Technology*, **33**(3):144–155, August 1984.

[7] A.F. Naguib, A. Paulraj, and T. Kailath, "Capacity Improvement with Base-Station Antenna Arrays in Cellular CDMA", *IEEE Trans. Vehicular Technology*, **43**(3):691–698, Aug. 1994.

[8] E. Lindskog, A. Ahlén, and M. Sternad, "Combined spatial and temporal equalization using an adaptive antenna array and a decision feedback equalization scheme", In *IEEE International Conference on Acoustics, Speech and Signal Processing*, Detroit, USA, May 1995.

[9] S. Talwar, M. Viberg, and A. Paulraj, "Blind Estimation of Multiple Co-Channel Digital Signals Using an Antenna Array", *IEEE SP Letters*, **1**:29–31, Feb. 1994.

[10] B.G. Agee, A.V. Schell, and W.A. Gardner, "Spectral Self-Coherence Restoral: A New Approach to Blind Adaptive Signal Extraction Using Antenna Arrays", *Proc. IEEE*, **78**:753–767, Apr. 1990.

[11] G. Raleigh, S.N. Diggavi, V.K. Jones, and A. Paulraj, A blind adaptive transmit antenna algorithm for wireless communication, In *Proc. ICC*, 1995.

[12] U. Forssén, J. Karlsson, B. Johannisson, M. Almgren, F. Lotse, and F. Kronestedt, "Adaptive Antenna Arrays for GSM900/DCS1800", In *Proc. IEEE Veh. Technol. Conf.*, pages 605–609, 1994.

[13] T. Trump and B. Ottersten, Estimation of nominal direction of arrival and angular spread using an array of sensors, Submitted to Signal Processing, Elsevier, 1994.

[14] B. Ottersten, R. Roy, and T. Kailath, "Signal Waveform Estimation in Sensor Array Processing", In *Proc. 23rd Asilomar Conf. Sig., Syst.,Comput.*, pages 787–791, Nov. 1989.

[15] D. Gerlach and A. Paulraj, Adaptive transmitting antenna arrays with feedback, *IEEE SP Letters*, **1**(10):150–152, October 1994.

[16] P. Mogensen, F. Frederiksen, J. Wigard, and S. Petersen, "A Research Study of Antenna Diversity and Data Receivers for DECT", In *Proc. Nordic Radio Symposium*, Saltsjöbaden, Sweden, April 1995.

37

FROM SINUSOIDS IN NOISE TO BLIND DECONVOLUTION IN COMMUNICATIONS

Dirk Slock

Eurecom Institute
Mobile Communications Dept.
B.P. 193, 06904 Sophia Antipolis Cedex, FRANCE

Dedicated to Professor Kailath on the occasion of his 60th birthday, with fond memories of an inspiring teacher.

ABSTRACT

Equalization for digital communications constitutes a very particular blind deconvolution problem in that the received signal is cyclostationary. Oversampling (OS) (w.r.t. the symbol rate) of the cyclostationary received signal leads to a stationary vector-valued signal (polyphase representation (PR)). OS also leads to a fractionally-spaced channel model and equalizer. The multichannel formulation also arises in mobile communications, when multiple receiving antennas are used. In the multi-channel case, channel and equalizer can be considered as an analysis and synthesis filter bank. Zero-forcing (ZF) equalization corresponds to a perfect-reconstruction filter bank. We show that in the multichannel case FIR ZF equalizers exist for a FIR channel. The noise-free multichannel power spectral density matrix has rank one and the channel can be found as the (minimum-phase) spectral factor. The multi-channel linear prediction of the noiseless received signal becomes singular eventually, reminiscent of the single-channel prediction of a sum of sinusoids. As a result, a ZF equalizer can be determined from the received signal second-order statistics by linear prediction in the noise-free case, and by using a Pisarenko-style modification when there is additive noise. Due to the singularity and the FIR assumption, the spectral factorization reduces to the triangular factorization of a finite covariance matrix. In the given data case, Music (subspace) or ML techniques can be applied. We present these developments by drawing the parallel with existing techniques for the sinusoids in noise subspace problem.

1 INTRODUCTION

Consider linear digital modulation over a linear channel with additive Gaussian noise so that the received signal can be written as

$$\mathrm{y}(t) \;=\; \sum_k a(k)\mathrm{h}(t-kT) \;+\; \mathrm{v}(t) \tag{37.1}$$

where the $a(k)$ are the transmitted symbols, T is the symbol period, $\mathrm{h}(t)$ is the combined impulse response of channel and transmitter and receiver filters, but is often called the channel response for simplicity. Assuming the $\{a(k)\}$ and $\{\mathrm{v}(t)\}$ to be (wide-sense) stationary, the process $\{\mathrm{y}(t)\}$ is (wide-sense) cyclostationary with period T. If the channel would be known, then one could pass the received signal through a matched filter and sample the output at the symbol rate. These samples would provide sufficient statistics for the detection of the transmitted symbols. If $\{\mathrm{y}(t)\}$ is sampled with period T, the sampled process is (wide-sense) stationary and its second-order statistics contain no information about the phase of the channel. Tong, Xu and Kailath [1] have proposed to oversample the received signal with a period $\Delta = T/m$, $m > 1$. In what follows, we assume $\mathrm{h}(t)$ to have a finite duration. Tong *et al.*have shown that the channel can be identified from the second-order statistics of the oversampled received signal.

We review a number of developments in this context that have been inspired by their work. We shall concentrate here on the blind identification of a multichannel,i.e. without the use of training sequences for the transmitted symbols. In terms of second-order statistics, we shall only consider the statistics of the stationary vector received signal corresponding to the multichannel, and not cyclic statistics of possibly oversampled signals. We shall also not consider the exploitation of other side information such as the distribution of the symbols. The case of multiple sources was treated in e.g. [2].

1.1 Linear Multichannels

The multiple discrete-time channels we consider here come about by representing a diversity channel at the symbol rate. The diversity considered here can be spectral diversity, in which case oversampling w.r.t. the symbol rate exploits excess bandwidth (see Fig. 37.1), or spatial diversity obtained through multiple receiving antennas (and/or multiple polarizations) in wireless communications, see Fig. 37.2. A third case of diversity arises if the symbol constellation is one-dimensional (e.g. PAM or BPSK) and the transmitted signal is modulated. In that case the baseband channel impulse response has a real and an imaginary component, whereas the input is purely real. Hence, working with real signals only, we get a one-input two-output system. We assume in all cases the channel to be FIR with duration approximately NT. To further develop the case of oversampling, consider the sampling rate $\frac{m}{T}$. The sampling instants for the received

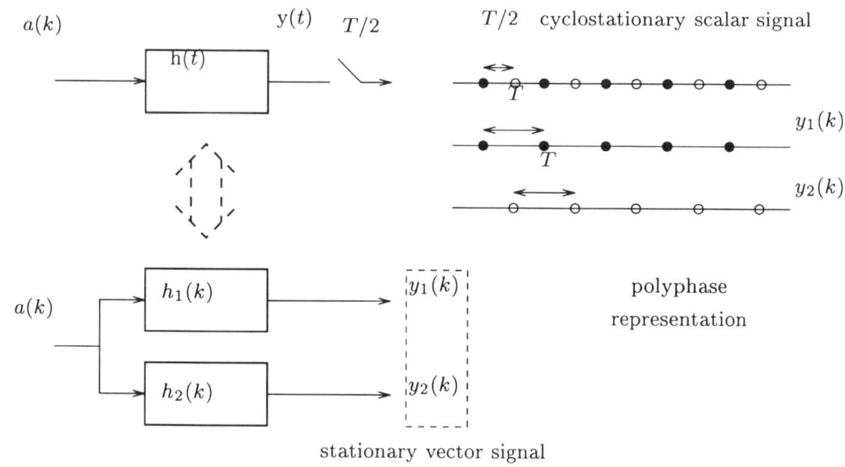

Figure 37.1 Polyphase and resulting vector channel representation of a $(T/2)$ oversampled received signal and channel.

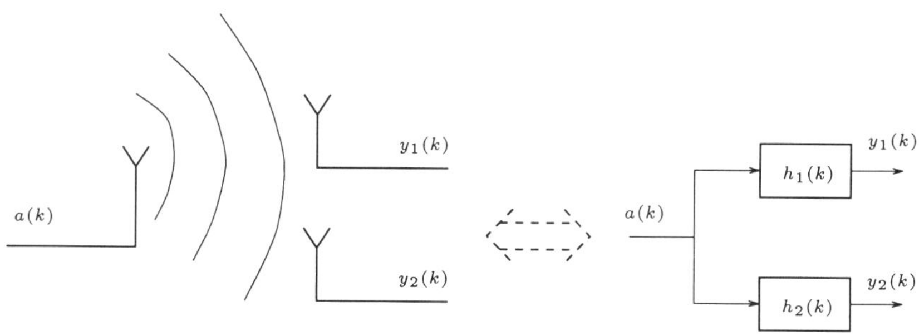

Figure 37.2 Vector channel representation of a signal received through multiple (2) antennas and sampled at the symbol rate.

signal in (37.1) are $t_0+T(k+\frac{j}{m})$ for integer k and $j=0,1,\ldots,m-1$. We introduce the polyphase description of the received signal: $y_j(k) = \mathrm{y}(t_0+T(k+\frac{j}{m}))$ for $j=0,1,\ldots,m-1$ are the m phases of received signal, and similarly for the channel impulse response and the additive noise. In the case of Fig. 37.1, the two polyphase components are the even and odd samples. In principle, it suffices to introduce a restricted $t_0 \in [0,T)$ to be fully general. However, we shall take $t_0 = t_0' + dT$ where $t_0' \in [0,T)$ and d is chosen as the smallest integer such that $\boldsymbol{h}(0) = \left[\mathrm{h}(t_0'+dT)\cdots \mathrm{h}(t_0'+(d+\frac{m-1}{m})T)\right]^T \neq 0$ (superscript ^{T} denotes transpose). The channel being causal implies that d is nonnegative;

d represents an inherent delay. The z-transform of the channel response at the sampling rate $\frac{m}{T}$ is $\mathrm{H}(z)=\sum_{j=1}^{m} z^{-(j-1)}\mathrm{H}_j(z^m)$.

The oversampled received signal can be represented in vector form at the symbol rate as

$$\begin{aligned} \boldsymbol{y}(k) &= \sum_{i=0}^{N-1} \boldsymbol{h}(i)a(k-i) + \boldsymbol{v}(k) = \boldsymbol{H}A_N(k) + \boldsymbol{v}(k)\ , \\ \boldsymbol{y}(k) &= \begin{bmatrix} y_1(k) \\ \vdots \\ y_m(k) \end{bmatrix}, \boldsymbol{v}(k) = \begin{bmatrix} v_1(k) \\ \vdots \\ v_m(k) \end{bmatrix}, \boldsymbol{h}(k) = \begin{bmatrix} h_1(k) \\ \vdots \\ h_m(k) \end{bmatrix} \\ \boldsymbol{H} &= [\boldsymbol{h}(N-1)\cdots\boldsymbol{h}(0)]\,, A_N(k) = \left[a(k-N+1)^H \cdots a(k)^H\right]^H \end{aligned} \tag{37.2}$$

where superscript ^{H} denotes Hermitian transpose. We formalize the finite duration NT assumption of the channel as follows (AFIR): $\boldsymbol{h}(0) \neq 0$, $\boldsymbol{h}(N-1) \neq 0$ and $\boldsymbol{h}(i) = 0$ for $i < 0$ or $i \geq N$.

A sequence of received signal samples can be represented as

$$\boldsymbol{Y}_L(k) = \mathcal{T}_L(\boldsymbol{H})\, A_{L+N-1}(k) + \boldsymbol{V}_L(k) \tag{37.3}$$

where $\boldsymbol{Y}_L(k) = [\boldsymbol{y}^H(k-L+1)\cdots\boldsymbol{y}^H(k)]^H$ and similarly for $\boldsymbol{V}_L(k)$, and $\mathcal{T}_M(\boldsymbol{x})$ is a (block) Toeplitz matrix with M (block) rows and $\left[\boldsymbol{x}\ 0_{r\times(M-1)s}\right]$ as first (block) row ($r \times s$ is the size of the blocks in $\boldsymbol{x}$). The generic identifiability of the channel (and the symbols) in the multichannel case becomes apparent by considering equation (37.3) in the noiseless case: $\boldsymbol{v}(k) \equiv 0$. In this case, (37.3) represents a set of mL equations in mN (in $\boldsymbol{H}$) plus $L+N-1$ (in $A_{L+N-1}(k)$) unknowns. This set of equations can in principle be solved for the unknowns if $mL \geq mN+L+N-1$, hence if $L \geq \frac{(m+1)N-1}{m-1}$. For any finite channel length N a data record length L can be found to satisfy this constraint if the number of channels $m \geq 2$. The system of equations is always underdetermined however for a single channel ($m=1$).

1.2 FIR Zero-Forcing (ZF) Equalization

In the multichannel representation, the Single Input Multiple Output (SIMO) vector channel transfer function from the single input $a(k)$ to the multiple outputs is $\mathbf{H}(z) = \sum_{i=0}^{N-1}\boldsymbol{h}(i)z^{-1}$ so that we can write for the vector received signal: $\boldsymbol{y}(k) = \mathbf{H}(q)\,a(k) + \boldsymbol{v}(k)$, where $q^{-1}a(k) = a(k-1)$. Consider now (in the oversampling context) a fractionally-spaced ($\frac{T}{m}$) equalizer of which the z-transform can also be decomposed into its polyphase components: $\mathrm{F}(z) = \sum_{j=1}^{m} z^{-(m-j)}\mathrm{F}_j(z^m)$. We assume the equalizer phases to be causal and FIR of length L: $\mathrm{F}_j(z) = \sum_{k=0}^{L-1} f_j(k)z^{-k}$, $j=1,\ldots,m$. The polyphase representation of the fractionally-spaced equalizer leads to a Multi

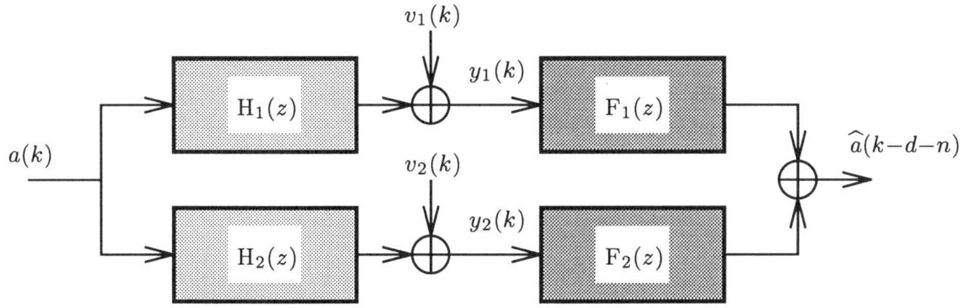

Figure 37.3 Polyphase representation of the T/m fractionally-spaced channel and equalizer for $m = 2$.

Input Single Output (MISO) system representation $\mathbf{F}(z) = \sum_{k=0}^{N-1} \boldsymbol{f}(k)z^{-k}$ with $\boldsymbol{f}(k) = [f_1(k) \cdots f_m(k)]$ and $\boldsymbol{F} = [\boldsymbol{f}(L{-}1) \cdots \boldsymbol{f}(0)]$, see Fig. 37.3 for $m = 2$. This MISO equalizer obvisouly applies to all multichannel formulations. The condition for the equalizer to be ZF is $\mathbf{F}(z)\mathbf{H}(z) = \sum_{j=1}^{m} \mathrm{F}_j(z)\mathrm{H}_j(z) = z^{-n}$ where $n \in \{0, 1, \ldots, N{+}L{-}2\}$. By equating equal powers of z^{-1}, we can write this in matrix form as

$$\boldsymbol{F}\ \mathcal{T}_L(\boldsymbol{H}) = [0 \cdots 0\ 1\ 0 \cdots 0] \tag{37.4}$$

where the 1 is in the $n{+}1st$ position from the end. (37.4) is a system of $L{+}N{-}1$ equations in Lm unknowns. To enable zero-forcing (ZF) equalization, we need to choose the equalizer length L such that the system of equations (37.4) is exactly or underdetermined. Hence

$$L \geq \underline{L} = \left\lceil \frac{N-1}{m-1} \right\rceil . \tag{37.5}$$

The block Toeplitz matrix $\mathcal{T}_L(\boldsymbol{H})$ is also a generalized Sylvester matrix. It can be shown that for $L \geq \underline{L}$ it has full column rank if $\mathbf{H}(z) \neq 0,\ \forall z$ or in other words if the $\mathrm{H}_j(z)$ have no zeros in common. Assuming $\mathcal{T}_L(\boldsymbol{H})$ to have full column rank, the nullspace of $\mathcal{T}_L^H(\boldsymbol{H})$ has dimension $L(m{-}1){-}N{+}1$. If we take the entries of any vector in this nullspace as equalizer coefficients, then the equalizer output is zero, regardless of the transmitted symbols. We shall call such an equalizer a blocking equalizer.

To find a ZF equalizer (corresponding to some delay n), it suffices to take an equalizer length equal to $\underline{L}$. We can arbitrarily fix $\underline{L}(m{-}1){-}N{+}1$ equalizer coefficients (e.g. take $\underline{L}(m{-}1){-}N{+}1$ equalizer phases of length $\underline{L}{-}1$ only, or fix the excess degrees of freedom by requiring minimal noise enhancement). The remaining $\underline{L}{+}N{-}1$ coefficients can be found from (37.4) if $\mathbf{H}(z) \neq 0,\ \forall z$. This shows that in the multichannel case, a FIR equalizer suffices for ZF equalization! With $m = N$ channels, the minimal required total number of equalizer coefficients N is found ($\underline{L} = 1$). In the multichannel case, FIR ZF equalization is an issue of finding an FIR synthesis bank given an FIR analysis bank so that the overall filterbank has the perfect reconstruction property. In [3], another interpretation of the ZF condition is given in the oversampled case that is similar to the Nyquist condition in the continuous-time case.

2 SINUSOIDS IN NOISE: A REVIEW

As can be seen from (37.3), the blind equalization problem in the multichannel case is a signal subspace estimation problem. That means that all the techniques that have been developed for subspace problems are applicable here also. Therefore we shall review these techniques in the context of a basic subspace problem: the sinusoids in noise problem. Afterwards we shall show that all the methods applicable to the sinusoids in noise problem are also applicable to the blind multichannel equalization problem.

2.1 Second-Order Statistics

Consider a signal x_k consisting of a sum of sinusoids

$$x_k = \sum_{i=1}^{M} A_i \cos(\omega_i k + \phi_i) \, y_k = x_k + v_k \ . \tag{37.6}$$

We shall assume that the amplitudes A_i and the frequencies $f_i = \omega_i/2\pi$ are deterministic unknowns, while the phases ϕ_i are unknown and random, independent of each other and uniformly distributed over $[0, 2\pi)$. Hence x_k is a stationary process with zero mean. The actual measured signal y_k is a noisy version of x_k with v_k being independent white noise with variance σ_v^2. But we first concentrate on the noise-free signal x_k. The support of the power spectral density function $S_{xx}(f) = \sum_{i=1}^{M} \frac{A_i^2}{2}(\delta(f-f_i) + \delta(f+f_i))$ has measure zero. The covariance matrix R_{XX} has rank $2M$ whenever its dimension exceeds $2M$. Hence, the rank remains finite, even if the dimension goes to infinity. The frequencies and phases can easily be determined from $S_{xx}(f)$.

2.2 Noise-free Prediction Problem

A sum of sinusoids satisfies a homogeneous difference equation. In particular

$$\mathrm{P}(q)\, x_k = 0\,, \quad \mathrm{P}(z) = \prod_{i=1}^{M} (1 - 2\cos\omega_i\, z^{-1} + z^{-2}) \ . \tag{37.7}$$

Hence, x_k is perfectly predictible from its previous $2M$ samples. $\mathrm{P}(z)$ and hence the ω_i can be found by linear prediction: this is *Prony*'s method. The normal equations governing the linear prediction problem are

$$P\,R_{XX} = [0 \cdots 0\; \sigma^2]\,, \quad \sigma^2 = 0 \tag{37.8}$$

where

$$\begin{aligned} R_{XX} &= \mathrm{E}\; X X^T\,, \quad X = [x_0 \cdots x_{2M}]^T \\ P &= [P_{2M} \cdots P_1\; P_0] \\ P_0 &= 1\,, \quad P_i = P_{2M-i}\,, \quad i = 0, \ldots, M-1\,. \end{aligned} \tag{37.9}$$

2.3 Signal and Noise Subspaces

The signal structure can be revealed by considering $X_k = \mathcal{V}\, S$:

$$\begin{bmatrix} x_0 \\ x_1 \\ \vdots \\ x_k \end{bmatrix} = \begin{bmatrix} 1 & 0 & \cdots & 1 & 0 \\ \cos\omega_1 & \sin\omega_1 & \cdots & \cos\omega_M & \sin\omega_M \\ \vdots & \vdots & \vdots & \vdots & \vdots \\ \cos\omega_1 k & \sin\omega_1 k & \cdots & \cos\omega_M k & \sin\omega_M k \end{bmatrix} \begin{bmatrix} A_1 \cos\phi_1 \\ -A_1 \sin\phi_1 \\ \vdots \\ A_M \cos\phi_M \\ -A_M \sin\phi_M \end{bmatrix} \tag{37.10}$$

where $\mathcal{V} = \mathcal{V}(\Omega)$ with $\Omega = [\omega_1 \cdots \omega_M]^T$ is a block Vandermonde matrix. Indeed, two consecutive block elements of a block column of $\mathcal{V}$ are proportional:

$$[\cos\omega_i k \quad \sin\omega_i k] = [\cos\omega_i(k-1) \quad \sin\omega_i(k-1)] \begin{bmatrix} \cos\omega_i & \sin\omega_i \\ -\sin\omega_i & \cos\omega_i \end{bmatrix}. \tag{37.11}$$

One calls

$$\begin{aligned} \mathit{Range}\,\{\mathcal{V}\} &= \text{SSS} &&= \text{signal subspace} \\ (\mathit{Range}\,\{\mathcal{V}\})^{\perp} &= \text{NSS} &&= \text{noise subspace} \end{aligned} \tag{37.12}$$

We get for the covariance structure of the noisy signal

$$Y_k = X_k + V_k = \mathcal{V}_k\, S + V_k \;\Rightarrow\; R_{YY} = \mathcal{V}\, R_{SS}\, \mathcal{V}^T + \sigma_v^2 I \tag{37.13}$$

where $R_{SS} = \frac{1}{2}\,\mathrm{diag}\,\{A_1^2, A_1^2, \ldots, A_M^2, A_M^2\}$. The signal component X_k of the measurement vector Y_k can only live in the signal subspace, whereas in the noise subspace only noise can be found.

Consider the eigendecomposition of R_{YY} ($\lambda_1 \geq \lambda_2 \geq \cdots$):

$$R_{YY} = \sum_{i=1}^{2M} \lambda_i V_i V_i^H + \sum_{i=2M+1}^{k+1} \lambda_i V_i V_i^H = V_{\mathcal{S}} \Lambda_{\mathcal{S}} V_{\mathcal{S}}^H + V_{\mathcal{N}} \Lambda_{\mathcal{N}} V_{\mathcal{N}}^H \tag{37.14}$$

where $V_{\mathcal{S}} = [V_1 \cdots V_{2M}]$, $V_{\mathcal{N}} = [V_{2M+1} \cdots V_{k+1}]$, $\Lambda_{\mathcal{S}} = \mathrm{diag}\,\{\lambda_1, \ldots, \lambda_{2M}\}$ and $\Lambda_{\mathcal{N}} = \sigma_v^2 I_{k+1-2M}$. Assuming $\mathcal{V}$ and R_{SS} to have full rank (all $\omega_i \in (0, \pi)$ are different, all $A_i > 0$), the sets of eigenvectors $V_{\mathcal{S}}$ and $V_{\mathcal{N}}$ are orthogonal:

$V_S^H V_{\mathcal{N}} = 0$, and $\lambda_i > \sigma_v^2$, $i = 1, \ldots, 2M$. The following equivalent descriptions of the signal and noise subspaces result:

$$\begin{aligned} \text{SSS} &= \textit{Range}\,\{V_S\} = \textit{Range}\,\{\mathcal{V}(\Omega)\} \\ \text{NSS} &= \textit{Range}\,\{V_{\mathcal{N}}\} \;:\; V_{\mathcal{N}}^T \mathcal{V} = 0\,. \end{aligned} \tag{37.15}$$

So far we have a parametric description of the SSS and we can compute both subspaces from the eigendecomposition of R_{YY}. We can also find a parametric description of the NSS. Indeed, we have $\mathrm{P}(q)\,\cos\omega_i k = 0$ and $\mathrm{P}(q)\,\sin\omega_i k = 0$, hence $\mathcal{G}(P)^T \mathcal{V} = 0$ where

$$\mathcal{G}(P)^T = \mathcal{T}_{k+1-2M}(P) = \begin{bmatrix} P_{2M} & \cdots & P_1 & P_0 & 0 & \cdots & 0 \\ 0 & P_{2M} & \cdots & P_1 & P_0 & \cdots & 0 \\ \vdots & \ddots & \ddots & & \ddots & \ddots & \vdots \\ 0 & \cdots & 0 & P_{2M} & \cdots & P_1 & P_0 \end{bmatrix} \tag{37.16}$$

is a Toeplitz matrix of full rank, equal to the dimension of the NSS. Hence

$$\text{NSS} = \textit{Range}\,\{\mathcal{G}(P)\}\,. \tag{37.17}$$

Note that both the SSS, $\mathcal{V}(\Omega)$, and the NSS, $\mathcal{G}(P)$, are parameterized by M independent parameters.

2.4 Frequency Estimation from Second-Order Statistics

Linear Prediction from Denoised Statistics

In the case of additive white noise, we can retrieve the noise-free prediction coefficients by replacing the monic constraint of linear prediction by a norm constraint in the minimization of the prediction error variance. We get the *Pisarenko* method

$$\begin{aligned} &\min_{\|P\|=1} P\,R_{YY}P^T = \min_{\|P\|=1} P\,R_{XX}P^T + \sigma_v^2 \\ &\Rightarrow\; P\,R_{XX} = [0\cdots 0]\,,\; P^T = V_{2M+1} \end{aligned} \tag{37.18}$$

This is the case of $k = 2M$: NSS dimension $= 1$. Equivalently, we can identify $\sigma_v^2 = \lambda_{2M+1}$ and apply linear prediction to the denoised statistics $R_{XX} = R_{YY} - \lambda_{2M+1} I_{2M+1}$.

Signal Subspace Fitting

We have two theoretically equivalent signal subspace descriptions: $\mathcal{V}(\Omega)$ and $V_S = V_S(R_{YY})$: both matrices have the same column space, hence one matrix can be transformed into the other one. V_S can be computed from the covariance matrix. By fitting $\mathcal{V}(\Omega)$ to V_S, we can determine Ω. With an estimated covariance matrix, V_S is approximate, so consider the following subspace fitting criterion

$$\min_{\Omega,T} \|\mathcal{V}(\Omega) - V_S T\|_F \tag{37.19}$$

where T is a square transformation matrix and the Frobenius norm is defined as $\|A\|_F^2 = \mathrm{tr} A^H A$ with tr denoting trace. The minimal value of the criterion is zero if V_S is exact. This criterion differs from the original subspace fitting strategy proposed in [4], which would propose $\min_{\Omega,T} \|\mathcal{V}(\Omega)T - V_S\|_F$ as criterion. We propose (37.19) because it leads to a simpler optimization problem. Both approaches can be made to be equivalent by the introduction of column space weighting. The criterion in (37.19) is separable. In particular, it is quadratic in T. Minimization w.r.t. T leads to $T = V_S^T \mathcal{V}$ and $\mathcal{V} - V_S T = P_{V_S}^{\perp} \mathcal{V}$ where $P_{V_S}^{\perp} = I - P_{V_S}$ and P_{V_S} is the projection matrix on the signal subspace (V_S). Hence

$$\begin{aligned} \min_T \|\mathcal{V}(\Omega) - V_S T\|_F^2 &= \|P_{V_S}^{\perp} \mathcal{V}\|_F^2 \\ &= \mathrm{tr}\, \mathcal{V}^T P_{V_S}^{\perp} \mathcal{V} = \mathrm{tr}\, \mathcal{V}^T P_{V_\mathcal{N}} \mathcal{V} = \|V_\mathcal{N}^T \mathcal{V}\|_F^2 \\ &= \sum_{i=2M+1}^{k} \|V_i^T \mathcal{V}\|^2 = \sum_{j=1}^{M} \sum_{i=2M+1}^{k} |V_i(\omega_j)|^2 \end{aligned} \tag{37.20}$$

where $V_i(\omega) = [1 \; e^{j\omega} \cdots e^{j\omega k}]\, V_i$ is the Fourier transform of the elements of V_i. The last expression in (37.20) needs to be minimized w.r.t. the ω_i. An approximate solution can be found as follows. Plot as a function of ω and find the ω_i as the abscissae of the M largest peaks of

$$\frac{1}{\sum_{i=2M+1}^{k} |V_i(\omega)|^2} \, . \tag{37.21}$$

This method is called MUSIC.

Noise Subspace Fitting

We have again two theoretically equivalent noise subspace descriptions: $\mathcal{G}(P)$ and $V_\mathcal{N} = V_\mathcal{N}(R_{YY})$. By fitting $\mathcal{G}(P)$ to $V_\mathcal{N}$, we can determine P. We introduce the following noise subspace fitting criterion

$$\min_{P,T} \|\mathcal{G}(P) - V_\mathcal{N} T\|_F \, . \tag{37.22}$$

Minimization w.r.t. T first leads again to $T = V_{\mathcal{N}}^T \mathcal{G}$ and $\mathcal{G} - V_{\mathcal{N}} T = P_{V_{\mathcal{N}}}^{\perp} \mathcal{G}$ and hence

$$\begin{aligned} \min_{P,T} \|\mathcal{G}(P) - V_{\mathcal{N}} T\|_F = \|P_{V_{\mathcal{N}}}^{\perp} \mathcal{G}\|_F^2 &= \operatorname{tr} \mathcal{G}^T P_{V_{\mathcal{N}}}^{\perp} \mathcal{G} = \operatorname{tr} \mathcal{G}^T P_{V_{\mathcal{S}}} \mathcal{G} \\ = \|V_{\mathcal{S}}^T \mathcal{G}\|_F^2 &= \sum_{i=1}^{2M} \|\mathcal{G}^T V_i\|^2 \end{aligned} \tag{37.23}$$

Due to the commutativity of convolution, we can write $\mathcal{G}^T V_i = \mathcal{H}_i P^T$ where $\mathcal{H}_i = \mathcal{H}(V_i)$ is Hankel. The symmetry of P can be expressed as $P = P J$ where J is the reverse identity matrix (ones on the main antidiagonal). We can assure the symmetry of the solution for P by explicitly expressing the symmetry of P. In this way, minimization of the last criterion in (37.23) w.r.t. P leads to

$$\min_{P} P \left[(\sum_{i=1}^{2M} \mathcal{H}_i^T \mathcal{H}_i) + J (\sum_{i=1}^{2M} \mathcal{H}_i^T \mathcal{H}_i) J \right] P^T \tag{37.24}$$

subject to $P_0 = 1$ or $\|P\| = 1$. Roughly this approach has been proposed recently in [5].

2.5 Frequency Estimation from Data: ML

With white Gaussian additive noise v_k, the likelihood function becomes the following least-squares criterion

$$\min_{\Omega,S} \|Y - \mathcal{V}(\omega) S\|^2 . \tag{37.25}$$

The criterion is again separable and minimization w.r.t. S first leads to $S = (\mathcal{V}^T \mathcal{V})^{-1} \mathcal{V}^T Y$. Hence

$$\min_{S} \|Y - \mathcal{V} S\|^2 = Y^T P_{\mathcal{V}}^{\perp} Y = Y^T P_{\mathcal{G}(P)} Y = P \mathcal{Y}^T (\mathcal{G}(P)^T \mathcal{G}(P))^{-1} \mathcal{Y} P^T \tag{37.26}$$

where we again exploited the commutativity of convolution: $\mathcal{G}(P)^T Y = \mathcal{H}(Y) P^T = \mathcal{Y} P^T$. Note that in (37.26), we went from the SSS parameterization in terms of Ω to the equivalent NSS parameterization in terms of P. The reason is that now a straightforward iterative procedure suggests itself, known as the Iterative Quadratic Maximum Likelihood (IQML) procedure [6]. At iteration n, we get the following quadratic criterion:

$$\min_{P^{(n)}} P^{(n)} \mathcal{Y}^T (\mathcal{G}(P^{(n-1)})^T \mathcal{G}(P^{(n-1)}))^{-1} \mathcal{Y} P^{(n)T} \tag{37.27}$$

subject to $P_0 = 1$ or $\|P\| = 1$. We could again incorporate the symmetry of P. The IQML iterations are not guaranteed to converge. However, with a consistent initialization (such as obtained from the second-order statistics based methods), only one iteration is required to get an Asymptotically Best Consistent (ABC) estimate for P and hence Ω.

2.6 Adaptive Notch Filtering

The model for the sum of sinusoids $\mathrm{P}(q)x_k = 0$ naturally leads to the following constrained ARMA representation for the measured signal y_k:

$$\mathrm{P}(q)y_k = \mathrm{P}(q)v_k \;\Rightarrow\; v_k = \frac{\mathrm{P}(q)}{\mathrm{P}(q/\rho)}\, y_k = \mathrm{N}(q)\, y_k \;\text{ as } \rho \lesssim 1 \tag{37.28}$$

where as indicated we can in principle recover the additive noise v_k from the measurements y_k using an infinitely sharp notch filter. This infinitely sharp notch filter is in practice approximated by the IIR notch filter $\mathrm{N}(z) = \mathrm{P}(z)/\mathrm{P}(z/\rho)$ with zeros $e^{\pm j\omega_i}$ and poles $\rho\, e^{\pm j\omega_i}$. When the coefficients of P in $\mathrm{N}(z)$ (the notches) are not properly chosen, the notch filter output will not equal (approximate) v_k but can be written in general as $\epsilon_k = \mathrm{N}(q)\, y_k = \mathrm{N}(q)x_k + \mathrm{N}(q)v_k$. The notch filter output variance is $(N(f) = \mathrm{N}(e^{j2\pi f}))$

$$\begin{aligned} \mathrm{E}\, \epsilon_k^2 &= \textstyle\int_{-\frac{1}{2}}^{\frac{1}{2}} |N(f)|^2 S_{xx}(f)df + \int_{-\frac{1}{2}}^{\frac{1}{2}} |N(f)|^2 S_{vv}(f)df \\ &= \sum_{i=1}^{M} \frac{A_i^2}{2}\, |N(f_i)|^2 + \sigma_v^2 \end{aligned} \tag{37.29}$$

where the expression for the second term is valid for infinitely sharp notches. We can find P by minimizing the notch filter output variance w.r.t. it since indeed

$$\min_P \; \mathrm{E}\, \epsilon_k^2 \;=\; \sigma_v^2 \,. \tag{37.30}$$

Due to the long transients of the notch filter, this approach lends itself to adaptive filtering. An adaptive notch filter can be obtained by applying any adaptive filtering strategy to the MMSE criterion. Remark in particular that the ML criterion is in fact the sum of squares of the additive noise $v_k = y_k - x_k$.

3 BLIND CHANNEL ESTIMATION

Now we shall draw the parallel of the previous approaches for the blind equalization problem.

3.1 Channel Identification from Second-order Statistics: Frequency Domain Approach

Consider the noise-free case and let the transmitted symbols be uncorrelated with variance σ_a^2. Then the power spectral density (psd) matrix of the stationary vector process $\boldsymbol{y}(k)$ is

$$\mathrm{S}_{\boldsymbol{yy}}(z) \;=\; \sigma_a^2 \mathbf{H}(z)\, \mathbf{H}^{\dagger}(z) \tag{37.31}$$

where $\mathbf{H}^{\dagger}(z) = \mathbf{H}^{H}(z^{-*})$. The following spectral factorization result can be found in [7]. Let $\mathbf{K}(z)$ be a $m \times 1$ rational transfer function that is causal and stable. Then $\mathbf{K}(z)$ is called minimum-phase if $\mathbf{K}(z) \neq 0,\ |z| > 1$. Let $\mathrm{S}_{\boldsymbol{yy}}(z)$ be a rational $m \times m$ spectral density matrix of rank 1. Then there exists a rational $m \times 1$ transfer matrix $\mathbf{K}(z)$ that is causal, stable , minimum-phase, unique up to a unitary constant, of (minimal) McMillan degree $\deg(\mathbf{K}) = \frac{1}{2}\deg(\mathrm{S}_{\boldsymbol{yy}})$ such that

$$\mathrm{S}_{\boldsymbol{yy}}(z) = \mathbf{K}(z)\,\mathbf{K}^{\dagger}(z)\,. \tag{37.32}$$

In our case, $\mathrm{S}_{\boldsymbol{yy}}$ is polynomial (FIR channel) and $\mathbf{H}(z)$ is minimum-phase since we assume $\mathbf{H}(z) \neq 0$, $\forall z$. Hence, the spectral factor $\mathbf{K}(z)$ identifies the channel

$$\mathbf{K}(z) = \sigma_a\, e^{j\phi}\, \mathbf{H}(z) \tag{37.33}$$

up to a constant $\sigma_a e^{j\phi}$. So the channel identification from second-order statistics is simply a multivariate MA spectral factorization problem.

Note that the psd matrix of the noise-free signal $\boldsymbol{y}_k$ is of rank one and hence is singular in the multichannel case $m > 1$. We recall that the input-output relation of the channel is

$$\boldsymbol{Y}_L(k) = \mathcal{T}_L(\boldsymbol{H})\, A_{L+N-1}(k). \tag{37.34}$$

Therefore, the structure of the covariance matrix of the received signal $\boldsymbol{y}(k)$ is

$$\mathrm{R}^{\boldsymbol{y}}_L = \mathrm{E}\,\boldsymbol{Y}_L(k)\boldsymbol{Y}^H_L(k) = \mathcal{T}_L(\boldsymbol{H})\,\mathrm{R}^a_{L+N-1}\mathcal{T}^H_L(\boldsymbol{H}) \tag{37.35}$$

where $\mathrm{R}^a_L = \mathrm{E}\,A_L(k)A^H_L(k) > 0$. When $mL > L+N-1$, $\mathrm{R}^{\boldsymbol{y}}_L$ is singular. If then L increases further by 1, the rank of $\mathrm{R}^{\boldsymbol{y}}_L$ increases by 1 and the dimension of its nullspace increases by $m-1$. In fact, $\frac{\text{rank}}{\text{dimension}} = \frac{L+N-1}{mL} \overset{L\to\infty}{\longrightarrow} \frac{1}{m}$.
So the channel can in principle be identified by spectral factorization, an iterative procedure that represents an infinite number of computations. We shall see however that due to the singularity and the FIR assumption the channel can be identified from the triangular factorization of a finite covariance matrix.

3.2 Noise-free Prediction Problem

Multichannel Linear Prediction

Consider now the problem of predicting $\boldsymbol{y}(k)$ from $\boldsymbol{Y}_L(k-1)$ The prediction error can be written as

$$\widetilde{\boldsymbol{y}}(k)|_{\boldsymbol{Y}_L(k-1)} = \boldsymbol{y}(k) - \widehat{\boldsymbol{y}}(k)|_{\boldsymbol{Y}_L(k-1)} = \boldsymbol{P}_L\boldsymbol{Y}_{L+1}(k) \tag{37.36}$$

with $\boldsymbol{P}_L = [\boldsymbol{P}_{L,L} \cdots \boldsymbol{P}_{L,1}\ \boldsymbol{P}_{L,0}]$, $\boldsymbol{P}_{L,0} = I_m$. Minimizing the prediction error variance leads to the following optimization problem

$$\min_{\boldsymbol{P}_L:\boldsymbol{P}_{L,0}=I_m} \boldsymbol{P}_L\mathrm{R}^{\boldsymbol{y}}_{L+1}\boldsymbol{P}^H_L = \sigma^2_{\widetilde{\boldsymbol{y}},L} \tag{37.37}$$

or hence

$$\boldsymbol{P}_L \mathrm{R}^{\boldsymbol{y}}_{L+1} \;=\; [0\cdots 0]\;\; \sigma^2_{\widetilde{\boldsymbol{y}},L}\,. \tag{37.38}$$

When $mL > L{+}N{-}1$, $\mathcal{T}_L\left(\boldsymbol{H}\right)$ has full column rank. Hence, using (37.34) and (37.36),

$$\begin{aligned}
\widetilde{\boldsymbol{y}}(k)|_{\boldsymbol{Y}_{L(k-1)}} = \widetilde{\boldsymbol{y}}(k)|_{A_{L+N-1}(k-1)} &= \boldsymbol{y}(k) - \widehat{\boldsymbol{y}}(k)|_{A_{L+N-1}(k-1)} \\
&= \sum_{i=0}^{N-1} \boldsymbol{h}(i)a(k{-}i) - \sum_{i=0}^{N-1} \boldsymbol{h}(i)\,\widehat{a}(k{-}i)|_{A_{L+N-1}(k-1)} \\
&= \sum_{i=0}^{N-1} \boldsymbol{h}(i)a(k{-}i) - \sum_{i=1}^{N-1} \boldsymbol{h}(i)a(k{-}i) - \boldsymbol{h}(0)\,\widehat{a}(k)|_{A_{L+N-1}(k-1)} \\
&= \boldsymbol{h}(0)\,\widetilde{a}(k)|_{A_{L+N-1}(k-1)}
\end{aligned} \tag{37.39}$$

Now let us consider the prediction problem for the transmitted symbols. We get similarly

$$\begin{aligned}
\widetilde{a}(k)|_{A_M(k-1)} &= a(k) - \widehat{a}(k)|_{A_M(k-1)} = \boldsymbol{Q}_M\, A_M(k{-}1), \text{ and} \\
\boldsymbol{Q}_M \mathrm{R}^a_{M+1} &= \left[0\cdots 0\,\sigma^2_{\widetilde{a},M}\right]
\end{aligned} \tag{37.40}$$

where $\boldsymbol{Q}_M = [\boldsymbol{Q}_{M,M}\cdots \boldsymbol{Q}_{M,1}\;1]$. We find from (37.39),(37.40)

$$\widetilde{\boldsymbol{y}}(k)|_{\boldsymbol{Y}_{L(k-1)}} = \boldsymbol{P}_L \mathcal{T}_{L+1}\left(\boldsymbol{H}\right) A_{L+N}(k) = \mathbf{h}(0)\,\boldsymbol{Q}_{L+N-1}A_{L+N}(k) \tag{37.41}$$

for all $A_{L+N}(k)$ and hence

$$\boldsymbol{P}_L \mathcal{T}_{L+1}\left(\boldsymbol{H}\right) \;=\; \mathbf{h}(0)\,\boldsymbol{Q}_{L+N-1}\,. \tag{37.42}$$

From (37.39), we also get

$$\sigma^2_{\widetilde{\boldsymbol{y}},L} \;=\; \boldsymbol{h}(0)\sigma^2_{\widetilde{a},L+N-1}\boldsymbol{h}^H(0)\,. \tag{37.43}$$

All this holds for $L \geq \underline{L}$. As a function of L, the rank profile of $\sigma^2_{\widetilde{\boldsymbol{y}},L}$ behaves like

$$\operatorname{rank}\left(\sigma^2_{\widetilde{\boldsymbol{y}},L}\right) \begin{cases} =\; 1 & ,\; L \geq \underline{L} \\ = m{-}\underline{m} \in \{2,3,\ldots,m\} & ,\; L = \underline{L}{-}1 \\ =\; m & ,\; L < \underline{L}{-}1 \end{cases} \tag{37.44}$$

where $\underline{m} = m\underline{L}{-}(\underline{L}{+}N{-}1) \in \{0,1,\ldots,m{-}2\}$ represents the degree of singularity of $\mathrm{R}^{\boldsymbol{y}}_{\underline{L}}$. Note that multichannel linear prediction corresponds to block triangular factorization of (some generalized) inverse of $\mathrm{R}^{\boldsymbol{y}}$. Indeed,

$$\boldsymbol{L}_L\, \mathrm{R}^{\boldsymbol{y}}_L \boldsymbol{L}_L^H = \boldsymbol{D}_L\,,\; (\boldsymbol{L}_L)_{i,j} = \boldsymbol{P}_{i-1,i-j}\,,\; (\boldsymbol{D}_L)_{i,i} = \sigma^2_{\widetilde{\boldsymbol{y}},i-1} \tag{37.45}$$

where $\boldsymbol{L}_L$ is block lower triangular and $\boldsymbol{D}_L$ is block diagonal. (A slight generalization to the singular case of) the multichannel Levinson algorithm can be used to compute the prediction quantities and hence the triangular factorization above in a fast way. In the case that $\mathrm{R}^{\boldsymbol{y}}_{\underline{L}}$ is singular, some precaution is necessary in the determination of the last block coefficient $P_{\underline{L},\underline{L}}$ (see [8]). Similar singularities will then arise at higher orders.

Uncorrelated Symbols

We shall now concentrate on the case in which the symbols $a(k)$ are uncorrelated. In this case the noise-free received signal is a singular multivariate MA process. Observe that for $L = \underline{L}$ we have

$$\boldsymbol{y}(k) + \sum_{i=1}^{\underline{L}} P_{\underline{L},i}\boldsymbol{y}(k-i) = \widetilde{\boldsymbol{y}}_{\underline{L}}(k) \;=\; \boldsymbol{h}(0)\,\widetilde{a}_{\underline{L}+N-1}(k) \;=\; \boldsymbol{h}(0)\,a(k) \qquad (37.46)$$

so that the prediction error is a singular white noise. This means that the noise-free received signal $\boldsymbol{y}(k)$ is also a singular multivariate AR process. Hence

$$\boldsymbol{P}_L = \left[\cdots 0 \;\; \boldsymbol{P}_{\underline{L}}\right] \;,\;\; \sigma^2_{\widetilde{\boldsymbol{y}},L} = \sigma^2_{\widetilde{\boldsymbol{y}},\underline{L}} \;,\;\; L > \underline{L}\,. \qquad (37.47)$$

Hence the factors $\boldsymbol{L}_L$ and $\boldsymbol{D}_L$ in the factorization (37.45) become block Toeplitz after $\underline{L}$ lines.

For $L = \underline{L}$, (37.43) allows us to find $\boldsymbol{h}(0)$ up to a scalar multiple. We see from (37.42) that $\frac{\boldsymbol{h}^H(0)}{\boldsymbol{h}^H(0)\boldsymbol{h}(0)}\boldsymbol{P}_{\underline{L}}$ is a zero-delay ZF equalizer. Given $\boldsymbol{h}(0)$ from (37.43) and $\boldsymbol{P}_{\underline{L}}$, we can solve for the channel impulse response $\boldsymbol{H}$ from (37.42). The channel can alternatively be found from

$$\boldsymbol{P}_{\underline{L}}\,\mathrm{E}\boldsymbol{Y}_{\underline{L}+1}(k)\boldsymbol{Y}^H_N(k+N-1) \;=\; \sigma^2_a\,\boldsymbol{h}(0)[\boldsymbol{h}^H(0)\cdots\boldsymbol{h}^H(N-1)] \qquad (37.48)$$

or from $\mathbf{P}_{\underline{L}}(z)\mathbf{H}(z) = \boldsymbol{h}(0) \;\Rightarrow\; \mathbf{H}(z) = \mathbf{P}^{-1}_{\underline{L}}(z)\,\boldsymbol{h}(0)$ using the lattice parameterization for $\mathbf{P}_{\underline{L}}(z)$ obtained with the Levinson algorithm.

In the uncorrelated symbols case, the prediction problem allows us also (in theory) to check whether the H_j have zeros in common. Indeed, the common factor colors the transmitted symbols (MA process) and hence once $\sigma^2_{\widetilde{\boldsymbol{y}},L}$ becomes of rank 1, its one nonzero eigenvalue $\sigma^2_{\widetilde{a},L+N-1}\boldsymbol{h}^H(0)\boldsymbol{h}(0)$ continues to decrease as a function of L since for a MA process, $\sigma^2_{\widetilde{a},L}$ is a decreasing function of L.

Correlated Symbols

Now consider the case in which the symbols $a(k)$ are correlated. Still, for $L = \underline{L}$, (37.43) allows us to find $\boldsymbol{h}(0)$ up to a scalar multiple. Let $\boldsymbol{h}^{\perp}$ be $m \times (m-1)$ of rank $m-1$ such that $\boldsymbol{h}^{\perp H}\boldsymbol{h}(0) = 0$, then

$$\boldsymbol{F}^b_{\underline{L}+1} \;=\; \boldsymbol{h}^{\perp H}\,\boldsymbol{P}_{\underline{L}} \qquad (37.49)$$

is a set of $m-1$ blocking equalizers since indeed $\boldsymbol{F}^b \boldsymbol{Y}_L(k) = 0$. We introduce a block-componentwise transposition operator ^{t}, viz.

$$\begin{aligned} \boldsymbol{H}^t &= [\boldsymbol{h}(N-1) \cdots \boldsymbol{h}(0)]^t = \left[\boldsymbol{h}^T(N-1) \cdots \boldsymbol{h}^T(0)\right] \\ \boldsymbol{F}_L^t &= [\mathbf{f}(L-1) \cdots \mathbf{f}(0)]^t = \left[\mathbf{f}^T(L-1) \cdots \mathbf{f}^T(0)\right] \end{aligned} \tag{37.50}$$

where ^{T} is the usual transposition operator. Due to the commutativity of convolution, we find

$$\boldsymbol{F}_L^b \, \mathcal{T}_L(\boldsymbol{H}_N) = 0 \iff \mathbf{H}_N^t \, \mathcal{T}_N\left(\boldsymbol{F}_L^{b\,t}\right) = 0 \,. \tag{37.51}$$

Now

$$\dim\left(Range^{\perp}\left\{\mathcal{T}_N\left(\boldsymbol{F}_{\underline{L}+1}^{b\,t}\right)\right\}\right) = 1 \tag{37.52}$$

so that we can identify the channel $\mathbf{H}_N^{t\,H}$ (up to scalar multiple) as the last right singular vector of $\mathcal{T}_N\left(\boldsymbol{F}_{\underline{L}+1}^{b\,t}\right)$ (a QR factorization would require less computations but might be less reliable numerically). From (37.42), one can furthermore identify $\boldsymbol{Q}_{\underline{L}+N-1}$ and via (37.40), this leads to the identification of the (Toeplitz) symbol covariance matrix $\mathrm{R}_{\underline{L}+N}^{a}$ up to a multiplicative scalar.

Modular Multichannel Linear Prediction

We noted previously that the consecutive multichannel linear prediction problems correspond to a block triangular factorization. They also correspond to Gram-Schmidt orthogonalization of the block components of the vector $\boldsymbol{Y}$. We can alternatively introduce sequential processing in the orthogonalization process and orthogonalize scalar component by scalar component the elements of the vector $\boldsymbol{Y}$. This leads to cyclic prediction filters and a true (non-block) triangular factorization. To make the orthogonalization process unique in the singular case, we need to introduce a convention. We shall assume that components of $\boldsymbol{Y}$ of which the orthogonalized versions are zero are not used in the orthogonalization of the further components (this corresponds to some minimum-norm choice). The consequence of this convention is that zeros will appear in the triangular factor (cyclic prediction filters). We get

$$\boldsymbol{L}_L' \, \mathrm{R}_L^{\boldsymbol{y}} \boldsymbol{L}_L'^H = \boldsymbol{D}_L' \tag{37.53}$$

where $\boldsymbol{L}_L'$ is a unit-diagonal lower triangular factor and $\boldsymbol{D}_L'$ is a diagonal matrix. After $m\underline{L}$ rows, both matrices become block Toeplitz again in the case of uncorrelated symbols. The steady-state diagonal elements of $\boldsymbol{D}_L'$ become in that case $\sigma_a^2 |h_1(0)|^2$ followed by $m-1$ zeros (since $\boldsymbol{h}(0) \neq 0$, we can w.l.o.g. assume that $h_1(0) \neq 0$). If we introduce a permutation matrix $\mathcal{P}$

$$\mathcal{P}\boldsymbol{L}_L' \, \mathrm{R}_L^{\boldsymbol{y}} \boldsymbol{L}_L'^H \mathcal{P}^H = \mathcal{P}\boldsymbol{D}_L' \mathcal{P}^H \Rightarrow \begin{bmatrix} \boldsymbol{L}'' \\ \mathcal{G} \end{bmatrix} \mathrm{R}_L^{\boldsymbol{y}} \begin{bmatrix} \boldsymbol{L}'' \\ \mathcal{G} \end{bmatrix}^H = \begin{bmatrix} \boldsymbol{D}'' & 0 \\ 0 & 0 \end{bmatrix} \tag{37.54}$$

so that the non-singular and singular parts get separated ($\boldsymbol{D}^{''}$ is non-singular). $\boldsymbol{L}^{''}$ eventually becomes block Toeplitz with $1 \times m$ blocks. Its repeated row then corresponds to a zero-delay ZF equalizer (up to a scalar multiple). $\mathcal{G} = \mathcal{G}(G)$ is block Toeplitz with $(m-1) \times m$ blocks and contains $m-1$ blocking equalizers parameterized by G. The number of elements in G is $mN-1$ (the number of degrees of freedom in $\boldsymbol{H}$ that can be determined blindly). Apart from the elements of G, $\mathcal{G}$ also contains 1's.

Whereas a modular multichannel Levinson algorithm (slightly adapted to handle singularities) can be used to find this factorization and the prediction quantities involved fast, similarly, a corresponding modular multichannel Schur algorithm [9] can be used to find the LDU factorization of $\mathrm{R}_L^{\boldsymbol{y}}$ itself:

$$\mathrm{R}_L^{\boldsymbol{y}} = \boldsymbol{U}_L^H \boldsymbol{D}_L^{'} \boldsymbol{U}_L \tag{37.55}$$

where we get the same diagonal factor and $\boldsymbol{U}_L$ is a unit-diagonal upper triangular matrix. Also $\boldsymbol{U}_L$ becomes block Toeplitz after $m\underline{L}$ rows. If we again introduce the permutation matrix $\mathcal{P}$, then

$$\begin{aligned}\mathrm{R}_L^{\boldsymbol{y}} &= \left(\boldsymbol{U}_L^H \mathcal{P}^H\right)\left(\mathcal{P}\boldsymbol{D}_L^{'}\mathcal{P}^H\right)\left(\mathcal{P}\boldsymbol{U}_L\right) \\ &= \begin{bmatrix}\boldsymbol{U}^{''} & \boldsymbol{U}^{'}\end{bmatrix}\begin{bmatrix}\boldsymbol{D}^{''} & 0 \\ 0 & 0\end{bmatrix}\begin{bmatrix}\boldsymbol{U}^{''} & \boldsymbol{U}^{'}\end{bmatrix}^H = \boldsymbol{U}^{''}\boldsymbol{D}^{''}\boldsymbol{U}^{''H}\end{aligned} \tag{37.56}$$

After a finite number of columns, $\boldsymbol{U}^{''}$ becomes block Toeplitz with $m \times 1$ blocks and the column that gets repeated contains (a multiple of) the channel impulse response for reasons that are related to the fact that $\boldsymbol{L}^{''}$ contains a ZF equalizer and to (37.48). This shows our earlier claim that the spectral factorization of $\mathrm{S}_{\boldsymbol{yy}}(z)$ can be replaced by a triangular factorization of a finite covariance matrix $\mathrm{R}^{\boldsymbol{y}}$. More details can be found in [10]

3.3 Signal and Noise Subspaces

Consider now the measured data with additive independent white noise $\boldsymbol{v}(k)$ with zero mean and assume $\mathrm{E}\,\boldsymbol{v}(k)\boldsymbol{v}^H(k) = \sigma_v^2 I_m$ with unknown variance σ_v^2 (in the complex case, real and imaginary parts are assumed to be uncorrelated, colored noise with known correlation structure but unknown variance could equally well be handled). A vector of L measured data can be expressed as

$$\boldsymbol{Y}_L(k) \;=\; \mathcal{T}_L\left(\boldsymbol{H}\right) A_{L+N-1}(k) + \boldsymbol{V}_L(k)\,. \tag{37.57}$$

Therefore, the structure of the covariance matrix of the received signal $\boldsymbol{y}(k)$ is

$$\mathrm{R}_L^{\boldsymbol{y}} = \mathrm{E}\,\boldsymbol{Y}_L(k)\boldsymbol{Y}_L^H(k) = \mathcal{T}_L\left(\boldsymbol{H}\right)\mathrm{R}_{L+N-1}^a \mathcal{T}_L^H\left(\boldsymbol{H}\right) + \sigma_v^2 I_{mL}\,. \tag{37.58}$$

Clearly, the column space of $\mathcal{T}_L\left(\boldsymbol{H}_N\right)$ is the signal subspace. Since $\mathcal{G}(G)\mathcal{T}_L(\boldsymbol{H}) = 0$, the column space of $\mathcal{G}^H$ is the noise subspace and G provides a linear parameterization for it.

Consider the eigendecomposition of $\mathrm{R}^{\boldsymbol{y}}_L$ of which the real positive eigenvalues are ordered in descending order:

$$\mathrm{R}^{\boldsymbol{y}}_L = \sum_{i=1}^{L+N-1} \lambda_i V_i V_i^H + \sum_{i=L+N}^{mL} \lambda_i V_i V_i^H = V_{\mathcal{S}} \Lambda_{\mathcal{S}} V_{\mathcal{S}}^H + V_{\mathcal{N}} \Lambda_{\mathcal{N}} V_{\mathcal{N}}^H \quad (37.59)$$

where $\Lambda_{\mathcal{N}} = \sigma_v^2 I_{(m-1)L-N+1}$ (see (37.58)). The sets of eigenvectors $V_{\mathcal{S}}$ and $V_{\mathcal{N}}$ are orthonormal: $V_{\mathcal{S}}^H V_{\mathcal{N}} = 0$, and $\lambda_i > \sigma_v^2$, $i = 1, \ldots, L+N-1$. We then have the following equivalent descriptions of the signal and noise subspaces

$$\begin{aligned} \text{SSS} &= \mathit{Range}\{V_{\mathcal{S}}\} = \mathit{Range}\{\mathcal{T}_L(\boldsymbol{H})\} \\ \text{NSS} &= \mathit{Range}\{V_{\mathcal{N}}\} = \mathit{Range}\{\mathcal{G}(G)^H\} \; . \end{aligned} \quad (37.60)$$

The noise subspace parameterization that we consider here is prediction based. It can perhaps be more easily expressed in the frequency domain by noting that

$$\mathbf{F}^b(z)\mathbf{H}(z) = 0 \, , \quad \mathbf{F}^b(z) = \boldsymbol{h}^{\perp H}\mathbf{P}(z) \, . \quad (37.61)$$

G above is not the same but is related to this choice of $\mathbf{F}^b(z)$. Another set of blocking equalizers and hence another linear parameterization of the noise subspace is channel based, e.g.

$$\mathbf{F}^b(z) = \begin{bmatrix} \mathrm{H}_2(z) & -\mathrm{H}_1(z) & & 0 \\ \vdots & & \ddots & \\ \mathrm{H}_m(z) & 0 & & -\mathrm{H}_1(z) \end{bmatrix} \quad (37.62)$$

Many other choices of $\mathbf{F}^b(z)$ are possible involving other pairs of channels. It is also possible to consider more than $m-1$ blocking equalizers, possibly involving up to all $\frac{m(m-1)}{2}$ possible pairs of channels.

3.4 Channel Estimation from Second-Order Statistics

Linear Prediction from Denoised Statistics

σ_v^2 can again be identified as the smallest eigenvalue of $\mathrm{R}^{\boldsymbol{y}}_L$. Replacing $\mathrm{R}^{\boldsymbol{y}}_L$ by $\mathrm{R}^{\boldsymbol{y}}_L - \sigma_v^2 I_{mL}$ gives us the covariance matrix for noise-free data, to which the prediction techniques discussed previously can be applied.

Signal Subspace Fitting

Consider now the following subspace fitting problem

$$\min_{\boldsymbol{H},T} \|\mathcal{T}_L(\boldsymbol{H}) - V_{\mathcal{S}}\, T\|_F \; . \quad (37.63)$$

The optimal transformation matrix T can again be found to be

$$T = V_{\mathcal{S}}^H \mathcal{T}_L(\boldsymbol{H}) \ . \tag{37.64}$$

Using (37.64) and the commutativity of the convolution operator, one can show that (37.63) is equivalent to

$$\begin{aligned} &\min_{\boldsymbol{H}^t} \boldsymbol{H}^t \left(\sum_{i=L+N}^{mL} \mathcal{T}_N\left(V_i^{H\,t}\right) \mathcal{T}_N^H\left(V_i^{H\,t}\right) \right) \boldsymbol{H}^{t\,H} \\ &= \min_{\boldsymbol{H}^t} \left[L \left\|\boldsymbol{H}^t\right\|_2^2 - \boldsymbol{H}^t \left(\sum_{i=1}^{L+N-1} \mathcal{T}_N\left(V_i^{H\,t}\right) \mathcal{T}_N^H\left(V_i^{H\,t}\right) \right) \boldsymbol{H}^{t\,H} \right] \end{aligned} \tag{37.65}$$

where V_i^H is considered a block vector with L blocks of size $1 \times m$. These optimization problems have to be augmented with a nontriviality constraint on $\boldsymbol{H}^t$. In case we choose the quadratic constraint $\left\|\boldsymbol{H}^t\right\|_2 = 1$, $\boldsymbol{H}^t$ is found as the minimum eigenvector of the first matrix in brackets in (37.65). This solution reflects orthogonalization of parameterized SSS and estimated NSS. Alternatively, the last term in (37.65) leads equivalently to

$$\max_{\left\|\boldsymbol{H}^t\right\|_2=1} \boldsymbol{H}^t \left(\sum_{i=1}^{L+N-1} \mathcal{T}_N\left(V_i^{H\,t}\right) \mathcal{T}_N^H\left(V_i^{H\,t}\right) \right) \boldsymbol{H}^{t\,H} \tag{37.66}$$

the solution of which is the eigenvector corresponding to the maximum eigenvalue of the matrix appearing between the brackets. This solution reflects really the attempt to fit parameterized and estimated SSS's.

Noise Subspace Fitting

Alternatively we may work with the parameterized noise subspace and consider the following subspace fitting approach

$$\min_{\boldsymbol{F}^b, T} \left\| \mathcal{T}\left(\boldsymbol{F}^b\right)^H - V_{\mathcal{N}}\, T \right\|_F \ . \tag{37.67}$$

One choice would be

$$\min_{G,T} \left\| \mathcal{G}\left(G\right)^H - V_{\mathcal{N}}\, T \right\|_F \ . \tag{37.68}$$

Again, two possible solutions can be obtained, depending on whether we attempt to orthogonalize the parameterized NSS to the estimated SSS or to fit it to the estimated NSS. The choice of $\boldsymbol{F}^b$ as in (37.62) corresponds to Xu's deterministic least-squares channel identification approach.

3.5 Channel Estimation from Data: ML

The transmitted symbols a_k are considered deterministic, the stochastic part is considered to come only from the additive Gaussian white noise. We assume the data $\boldsymbol{Y}_M(k)$ to be available. The maximization of the likelihood function boils down to the following least-squares problem

$$\min_{\boldsymbol{H}, A_{M+N-1}(k)} \|\boldsymbol{Y}_M(k) - \mathcal{T}_M(\boldsymbol{H})\, A_{M+N-1}(k)\|_2^2 \,. \tag{37.69}$$

The optimization problem in (37.69) is separable. Eliminating $A_{M+N-1}(k)$ in terms of $\boldsymbol{H}$, we get

$$\min_{\boldsymbol{H}} \left\| P^{\perp}_{\mathcal{T}_M(\boldsymbol{H})} \boldsymbol{Y}_M(k) \right\|_2^2 \tag{37.70}$$

subject to a nontriviality constraint on $\boldsymbol{H}$. In order to find an attractive iterative procedure for solving this optimization problem, we should work with a minimal parameterization of the noise subspace, which we have obtained before. Indeed,

$$P^{\perp}_{\mathcal{T}_M(\boldsymbol{H})} = P_{\mathcal{G}_M^H(G)} \,. \tag{37.71}$$

The number of degrees of freedom in $\boldsymbol{H}$ and G is both $mN-1$ (the proper scaling factor cannot be determined). So $\boldsymbol{H}$ can be uniquely determined from G and vice versa. Hence, we can reformulate the optimization problem in (37.70) as

$$\min_{G} \left\| P_{\mathcal{G}_M^H(G)} \boldsymbol{Y}_M(k) \right\|_2^2 \,. \tag{37.72}$$

Due to the (almost) block Toeplitz character of $\mathcal{G}_M$, the product $\mathcal{G}_M \boldsymbol{Y}_M(k)$ represents a convolution. Due to the commutativity of convolution, we can write $\mathcal{G}_M(G)\boldsymbol{Y}_M(k) = \mathcal{Y}_N(\boldsymbol{Y}_M(k))[1 \quad G^H]^H$ for some properly structured $\mathcal{Y}_N(\boldsymbol{Y}_M(k))$. This leads us to rewrite (37.72) as

$$\min_{G} \begin{bmatrix} 1 \\ G \end{bmatrix}^H \mathcal{Y}_N^H(\boldsymbol{Y}_M(k)) \left(\mathcal{G}_M(G)\, \mathcal{G}_M^H(G)\right)^{-1} \mathcal{Y}_N(\boldsymbol{Y}_M(k)) \begin{bmatrix} 1 \\ G \end{bmatrix} \tag{37.73}$$

This optimization problem can now easily be solved iteratively in the classical IQML style. An initial estimate may be obtained from the subspace fitting approach discussed above. Such an initial estimate is consistent and hence one iteration of (37.73) will be sufficient to generate an estimate that is asymptotically equivalent to the global optimizer of (37.73). Cramer-Rao bounds have been obtained and analyzed in [3]. The choice of the noise subspace parameterization in (37.62) using all pairs of channels leads to Yingbo Hua's ML method. More discussion on the ML method can be found in [11],[12].

3.6 Constrained IIR Filter DFE

Here we consider an equalizer structure with decision feedback. The approach is in fact a multichannel extension of the adaptive notch filter approach for sinusoids in noise. As a consequence, the method will continue to work well even if the additive noise and/or the transmitted symbols are colored.

Let $\mathbf{P}_{\widehat{\boldsymbol{y}}_L}(z)$ and $\mathbf{P}_{\widetilde{\boldsymbol{y}}_L}(z)$ be the z-transforms of the forward prediction and prediction error filters (of the noise-free case) so that $\mathbf{P}_{\widetilde{\boldsymbol{y}}_L}(z) = I_m - z^{-1}\mathbf{P}_{\widehat{\boldsymbol{y}}_L}(z)$. To alleviate the notation, $\mathbf{P}(z)$ will continue to represent $\mathbf{P}_{\widetilde{\boldsymbol{y}}_L}(z)$ (as before). Since $\mathbf{P}(z)\mathbf{H}(z) = \boldsymbol{h}(0)$, the noise-free received vector signal $\boldsymbol{y}(k) = \mathbf{H}(q)a_k$, which is a multichannel MA process, is also a (singular) multichannel AR process: $\mathbf{P}(q)\boldsymbol{y}(k) = \boldsymbol{h}(0)a_k$. For the noisy received signal $\boldsymbol{y}(k) = \mathbf{H}(q)a_k + \boldsymbol{v}(k)$, we get

$$\mathbf{P}(q)\boldsymbol{y}(k) = \boldsymbol{h}(0)a_k + \mathbf{P}(q)\boldsymbol{v}(k) \tag{37.74}$$

which is a constrained multichannel ARMA process, apart from the term $\boldsymbol{h}(0)a_k$, which will require detection. In the scalar case, the prediction error filter is minimum-phase. For the multichannel case, the extension is that $\det[\mathbf{P}(z)]$ is minimum-phase, even in the singular case. So we can recover $\boldsymbol{v}(k)$ as follows:

$$\boldsymbol{v}(k) = \mathbf{P}^{-1}(q)[\mathbf{P}(q)\boldsymbol{y}(k) - \boldsymbol{h}(0)a_k]\ . \tag{37.75}$$

This can be more straightforwardly implemented by the following procedure

$$\begin{cases} \boldsymbol{s}(k) = \mathbf{P}_{\widetilde{\boldsymbol{y}}_L}(q)\boldsymbol{y}(k) - \mathbf{P}_{\widehat{\boldsymbol{y}}_L}(q)\widehat{\boldsymbol{v}}(k-1) \\ \widehat{a}_k = \mathrm{dec}[\frac{\boldsymbol{h}^H(0)}{\boldsymbol{h}^H(0)\boldsymbol{h}(0)}\boldsymbol{s}(k)] \\ \widehat{\boldsymbol{v}}(k) = \boldsymbol{s}(k) - \boldsymbol{h}(0)\widehat{a}_k \end{cases} \tag{37.76}$$

where dec denotes the decision operation, whose argument is ideally $a_k + \frac{\boldsymbol{h}^H(0)}{\boldsymbol{h}^H(0)\boldsymbol{h}(0)}\boldsymbol{v}(k)$. Various algorithms are now possible to adapt the coefficients $\mathbf{P}_{\widehat{\boldsymbol{y}}_L}$ such as the Recursive Prediction Error Method and its simplifications.

REFERENCES

[1] L. Tong, G. Xu, and T. Kailath. "A New Approach to Blind Identification and Equalization of Multipath Channels". In *Proc. of the 25th Asilomar Conference on Signals, Systems & Computers*, pages 856–860, Pacific Grove, CA, Nov. 1991.

[2] D.T.M. Slock. "Blind Joint Equalization of Multiple Synchronous Mobile Users Using Oversampling and/or Multiple Antennas". In *Proc. 28th Asilomar Conf. on Signals, Systems and Computers*, pages 1154–1158, Pacific Grove, CA, Oct. 31 - Nov. 2 1994.

[3] D.T.M. Slock and C.B. Papadias. "Blind Fractionally-Spaced Equalization Based on Cyclostationarity". In *Proc. Vehicular Technology Conf.*, pages 1286–1290, Stockholm, Sweden, June 1994.

[4] M. Viberg and B. Ottersten. "Sensor Array Procesing Based on Subspace Fitting". *IEEE Trans. Acoust., Speech and Sig. Proc.*, ASSP-39(5):1110–1121, May 1991.

[5] D. Kundu and A. Mitra. "Estimating the parameters of exponentially damped/undamped sinusoids in noise: A non-iterative approach". *Signal Processing (EURASIP)*, 46:363–368, 1995.

[6] Y. Bresler and A. Macovski. "Exact Maximum Likelihood Parameter Estimation of Superimposed Exponential Signals in Noise". *IEEE Trans. Acoust., Speech and Sig. Proc.*, ASSP-34:1081–1089, Oct. 1986.

[7] Y.A. Rozanov. *Stationary Random Processes.* Holden-Day, San Fransisco, 1967. *Translated by* A. Feinstein.

[8] B. Porat. *Contributions to the Theory and Applications of Lattice Filters.* PhD thesis, Stanford University, Stanford, CA, Aug. 1982.

[9] K. Gallivan, S. Thirumalai, and P. Van Dooren. "A Block Toeplitz Look-Ahead Schur Algorithm". In *Proc. 3rd International Workshop on SVD and Signal Processing*, Leuven, Belgium, Aug. 22-25 1994.

[10] D.T.M. Slock. "Subspace Techniques in Blind Mobile Radio Channel Identification and Equalization using Fractional Spacing and/or Multiple Antennas". In *Proc. 3rd International Workshop on SVD and Signal Processing*, Leuven, Belgium, Aug. 22-25 1994.

[11] D.T.M. Slock and C.B. Papadias. "Further Results on Blind Identification and Equalization of Multiple FIR Channels". In *Proc. ICASSP 95 Conf.*, Detroit, Michigan, May 1995.

[12] D.T.M. Slock. "Spatio-Temporal Training-Sequence-Based Channel Equalization and Adaptive Interference Cancellation". In *Proc. ICASSP 96 Conf.*, Atlanta, Georgia, May 1996.

Index

D

N